Progress in Computer Science and Applied Logic
Volume 20

Editor

John C. Cherniavsky, National Science Foundation

Associate Editors

Robert Constable, Cornell University
Jean Gallier, University of Pennsylvania
Richard Platek, Cornell University
Richard Statman, Carnegie-Mellon University

Cryptography and Computational Number Theory

Kwok-Yan Lam
Igor Shparlinski
Huaxiong Wang
Chaoping Xing
Editors

Springer Basel AG

Editors:

Kwok-Yan Lam
Department of Computer Science
National University of Singapore
2 Science Drive 2
Singapore 117543
e-mail: lamky@comp.nus.edu.sg

Igor Shparlinski
Department of Computing
Macquarie University
NSW 2109
Australia
e-mail: igor@mpce.mq.edu.au

Huaxiong Wang
Department of Computer Science
University of Wollongong
NSW 2522
Australia
e-mail: huaxiong@uow.edu.au

Chaoping Xing
Department of Mathematics
National University of Singapore
2 Science Drive 2
Singapore 117543
e-mail: matxcp@nus.edu.sg

2000 Mathematics Subject Classification 11Yxx, 11Txx, 94Axx, 68P25

A CIP catalogue record for this book is available from the Library of Congress,
Washington D.C., USA

Deutsche Bibliothek Cataloging-in-Publication Data

Cryptography and computational number theory : workshop in Singapore 1999 /
Kwok-Yan Lam ..., ed. – Basel ; Boston ; Berlin : Birkhäuser, 2001
  (Progress in computer science and applied logic ; Vol. 20)

ISBN 978-3-0348-9507-1   ISBN 978-3-0348-8295-8 (eBook)
DOI 10.1007/978-3-0348-8295-8

© 2001 Springer Basel AG
Originally published by Birkhäuser Verlag, Basel in 2001
Member of the BertelsmannSpringer Publishing Group
Printed on acid-free paper produced of chlorine-free pulp. TCF ∞

9 8 7 6 5 4 3 2 1

Contents

Cryptography

Preface

This volume contains the refereed proceedings of the Workshop on Cryptography and Computational Number Theory, CCNT'99, which has been held in Singapore during the week of November 22–26, 1999.

The workshop was organized by the Centre for Systems Security of the National University of Singapore. We gratefully acknowledge the financial support from the Singapore National Science and Technology Board under the grant number RP960668/M.

The idea for this workshop grew out of the recognition of the recent, rapid development in various areas of cryptography and computational number theory. The event followed the concept of the research programs at such well-known research institutions as the Newton Institute (UK), Oberwolfach and Dagstuhl (Germany), and Luminy (France). Accordingly, there were only invited lectures at the workshop with plenty of time for informal discussions.

It was hoped and successfully achieved that the meeting would encourage and stimulate further research in information and computer security as well as in the design and implementation of number theoretic cryptosystems and other related areas. Another goal of the meeting was to stimulate collaboration and more active interaction between mathematicians, computer scientists, practical cryptographers and engineers in academia, industry and government.

Talks concerning many different aspects of cryptography and computational number theory such as theory, techniques, applications and practical experiences were given. Some other related areas of number theory and computer science were covered as well. These include but are not limited to talks devoted to

- o new cryptographic systems and protocols;
- o new attacks on the existing cryptosystems;
- o new cryptographic paradigms such as visual and audio cryptography;
- o pseudorandom number generator and stream cipher;
- o primality proving and integer factorization;
- o fast algorithms;
- o cryptographic aspects of the theory of elliptic and higher genus curves;
- o polynomials over finite fields;
- o analytical number theory.

Some of the talks, and their associated papers, were surveys giving comprehensive state-of-the-art outlines of some number theoretic research areas of significance to cryptography. Some were descriptions of new original results and ideas for which this workshop provided the first forum where they were publicly presented.

The contents of this volume reflects the whole variety of the topics which have been considered at the workshop. We believe it will provide a valuable contribution to and stimulate further progress of cryptography and computational number theory.

The Organizers:

Kwok-Yan Lam (National University of Singapore)
Igor Shparlinski (Macquarie University, Australia)
Huaxiong Wang (University of Wollongong, Australia)
Chaoping Xing (National University of Singapore)

Computational
Number Theory

Progress in Computer Science and Applied Logic, Vol. 20
© 2001 Birkhäuser Verlag Basel/Switzerland

On the Dimension and the Number
of Parameters of a Unirational Variety

Cesar Alonso, Jaime Gutierrez, and Rosario Rubio

Abstract. In this paper we study the relation between the dimension of a parametric variety and the number of parameters. We present an algorithm to reparameterize a variety in order to obtain a parameterization where the number of parameters equals the dimension of the variety.

1. Introduction

Let $\mathbb{K}$ be an algebraically closed field of characteristic zero and $T = (T_1, \ldots, T_m)$ indeterminates over $\mathbb{K}$. Following the notation of [5], given n rational functions in $\mathbb{K}(T)$

$$F_1(T) = \frac{F_{1N}(T)}{F_{1D}(T)}, \ldots, F_n(T) = \frac{F_{nN}(T)}{F_{nD}(T)};$$

we consider $W = \bigcup_{1 \leq i \leq n} \mathbf{V}(F_{iD}(T)) \subset \mathbb{K}^m$, where $\mathbf{V}(F_{iD}(T))$ is the zero set of the polynomial $F_{iD}(T) \in \mathbb{K}[T]$, and the map $\varphi \colon (\mathbb{K}^m - W) \to \mathbb{K}^n$ such that:

$$\varphi(t_1, \ldots, t_m) = (F_1(t_1, \ldots, t_m), \ldots, F_n(t_1, \ldots, t_m)).$$

An affine variety $V \subseteq \mathbb{K}^n$ is a parametric or unirational variety if there exists a collection of rational functions $\mathcal{F} = \{F_1(T), \ldots, F_n(T)\}$ such that V is the Zariski closure of $\varphi(\mathbb{K}^m - W)$. The set $\mathcal{F}$ is a parameterization of V.

Most affine varieties cannot be parameterized in the sense described above. In general, it is difficult to tell whether a given variety is unirational or not. It is well known that any parametric variety is irreducible; in fact, the coordinate ring of V is isomorphic to $\mathbb{K}[F_1(T), \ldots, F_n(T)]$, that is,

$$\mathbb{K}[x_1, \ldots, x_n]/\mathbf{I}(V) \simeq \mathbb{K}[F_1(T), \ldots, F_n(T)],$$

where $\mathbf{I}(V)$ is the polynomial ideal of V. So, the dimension of V is the transcendence degree of $\mathbb{K}(F_1(T), \ldots, F_n(T))$ over $\mathbb{K}$, that is,

$$d = \dim(V) = trans.deg_{\mathbb{K}}(\mathbb{K}(F_1(T), \ldots, F_n(T))).$$

A parameterization $\mathcal{F}$ is called *faithful* if $\mathbb{K}(F_1(T), \ldots, F_n(T)]) = \mathbb{K}(T)$. In geometric terms, it means there exist a one to one map from points of the variety to values of the parameters T, except an algebraic set of dimension $d - 1$. If the parameterization is not faithful, naturally we would ask whether we can reparameterize them so that new parameterization is faithful. In general cases, the answer is negative. However, in the case of algebraic curves $d = 1$, the existence of a faithful reparametrization for the original one is guaranteed by Lüroth's theorem. The papers [11] and [3] contain methods to reparameterize. In the case of algebraic surface $d = 2$ and if the ground field $\mathbb{K}$ is the complex number, then there always exists a faithful parameterization [2]. If $d > 2$, then there exist parameterization which are non-faithful and cannot be reparameterized [1].

We say that a parameterization $\mathcal{F}$ of V is *simple* if the number of the parameters m coincides with the dimension of V, that is, if $d = m$. Obviously, any faithful parameterization is simple, but not conversely. In most cases, the parameters in a set of parametric equations are independent, that is, $d = m$. But, there are not simple parameterizations. Consider the following example

$$
\begin{aligned}
F_1 &= T_1 + T_2^2 + 1, \\
F_2 &= T_1^3 + 3T_1T_2^2 + 3T_1T_2^4 + T_2^6 - T_1 - T_2^2, \\
F_3 &= T_1^2 + 2T_1T_2^2 + T_2^4 - 2T_1 - 2T_2^2 - 1.
\end{aligned}
$$

At the first sight, one might think that it represents a surface $d = 2$, in the space. Actually, it represents a curve $d = 1$, in the space; because if $S = T_1 + T_2^2$, then the above parametric equations become

$$
F_1 = S + 1, \quad F_2 = S^3 - S, \quad F_3 = S^2 - 2S - 1.
$$

For this example, each point of the curve corresponds to infinitely many values of T_1 and T_2. From computer aided geometric design (CAGD) point of view, such bad parameterization should be avoided, if it is possible. Now the question is: does every unirational variety have a simple parameterization? The answer to this question is positive and we will see that it is a corollary of the following classical result (see [9]):

Theorem 1.1. *Let $x_1, \ldots, x_n$ be algebraically independent elements over an infinite field $\mathbb{K}$. If $\mathbb{F}$ is a field such that $\mathbb{K} \subset \mathbb{F} \subset \mathbb{K}(x_1, \ldots, x_n)$. Then there are algebraically independent elements $y_1, \ldots, y_d$ over $\mathbb{K}$ such that $\mathbb{F} \subset \mathbb{K}(y_1, \ldots, y_d)$, where $d = trans.deg_{\mathbb{K}}(\mathbb{F})$.*

On the other hand, if the parameterization is not simple, naturally we would ask whether we can reparameterize them so that new parameterization is simple. Regarding algorithms for this last problem, we can mention the work [4]. They presented a method based on characteristic set theory.

In this introduction we have provided a coherent framework of this problem from the mathematical point of view. We also present a method to compute a simple parameterization of a unirational variety from any parameterization, based on the above classical result. We will show that the proof of Theorem 1.1 given by [10] is constructive and this supplies an algorithm to embed any unirational field $\mathbb{F}$, $\mathbb{K} \subset \mathbb{F} \subset \mathbb{K}(x_1, \ldots, x_m)$, with transcendence degree d over $\mathbb{K}$ into $\mathbb{K}(x_{i_1}, \ldots, x_{i_d})$.

2. Embedding $\mathbb{F}$ in $\mathbb{K}(x_{i_1}, \ldots, x_{i_d})$

In this section we will show a constructive version of Theorem 1.1. The algorithm we are presenting is based on the proof in [10]. We will provide all details from the computational point of view.

Algorithm 2.1.

    Input: $f_1, \ldots, f_n \in \mathbb{K}(x_1, \ldots, x_m)$.
    Output: $\Phi : \mathbb{K}(f_1, \ldots, f_n) \longrightarrow \mathbb{K}(x_{i_1}, \ldots, x_{i_d})$ *injective homomorphism, with*
    $d = trans.deg_{\mathbb{K}}(\mathbb{K}(f_1, \ldots, f_n))$.
    Steps:
    A: *Compute rational functions* $\overline{f_1}, \ldots, \overline{f_n}$ *such that*
        A.1: $\mathbb{K}(\overline{f_1}, \ldots, \overline{f_n}) = \mathbb{K}(f_1, \ldots, f_n)$.
        A.2: $\overline{f_1}, \ldots, \overline{f_d}$ *are algebraically independent over* $\mathbb{K}$.
        A.3: $\overline{f_{d+1}}, \ldots, \overline{f_n}$ *are integer over* $\mathbb{K}[\overline{f_1}, \ldots, \overline{f_d}]$.
        A.4: *If* $d = m$, *then return* $\Phi = id$.
    B: *Rewrite variables* $x_1, \ldots, x_m$ *such that*
        B.1: $x_{d+1}, \ldots, x_m$ *are algebraically independent over* $\mathbb{K}(\overline{f_1}, \ldots, \overline{f_d})$.
        B.2: $x_1, \ldots, x_d$ *are algebraic over* $\mathbb{K}(\overline{f_1}, \ldots, \overline{f_d}, x_{d+1}, \ldots, x_m)$.
    C: *Let* $P_i(\overline{f_1}, \ldots, \overline{f_d}, x_{d+1}, \ldots, x_m, Z) \in \mathbb{K}[\overline{f_1}, \ldots, \overline{f_d}, x_{d+1}, \ldots, x_m, Z]$ *be a nonconstant polynomial such that* $P_i(\overline{f_1}, \ldots, \overline{f_d}, x_{d+1}, \ldots, x_m, x_i) = 0$, *for* $i = 1, \ldots, d$. *Let* $f(x_1, \ldots, x_m)$ *be a common denominator of* $\overline{f_1}, \ldots, \overline{f_d}$
    and write $P_i = \dfrac{\widetilde{P_i}}{f^{r_i}}$.
        Take $\nu = \max\{\deg \widetilde{P_i}, \deg f, m\} + 1$.
    D: *Let* φ *be the monomorphism*

$$\varphi : \begin{array}{ccc} \mathbb{K}(f_1, \ldots, f_n) & \longrightarrow & K(x_1, \ldots, x_{m-1}). \\ f_i(x_1 \ldots, x_m) & \longmapsto & f_i(x_1, \ldots, x_{m-1}, x_1^{\nu}) \end{array}$$

    Let $\Phi = \varphi \circ id$.
    E: *If* $m - 1 = d$ *then return* Φ. *Otherwise repeat* **B-E** *for* $\Phi(\overline{f_1}), \ldots, \Phi(\overline{f_n})$.

Theorem 2.2. *Algorithm 2.1 is correct and each step of it can be effectively computed.*

Proof. The choice of the natural number ν, is the clue for the injectiveness of the function φ. First of all we will see that the function

$$\varphi : \begin{array}{ccc} \mathbb{K}[\overline{f_1}, \ldots, \overline{f_d}] & \longrightarrow & K(x_1, \ldots, x_{m-1}) \\ f_i'(x_1 \ldots, x_m) & \longmapsto & f_i(x_1, \ldots, x_{m-1}, x_1^{\nu}) \end{array}$$

is a monomorphism.

Since $\nu > \deg f$, the function φ is well defined: $f(x_1, \ldots, x_{m-1}, x_1^{\nu}) \neq 0$. Suppose that

$$P_1 = \sum_{u,v} a_{uv}(\overline{f_1}, \ldots, \overline{f_d}, x_{d+1}, \ldots, x_{m-1}) x_m^u Z^v$$

and for $2 \leq i \leq t$,

$$P_i = \sum_j b_{ij}(\overline{f_1}, \ldots, \overline{f_d}, x_{d+1}, \ldots, x_m)Z^j.$$

The coefficients a_{uv} and b_{ij} can be written as rational functions of the form $\dfrac{g(x_1 \ldots, x_m)}{f(x_1, \ldots, x_m)^r}$, for some polynomial $g(x_1 \ldots, x_m)$ and for some positive integer number r. Since we chose $\nu > \deg \widetilde{P_i}$, $\deg f$, we have

$$g(x_1, \ldots, x_{m-1}, x_1^\nu)f(x_1, \ldots, x_{m-1}, x_1^\nu) \neq 0.$$

We also imposed $\nu > m$, so if $(u, v) \neq (u', v')$ then $\nu u + v \neq \nu u' + v'$. Therefore,

$$P_1' = \sum_{u,v} a_{uv}(\varphi(\overline{f_1}), \ldots, \varphi(\overline{f_d}), x_{d+1}, \ldots, x_{m-1})Z^{\nu u+v} \neq 0$$

and for $2 \leq i \leq t$,

$$P_i' = \sum_j b_{ij}(\varphi(\overline{f_1}), \ldots, \varphi(\overline{f_d}), x_{d+1}, \ldots, x_{m-1}, x_1^\nu)Z^j \neq 0.$$

Moreover, P_i' vanishes for $Z = x_i$, thus $x_1, \ldots, x_d$ are algebraic over the function field $\mathbb{K}(\varphi((\overline{f_1}), \ldots, \varphi(\overline{f_d}), x_{d+1}, \ldots, x_{m-1})$. As a consequence, the elements $\varphi(\overline{f_1}), \ldots, \varphi(\overline{f_d}), x_{d+1}, \ldots, x_{m-1}$ are algebraically independent over $\mathbb{K}$, otherwise $trans.deg_\mathbb{K}(\mathbb{K}(x_1, \ldots, x_{m-1})) < m - 1$. In particular, $\varphi(\overline{f_1}), \ldots, \varphi(\overline{f_d})$ are algebraically independent over $\mathbb{K}$ and φ is injective.

We have just proved that $\varphi : \mathbb{K}[\overline{f_1}, \ldots, \overline{f_d}] \longrightarrow \mathbb{K}(x_1, \ldots, x_{m-1})$ is an injective morphism. Let us show that $\varphi : \mathbb{K}[\overline{f_1}, \ldots, \overline{f_n}] \longrightarrow \mathbb{K}(x_1, \ldots, x_{m-1})$ is also a monomorphism.

Let P be its kernel. P is a prime ideal in $\mathbb{K}[f_1, \ldots, f_n]$ and intersecting with $\mathbb{K}[\overline{f_1}, \ldots, \overline{f_d}]$ is (0). Since the domain $\mathbb{K}[\overline{f_1}, \ldots, \overline{f_n}]$ is integral over $K[\overline{f_1}, \ldots, \overline{f_d}]$, by Cohen–Seidenberg Theorem, P must be (0).

Finally, we observe the injection $\mathbb{K}[f_1, \ldots, f_n] \hookrightarrow \mathbb{K}(x_1, \ldots, x_{m-1})$ induces an injection $\mathbb{K}(f_1, \ldots, f_n) \to \mathbb{K}(x_1, \ldots, x_{m-1})$.

After step **D**, functions $\varphi(\overline{f_1}), \ldots, \varphi(\overline{f_n})$ and $f_1, \ldots, f_n$ have the same properties: **A.1**, **A.2** and **A.3**. So the algorithm works.

Every step at the algorithm can be computed.

Step **A** is Noether Normalization Lemma, most of the classical proofs of this result are constructive. For instance, the paper (cf. [12]) shows that most linear combinations of coordinates $\overline{f_i} = \sum_{j=1}^m a_{ij}f_j$ obtain bases with this properties. Other constructive proof can be found in [8].

Steps **B** and **C** require to compute minimal polynomials over unirational fields. In [13] a method has been presented which uses Gröbner Bases computation to do that.

The complexity is dominated by computing the transcendence degree and is exponential in the number of variables and polynomial in the rest of the data. □

3. Reparameterizing unirational varieties

In this section we present a method to reparameterize a parametric variety. This method is based on Theorem 1.1 and the following proposition

Proposition 3.1. *Let $F_1(T), \ldots, F_n(T)$ be a parameterization of a variety V of dimension d. Let $\overline{F}_1(T), \ldots, \overline{F}_d(T)$ in $A = \mathbb{K}[F_1(T), \ldots, F_n(T)]$, algebraically independent over $\mathbb{K}$ and such that A is an integer extension of $\mathbb{K}[\overline{F}_1(T), \ldots, \overline{F}_d(T)]$. Then*

$$\mathbb{K}[F_1(T), \ldots, F_n(T)] = \mathbb{K}[\overline{F}_1(T), \ldots, \overline{F}_d(T), F_{d+1}(T), \ldots, F_n(T)].$$

Proof. Write $A = \mathbb{K}[x_1, \ldots, x_n]/\mathbf{I}(V)$. Following the proof of the Noether Normalization Lemma in [8], we get a subalgebra $K[y_1, \ldots, y_n]$ with coordinate changes $y_i = x_i + \sum_{j>d} a_{ij}x_j$, $1 \leq i \leq d$, $a_{ij} \in K$; and $y_i = x_i$, for $i > d$. Thus we have $\overline{F}_i(X) = F_i(X) + \sum_{j>d} a_{ij}F_j(X)$, $1 \leq j \leq d$, satisfying the hypothesis. □

From Algorithm 2.1 and Proposition 3.1, we have a method to reduce the number of parameters in the parameterization of a parametric variety V, when the number of them is greater than the dimension. Suppose we have a parameterization $F_1(T), \ldots, F_n(T)$. Performing Algorithm 2.1 to $\mathbb{K}(F_1(T), \ldots, F_n(T))$, we get the parameterization $G_i(T) = \Phi(F_i(T)) \in \mathbb{K}(T_1, \ldots, T_d)$ for $i = 1, \ldots, n$.

We illustrate this algorithm with an example.

Example 3.2. *We consider the variety given by the following parameterization ([4]):*

$$F_1 = \frac{T_1 + T_2}{T_1 - T_2}, \quad F_2 = \frac{2T_2^2 + 2T_1^2}{(T_1 - T_2)^2}, \quad F_3 = \frac{2T_2^3 + 6T_1^2T_2}{(T_1 - T_2)^3}.$$

A. *First of all, we calculate the reduced Gröbner basis of the polynomial ideal I generated by*

$$\{(T_1 - T_2)y - 1, \quad (T_1 - T_2)x_1 - T_1 - T_2,$$
$$(T_1 - T_2)^2 x_2 - 2T_2^2 - 2T_1^2, \quad (T_1 - T_2)^3 x_3 - 2T_2^3 - 6T_1^2T_2\}.$$

where y is an extra variable. We consider the lexicographical order $y > T_1 > T_2 > x_3 > x_2 > x_1$. A Gröbner bases of $\mathbf{I}(V)$ (see [6]) is the intersection of the Gröbner basis of I with $K[x_1, x_2, x_3]$. Thus,

$$GB(I) = \{T_1x_1 - T_1 - T_2x_1 - T_2,$$
$$yT_1^3 - \tfrac{1}{8}x_1^3 - \tfrac{3}{8}x_1^2 - \tfrac{3}{8}x_1 - \tfrac{1}{8},$$
$$yT_1^2T_2 - \tfrac{1}{8}x_1^3 - \tfrac{1}{8}x_1^2 + \tfrac{1}{8}x_1 + \tfrac{1}{8},$$
$$yT_1T_2^2 - \tfrac{1}{8}x_1^3 + \tfrac{1}{8}x_1^2 + \tfrac{1}{8}x_1 - \tfrac{1}{8},$$
$$yT_2^3 - \tfrac{1}{8}x_1^3 + \tfrac{3}{8}x_1^2 - \tfrac{3}{8}x_1 + \tfrac{1}{8},$$
$$x_3 - x_1^3 + 1,$$
$$x_2 - x_1^2 - 1\}.$$

Then $\mathbf{I}(V)$ is generated by $\{x_3 - x_1^3 + 1, \quad x_2 - x_1^2 - 1\}$. We can conclude that the dimension is 1, since in its Gröbner bases there exists a polynomial with *principal variable x_3 and another with principal variable x_2 (see [7]). Therefore our variety is a curve.*

Applying Logar's algorithm [8], we observe that x_3, x_2 are integer over $\mathbb{K}[x_1]$, thus $\overline{F}_1(T) = F_1(T)$. That is, $\mathbb{K}[F_2(T), F_3(T)]$ is integer over $\mathbb{K}[F_1(T)]$.

B. *To find the parameter which is algebraically independent over $\mathbb{K}(F_1(T))$, we compute the reduced Gröbner basis of I, with respect to the lexicographical ordering $y > x_3 > x_2 > T_2 > T_1 > x_1$:*

$$GB(I) = \{T_2 x_1 + T_2 - T_1 x_1 + T_1, \ yT_2^3 - \tfrac{1}{8}x_1^3 + \tfrac{3}{8}x_1^2 - \tfrac{3}{8}x_1 + \tfrac{1}{8},$$
$$yT_2^2 T_1 - \tfrac{1}{8}x_1^3 + \tfrac{1}{8}x_1^2 + \tfrac{1}{8}x_1 - \tfrac{1}{8}, \ yT_2 T_1^2 - \tfrac{1}{8}x_1^3 - \tfrac{1}{8}x_1^2 + \tfrac{1}{8}x_1 + \tfrac{1}{8},$$
$$yT_2^3 - \tfrac{1}{8}x_1^3 - \tfrac{3}{8}x_1^2 - \tfrac{3}{8}x_1 - \tfrac{1}{8}, \ x_3 - x_1^3 + 1, \ ; x_2 - x_1^2 - 1\}.$$

We note, there is not polynomial in the variables T_1, x_1; so T_1 is algebraically independent over $\mathbb{K}(F_1(T))$.

C. *The algebraic relation between the variables F_1, T_2 and T_1 is given by the polynomial $P_1(x_1, T_1, Z) = Z(x_1 + 1) - T_1(x_1 - 1)$, which appears in the above Gröbner basis, that is $P_1(F_1, T_1, T_2) = T_2(F_1 + 1) - T_1(F_1 - 1) = 0$. Now, we take $\nu = 2$.*

D. *Performing the substitution T_1 by T_2^ν in the initial parameterization, we get a parameterization with one parameter:*

$$G_1(T_2) = \frac{T_2^2 + T_2}{T_2^2 - T_2}, \quad G_2(T_2) = \frac{2(T_2^2 + T_2^4)}{(T_2^2 - T_2)^2}, \quad G_3(T_2) = \frac{2T_2^3 + 6T_2^5}{(T_2^2 - T_2)^3}.$$

Simplifying:

$$G_1 = \frac{T_2 + 1}{T_2 - 1}, \quad G_2 = \frac{2(1 + T_2^2)}{(T_2 - 1)^2}, \quad G_3 = \frac{2(1 + 3T_2^2)}{(T_2 - 1)^3}.$$

References

[1] M. Artin and D. Mumford, *Some elementary examples of Unirational Varieties which are non-rational*, Proc. London Math. Soc., **25** (1972), 75–95.

[2] A. Castelnuovo, *Sulla rationalità della involuzioni pinae*, Math. Ann., **44** (1894), 125–155.

[3] A. Cesar, J. Gutierrez and T. Recio, *Reconsidering algorithms for real parametric curves.* Applicable Algebra in Eng. Comm. and Computing, **6** (1995), 345–352.

[4] S. C. Chou, X. S. Gao and Z. M. Li, *Computations with rational parametric equations*, Proc. Computer mathematics, Nankai Ser. Pure Appl. Math. Theor. Phys. World Sci. Publishing, **5** (1993), 86–111.

[5] D. Cox, J. Little and D. O'Shea, *Ideals, Varieties, and Algorithms*, Undergraduate Texts in Mathematics, Springer-Verlag, 1996.

[6] P. Gianni, B. Trager and G. Zacharias, *Gröbner bases and primary decomposition of polynomial ideals*, J. Symbolic Comp., **6** (1988), 149–167.

[7] H. Kredel and V. Weispfenning, *Computing dimension and independent sets for polynomial ideals.* J. Symbolic Comput., **6** (1988), 213–248.

[8] A. Logar, *A computational proof of the Noether Normalization Lemma,* Proc. the 8th Symp. on Appl. Algebra, Algebraic Algorithms, and Error-Correcting Codes, Lectures Notes in Comput. Sci., **357** (1988), 259–273.

[9] M. Nagata, *Theory of Commutative Fields,* Translations of Mathematical Monographs, Amer. Math. Soc., **125** (1993).

[10] M. Ojanguren, *The Witt group and the problem of Lüroth,* Università di Pisa. Dipartimento di Matematica. Dottorado di recerca in matematica. Ets editrice Pisa, 1990.

[11] T. Sederberg, *Improperly parametrized rational curves,* C. Aided Geometric Design **3** (1986), 67–75.

[12] A. Seidenberg, *Constructions in algebra,* Trans. Amer. Math. Soc. **197** (1974), 273–313.

[13] M. Sweedler, *Using Groebner bases to determine the algebraic and trancendental nature of field extensions: return of the killer tag variables,* Proc. the 10th Symp. on Appl. Algebra, Algebraic Algorithms, and Error-Correcting Codes, Lectures Notes in Comput. Sci., **673** (1993), 66–75.

Departamento de Informática, Universidad de Oviedo
Sedes Departamentales Oeste, edif. 1, Campus de Viesques
E-33271 Gijon, Spain
E-mail address: calonso@aic.uniovi.es

Department of Mathematics, Statistics and Computation, University of Cantabria
Avenida de los Castros, s/n
E-39071 Santander, Spain
E-mail address: jaime@matesco.unican.es

Department of Mathematics, Statistics and Computation, University of Cantabria
Avenida de los Castros, s/n
E-39071 Santander, Spain
E-mail address: sarito@matesco.unican.es

Progress in Computer Science and Applied Logic, Vol. 20
© 2001 Birkhäuser Verlag Basel/Switzerland

On Elements of High Order in Finite Fields

Alessandro Conflitti

Abstract. We provide a more careful analysis of a construction of elements high order in finite fields which has recently been proposed by S. Gao. In particular, we improve and generalize one of his results.

Let $\mathbb{F}_q$ be a finite field of q elements. We improve and generalize a lower bound on the order of elements constructed in Theorem 1.1 of [2]. This generalization can probably be combined with some ideas of [1, 6] to obtain new constructions of elements of high order. Different approaches to constructing elements of high order (sometimes exponentially large) have been developed in [3, 4, 5].

We recall that the *order* $\operatorname{ord}\alpha$ of an element $\alpha \in \mathbb{F}_{q^n}$ is the smallest positive integer t such that $\alpha^t = 1$.

In [2] the following result has been proved. Consider $\alpha \in \mathbb{F}_{q^n}$ of degree n over $\mathbb{F}_q$ and a root of $X^{\overline{m}} - g(X)$, where $\overline{m}$ is the smallest power of q greater than or equal to n and $g(X) \in \mathbb{F}_q[X]$ with $\deg g(X) \leq 2\log_q n$ and $g(X) \neq aX^k$ or $aX^{p^l} + b$ for any $a, b \in \mathbb{F}_q$, $k, l \geq 0$, where p is the characteristic of $\mathbb{F}_q$. Then

$$\operatorname{ord}\alpha \geq n^{\frac{\log_q n}{4\log_q(2\log_q n)} - \frac{1}{2}}. \tag{1}$$

The proof of (1) is based on the following statement.

Lemma 1 (Theorem 1.4 of [2]). *If $f \in \mathbb{F}_q[X]$ let us define the sequence of polynomials $f^{(0)}(X) = X$,*

$$f^{(j)}(X) = f\left(f^{(j-1)}(X)\right), \qquad j = 1, 2, \ldots .$$

Suppose that $f(X)$ is not a monomial nor a binomial of the form $aX^{p^l} + b$ where p is the characteristic of $\mathbb{F}_q$. Then the polynomials

$$f(X), f^{(2)}(X), \ldots, f^{(n)}(X), \ldots$$

are multiplicatively independent in $\mathbb{F}_q[X]$, that is, if

$$(f(X))^{k_1} \left(f^{(2)}(X)\right)^{k_2} \cdots \left(f^{(n)}(X)\right)^{k_n} = 1,$$

for any integers $n \geq 1, k_1, k_2, \ldots, k_n$, then $k_1 = k_2 = \ldots = k_n = 0$.

Here we give a more careful analysis of the arguments of [2] and we prove the following improvement and generalization of (1).

Theorem 2. *Let $\alpha \in \mathbb{F}_{q^n}$ be of degree n over $\mathbb{F}_q$ and a root of $X^m - g(X)$, where m is a power of q with $m \geq n$ and $g(X) \in \mathbb{F}_q[X]$ is a polynomial of degree $\deg g = d < n$ such that $g(X) \neq aX^k$ and $g(X) \neq aX^{p^l} + b$ for $a, b \in \mathbb{F}_q$ and integers $k, l \geq 0$, where p is the characteristic of $\mathbb{F}_q$. Then*

$$\operatorname{ord} \alpha \geq \left(\frac{nd}{\log_d^2 n} \right)^{\frac{1}{2} \log_d n}.$$

Proof. If α is a root of $X^m - g(X)$ then since m is a power of q, applying iteratively the Frobenius automorphism we have

$$\alpha^{m^j} = g^{(j)}(\alpha) \qquad j \in \mathbb{N},$$

where as in the statement of Lemma 1, $g^{(j)}(X)$ is the polynomial obtained by composing $g(X)$ with itself j times.

Fix some integer $t \geq 1$ and define the set

$$S_t = \left\{ \sum_{j=0}^{t-1} a_j m^j : 0 \leq a_j \leq \mu_j \right\}$$

where the integers μ_j are defined by the inequalities

$$\frac{n}{td^j} - 1 \leq \mu_j < \frac{n}{td^j}, \qquad j = 0, \dots, t-1.$$

We show that α^a are distinct elements in $\mathbb{F}_{q^n}$ for all $a \in S_t$, thus α has order at least

$$\#S_t = \prod_{j=0}^{t-1} (\mu_j + 1) \geq \prod_{j=0}^{t-1} \frac{n}{td^j} = \left(\frac{n}{t} \right)^t \prod_{j=0}^{t-1} d^{-j} = \left(\frac{n}{t} \right)^t d^{-\frac{t(t-1)}{2}} \qquad (2)$$

Let $a, b \in S_t$ be such that $\alpha^a = \alpha^b$. Write

$$a = \sum_{j=0}^{t-1} a_j m^j \qquad \text{and} \qquad b = \sum_{j=0}^{t-1} b_j m^j$$

where $0 \leq a_j, b_j \leq \mu_j$ for $j = 0, \dots, t-1$. Then $\alpha^a = \alpha^b$ is equivalent to

$$\prod_{j=0}^{t-1} \left(\alpha^{m^j} \right)^{a_j} = \prod_{j=0}^{t-1} \left(\alpha^{m^j} \right)^{b_j}.$$

So we get

$$\prod_{j=0}^{t-1} \left(g^{(j)}(\alpha) \right)^{a_j} = \prod_{j=0}^{t-1} \left(g^{(j)}(\alpha) \right)^{b_j}.$$

Let us define the polynomials

$$h_1(X) = \prod_{a_j > b_j} \left(g^{(j)}(X) \right)^{a_j - b_j} \qquad \text{and} \qquad h_2(X) = \prod_{b_j > a_j} \left(g^{(j)}(X) \right)^{b_j - a_j}.$$

Then $h_1(\alpha) = h_2(\alpha)$ and since α has degree m and $g^{(k)}(X)$ has degree d^k, so $h_1(X)$ and $h_2(X)$ are of degree at most

$$\sum_{j=0}^{t-1} \mu_j d^j < n,$$

thus $h_1(X)$ must be equal to $h_2(X)$. Therefore

$$\prod_{j=0}^{t-1} \left(g^{(j)}(X)\right)^{a_j - b_j} = 1$$

and by Lemma 1, the polynomials $g^{(j)}(X)$ are multiplicatively independent in $\mathbb{F}_q[X]$. So $a_j = b_j$ for $0 \le j \le t-1$, and thus $a = b$.

Therefore, by (2)

$$\operatorname{ord} \alpha \ge \left(\frac{n}{t}\right)^t d^{-\frac{t(t-1)}{2}} = \left(\frac{n\sqrt{d}}{t\sqrt{d^t}}\right)^t$$

for any integer $t \ge 1$. Remarking that the function

$$\psi(z) = \left(\frac{n}{z}\right)^z d^{-\frac{z(z-1)}{2}}$$

monotonically decreases for $z \ge \log_d n$ and selecting

$$t = \lfloor \log_d n \rfloor,$$

we obtain

$$\operatorname{ord} \alpha \ge \left(\frac{n}{\log_d n}\right)^{\log_d n} d^{-\frac{\log_d n (\log_d n - 1)}{2}}$$

and the desired result follows. $\qquad\square$

In [2] some heuristic arguments have been given, showing that there exists a polynomial $g(X) \in \mathbb{F}_q[X]$ of degree $d \le 2\log_q m$ satisfying the condition of Theorem 2 with $n = m$, thus such that $X^m - g(X)$ is irreducible. In this case, that is, for $n = m$ and $d \le 2\log_q m$, the lower bound of Theorem 2 is about the square of the lower bound (1).

It would be interesting to find new explicitly given classes of elements $\alpha \in \mathbb{F}_{q^n}$ of large order.

Open Question 3. *Extend the lower bound of Theorem 2 to roots of polynomials of the form $X^m h(X) + g(X) \in \mathbb{F}_q[X]$ where m is a power of q and $h(X), g(X) \in \mathbb{F}_q[X]$ are polynomials of degree at most d.*

In particular, one should obtain an analogue of Lemma 1 for rational functions (which could be of independent interest).

14 A. Conflitti

Acknowledgment

The author would like to thank Igor Shparlinski for suggesting this problem and his helpful advice, and Francesco Pappalardi for his useful remarks during the final preparation of this paper.

References

[1] I. F. Blake, S. Gao and R. Mullin, *Normal and self-dual normal bases from factorization of $cx^{q+1} + dx^q - ax - b$*, SIAM J. Discr. Math. **7** (1994), 499–512.

[2] S. Gao, *Elements of provable high orders in finite fields*, Proc. Amer. Math. Soc., **127** (1999), 1615–1623.

[3] J. von zur Gathen and I. Shparlinski, *Orders of Gauss periods in finite fields*, Appl. Algebra in Engin., Commun. and Computing, **9** (1998), 15–24.

[4] J. von zur Gathen and I. Shparlinski, *Constructing elements of large order in finite fields and Gauss periods*, Proc. the 13th Symp. on Appl. Algebra, Algebraic Algorithms, and Error-Correcting Codes, Hawaii, 1999, Lect. Notes in Comp. Sci., **1719** (1999), Springer-Verlag, Berlin, 404–409.

[5] J. von zur Gathen and I. Shparlinski, *Gauss periods in finite fields*, Proc. 5th Conference of Finite Fields and their Applications, Augsburg, 1999, (to appear).

[6] V. M. Sidel'nikov, *On the normal bases of a finite field*, Matem. Sbornik, **133** (1987), 497–507 (in Russian), translation in Math. USSR Sbornik, **61** (1988), 485–494.

Dipartimento di Matematica,
Università degli Studi "Roma Tre",
Largo San Leonardo Murialdo 1,
I-00146 Roma, Italy
E-mail address: confliti@mat.uniroma3.it

Progress in Computer Science and Applied Logic, Vol. 20
© 2001 Birkhäuser Verlag Basel/Switzerland

Counting the Number of Points on Affine Diagonal Curves

Cunsheng Ding, David R. Kohel, and San Ling

Abstract. The number of points on affine diagonal curves $aX^m + bY^n = c$ over finite fields can be computed in terms of cyclotomic numbers. The approach of Berndt, Evans and Williams [1] is to express the number of points in terms of generalized Jacobi sums, then to relate the Jacobi sums $J_r(\chi^u, \chi^v)$ to cyclotomic numbers. In this article we present the direct elementary method for the number of points on the affine curves $aX^m + bY^n = c$ over finite fields in terms of cyclotomic numbers. This approach is applicable when explicit formulas are already known for cyclotomic numbers, and circumvents the use of Jacobi sums. It generalizes to the determination of the number of points on affine diagonal hypersurfaces of higher dimension. The curves for which this method applies includes examples of elliptic and hyperelliptic curves which are of interest for public-key cryptosystems, coding theory and the design and analysis of sequences.

1. Introduction

Public-key cryptosystems play an important role in information and system security [21]. Elliptic and hyperelliptic curves have been successfully employed to construct public-key cryptosystems [12, 13, 19, 20]. Counting the number of points on these curves is necessary for the construction of such cryptosystems [14, 18]. Curves have also important applications in sequences and coding theory [24, 25].

Let α be a generating element of $\mathrm{GF}(q)^*$, and let e be a positive divisor of $q - 1$. The cyclotomic classes $C_i^{(e)}$ of order e are defined with respect to α as $C_i^{(e)} = \alpha^i C_0^{(e)}$ for $i = 0, \ldots, e - 1$, where $C_0^{(e)} = \{\beta^e \mid \beta \in \mathrm{GF}(q)^*\}$ is the set of e-th residues in $\mathrm{GF}(q)$. The corresponding *cyclotomic numbers* of order e are defined by

$$(i, j)_e = \left| (C_i^{(e)} + 1) \cap C_j^{(e)} \right|, \quad 0 \le i, j \le e - 1.$$

Cyclotomic numbers were introduced by Gauss [11, §358] in his treatment of the number of solutions (x, y) of

$$aX^3 \equiv bY^3 + 1 \pmod{p}.$$

In the course of his general study he also defines the *periods* [11, §343], which we now refer to as Gaussian periods. The study of relations, properties and formulas for cyclotomic numbers and Gaussian periods is referred to as *cyclotomy*.

Cyclotomy finds applications in Waring's problems [5], the construction of difference sets [26, 27, 29] and almost difference sets [3], coding theory [7, 8, 9, 16, 17], and cryptography [3]. Dickson [6] used cyclotomy to derive results on the number of solutions of the diagonal surfaces $aX^m + bY^m + cZ^m = d$.

The connection between the number of points on affine diagonal curves and cyclotomy is generally expressed in terms of the theory of Jacobi sums. Berndt et al. [1], for example, express the number of points on an affine diagonal curve $aX^m + bY^n = c$ in terms of cyclotomic numbers via a two step process. First the number of points is expressed in terms of generalized Jacobi sums, then they show that the collection of Jacobi sums $J_r(\chi^u, \chi^v)$ can be expressed in terms of cyclotomic numbers. In this paper we present the direct elementary method for the calculation of the number of points on the affine diagonal curves $aX^m + bY^n = c$ over finite fields in terms of cyclotomic numbers. This applies in particular when m and n divide an exponent e for which explicit formulas are known for cyclotomic numbers. If one of the exponents $m = 2$, and $n > 4$, then the corresponding curve is hyperelliptic, and when (m, n) is one of $(2, 3)$, $(2, 4)$, or $(3, 3)$, the curve is elliptic. In the latter cases the curves have complex multiplication, and the determination of the number of points over a finite field does not require any of the sophisticated methods of Schoof, Atkin, and Elkies (see [22, 23, 10]). This direct approach circumvents the need to compute or analyse Jacobi sums, and provides an effective means of computing the number of points on more general diagonal hypersurfaces in terms of cyclotomic numbers.

2. Cyclotomy and Affine Diagonal Curves

In this section we use cyclotomic numbers to express the number of points on the curve

$$aX^m + bY^n = 1 \tag{1}$$

over $GF(q)$, where a and b are in $GF(q)^*$, in terms of the cyclotomic numbers of certain order. It is straightforward to see that when $m = n \mid q - 1$, then the number of points (x, y) on (1) with $xy \neq 0$ is m^2 times a cyclotomic number.

Given a curve $aX^m + bY^n = c$, we first write $m = m_1 m_2$ and $n = n_1 n_2$, where $m_2 = \gcd(q - 1, m)$ and $n_2 = \gcd(q - 1, n)$. Then there exist integers r and s, relatively prime to $q - 1$, such that $rm \equiv m_2 \bmod q - 1$ and $sn \equiv n_2 \bmod q - 1$. Since the maps $\alpha \mapsto \alpha^r$ and $\alpha \mapsto \alpha^s$ are automorphisms of $GF(q)^*$ sending k-th residues to k-residues, the curve $aX^{mr} + bY^{ns} = 1$, or equivalently the curve $aX^{m_2} + bY^{n_2} = 1$, has the same number of points as the curve (1). We can thus reduce to the case that m and n both divide $q - 1$. Henceforth, we assume that m and n divide $q - 1$. We set e equal to the least common multiple of m and n, which clearly also divides $q - 1$.

The genus of the curve (1) is known to be

$$\frac{(m-1)(n-1) - \gcd(m,n) + 1}{2}.$$

For exponents (m, n) with $1 < m \leq n$, the curve has genus zero if and only if $(m, n) = (2, 2)$ and has genus one if and only if (m, n) is in $\{(2,3),(2,4),(3,3)\}$. Of particular interest are the hyperelliptic curves, where $m = 2$ and $n > 4$.

As before, fix a primitive element α of $\mathrm{GF}(q)$, and define cyclotomic classes $C_i^{(m)}$ of order m. Now we consider the relation among the cyclotomic classes of orders m and e.

Lemma 2.1. *Suppose that $e = mr$. Then $C_j^{(m)}$ is the disjoint union of the r cyclotomic classes $C_{im+j}^{(e)}$ for $0 \leq i < r$.*

Proof. It is straightforward to see that $C_0^{(m)} = \bigcup_{i=0}^{r-1} C_{im}^{(e)}$. Hence $C_j^{(m)} = \alpha^j C_0^{(m)}$ has the form indicated. □

Let $N(a, b)$ denote the number of points on the curve (1), and define $\delta_m(c)$ to be the number of solutions of the equation $cX^m = 1$. It is clear that $\delta_m(c)$ equals m if c is an m-th residue and is zero otherwise.

Theorem 2.2. *Let r and s be the integers such that $e = mr = ns$, and define h and k to be integers such that $-a$ and b lie in $C_h^{(e)}$ and $C_k^{(e)}$, respectively. Then*

$$N(a, b) = \delta_m(a) + \delta_n(b) + mn \sum_{i=0}^{r-1} \sum_{j=0}^{s-1} (in + h, jm + k)_e.$$

Proof. It is clear that $N(a, b) - \delta_m(a) - \delta_n(b)$ is the number of points (x, y) on (1) such that $xy \neq 0$. Since x^m takes on each element of $C_0^{(m)}$ exactly m times as x ranges over $\mathrm{GF}(q)^*$, it follows that

$$
\begin{aligned}
N(a,b) - \delta_m(a) - \delta_n(b) &= mn \left| \left(-a\, C_0^{(m)} + 1 \right) \bigcap b\, C_0^{(n)} \right| \\
&= mn \left| \left(\bigcup_{i=0}^{s-1} -a\, C_{im}^{(e)} + 1 \right) \bigcap \left(\bigcup_{j=0}^{s-1} b\, C_{jn}^{(e)} \right) \right| \\
&= mn \left| \left(\bigcup_{i=0}^{r-1} C_{im+h}^{(e)} + 1 \right) \bigcap \left(\bigcup_{j=0}^{s-1} C_{jn+k}^{(e)} \right) \right| \\
&= mn \sum_{i=0}^{r-1} \sum_{j=0}^{s-1} (im + h, jn + k)_e,
\end{aligned}
$$

which completes the proof. □

In the next section we apply the theorem to the determination of the number of points on curves which have the form (1).

3. Examples and Computations

Theorem 2.2 shows that the number of points on curves of form (1) can be calculated when cyclotomic numbers of order e are known. We now do some specific computations to illustrate this idea.

Example 3.1. We consider the genus one curve

$$aX^2 + bY^4 = 1,$$

over $\mathrm{GF}(q)$, where $q \equiv 1 \bmod 4$. By definition $\delta_2(a) = 2$ if $a \in C_0^{(4)} \cup C_2^{(4)}$, and $\delta_2(a) = 0$ otherwise. Similarly $\delta_4(b) = 4$ if $b \in C_0^{(4)}$, and $\delta_4(b) = 0$ otherwise. To calculate the number of points on this curve, we apply the known formulas for the cyclotomic numbers of order 4.

It has been proven [26] that the 16 possible cyclotomic numbers $(h, k)_4$ are determined by the decomposition $q = u^2 + 4v^2$, where $u \equiv 1 \bmod 4$ and the sign of v is dependent on the choice of the primitive root used to define the cyclotomic classes. There are at most five distinct cyclotomic numbers of order 4. The relations of these numbers are given in Table 1, and the values A, B, C, D and E are given by Table 2.

$h\backslash k$	0	1	2	3
0	A	B	C	D
1	B	D	E	E
2	C	E	C	E
3	D	E	E	B

when $q \equiv 1 \bmod 8$

$h\backslash k$	0	1	2	3
0	A	B	C	D
1	E	E	D	B
2	A	E	A	E
3	E	D	B	E

when $q \equiv 5 \bmod 8$

TABLE 1. The relations of the cyclotomic numbers of order 4.

	$q \equiv 1 \bmod 8$	$q \equiv 5 \bmod 8$
$16A$	$q - 11 - 6u$	$q - 7 + 2u$
$16B$	$q - 3 + 2u + 8v$	$q + 1 + 2u - 8v$
$16C$	$q - 3 + 2u$	$q + 1 - 6u$
$16D$	$q - 3 + 2u - 8v$	$q + 1 + 2u + 8v$
$16E$	$q + 1 - 2u$	$q - 3 - 2u$

TABLE 2. The values of the cyclotomic numbers of order 4.

Let $(-a, b) \in C_h^{(2)} \times C_k^{(4)}$. From Theorem 2.2 and the tables of cyclotomic numbers and their relations, it follows that the number of points $N(a, b)$ on the curves $aX^2 + bY^4 = 1$ are those given by Table 3. □

(h,k)	$q \equiv 1 \bmod 8$	$q \equiv 5 \bmod 8$
$(0,0)$	$q - 1 - 2u$	$q - 1 + 2u$
$(0,1)$	$q + 1 + 4v$	$q + 1 - 4v$
$(0,2)$	$q - 1 + 2u$	$q - 1 - 2u$
$(0,3)$	$q + 1 - 4v$	$q + 1 + 4v$
$(1,0)$	$q + 1 + 2u$	$q + 1 - 2u$
$(1,1)$	$q - 1 - 4v$	$q - 1 + 4v$
$(1,2)$	$q + 1 - 2u$	$q + 1 + 2u$
$(1,3)$	$q - 1 + 4v$	$q - 1 - 4v$

TABLE 3. The number of points on $aX^2 + bY^4 = 1$.

Remark 3.2. Since the genus of the curve is one, the number of points of the projective model $\mathcal{C} : aX^2 Z^2 + bY^4 = Z^4$ of the curve of Example 3.1 must satisfy the Hasse bound

$$||\mathcal{C}(\mathrm{GF}(q))| - q - 1| \leq 2\lfloor \sqrt{q} \rfloor.$$

If $u = 1$ or $v = \pm 1$ remains fixed, then for $q = u^2 + 4v^2$ sufficiently large, one verifies from Table 3 that there exist curves $\mathcal{C}$ of this form attaining the maximal possible points for this genus.

Example 3.3. We consider the genus 2 hyperelliptic curve

$$X^2 = Y^6 + 1.$$

over $\mathrm{GF}(q)$, where $q \equiv 7 \bmod 12$. In the notation of Theorem 2.2 we have $m = 2$, $n = 6$, $e = 6$, and $a = -b = 1$, so find $\delta_2(a) = 2$ and $\delta_6(b) = 0$. To calculate the number of points on this curve, we apply known formulas for cyclotomic numbers of order 6 (see [26]). The relations of these numbers are given in Table 4.

$h\backslash k$	0	1	2	3	4	5
0	A	B	C	D	E	F
1	G	H	I	E	C	I
2	H	J	G	F	I	B
3	A	G	H	A	G	H
4	G	F	I	B	H	J
5	H	I	E	C	I	G

TABLE 4. The relations of the cyclotomic numbers of order 6.

For $q \equiv 7 \bmod 12$, the 36 cyclotomic numbers are functions of a representation $q = u^2 + 3v^2$, where $u \equiv 1 \bmod 3$ and the sign of v is dependent on the choice of primitive root used to define the cyclotomic classes.

Let α be the primitive element of $\mathrm{GF}(q)$ employed to define the cyclotomic classes of order 6, and let $2 = \alpha^m$. The values of the 10 basic constants are given in Table 5. By Theorem 2.2 and the above cyclotomic numbers of order 6, we find

	$m \equiv 0 \bmod 3$	$m \equiv 1 \bmod 3$	$m \equiv 2 \bmod 3$
$36A$	$q - 11 - 8u$	$q - 11 - 2u$	$q - 11 - 2u$
$36B$	$q + 1 - 2u + 12v$	$q + 1 - 2u - 12v$	$q + 1 + 4u$
$36C$	$q + 1 - 2u + 12v$	$q + 1 - 8u + 12v$	$q + 1 - 2u + 12v$
$36D$	$q + 1 + 16u$	$q + 1 + 10u + 12v$	$q + 1 + 10u - 12v$
$36E$	$q + 1 - 2u - 12v$	$q + 1 - 2u - 12v$	$q + 1 - 8u - 12v$
$36F$	$q + 1 - 2u - 12v$	$q + 1 + 4u$	$q + 1 - 2u + 12v$
$36G$	$q - 5 + 4u + 6v$	$q - 5 + 4u + 6v$	$q - 5 - 2u + 6v$
$36H$	$q - 5 + 4u - 6v$	$q - 5 - 2u - 6v$	$q - 5 + 4u - 6v$
$36I$	$q + 1 - 2u$	$q + 1 + 4u$	$q + 1 + 4u$
$36J$	$q + 1 - 2u$	$q + 1 - 8u + 12v$	$q + 1 - 8u - 12v$

TABLE 5. The values of the cyclotomic numbers of order 6.

$$N(1, -1) = 2 + 12 \sum_{i=0}^{2} (0, 2i + 3)_6$$
$$= 2 + 12(E + A + C) = q - 1 - 4u$$

for the number of points on the curve $X^2 = Y^6 + 1$.   □

Let $F(X)$ and $G(X)$ be permutation polynomials for $\mathrm{GF}(q)$. Then Theorem 2.2 also applies to curves of the form

$$aF(X)^m + bG(Y)^n = 1.$$

We indicate in the next example how this can be applied to point counting on curves of a more exotic form.

Example 3.4. Let q be of the form $30t + 7$. Then $5X^5 + 5cX^3 + c^2 X$ is a permutation polynomial of $\mathrm{GF}(q)$ [15, p. 352]. Let $F(X) = 5X^5 + 5X^3 + X$ and $G(Y) = 5Y^5 - 5Y^3 + Y$. Then both $F(X)$ and $G(Y)$ are permutation polynomials of $\mathrm{GF}(q)$. Then the number of points on the curve

$$125X^{15} + 375X^{13} + 450X^{11} + 275X^9 + 90X^7 + 15X^5 + X^3 +$$
$$25Y^{10} - 50Y^8 + 35Y^6 - 10Y^4 + Y^2 = 1$$

can be computed with the help of cyclotomic numbers of order 6.   □

It should be noted that in each of the above examples a specific equation was treated. With a minimal amount of additional work, by Theorem 2.2 the complete set of cyclotomic numbers $(i, j)_{m,n}$ of *mixed order* m, n could be determined, reducing the formulas for cyclotomic numbers of order e to explicit mixed order formulas for all divisors m, n of e.

4. Generalization

The concepts of cyclotomic numbers have natural generalizations to higher dimensions, and for many fixed exponents and dimensions it is possible to find explicit formulas for these values. This permits a cyclotomic approach to the study of general diagonal hypersurfaces

$$a_1 X_1^{e_1} + a_2 X_2^{e_2} + \cdots + a_n X_n^{e_n} = 1 \tag{2}$$

over $GF(q)^n$. In fact these equations are *quotients* of the hypersurface

$$a_1 Z_1^e + a_2 Z_2^e + \cdots + a_n Z_n^e = 1, \tag{3}$$

where e_i divides e for each i, by the map $(z_1, \ldots, z_n) \longmapsto (z_1^{e/e_1}, \ldots, z_n^{e/e_n})$. There exist other quotients, and it is an interesting problem to determine the number of points on these general quotients.

To illustrate how the number of solutions of the general diagonal hypersurface (2) is determined by cyclotomic numbers, we consider the special case

$$X_1^2 + X_2^4 + X_3^2 + X_4^4 = 1 \tag{4}$$

over the field $GF(q)$, where $q \equiv 1 \bmod 4$.

We denote the number of points on $X^2 + Y^4 = a$ by $N(a)$. Then $N(0) = 2q - 1$, and otherwise, when $a \in C_k^{(4)}$, it follows from the computation in Example 3.3 that $N(a)$ is given by

k	$q \equiv 1 \bmod 8$	$q \equiv 5 \bmod 8$
0	$q - 1 + 2u$	$q - 1 - 2u$
1	$q - 1 - 4v$	$q - 1 + 4v$
2	$q - 1 - 2u$	$q - 1 + 2u$
3	$q - 1 + 4v$	$q - 1 - 4v$

where $q = u^2 + 4v^2$. Let $h = (q - 1)/2 \bmod 4$ so that $C_h^{(4)}$ is the cyclotomic class of -1. Then the number N of points on the hypersurface (4) is given by

$$
\begin{aligned}
N &= \sum_{a \in GF(q)} N(a) N(1 - a) \\
&= 2N(0)N(1) + \sum_{i=0}^{3} \sum_{j=0}^{3} |C_i^{(4)} \cap (1 - C_j^{(4)})| N(\alpha^i) N(\alpha^j) \\
&= 2N(0)N(1) + \sum_{i=0}^{3} \sum_{j=0}^{3} |C_{i+h}^{(4)} \cap (C_j^{(4)} - 1)| N(\alpha^i) N(\alpha^j) \\
&= 2N(0)N(1) + \sum_{i=0}^{3} \sum_{j=0}^{3} (i + h, j)_4 N(\alpha^i) N(\alpha^j).
\end{aligned}
$$

From this formula, we find that when $q \equiv 1 \bmod 8$ we have

$$N = q^3 - q(4u + 1) - 2(u^3 + u^2 + 4uv^2 + 4v^2),$$

for the number of points on the hypersurface (4). For example, when $q = 17$, we have $N = 4760$. When $q \equiv 5 \bmod 8$, we derive a similar formula.   □

The approach given in Berndt et al. [1] requires the computation of generalized Jacobi sums $J_r(\chi_1^{n_1}, \chi_2^{n_2}, \chi_3^{n_3}, \chi_4^{n_4})$. This example shows, for small exponents e, how the approach through cyclotomic numbers suffices to compute points on general diagonal hypersurfaces.

5. Concluding Remarks

The classical approach to the study of point counting on curves over finite fields is to express the number of points in terms of character sums. The number of points on the curve (1) and more generally the diagonal hypersurface (2), can be expressed in terms of Jacobi or Gaussian sums (see [15, Section 6.3] and [1]). In 1934 Davenport and Hasse [4] gave theoretical characterizations of these sums, which is the foundation for most of the present-day explicit formulas for cyclotomic numbers. Via a more computationally sophisticated algorithm, Buhler and Koblitz [2] recently showed, at least for prime exponents e, that it is possible to apply this characterization directly to compute the number of points on certain hyperelliptic curves of the form (1) in polynomial time. In contrast, the present approach makes use of elementary formulas with simple implementation, to treat the same hyperelliptic curves.

The use of elliptic curves has become central to public-key cryptography in the last years, and considerable attention has been given to the subject of hyperelliptic curves in cryptography. Beyond the examples of elliptic and hyperelliptic curves for which the present method can be applied, the general class of diagonal curves may be of future interest for cryptosystems because of their rich structure and amenability to rapid point counting algorithms. In addition, curves of the form (1) with relative small exponents may be useful in constructing error correcting codes and sequences. Detailed information about curves and their applications in coding theory, sequences and cryptography can be found in [24, 25].

Acknowledgments: The authors thank the reviewer for helpful comments that improved the presentation of this paper.

References

[1] B. C. Berndt, R. J. Evans, and K. S. Williams, *Gauss and Jacobi sums,* John Wiley & Sons, Inc., 1998.

[2] J. Buhler and N. Koblitz, *Lattice basis reduction, Jacobi sums, and hyperelliptic cryptosystems,* Bull. Austral. Math. Soc., **58** (1998), 147–154.

[3] T. Cusick, C. Ding, and A. Renvall, *Stream ciphers and number theory,* North-Holland Mathematical Library, Elsevier/North-Holland, 1998

[4] H. Davenport and H. Hasse, *Die Nullstellen der Kongruenzzetafunktionen in gewissen zyklischen Fällen,* J. Reine Angew. Math., **172** (1934), 115–182.

[5] L. E. Dickson, *Cyclotomy, higher congruences, and Waring's problem*, Amer. J. Math., **57** (1935), 391–424, and 463–474.

[6] L. E. Dickson, *Congruences involving only e-th powers*, Acta Arith., **1** (1936), 161–167.

[7] C. Ding and T. Helleseth, *New generalized cyclotomy and its applications*, Finite Fields and Their Applications, **4** (1998), 140–166.

[8] C. Ding and T. Helleseth, *Generalized cyclotomic codes of length $p_1^{e_1} \cdots p_t^{e_t}$*, IEEE Trans. Information Theory, **45(2)** (1999), 467–474.

[9] C. Ding and V. Pless, *Cyclotomy and duadic codes of prime lengths*, IEEE Trans. Information Theory, **45(2)** (1999), 453–466.

[10] N. Elkies, *Elliptic and modular curves over finite fields and related computational issues*, in: Computational Perspectives on Number Theory, A conference in honor of A.O.L. Atkin, 1998.

[11] C. F. Gauss, *Disquisitiones Arithmeticae*, Leipzig, 1801. English translation, Yale, New Haven, 1966.

[12] N. Koblitz, *Elliptic curve cryptosystems*, Math. Comp., **48** (1987), 203–209.

[13] N. Koblitz, *Hyperelliptic curve cryptosystems*, J. Cryptology, **1** (1989), 139–150.

[14] N. Koblitz, *CM curves with good cryptographic properties*, Advances in Cryptology — Proc. Crypto'91, Lecture Notes in Comput. Sci., Vol. 576, Springer-Verlag, Berlin, 1992, 279–287.

[15] R. Lidl, H. Niederreiter, *Finite fields*, Encyclopedia of Mathematics and Its Applications, Vol. 20, Addison-Wesley, 1983; Second edition, Cambridge University Press, 1997.

[16] F. J. MacWilliams, *Cyclotomic numbers, coding theory and orthogonal polynomials*, Discrete Mathematics, **3** (1972), 133–151.

[17] R. J. McEliece and H. Rumsey, Jr., *Euler products, cyclotomy, and coding*, J. of Number Theory, **4** (1972), 302–311.

[18] A. J. Menezes, S. A. Vanstone, and R. J. Zuccherato, *Counting points on elliptic curves over F_{2^m}*, Math. Comp., **60** (1993), 407–420.

[19] V. Miller, *Uses of elliptic curves in cryptography*, Advances in Cryptology–Proc. Crypto'85, Lecture Notes in Comp. Sci., Vol. 218, Springer-Verlag, Berlin, 1986, 417–426.

[20] T. Okamoto and K. Sakurai, *Efficient algorithms for the construction of hyperelliptic cryptosystems*, Advances in cryptology – Proc. Crypto'91, Lecture in Comput. Sci., Vol. 576. Springer-Verlag, Berlin, 1991, 267–278.

[21] A. Salomaa, *Public-key cryptography*, 2nd Ed., Springer-Verlag, 1996.

[22] R. Schoof, *Elliptic curves over finite fields and the computation of square roots mod p*, Math. Comp., **44** (1985), 483–494.

[23] R. Schoof, *Counting points on elliptic curves over finite fields*, J. Théorie des Nombres de Bordeaux., **7** (1995), 219–254.

[24] I. E. Shparlinski, *Computational and algorithmic problems in finite fields*, Mathematics and Its Applications, **88** (1992), Kluwer Academic Publishers, Boston, London.

[25] I. E. Shparlinski, *Finite fields: theory and computation – the meeting point of number theory, computer science, coding theory and cryptography*, Mathematics and Its Applications, **477** (1999), Kluwer Academic Publishers, Boston, London.

[26] T. Storer, *Cyclotomy and difference sets*, Markham, Chicago, 1967.

[27] T. Storer, *Cyclotomies and difference sets modulo a product of two distinct odd primes*, Michigan Math. J., **14** (1967), 117–127.

[28] A. L. Whiteman, *The cyclotomic numbers of order twelve*, Acta Arith., **6** (1960), 53–76.

[29] A. L. Whiteman, *A family of difference sets*, Illinois J. Math., **6** (1962), 107–121.

Cunsheng Ding
Department of Computer Science
National University of Singapore
S16, Room 05-08, 3 Science Drive 2
Singapore 117543
E-mail address: dingcs@comp.nus.edu.sg

David R. Kohel
School of Mathematics and Statistics
Carslaw Building, F7, University of Sydney
Sydney, NSW 2006, Australia
E-mail address: kohel@maths.usyd.edu.au

San Ling
Department of Mathematics
National University of Singapore
2 Science Drive 2, Singapore 117543
E-mail address: matlings@nus.edu.sg

Progress in Computer Science and Applied Logic, Vol. 20
© 2001 Birkhäuser Verlag Basel/Switzerland

Small Values of the Carmichael Function and Cryptographic Applications

John B. Friedlander, Carl Pomerance, and Igor E. Shparlinski

Abstract. We outline some cryptographic applications of the recent results of the authors about small values of the Carmichael function and the period of the power generator of pseudorandom numbers. Namely, we show rigorously that almost all randomly selected RSA moduli are safe against the so-called cycling attack and we also provide some arguments in support of the reliability of the timed-release crypto scheme, which has recently been proposed by R. L. Rivest, A. Shamir and D. A. Wagner.

1. Introduction

For an integer $n \geq 1$ we define the *Carmichael function* $\lambda(n)$ as the largest possible order of elements of the unit group in the residue ring modulo n. More explicitly, for a prime power p^k we write

$$\lambda\left(p^k\right) = \begin{cases} p^{k-1}(p-1), & \text{if } p \geq 3 \text{ or } k \leq 2; \\ 2^{k-2}, & \text{if } p = 2 \text{ and } k \geq 3; \end{cases}$$

and finally,

$$\lambda(n) = \mathrm{lcm}\left(\lambda(p_1^{k_1}), \ldots, \lambda(p_\nu^{k_\nu})\right),$$

where

$$n = p_1^{k_1} \ldots p_\nu^{k_\nu}$$

is the prime number factorization of n.

Various upper and lower bounds for $\lambda(n)$ have been obtained in [6]. In particular, good bounds were given there for the normal order of $\lambda(n)$, namely estimates for the size of $\lambda(n)$ for most numbers n. In [9] a new result was obtained about the distribution of the exceptional numbers n for which $\lambda(n)$ is abnormally small, which in turn was used to obtain lower bounds for the period of the *power generator* pseudorandom sequence. This sequence of integers (u_n), satisfies the recurrence relation

$$u_n \equiv u_{n-1}^e \pmod{m}, \quad 0 \leq u_n \leq m-1, \quad n = 1, 2, \ldots, \qquad (1)$$

with the *initial value* $u_0 = \vartheta$ (an integer coprime to m) and *exponent* e (an integer at least 2). Besides being a source of pseudorandom numbers, this generator has numerous cryptographic applications and has been extensively studied in the literature, see [2, 3, 4, 5, 7, 8, 10, 11, 13, 16, 18, 24, 25].

It is easy to see that the power generator sequence (1) is eventually periodic and that the largest possible period for a given modulus m is $\lambda(\lambda(m))$. Thus, we have the connection to Carmichael's function. In two special cases $\gcd(e, \varphi(m)) = 1$, where $\varphi(m)$ is the Euler function, and $e = 2$ the sequence (1) is known as the *RSA generator* and as the *Blum–Blum–Shub generator*, respectively. It is easy to see that if $\gcd(e, \varphi(m)) = 1$ then the sequence is purely periodic.

It has been shown in [9] that if the modulus m of the power generator is the product pl of two primes, the case of greatest importance for cryptographic applications, then, provided that the parameters ϑ, e, p, l are selected at random, with overwhelming probability the period t is greater than $m^{1-\varepsilon}$. Thus one can apply the results of [10] to establish the uniformity of distribution of this generator. This application was essentially the original motivation of [9].

Here we give a brief outline of these results and describe further crypto-graphic applications of two types which have been only briefly mentioned in [9]. Namely, we consider the feasibility of the *cycling attack* on RSA and a variation thereof, and also the possibility of finding a "short-cut" in the timed-release crypto scheme which has recently been proposed by R. L. Rivest, A. Shamir and D. A. Wagner [22].

We use $\log x$ to denote the natural logarithm of x and $\mathcal{P}$ to denote the set of primes.

2. Small Values of the Carmichael Function and the Period of the Power Generator

Here we outline some relevant results from [9]. First of all it has been shown there that the set of integers n for which $\lambda(n)$ is small is rather thin.

Theorem 2.1. *For all sufficiently large numbers N and for $\Delta \geq (\log \log N)^3$, the number of positive integers $n \leq N$ with*

$$\lambda(n) \leq n \exp(-\Delta)$$

is at most $N \exp\left(-0.69 \left(\Delta \log \Delta\right)^{1/3}\right)$.

Theorem 2.1 was applied in [9] to obtain a lower bound for the largest possible period of the power generator.

Theorem 2.2. *For Q sufficiently large and for any $\Delta \geq 2 (\log \log Q)^3$ the number of pairs $(p, l) \in \mathcal{P}^2$, $1 < p < l \leq Q$, with*

$$\lambda(\lambda(pl)) < Q^2 \exp(-\Delta)$$

is at most $Q^2 \exp\left(-0.16 \left(\Delta \log \Delta\right)^{1/3}\right)$.

Using Theorem 2.2 and some bounds for the number of choices of the initial value ϑ and the exponent e for which the period of the sequence (u_n) given by (1) is small, it is shown that if p, l, ϑ and e are selected at random then it is likely that the period of the corresponding generator modulo $m = pl$ is close to m. More precisely, we proved:

Theorem 2.3. *For Q sufficiently large, for any $\Delta \geq 6 \left(\log \log Q \right)^3$, and for all pairs $(p, l) \in \mathcal{P}^2$, $1 < p < l \leq Q$, except at most $Q^2 \exp \left(-0.1 \left(\Delta \log \Delta \right)^{1/3} \right)$ of them, the following statement holds. For all pairs (ϑ, e) with*

$$1 \leq \vartheta \leq m - 1, \qquad 1 \leq e \leq \lambda(m), \qquad \gcd(\vartheta, m) = \gcd(e, \lambda(m)) = 1,$$

where $m = pl$, except at most $m\lambda(m) \exp \left(-0.2\Delta \right)$ of them, the period t of the sequence (u_n) given by (1) satisfies

$$t \geq Q^2 \exp \left(-\Delta \right).$$

Finally the case $e = 2$, that is, the Blum–Blum–Shub generator, was considered as well, although fixing the value of e makes the problem much more difficult and so in this case the result is weaker. It guarantees that, if other parameters, that is, p, l and ϑ, are selected at random, then the period is likely to be of order almost $m^{1/2}$ at least.

Theorem 2.4. *Given $\varepsilon > 0$, there exist positive constants c, γ such that for Q sufficiently large, there are more than $cQ^2 / \left(\log Q \right)^4$ pairs $(p, l) \in \mathcal{P}^2$, $p < l \leq Q$, such that for all integers ϑ with*

$$1 \leq \vartheta \leq m - 1 \qquad and \qquad \gcd(\vartheta, m) = 1,$$

where $m = pl$, except at most $m^{1-\gamma}$ of them, the period t of the sequence (u_n) given by (1) with $e = 2$ satisfies

$$t \geq cQ^{1-\varepsilon}.$$

3. Cycling Attacks on RSA

The first application we discuss is the conclusion that the so-called *cycling attack* on the RSA cryptosystem has a negligible chance to be efficient. Despite the common belief that this should be the case, no rigorous proof of the statement has been given. The attack is based on the observation that the power generator (1) can be considered as a sequence of consecutive RSA encryptions starting with the "message" u_0. Thus, if the period is t then after $t - 1$ iterations of the encrypted message u_1 we obtain $u_t \equiv u_0 \pmod{m}$ and if t is small then this is an efficient procedure. (Note that in RSA the encryption exponent e is coprime to $\varphi(m)$, so the sequence (1) is purely periodic.) Even more, if t is small then, because it is very likely that the periods t_p and t_l of this sequence modulo p and l are distinct, after at most $\min\{t_p, t_l\}$ iterations this attack may produce a complete factorization of m. Indeed, it is very likely that

$$\gcd\left(u_{t_p} - u_0, m\right) = p \qquad \text{and} \qquad \gcd\left(u_{t_l} - u_0, m\right) = l.$$

This attack, as well as various ways of protecting against it, have been discussed in the literature, see [3, 17, 18, 21, 23]. In particular, so-called *safe primes* have been introduced. R. L. Rivest and R. D. Silverman [23] present heuristic arguments which show that randomly selected primes p and l are already likely to be strong against this attack so that it is not so necessary to make special choices. Theorem 2.3 implies a more precise and completely rigorous statement which basically means that for a random selection of parameters the expected complexity of this attack is about $m^{1/2}$, that is, of the same magnitude as the trial-division factorization algorithm. Indeed, obviously $t_p(l-1) \geq t$ and $t_l(p-1) \geq t$; thus when t is of order m and $p \sim l \sim m^{1/2}$ we obtain that $\min\{t_p, t_l\}$ is of order $m^{1/2}$. In fact t_p and t_l can be studied independently using Theorem 2.1 directly, thus getting slightly more precise results.

Furthermore, in [12] some modifications of the cycling attack have been introduced. One of these is based on the properties of the sequence

$$u_n \equiv u_{n-1}^e u_{n-2}^{-1} \pmod{m}, \quad 0 \leq u_n \leq m-1, \qquad n = 2, 3, \ldots, \qquad (2)$$

with the *initial values* $u_0 = \vartheta^2$, $u_1 = \vartheta^e$, where ϑ is an integer coprime to m, and the *exponent* $e \neq 0, \pm 2$. Thus, the sequence of exponents is running through the Lucas sequence $\alpha^j + \beta^j$ where α, β are the roots of the polynomial $x^2 - ex + 1$. The period of the sequence (2) and thus the success of the corresponding cycling attack depends on the behaviour of the function $\Omega_D(n)$, defined in [12], where $D = e^2 - 4$. This function is defined in a somewhat similar way to the Carmichael function. Namely,

$$\Omega_D\left(2^k\right) = \begin{cases} 2^{k-1}, & \text{if } D \text{ is even;} \\ 3 \cdot 2^{k-1}, & \text{if } D \text{ is odd.} \end{cases}$$

For a prime $p \geq 3$,

$$\Omega_D\left(p^k\right) = \begin{cases} (p - (D/p)) \, p^{k-1}, & \text{if } (D/p) \neq 0; \\ 2p^{k-1}, & \text{if } (D/p) = 0; \end{cases}$$

where (D/p) is the quadratic character of D modulo p. Finally, as in the case of the Carmichael function we put

$$\Omega_D(n) = \mathrm{lcm}\left(\Omega_D(p_1^{k_1}), \ldots, \Omega_D(p_\nu^{k_\nu})\right),$$

where

$$n = p_1^{k_1} \ldots p_\nu^{k_\nu}$$

is the prime number factorization of n.

The method of [9] can be used to obtain complete analogues of the results of Section 2 for the function $\Omega_D(n)$ and thus to also rule out the feasibility of this modified attack for randomly selected RSA moduli.

4. Security of Timed-Release Crypto

R. L. Rivest, A. Shamir and D. A. Wagner [22] have recently introduced the notion of *timed-release crypto* which applies to the following scenario.

Assume one wants to send a message μ which is supposed to be read T seconds later. The simplest way to arrange this is to encrypt the message using any reliable (public/private key) cryptosystem using a secret key K and then, T seconds later, send the key K. However, this solution cannot be used if for some reason the sender is no longer on-line in T seconds, which may especially be true if the actual delay is several years. See the web site

http://www.lcs.mit.edu/research/demos/cryptopuzzle0499

for such an example.

In [22] the following elegant solution has been proposed. As before, use any (public/private key) cryptosystem and encrypt it with a private key K, getting the encrypted message $E(\mu, K)$. Select an RSA modulus $m = pl$, an integer ϑ and an exponent e. We remark that in the original paper [22] only the value $e = 2$ has been considered but it may be more convenient to use a random value of e (to take advantage of the difference in the strengths of Theorem 2.3 and Theorem 2.4). Now the sender evaluates $U \equiv \vartheta^{e^s} \pmod{m}$, $0 \le U < m$, by computing

$$f \equiv e^s \pmod{\varphi(m)}, \qquad 0 \le f \le \varphi(m),$$

in $O(\log s)$ steps and then by computing

$$\vartheta^{e^s} \equiv \vartheta^f \pmod{m}$$

in another $O(\log m)$ steps (using repeated squaring in both cases). Finally, the sender computes $L \equiv K + \vartheta^{e^s} \pmod{m}$, $0 \le L < m$, and sends the 6-tuple

$$[E(\mu, K), m, L, \vartheta, e, s].$$

To decrypt the message μ the receiver must recover K, that is, the receiver must compute $U \equiv \vartheta^{e^s} \pmod{m}$. It seems that the only way to compute U is to make s consecutive exponentiations

$$u_i \equiv u_{i-1}^e \pmod{m}$$

with $u_0 = \vartheta$, until it reaches $U = u_s$. Hopefully this is the only way without the knowledge of $\varphi(m)$, moreover, there seems to be no advantage to the use of parallel computation; see the discussion of this issue in [22]. Thus, adjusting s and taking into account the expected computer performance one can predict the "breaking" time. Because there is no reasonable way to parallelize the procedure this time does not depend on the total computational power, only on the performance of a single computer, which is more predictable.

However, if the period of the sequence $u_n \equiv \vartheta^{e^n} \pmod{m}$ is small, say $t < s$, then one can compute U in about t steps rather than in s steps, thus reading the message earlier than expected. This issue has been discussed at the end of Section 2.1 of [22]. Our results, namely Theorem 2.3 for a randomly selected e

and, in weaker form, Theorem 2.4 for $e = 2$, provide rigorous support for the assumption that premature decryption will occur only extremely rarely.

5. Remarks and Open Problems

We have already mentioned that the original motivation of [9] was the application thereof to some uniformity of distribution results in [10]. For example, Theorem 2.3 combined with Theorem 3.2 of [10] produces the bound $O(m^{-1/8+\varepsilon})$ for the discrepancy of the power generator (1) modulo $m = pl$ for almost all values of the parameters. On the other hand, Theorem 2.4 is unfortunately too weak to guarantee such a result when the exponent e is fixed, say $e = 2$, and only the other parameters are selected at random. In order for the results of [10] to be nontrivial one needs the period to be of size at least $m^{3/4+\delta}$ for some fixed $\delta > 0$. Thus there is a gap roughly between $m^{1/2}$ (period estimates) and $m^{3/4}$ (discrepancy estimates) which would be very interesting to close. Probably pushing from the upper end looks more promising; we are quite pessimistic about beating the $m^{1/2}$-threshold.

Another challenging problem is to study the period length and distribution of some bit configurations extracted from the values of u_n. For example, in many cryptographic applications it is only the binary sequence (ξ_n) of the rightmost bits of (u_n) which is used. It has been shown in [10] that if the period of the Blum–Blum–Shub generator exceeds $m^{3/4+\delta}$, for some fixed $\delta > 0$, then the period of (ξ_n) grows as m^γ with some $\gamma > 0$, and if the period is close to m then one can take any $\gamma < 1/24$. Unfortunately, Theorem 2.4 is too weak to produce a useful combination with this result and no other rigorous results about the period of the sequence (ξ_n) seem to be known.

On the other hand, there is another way to guarantee that the period of (ξ_n) is large enough. Indeed, this period is obviously a divisor of the period of (u_n). Thus, if $\lambda(\lambda(pl))$ has no small prime divisors (except 2) then obviously the period of (ξ_n) is not too small (or is just 2 which can easily be detected and avoided). If one is willing to accept heuristic results then, as a simple consequence of the well-known conjecture about prime k-tuplets, see [1], there are in every interval $(x, 2x)$ with large x, at least $c_1 x/(\log x)^3$ primes p such that $q = (p - 1)/2$ and $r = (q - 1)/2$ are both also prime. That is, we request that r, $q = 2r + 1$, and $p = 4r + 3$ are prime. It follows on pairing such primes p that there are at least $c_2 Q^2/(\log Q)^6$ pairs of primes (p, l) with $Q/2 < p < l \leq Q$ for which $\lambda(\lambda(pl)) = 2rs$ where $r = p/4 + O(1)$ and $s = l/4 + O(1)$ are primes. Thus the period of (ξ_n) is either 2 or at least $2^{-5/2} m^{1/2}$. However, obtaining unconditional results about the arithmetic structure of the values of the Carmichael function (as is the case for the Euler function) is a notoriously hard problem.

Finally, from among a number of ingredients in the proofs of the results presented in Section 2 we draw attention to an estimate for the number $S_K(M)$ of elements g in the unit group $\mathbb{Z}_M^*$ of the residue ring modulo M whose multiplicative order $\mathrm{ord}_M g \leq \lambda(M)/K$. Namely, if $\tau(m)$ denotes the number of distinct positive integer divisors of m, then

$$S_K(M) \leq \varphi(M)\tau(\lambda(M))/K.$$

Results of this type, although in a weaker form, have appeared in several papers devoted to various cryptographic problems. In particular, the above bound is a direct improvement of Proposition 1 from [14]. It can also be useful for the purposes of the works [15, 20], and possibly some others.

We also note the following result from [19] which can be helpful for various cryptographic applications (in particular for the purposes of the work [20]). Let $E(M)$ be the expected number of elements chosen uniformly and independently at random in $\mathbb{Z}_M^*$ to generate the whole group $\mathbb{Z}_M^*$. Then

$$E(M) < \begin{cases} \nu(M) + \sigma, & \text{if } M \equiv 1 \pmod{2}, \\ \nu(M) + \sigma - 1, & \text{if } M \equiv 2 \pmod{4}, \\ \nu(M) + \rho, & \text{if } M \equiv 4 \pmod{8}, \\ \nu(M) + \rho + 1, & \text{if } M \equiv 0 \pmod{8}, \end{cases}$$

where $\nu(m)$ is the number of distinct prime divisors of m and $\sigma = 2.118\ldots$ and $\rho = 1.742\ldots$ are some explicitly given constants. In fact that paper contains a much more general statement about arbitrary abelian groups.

Acknowledgement. We thank Ron Rivest for his interest and for helpful references. J. F. was supported in part by NSERC and an NEC grant to the Institute for Advanced Study. I. S. was supported in part by ARC.

References

[1] A. Balog, *The prime k-tuplets conjecture on average*, Analytic Number Theory, Progress in Mathematics **85**, Birkhäuser, Boston, 1990, 47–75.

[2] L. Blum, M. Blum and M. Shub, *A simple unpredictable pseudorandom number generator*, SIAM J. Comp., **15** (1986), 364–383.

[3] J. J. Brennan and B. Geist, *Analysis of iterated modular exponentiation: The orbit of $x^\alpha \bmod N$*, Designs, Codes and Cryptography, **13** (1998), 229–245.

[4] T. W. Cusick, *Properties of the $x^2 \bmod N$ pseudorandom number generator*, IEEE Trans. Inform. Theory, **41** (1995), 1155–1159.

[5] T. W. Cusick, C. Ding and A. Renvall, *Stream Ciphers and Number Theory*, Elsevier, Amsterdam, 1998.

[6] P. Erdős, C. Pomerance and E. Schmutz, *Carmichael's lambda function*, Acta Arith., **58** (1991), 363–385.

[7] R. Fischlin and C. P. Schnorr, *Stronger security proofs for RSA and Rabin bits*, Lect. Notes in Comp. Sci., Springer-Verlag, Berlin, **1233** (1997), 267–279.

[8] J. B. Friedlander, D. Lieman and I. E. Shparlinski, *On the distribution of the RSA generator*, Proc. Intern. Conf. on Sequences and their Applications (SETA'98), Singapore, Springer-Verlag, London, 1999, 205–212.

[9] J. B. Friedlander, C. Pomerance and I. E. Shparlinski, *Period of the power generator and small values of Carmichael's function*, Math. Comp., (to appear).

[10] J. B. Friedlander and I. E. Shparlinski, *On the distribution of the power generator,* Math. Comp., (to appear).

[11] F. Griffin and I. E. Shparlinski, *On the linear complexity profile of the power generator,* Trans. IEEE on Information Theory, (to appear).

[12] M. Gysin and J. Seberry, *Generalised cycling attacks on RSA and strong RSA primes,* Proc. 4th Australasian Conf. on Information Security and Privacy (ACISP'99), Lect. Notes in Comp. Sci., vol. 1587, Springer-Verlag, Berlin, 1999, Wollongong, 149–163.

[13] J. Håstad and M. Näslund, *The security of individual RSA bits,* Proc. 39th IEEE Symp. on Foundations of Comp. Sci., 1998, 510–519.

[14] J. Håstad, A. W. Schrift and A. Shamir, *The discrete logarithm modulo a composite hides $O(n)$ bits,* J. Comp. and Syst. Sci., **47** (1993), 376–404.

[15] M. Liskov and R. D. Silverman, *A statistical limited-knowledge proof for secure RSA keys,* IEEE Working Group P1363: Research Contributions, 1998, 1–14 (available from http://grouper.ieee.org/groups/1363/Research/).

[16] J. C. Lagarias, *Pseudorandom number generators in cryptography and number theory,* Proc. Symp. in Appl. Math., Amer. Math. Soc., Providence, RI, **42** (1990), 115–143.

[17] U. M. Maurer, *Fast generation of prime numbers and secure public-key cryptographic parameters,* J. Cryptology, **8** (1995), 123–155.

[18] A. J. Menezes, P. C. van Oorschot and S. A. Vanstone, *Handbook of Applied Cryptography,* CRC Press, Boca Raton, FL, 1996.

[19] C. Pomerance, *The expected number of random elements to generate a finite abelian group,* Preprint, 2000.

[20] G. Poupard and J. Stern, *Short proofs of knowledge for factoring,* Proc. Inter. Workshop on Practice abnd Theory of Public Key Cryptography (PKC'2000), Lect. Notes in Comp. Sci., Springer-Verlag, Berlin, **1751** (2000), 147–166.

[21] R. L. Rivest, *Remarks on a proposed cryptanalytic attack on the M.I.T. public-key cryptosystem,* Cryptologia, **2** (1978), 62–65.

[22] R. L. Rivest, A. Shamir and D. A. Wagner, *Time-lock puzzles and timed-release crypto,* Preprint, 1996, 1–9.

[23] R. L. Rivest and R. D. Silverman, *Are "strong" primes needed for RSA?,* Preprint, 1999, 1–23.

[24] I. E. Shparlinski, *On the linear complexity of the power generator,* Designs, Codes and Cryptography, (to appear).

[25] D. R. Stinson, *Cryptography: Theory and Practice,* CRC Press, Boca Raton, FL, 1995.

Dept. of Mathematics, University of Toronto,Toronto, Ontario M5S 3G3, Canada
E-mail address: frdlndr@math.toronto.edu

Dept. of Fundamental Mathematics, Bell Labs, Murray Hill, NJ 07974-0636, USA
E-mail address: carlp@research.bell-labs.com

Dept. of Computing, Macquarie University, Sydney, NSW 2109, Australia
E-mail address: igor@ics.mq.edu.au

Progress in Computer Science and Applied Logic, Vol. 20
© 2001 Birkhäuser Verlag Basel/Switzerland

Density Estimates Related to Gauß Periods

Joachim von zur Gathen and Francesco Pappalardi

Abstract. Given two integers q and k, for any prime r not dividing q with $r \equiv 1 \bmod k$, we denote by $\text{ind}_r(q)$ the index of $q \bmod r$. In [2] the question was raised of calculating the density of the primes r for which $\text{ind}_r(q)$ and $(r-1)/k$ are coprime; this is the condition that the Gauß period in $\mathbb{F}_{q^{(r-1)/k}}$ defined by these data be normal over $\mathbb{F}_q$. We assume the Generalized Riemann Hypothesis and calculate a formula for this density for all q and k. We prove unconditionally that our formula is an upper bound for the density and then express it as an Euler product. Finally we apply the results to characterize the existence of a special type of Gauß periods.

1. Introduction

Let q and k be integers with $|q| > 1$ and $k > 0$. For any prime r not dividing q, we define the index of $q \bmod r$ as $\text{ind}_r(q) = [\mathbb{F}_r^* : \langle q \bmod r \rangle]$, so that $\text{ind}_r(q) = (r-1)/\text{ord}_r(q)$. If $r \equiv 1 \bmod k$, we also set

$$g_{q,k}(r) = \gcd\left(\text{ind}_r(q), (r-1)/k\right).$$

Finally we let $M_{q,k}(x)$ be the number of primes $r \equiv 1 \bmod k$ up to x for which $g_{q,k}(r) = 1$.

The interest in this quantity comes from the construction of normal Gauß periods in $\mathbb{F}_{q^n}$ over $\mathbb{F}_q$, where $q \in \mathbb{N}$ is a prime power. If $n = (r-1)/k$, $g_{q,k}(r) = 1$, $\beta \in \mathbb{F}_{q^{r-1}}$ is a primitive r-th root of unity, $K \subseteq \mathbb{F}_r^*$ is the unique subgroup of order k, and $\alpha = \sum_{i \in K} \beta^i$, then (n, k) is called in [2] a *Gauß pair* (over $\mathbb{F}_a$), and indeed the *Gauß period* α generates a normal basis for $\mathbb{F}_{q^n}$ over $\mathbb{F}_q$. It was noted a few years ago that such a normal basis is useful for fast exponentiation in finite fields, which in turn has various cryptographic applications. Theory and applications of this, including implementations, are discussed in [2], [3], [4], [5], [6], [7]. A survey of these results is in [8]. In particular, two elements of $\mathbb{F}_{q^n}$ represented in such a basis can be multiplied at essentially the same cost as multiplying two polynomials of degree nk over $\mathbb{F}_q$.

Therefore a natural question is: given q and n as above, what is the smallest k such that (n, k) is a Gauß pair over $\mathbb{F}_q$?

In this paper we turn this question around and ask: given q and a (small) k, for how many n is (n, k) a Gauß pair over $\mathbb{F}_q$?

The paper [1] gives a generalization of Gauß periods, where basically the prime r is replaced by an arbitrary integer; our considerations only apply to the classical case as treated by Gauß, where $r = nk + 1$ is prime.

For $k = 1$, it is clear that $g_{q,k}(r) = 1$ if and only if $\mathrm{ind}_r(q) = 1$, and this happens exactly when q is a primitive root modulo r. Hence $M_{q,1}(x)$ is the number of primes r up to x for which q is a primitive root modulo r; the famous Artin Conjecture for primitive roots states that the set of these primes has a positive density unless q is a square or equals -1. In 1965, C. Hooley [11] proved that the Generalized Riemann Hypothesis implies the asymptotic formula

$$M_{q,1}(x) = \left(\delta_q + \mathrm{O} \left(\frac{\log \log x + \log q}{\log x} \right) \right) \frac{x}{\log x}$$

uniformly with respect to q, where δ_q depends only upon q. Unconditionally, the work of Gupta and Murty [9] and of Heath-Brown [10] provides evidence for the Artin Conjecture.

Our question can be considered as a natural generalization of Hooley's famous result. This generalization is meaningful also if q is a square.

For $r \in \mathbb{N}$, we let $\zeta_r \in \mathbb{C}$ be a primitive rth root of unity. We will prove the following results.

Theorem 1.1. *Let q and k be integers with $|q| > 1$ and $k > 0$, and for $m \in \mathbb{N}$ set $K_m = \mathbb{Q}(\zeta_{km}, q^{1/m})$ and $n_m = [K_m : \mathbb{Q}]$, and*

$$\delta_{q,k} = \sum_{1 \leq m} \frac{\mu(m)}{n_m}.$$

Then there exists $c_{q,k} \in \mathbb{R}$ that depends only on q and k such that

$$M_{q,k}(x) \leq \left(\delta_{q,k} + \frac{c_{q,k}}{\log \log x} \right) \frac{x}{\log x}.$$

If the Generalized Riemann Hypothesis holds for all these fields K_m, then

$$M_{q,k}(x) = \left(\delta_{q,k} + \mathrm{O} \left(\frac{\log \log x}{\log x} \right) \right) \frac{x}{\log x}.$$

Next we express the densities as Euler products. The parameter l in the products below ranges over the primes. We let

$$A = \prod_{l \text{ prime}} \left(1 - \frac{1}{l(l-1)} \right) \approx 0.373956$$

be Artin's constant, and μ the Möbius function.

Theorem 1.2. *With the notation of Theorem 1.1, we write $q = b^h$ and $b = b_1^2 b_2$ with integers b, b_1, b_2, and h, where b is not a perfect power and b_2 is squarefree,* *set*

$$b_3 = \begin{cases} 4b_2/\gcd(4b_2,k) & \text{if } b_2 \equiv 2,3 \bmod 4, \\ b_2/\gcd(b_2,k) & \text{if } b_2 \equiv 1 \bmod 4, \end{cases}$$

write $b_3 = \alpha b_4$ with α a power of two and b_4 odd, so that the values of α are given by the following table:

	$2\nmid k$	$2\|k$	$4\|k$	$8 \mid k$
$b_2 \equiv 1 \bmod 4$	1	1	1	1
$b_2 \equiv 3 \bmod 4$	4	2	1	1
$b_2 \equiv 2 \bmod 4$	8	4	2	1

Furthermore, we set

$$A_{h,k} = \frac{A}{k} \prod_{l\|k} \left(1 + \frac{l}{l^2 - l - 1}\right) \prod_{\substack{l\mid h \\ l\nmid k}} \left(1 - \frac{l-1}{l^2 - l - 1}\right).$$

Then we have

$$\delta_{q,k} = A_{h,k} \cdot \left(1 - \frac{\mu(b_4 \cdot \gcd(h,2)^2) \cdot |\mu(\alpha)|}{2\gcd(2,k) - 1} \prod_{\substack{l\mid b_4 \\ l\nmid h}} \frac{1}{l^2 - l - 1} \prod_{\substack{l\mid b_4 \\ l\mid h}} \frac{1}{l - 2}\right), \qquad (1)$$

and $A_{h,k} = 0$ if and only if h is even and k is odd.

Finally we apply the above results to the problem of Gauß pairs.

Corollary 1.3. *Let p be a prime, h and k be positive integers, $q = p^h$, and assume that the GRH holds for all fields K_m of Theorem 1.1.*

(i) *$\delta_{q,k} = 0$ if and only if at least one of the following two conditions is satisfied:*
   (a) *$2 \mid h$ and $2 \nmid k$,*
   (b) *$2 \nmid k$, $p \mid k$, and $p \equiv 1 \bmod 4$.*
(ii) *If $\delta_{q,k} = 0$, then there is no Gauß pair (n,k) over $\mathbb{F}_q$.*

Proof. (i) We write (1) as $\delta_{q,k} = A_{h,k} \cdot B$, so that

$$\delta_{q,k} = 0 \iff A_{h,k} = 0 \text{ or } B = 0 \iff (2 \mid h \text{ and } 2 \nmid k) \text{ or } B = 0,$$

using Theorem 1.2. Furthermore,

$$B = 0 \iff \mu(b_4)|\mu(\alpha)| = (2\gcd(2,k) - 1)\prod_{\substack{l\mid b_4 \\ l\nmid h}} (l^2 - l - 1) \prod_{\substack{l\mid b_4 \\ l\mid h}} (l - 2).$$

The left-hand side has absolute value 1, and the right-hand side is positive, since b_4 is odd. They are equal if and only if both are equal to 1. If that is the case, then $b_4 = 1$, since otherwise it would have at least two distinct prime factors, by

$\mu(b_4) = 1$, and then one of the factors on the right-hand side would be greater than 1. Since $|\mu(\alpha)| = 1$ if and only if $\alpha \leq 2$, we have

$$\begin{aligned} B = 0 &\iff \alpha \leq 2, 2 \nmid k, b_4 = 1 \\ &\iff 2 \nmid k, \alpha = 1, b_3 = b_4 = 1, b_2 \equiv 1 \bmod 4 \\ &\iff 2 \nmid k, p \mid k, p \equiv 1 \bmod 4, \end{aligned}$$

since $b_2 = b = p$.

(ii) Since $\delta_{q,k} = 0$, either (a) or (b) holds. From (a) we find that $\mathrm{ind}_r(q)$ and $(r-1)/k$ are both even, so that $g_{q,k}(r)$ is even, for all odd primes r, and thus there is no Gauß pair (n, k) over $\mathbb{F}_q$. So now we assume that (b) holds, and let r be an odd prime with $r \equiv 1 \bmod k$. Then $(r-1)/k$ is even. Since p divides k, we also have $r \equiv 1 \bmod p$. We may assume that h is odd, since otherwise (a) holds. Then the quadratic reciprocity law gives the following for the Legendre symbol

$$\left(\frac{q}{r}\right) = \left(\frac{p^h}{r}\right) = \left(\frac{p}{r}\right) = \left(\frac{r}{p}\right) = \left(\frac{1}{p}\right) = 1.$$

Thus q is a square modulo r and $\mathrm{ind}_r(q)$ is even. Therefore again $g_{q,k}(r)$ is even, and there is no Gauß pair, as claimed. $\square$

In particular, for q and k as in Corollary 1.3, the set of primes r for which $((r-1)/k, k)$ is a Gauß pair over $\mathbb{F}_q$ is either empty or has the positive density $\delta_{q,k}$.

Wassermann proves in [14] an existence result starting from a different set of parameters. His Theorem 3.3.4 states that for any given integers h, n and a prime p, there exists a Gauß pair (n, k) over $\mathbb{F}_{p^h}$ if and only if $\gcd(h, n) = 1$ and

$$2p \nmid n \text{ if } p \equiv 1 \bmod 4,$$
$$4p \nmid n \text{ if } p \equiv 2, 3 \bmod 4.$$

2. Proof of the Theorems

The following lemma is the Chebotarev Density Theorem. The proof of the two versions that we state here is due to Lagarias and Odlyzko [12].

Lemma 2.1. *Suppose that L is a Galois extension of $\mathbb{Q}$ with absolute discriminant d_L and degree n_L over $\mathbb{Q}$, and define*

$$\pi(x, L\colon \mathbb{Q}) = \#\{p \leq x\colon p \text{ is unramified and splits completely in } L\}.$$

If the Generalized Riemann Hypothesis holds for the Dedekind zeta function of L, then

$$\pi(x, L\colon \mathbb{Q}) = \frac{1}{n_L} \operatorname{li}(x) + \mathrm{O}(x^{1/2} \log(x \cdot d_L^{1/n_L})).$$

In general (unconditionally) there exists absolute constants C_1 and B such that for

$$\sqrt{\log x} \geq C_1\, n_L^{1/2} \max\{\log |d_L|, |d_L|^{1/n_L}\}, \tag{2}$$

one has

$$\pi(x, L : \mathbb{Q}) = \frac{1}{n_L} \operatorname{li}(x) + O(x \exp(-B n_L^{-1/2} \sqrt{\log x})). \qquad \square$$

Proof of Theorem 1.1. The argument is similar to the original one of Hooley, therefore we only mention the main steps.

We start by noticing that the condition for a prime $l \neq p$ to divide the index $\operatorname{ind}_p(q)$ is equivalent to p splitting completely in $\mathbb{Q}(\zeta_l, q^{1/l})$, while the condition that l divides $(p-1)/k$ is equivalent to p splitting completely in the cyclotomic field $\mathbb{Q}(\zeta_{lk})$. Since a prime splits completely in two extensions if and only if it splits completely in the compositum, by the inclusion–exclusion principle we gather that

$$M_{q,k}(x) = \sum_{1 \leq m} \mu(m) \pi(x, \mathbb{Q}(\zeta_{km}, q^{1/m}) : \mathbb{Q}).$$

We now consider the set $S(y)$ of those squarefree "y-smooth" integers $m \geq 1$ all of whose prime divisors are less than a (sufficiently small) parameter y. We note that $S(y)$ has $2^{\pi(y)}$ elements, and if $m \in S(y)$, then $m \leq P(y)$, where $P(y)$ denotes the product of the primes up to y.

Furthermore, we let N and D denote the degree and the discriminant of K_m over $\mathbb{Q}$. Then $\sqrt{N} \leq \sqrt{km} \leq \sqrt{k} P(y)$, $\log D \ll N \log N \ll y P(y)^2$, and $D^{1/N} \ll N \prod_{l|D} l \ll P(y)^3$, where the implied constants depend on a and k. By choosing y such that $P(y) = C_2 (\log x)^{1/8}$ for some constant C_2, we can use the unconditional part of Lemma 2.1. The inclusion–exclusion principle then yields the (unconditional) upper bound

$$
\begin{aligned}
M_{q,k}(x) &\leq \sum_{m \in S(y)} \mu(m) \pi(x, \mathbb{Q}(\zeta_{km}, q^{1/m}) : \mathbb{Q}) \\
&= \sum_{m \in S(y)} \mu(m) \left\{ \frac{\operatorname{li}(x)}{n_m} + O\left(x \exp(-C_3 \sqrt{(\log x)/n_m}) \right) \right\} \\
&= \left(\delta_{q,k} + O\left(\sum_{m > y} \frac{1}{m\varphi(m)} \right) \right) \operatorname{li}(x) + O\left(2^{\pi(y)} x \exp\left(-C_4 \frac{\sqrt{\log x}}{P(y)} \right) \right) \\
&= \left(\delta_{q,k} + O\left(\frac{1}{y} \right) \right) \frac{x}{\log x} + O\left(x \exp\left(-C_5 (\log x)^{3/8} \right) \right) \\
&= \left(\delta_{q,k} + O\left(\frac{1}{\log \log x} \right) \right) \frac{x}{\log x},
\end{aligned}
$$

where we used the fact that $\varphi(m) m \ll n_m$. This proves the second part of Theorem 1.1. We note that the method of A. I. Vinogradov [13] could be used here to establish a sharper error term.

For the second claim we note that

$$
\begin{aligned}
M_{q,k}(x) &\leq \sum_{m \in S(y)} \mu(m) \pi(x, \mathbb{Q}(\zeta_{km}, q^{1/m}) : \mathbb{Q}) \\
&\leq M_{q,k}(x) + \# \{ p \leq x : \exists l \geq y \quad l \mid g_{q,k} \}.
\end{aligned}
$$

Therefore

$$M_{q,k}(x) = \sum_{m \in S(y)} \mu(m)\pi(x, \mathbb{Q}(\zeta_{km}, q^{1/m}) : \mathbb{Q}) + O\left(\# \{p \le x : \exists l \ge y \quad l \mid g_{q,k}\}\right).$$

The main term is estimated using the version of the Chebotarev Density Theorem in Lemma 2.1 dependent on the Generalized Riemann Hypothesis which leads to a choice of $y = \frac{1}{6} \log x$. The error term can be handled exactly as in Hooley's case, ignoring the condition that $l \mid (p-1)/k$. $\square$

For the proof of Theorem 1.2, we need the following two lemmas. We will have an integer h, and for an integer m we set

$$\hat{m} = m/\gcd(h, m).$$

Lemma 2.2. *Let* $q, k, m \in \mathbb{Z}$ *with* $m, k > 0, |q| > 1$, *and* m *squarefree. We write* $q = b^h$ *with* b *not a perfect power,* $b = b_1^2 b_2$ *with* b_2 *squarefree, and set*

$$\varepsilon = \begin{cases} 2 & \text{if } 2 \mid \hat{m}, b_2 \mid mk, \text{ and } b_2 \equiv 1 \bmod 4, \\ 2 & \text{if } 2 \mid \hat{m}, 4b_2 \mid mk, \text{ and } b_2 \not\equiv 1 \bmod 4, \\ 1 & \text{otherwise.} \end{cases}$$

Then $n_m = \varphi(km) \cdot \left[\mathbb{Q}(\zeta_{km}, q^{1/m}) : \mathbb{Q}\right] = \varphi(km)\hat{m}/\varepsilon$.

Proof. First we note that $\mathbb{Q}(\zeta_{km}, q^{1/m}) = \mathbb{Q}(\zeta_{km}, b^{1/\hat{m}})$. Since $[\mathbb{Q}(b^{1/\hat{m}}) : \mathbb{Q}] = \hat{m}$ and $[\mathbb{Q}(b^{1/\hat{m}})(\zeta_{km}) : \mathbb{Q}(b^{1/\hat{m}})]$ is a divisor of $\varphi(km)$, from the identity

$$[\mathbb{Q}(\zeta_{km}, b^{1/\hat{m}}) : \mathbb{Q}(\zeta_{km})] \cdot [\mathbb{Q}(\zeta_{km}) : \mathbb{Q}] = [\mathbb{Q}(b^{1/\hat{m}}, \zeta_{km}) : \mathbb{Q}(b^{1/\hat{m}})] \cdot [\mathbb{Q}(b^{1/\hat{m}}) : \mathbb{Q}]$$

we deduce that

$$n_m = \varphi(km)\left[\mathbb{Q}(\zeta_{km}, b^{1/\hat{m}}) : \mathbb{Q}(\zeta_{km})\right] = \varphi(km)\frac{\hat{m}}{d}$$

for some divisor d of $\hat{m}$. We claim that d is 1 or 2. Indeed, if l is a prime dividing d, then we have extensions

$$\mathbb{Q}(\zeta_{km}) \subseteq \mathbb{Q}(\zeta_{km}, b^{1/l}) \subseteq \mathbb{Q}(\zeta_{km}, b^{1/\hat{m}}).$$

Since $\hat{m}$ is squarefree, l does not divide $\hat{m}$, hence $\mathbb{Q}(\zeta_{km}, b^{1/l}) = \mathbb{Q}(\zeta_{km})$ and $b^{1/l} \in \mathbb{Q}(\zeta_{km})$. Therefore we have an inclusion of Abelian extensions $\mathbb{Q}(b^{1/l}) \subseteq \mathbb{Q}(\zeta_{km})$ of $\mathbb{Q}$. This can only happen when l is 1 or 2.

Furthermore $\mathbb{Q}(\sqrt{b}) = \mathbb{Q}(\sqrt{b_2})$, so that $d = 2$ if and only if $\hat{m}$ is even and $\sqrt{b_2} \in \mathbb{Q}(\zeta_{km})$.

The quadratic subfields of $\mathbb{Q}(\zeta_{km})$ are

$$\begin{cases} \mathbb{Q}(\sqrt{(\frac{-1}{D})|D|}) : D \mid km, D \text{ odd squarefree} \end{cases} \quad \text{if } 4 \nmid km,$$
$$\begin{cases} \mathbb{Q}(\sqrt{D}) : D \mid km, D \text{ odd squarefree} \end{cases} \qquad\quad \text{if } 4 \| km,$$
$$\begin{cases} \mathbb{Q}(\sqrt{D}) : D \mid km, D \text{ squarefree} \end{cases} \qquad\qquad\; \text{if } 8 \mid km.$$

In the first case, $d = 2$ if and only if $b_2|km$ and $b_2 \equiv 1 \bmod 4$, and in the second case, $d = 2$ if and only if b_2 is odd and divides km, and in the third case $d = 2$ if and only if $b_2|km$.

Finally, $d = \varepsilon$ and hence the claim. $\qquad\square$

Lemma 2.3. *Let $A_{h,k}$ be as in the statement of Theorem 1.2 and $t \in \mathbb{N}$. Then*

$$A_{h,k} = \sum_{1 \le m} \frac{\mu(m)}{\varphi(km)\hat{m}} = \frac{1}{\varphi(k)} \prod_{l \text{ prime}} \left(1 - \frac{\gcd(l,h)\varphi(\gcd(l,k))}{l\gcd(l,k)(l-1)}\right),$$

$$\sum_{\substack{1 \le m \\ \gcd(m,t)=1}} \frac{\mu(m)}{\varphi(km)\hat{m}} = \frac{1}{\varphi(k)} \prod_{l \nmid t} \left(1 - \frac{\varphi(\gcd(l,k))}{(l-1)\hat{l}\gcd(l,k)}\right). \tag{3}$$

Proof. We have

$$\sum_{1 \le m} \frac{\mu(m)}{\varphi(km)\hat{m}} = \sum_{d|k} \sum_{\substack{1 \le m \\ \gcd(m,k)=d}} \frac{\mu(m)}{\varphi(km)\hat{m}}$$

$$= \left(\sum_{\substack{1 \le m \\ \gcd(m,k)=1}} \frac{\mu(m)}{\varphi(km)\hat{m}}\right) \cdot \left(\sum_{d|k} \frac{\mu(d)}{d\hat{d}}\right)$$

$$= \frac{1}{\varphi(k)} \prod_{l \nmid k} \left(1 - \frac{1}{\hat{l}(l-1)}\right) \prod_{l|k} \left(1 - \frac{1}{\hat{l}l}\right),$$

since if $d \mid k$, then $\varphi(kmd) = d\varphi(km)$, and the claim is easily deduced. The second part is proven similarly. $\qquad\square$

Let us now prove Theorem 1.2.

If h is even, then $\hat{m}$ is odd for any squarefree m, and this implies that $n_m = \varphi(km)\hat{m}$. Therefore by Lemma 2.3, we have that $\delta_{a,k} = A_{h,k}$. We now assume that h is odd (so that $\hat{m}$ is even if and only if m is), and consider b_3, b_4, and α as in the theorem. We note that $\gcd(b_4,k) = 1$. Furthermore, for any squarefree m, ε as defined in Lemma 2.2 equals 2 if and only if $\alpha \le 2$ and $2b_4|m$.

Therefore, if $\alpha \ge 4$, then $\delta_{q,k} = A_{h,k}$. If $\alpha \le 2$, then

$$\delta_{q,k} = \sum_{2b_4 \nmid m} \frac{\mu(m)}{\varphi(km)\hat{m}} + 2\sum_{2b_4|m} \frac{\mu(m)}{\varphi(km)\hat{m}} = A_{h,k} + \frac{\mu(2b_4)}{2\hat{b_4}\varphi(b_4)} \sum_{\gcd(m,2b_4)=1} \frac{\mu(m)}{\varphi(2km)\hat{m}}.$$

By applying the multiplicative property (3) to the last sum above (with $t = 2b_4$ and $2k$ instead of k), we have

$$\delta_{q,k} = A_{h,k} - \frac{\mu(b_4)}{2\hat{b_4}\varphi(b_4)\varphi(2k)} \prod_{l \nmid 2b_4} \left(1 - \frac{\varphi(\gcd(k,l))}{(l-1)\hat{l}\gcd(l,k)}\right).$$

In the inner product we write $\gcd(k, l)$ instead of $\gcd(2k, l)$, since l is odd. Now, we can factor out $A_{h,k}$ as follows. We multiply and divide the inner product by $\prod_{l|2b_4} \left(1 - \frac{\varphi((k,l))}{\hat{l}\gcd(l,k)(l-1)}\right)$, and obtain:

$$
\begin{aligned}
\delta_{q,k} = {}& A_{h,k} - \frac{\mu(b_4)}{2\hat{b}_4\varphi(b_4)\varphi(2k)} \prod_{l} \left(1 - \frac{\varphi(\gcd(k,l))}{\hat{l}\gcd(l,k)(l-1)}\right) \\
& \cdot \prod_{l|2b_4} \left(1 - \frac{\varphi(\gcd(k,l))}{\hat{l}\gcd(l,k)(l-1)}\right)^{-1} \\
= {}& A_{h,k} \left(1 - \frac{\mu(b_4)}{2\hat{b}_4\varphi(b_4)} \frac{\varphi(k)}{\varphi(2k)} \prod_{l|2b_4} \left(\frac{\hat{l}\gcd(l,k)(l-1)}{\hat{l}\gcd(l,k)(l-1) - \varphi(\gcd(k,l))}\right)\right).
\end{aligned}
$$

It is easy to see that $\gcd(2, k)\varphi(k) = \varphi(2k)$ and $\hat{2} = 2$. If $l \mid b_4$, then $\gcd(l, k) = 1$, since $\gcd(b_4, k) = 1$. Therefore

$$
\begin{aligned}
\delta_{q,k} = {}& A_{h,k} \left(1 - \frac{\mu(b_4)}{2\hat{b}_4\varphi(b_4)} \frac{\varphi(k)}{\varphi(2k)} \prod_{l|2} \left(\frac{\hat{l}\gcd(l,k)(l-1)}{\hat{l}\gcd(l,k)(l-1) - \varphi(\gcd(k,l))}\right)\right. \\
& \left.\cdot \prod_{l|b_4} \left(\frac{\hat{l}(l-1)}{\hat{l}(l-1) - 1}\right)\right) \\
= {}& A_{h,k} \left(1 - \frac{\mu(b_4)}{2\hat{b}_4\varphi(b_4)} \frac{\varphi(k)}{\varphi(2k)} \frac{\hat{2}\gcd(2,k)}{\hat{2}\gcd(2,k) - 1}\right. \\
& \left.\cdot \prod_{l|b_4} \left(\hat{l}(l-1)\right) \prod_{l|b_4} \frac{1}{\hat{l}(l-1) - 1}\right) \\
= {}& A_{h,k} \left(1 - \frac{\mu(b_4)}{2\gcd(2,k) - 1} \prod_{l|b_4} \frac{1}{\hat{l}(l-1) - 1}\right).
\end{aligned}
$$

Finally we can combine the three cases h even, h odd and $\alpha \geq 4$, and h odd and $\alpha \leq 2$, in a single formula as

$$
\delta_{a,k} = A_{h,k} \left(1 - \frac{\mu(b_4 \cdot \gcd(h,2)^2)|\mu(\alpha)|}{2\gcd(2,k) - 1} \prod_{l|b_4} \frac{1}{\hat{l}(l-1) - 1}\right). \qquad \square
$$

Acknowledgements. The authors would like to thank Hans Roskam for having pointed out and corrected a mistake in Lemma 2.2 of the original version of the paper.

References

[1] S. Feisel, J. von zur Gathen, and M. A. Shokrollahi, *Normal bases via general Gauß periods*, Mathematics of Computation, **68**(225) (1999), 271–290.

[2] S. Gao, J. von zur Gathen, and D. Panario, *Gauß periods and fast exponentiation in finite fields*, In *Proceedings of LATIN '95,* Valparaíso, Chile, number 911 in Lecture Notes in Computer Science, 311–322, Springer-Verlag, 1995.

[3] S. Gao, J. von zur Gathen, and D. Panario, *Gauß periods: orders and cryptographical applications*, Mathematics of Computation, **67**(221) (1998), 343–352, Microfiche supplement.

[4] S. Gao, J. von zur Gathen, D. Panario, and V. Shoup, *Algorithms for exponentiation in finite fields*, J. of Symbolic Computation **29**(6) (2000), 879–889.

[5] S. Gao and H. W. Lenstra, Jr. Optimal normal bases. *Designs, Codes, and Cryptography,* **2** (1992), 315–323.

[6] J. von zur Gathen and M. Nöcker, *Exponentiation in finite fields: Theory and practice*, In Teo Mora and Harold Mattson, editors, *Applied Algebra, Algebraic Algorithms and Error-Correcting Codes: AAECC-12,* Toulouse, France, number 1255 in Lecture Notes in Computer Science, 88–113, Springer-Verlag, 1997.

[7] J. von zur Gathen and M. Nöcker, *Computing special powers in finite fields: Extended abstract*, In Sam Dooley, editor, *Proceedings of the 1999 International Symposium on Symbolic and Algebraic Computation ISSAC '99,* Vancouver, Canada, 83–90, ACM Press, 1999.

[8] Joachim von zur Gathen and Igor Shparlinski, *Gauß periods in finite fields*, In *Proceedings Fq5* (2000), Birkhäuser Verlag, Basel. To appear.

[9] R. Gupta and M. R. Murty, *A remark on Artin's conjecture*, Invent. Math., **78** (1), (1984), 127–130.

[10] D. R. Heath-Brown, *Artin's conjecture for primitive roots*, Quart. J. Math. Oxford Ser. (2), **37**, (1986), 27–38.

[11] C. Hooley, *On Artin's conjecture*, J. für Angew. und Reine Math. **225**, (1967), 209–220.

[12] J. C. Lagarias and A. M. Odlyzko, *Effective versions of the Chebotarev Density Theorem in algebraic number fields*, in: *Algebraic Number Theory*, Ed. A. Fröhlich, Academic Press, New York 1977, 409–464.

[13] A. I. Vinogradov, *Artin L–series and his conjectures*, Proc. Steklov Inst. Math. **112** (1971), 124–142.

[14] Alfred Wassermann, *Zur Arithmetik in endlichen Körpern*, Bayreuther Math. Schriften, **44**, 147–251, 1993.

Fachbereich Mathematik-Informatik, Universität Paderborn
D–33095 Paderborn, Germany
E-mail: gathen@upb.de

Dipartimento di Matematica, Università Roma Tre
Largo S. L. Murialdo, 1, I–00146, Roma, Italy
E-mail: pappa@mat.uniroma3.it

Progress in Computer Science and Applied Logic, Vol. 20
© 2001 Birkhäuser Verlag Basel/Switzerland

The Distribution of the Coefficients of Primitive Polynomials over Finite Fields

Wenbao Han

Abstract. In this report, we discuss Hansen-Mullen's conjecture on the distributions of primitive polynomials over finite field in two cases: 1). second coefficient and odd characteristic; 2). second coefficient and even characteristic.

1. Introduction

Let $\mathbf{F}_q$ be a finite field with $q = p^l$ elements where p is a prime number and l is a positive integer. A monic polynomial $f(x) \in \mathbf{F}_q[x]$ of degree n is called a *primitive polynomial* if the least positive integer e such that $f(x)|x^e - 1$ is $q^n - 1$. It is well known that $f(x)$ is irreducible over $\mathbf{F}_q[x]$. If ξ is a root of $f(x)$ in $\mathbf{F}_{q^n}$, then ξ is a primitive element of $\mathbf{F}_{q^n}$, namely a generator of the multiplicative group $\mathbf{F}_{q^n}^*$ of $\mathbf{F}_{q^n}$. Davenport [4] and Carlitz studied the properties of primitive elements in the sixties. Davenport's results are merely asymptotic, whereas he obtained complete results for the question studied for all *prime* fields. Recently, because of the applications of finite fields in cryptography, coding theory, designing Costas arrays, etc., various properties of primitive elements have been investigated again with complete results being sought (see Cohen's survey article [2]). For $f(x) = x^n + a_1 x^{n-1} + \cdots + a_n \in \mathbf{F}_q[x]$, we call a_i i-th coefficient of $f(x)$. In [9], Hansen and Mullen conjectured that with the three non-trivial exceptions

$$(q,\ n,\ i,\ a) = (4,\ 3,\ 1,\ 0),\ (4,\ 3,\ 2,\ 0),\ (2,\ 4,\ 2,\ 1),$$

there is a primitive polynomial of degree n with the i-th coefficient a prescribed $(0 < i < n)$ in advance.

For $i = 1$ this follows from Davenport [4], Cohen [1], Jungnickel and Vanstone [10], Moreno [13] etc. In this report, we investigate the distributions of primitive polynomials over finite fields in two cases:

1. The index $i = 2$, p is odd;
2. The index $i = 2$, $p = 2$.

These will be very important in coding theory, cryptography and so on. First we discuss Hansen-Mullen's conjecture for $i = 2$. We prove that if $n > 6$ and q odd, there exists a primitive polynomial with the first and second coefficients prescribed in advance; consequently Hansen-Mullen conjecture holds for $i = 2$ and $n > 6$. For $i = 2$ and q even, we first give estimates of two exponential sums over finite fields, associated with the so-called q-bilinear polynomials, these being sums for which Weil's estimates fail to work (see [16], pp45 Theorem 2G). Then we use our estimates and Cohen Sieve Method to discuss Hansen-Mullen's conjecture for $i = 2$ and q even. We prove that if $(n, a) \neq (4, 0), (5, 0), (6, 0)$, there always exists a primitive polynomial of degree $n \geq 4$ with the second coefficient $a \in \mathbf{F}_q$, see [8].

2. Characters and Exponential Sums

In study of the properties of primitive polynomials, the exponential sums and characters over finite fields are important tools. For this reason, we review a few basic facts about the characters over finite fields first.

Lemma 2.1. *Let* $\xi \in \mathbf{F}_{q^n}^*$, *then*

$$\sum_{d|q^n-1} \frac{\mu(d)}{\varphi(d)} \sum_{\chi_d} \chi_d(\xi) = \begin{cases} \frac{q^n-1}{\varphi(q^n-1)}, & \text{if } \xi \text{ is a primitive element of } \mathbf{F}_{q^n}; \\ 0, & \text{otherwise}, \end{cases}$$

where $\mu(d)$ *is Möbius function,* $\varphi(d)$ *is Euler's function and* χ_d *runs through all* d-th order multiplicative characters of $\mathbf{F}_{q^n}$.

There is a more general result analogous to Lemma 2.1 of the following. Let $e \mid q^n - 1$, we call ξ *not any kind of* e-th power in $\mathbf{F}_{q^n}$ if $\xi = \rho^d$, $\rho \in \mathbf{F}_{q^n}$, $d|e$ only if $d = 1$.

Lemma 2.2. *Let* $\xi \in \mathbf{F}_{q^n}^*, e \mid q^n - 1$, *then*

$$\sum_{d|e} \frac{\mu(d)}{\varphi(d)} \sum_{\chi_d} \chi_d(\xi) = \begin{cases} \frac{e}{\varphi(e)} & \text{if } \xi \text{ is not any kind of } e\text{-th power}; \\ 0 & \text{otherwise}, \end{cases}$$

where χ_d *runs through all* $\varphi(d)$ *multiplicative characters of* $\mathbf{F}_{q^n}$ *of exact order* d, $\mu(d)$ *is the Möbius function and* $\varphi(d)$ *is Euler function.*

Let Tr be the absolute trace from $\mathbf{F}_q$ to $\mathbf{F}_p$. Then the mapping $\psi : x \rightarrow e^{2\pi i Tr(x)/p}$ $(x \in \mathbf{F}_q)$ is an additive character of $\mathbf{F}_q$, called the *canonical additive character*. We define $\psi_a(x) = \psi(ax)$, a (fixed) , $x \in \mathbf{F}_q$. Then the $\psi_a's$ $(a \in \mathbf{F}_q)$ are all additive characters of $\mathbf{F}_q$. We also observe that $\psi_a(x) = \psi(aT(x))$ is an additive character of $\mathbf{F}_{q^n}$.

Lemma 2.3. *Let* $\xi \in \mathbf{F}_q$. *Then*

$$\sum_{a \in \mathbf{F}_q} \psi_a(\xi) = \begin{cases} q & \text{if } \xi = 0, \\ 0 & \text{if } \xi \neq 0. \end{cases}$$

We still need an estimate on twisted exponential sums. Thanks to Weil, we have the following result.

Lemma 2.4. (Weil's Theorem) [20] *Let χ be a d-th order multiplicative character and λ an additive character of $\mathbf{F}_q$. Let $g(x), h(x) \in \mathbf{F}_q[x], m = \deg g(x), r = \deg h(x)$. If $(m, d) = (r, q) = 1$, then*

$$| \sum_{x \in \mathbf{F}_q} \chi(g(x))\lambda(h(x))| \leq (m + r - 1)\sqrt{q}.$$

We see that Weil's Theorem fails to work when the degrees of $g(x)$ and $h(x)$ are large. In the following, we will improve Weil's estimate for some special cases, that is, the bilinear polynomials.

Let $T(x)$ be the trace from $\mathbf{F}_{q^n}$ to $\mathbf{F}_q$. A polynomial $b(x) \in \mathbf{F}_{q^n}[x]$ is called a *q-bilinear polynomial* if

$$b(x) = \sum_{0 \leq i,\, j < n} a_{ij}\, x^{q^i + q^j}, \quad a_{ij} \in \mathbf{F}_{q^n}.$$

For $b(x)$, we define a bilinear form

$$B(x,\, y) = T(b(x + y)) - T(b(x)) - T(b(y)).$$

If we consider $\mathbf{F}_{q^n}$ as a vector space over $\mathbf{F}_q$, it is easy to see that $B(x,\, y)$ is a symmetric bilinear form over $\mathbf{F}_q$. Let V^0 be the orthogonal subspace of $\mathbf{F}_{q^n}$ associated with $B(x, y)$, that is, $V^0 = \{x \in \mathbf{F}_{q^n} \mid B(x,\, y) = 0, \, \forall y \in \mathbf{F}_{q^n}\}$. We let d_0 denote the dimension of V^0 over $\mathbf{F}_q$. It is obvious that $d_0 \leq n$. In fact, $d_0 < n$ if $b(x) \neq 0$. Now we compute d_0 for the following special case which will occur in next section. From now on we fix $u \in \mathbf{F}_{q^n}$ such that $T(u) = 1$. Let

$$\begin{cases} a(x) &= \sum_{0 \leq i < j < n} x^{q^i + q^j} \\ A(x) &= ua(x). \end{cases}$$

Clearly, $a(x)$ and $A(x)$ are q-bilinear polynomials. Since $a(x)$ can be viewed as a function from $\mathbf{F}_{q^n}$ to $\mathbf{F}_q$, then $T(A(x)) = T(u\, a(x)) = a(x)\, T(u) = a(x)$.

Proposition 2.5. *Let V^0 be the orthogonal subspace of $\mathbf{F}_{q^n}$ associated with $A(x)$. Then*

$$V^0 = \begin{cases} \mathbf{F}_q & \text{if } p \mid n - 1; \\ 0 & \text{if } p \nmid n - 1. \end{cases}$$

Let ψ be the canonical additive character of $\mathbf{F}_q$, $b(x)$ be a q-bilinear polynomial over $\mathbf{F}_{q^n}$. Define

$$S(b(x)) = \sum_{x \in \mathbf{F}_{q^n}} \psi(T(b(x))).$$

Theorem 2.6. *(i). When q is odd, $|S(b(x))| = q^{\frac{n+d_0}{2}}$.*
(ii). When q is even,

$$|S(b(x))| = \begin{cases} q^{\frac{n+d_0}{2}} & \text{if } T(b(x)) = 0, \, \forall x \in V^0; \\ 0 & \text{otherwise.} \end{cases}$$

Carlitz [3] gave the exact value for $S(b(x))$ when q is even and $b(x)$ of degree 3, see also [5]. We note that while Theorem 2.6 can be deduced from Weil's deep general results from [20] we have given a simple proof of this result which will be of use. We now discuss the mixed exponential sums associated with $b(x)$. Let χ be a multiplicative character of $\mathbf{F}_{q^n}$. Define

$$\hat{S}(b(x),\,\chi) = \sum_{x \in \mathbf{F}_{q^n}} \psi(T(b(x)))\chi(x).$$

The following result is new.

Theorem 2.7. *Let ψ be the canonical additive character of $\mathbf{F}_q$, and let χ be a multiplicative character of $\mathbf{F}_{q^n}$ $(d > 1)$ of order d. Then*

$$|\hat{S}(b(x),\,\chi)| \le \sqrt{q^n - \sigma q^{d_0} + 2(q^n - q^{d_0})\sqrt{q^n}}$$

where d_0 is the dimension of the orthogonal subspace V^0 associated with $b(x)$ over $\mathbf{F}_q$ and

$$\sigma = \begin{cases} 0 & \text{if } q \text{ is even and there exists } z \in V^0 \text{ such that } T(b(z)) \ne 0; \\ \chi(-1) & \text{otherwise.} \end{cases}$$

3. Second Coefficient of Primitive Polynomials

3.1. Odd Characteristic

First of all, we give a lemma from which the second coefficient of an irreducible polynomial can be represented by the traces of a root and the square of a root. Then Hansen-Mullen's conjecture reduces to the existence of primitive element solutions of some equation associated with the trace from $\mathbf{F}_{q^n}$ to $\mathbf{F}_q$.

Lemma 3.1. *Let $f(x) = x^n + a_1 x^{n-1} + \cdots + a_n$ be an irreducible polynomial over $\mathbf{F}_q, \xi$ be a root of $f(x)$ in $\mathbf{F}_{q^n}, q$ odd. Then $a_2 = \frac{1}{2}(T(\xi)^2 - T(\xi^2))$, where $T(x)$ is the trace from $\mathbf{F}_{q^n}$ to $\mathbf{F}_q$.*

Proof. Since $f(x)$ is irreducible, $\xi, \xi^q, \cdots, \xi^{q^{n-1}}$ are all roots of $f(x)$ in $\mathbf{F}_{q^n}$. Therefore

$$f(x) = (x - \xi)(x - \xi^q) \cdots (x - \xi^{q^{n-1}})$$

and

$$a_2 = \sum_{0 \le i < j < n} \xi^{q^i} \xi^{q^j}$$

$$= \frac{1}{2} \sum_{0 \le i,j < n; i \ne j} \xi^{q^i} \xi^{q^j}$$

$$= \frac{1}{2} T(\xi^{1+q} + \xi^{1+q^2} + \cdots + \xi^{1+q^{n-1}})$$

$$= \frac{1}{2} T(\xi T(\xi) - \xi^2)$$

$$= \frac{1}{2} (T(\xi)^2 - T(\xi^2)).$$

This finishes the proof of the lemma.   □

By Lemma 3.1, the existence of primitive element solutions of the equation $T(x)^2 - T(x^2) = c$ for $c \in \mathbf{F}_q$ yields Hansen-Mullen's conjecture in the case of the second coefficient. But we prefer to consider the following system of equations to give a stronger conclusion:

$$\begin{cases} T(x) & = a, \\ T(x^2) & = b, \end{cases} \tag{1}$$

where $a, b \in \mathbf{F}_q$. If (1) has a primitive element solution ξ in $\mathbf{F}_{q^n}$, let $f(x)$ be the minimal polynomial of ξ over $\mathbf{F}_q$. Then the first and second coefficients of $f(x)$ are a and $\frac{1}{2}(a^2 - b)$. Furthermore, $f(x)$ is a primitive polynomial of degree n. Hence there exists a primitive polynomial with the first coefficient a and second coefficient $\frac{1}{2}(a^2 - b)$. So we need to discuss the existence of the primitive element solutions of (1). Let $N_{q,n}(a, b)$ denote the number of the primitive element solutions of (1) in $\mathbf{F}_{q^n}$ and $Q = \frac{q^n - 1}{q - 1}$. Now we can prove our main result.

Theorem 3.2. *(i)*

$$N_{q,n}(0,0) \ge \frac{\varphi(q^n - 1)}{q^2(q^n - 1)} \{q^n - q - (q - 1)q(\sqrt{q^n} + 1)$$
$$-(2^{\omega(Q)} - 1)(q - 1)(2q + 1)\sqrt{q^n}\}.$$

(ii) $a \ne 0$.

$$N_{q,n}(a,0) \ge \frac{\varphi(q^n - 1)}{q^2(q^n - 1)} \{q^n - 2(q - 1)$$
$$+(2q - 1)\sqrt{q^n} - 2^{\omega(Q)}(4q - 3)\sqrt{q^n}$$
$$-(2^{\omega(q^n - 1)} - 2^{\omega(Q)})(2q - 1)\sqrt{q^{n+1}}\}.$$

(iii) $b \neq 0$.

$$N_{q,n}(a,b) \geq \frac{\varphi(q^n - 1)}{q^2(q^n - 1)}\{q^n - \delta q + q(\sqrt{q} + 1)(\sqrt{q^n} + 1)$$
$$-(2^{\omega(Q)} - 1)(2(\sqrt{q} + 1)q + \delta q - 2\delta + 1)\sqrt{q^n}$$
$$-(2^{\omega(q^n - 1)} - 2^{\omega(Q)})(4q + 1 - \delta)\sqrt{q^{n+1}}\},$$

where $\omega(m)$ *is the number of the distinct prime factors of* m *and*

$$\delta = \begin{cases} 0, & \text{if } a \neq 0, \\ 1, & \text{if } a = 0. \end{cases}$$

We only give a sketch of the proof of the theorem above. By Lemma 2.1 and Lemma 2.3, we have

$$N_{q,n}(a,b) = \frac{q^n - 1}{q^2(q^n - 1)} \sum_{\xi \in F_{q^n}^*} \sum_{c_1 \in F_q} \psi_{c_1}(T(\xi) - a) \sum_{c_2 \in F_q} \psi_{c_2}(T(\xi^2) - b)$$
$$\times \sum_{d|q^n - 1} \frac{\mu(d)}{\varphi(d)} \sum_{\chi_d} \chi_d(\xi)$$
$$= \frac{\varphi(q^n - 1)}{q^2(q^n - 1)} \sum_{d|q^n - 1} \frac{\mu(d)}{\varphi(d)} \sum_{\chi_d} \sum_{c_1, c_2 \in F_q}$$
$$\times \sum_{\xi \in F_{q^n}^*} \psi(T(c_1\xi + c_2\xi^2) - c_1 a - c_2 b)\chi_d(\xi).$$

In the above, χ_d runs through all d-th order multiplicative characters of $\mathbf{F}_{q^n}$. By a complicated analysis, we can obtain the estimate above. Now we give a simple proposition to conclude when $N_{q,n}(a,b) > 0$.

Proposition 3.3. *(i). If* $q^{\frac{n}{2} - 2} \geq 2^{\omega(Q)}$, *then* $N_{q,n}(0,0) > 0$. *(ii). Let* $(a,b) \neq (0,0)$. *If* $q^{\frac{n}{2} - \frac{3}{2}} \geq (\frac{13}{3})2^{\omega(q^n - 1)}$, *then* $N_{q,n}(a,b) > 0$.

It is an easy consequence of Theorem 3.2.

Proposition 3.4. *(i). Let* $n \geq 5$. *Then* $N_{q,n}(0,0) > 0$ *for* q^n *large enough. (ii). Let* $n \geq 4, (a,b) \neq (0,0)$. *Then* $N_{q,n}(a,b) > 0$ *for* q^n *large enough.*

This shows that Hansen-Mullen's conjecture for $i = 2$ holds if $n \geq 4$ and q^n is large enough. Further by detailed analysis and computations of Proposition 3.3, we can prove

Proposition 3.5. *Let* $n \geq 7$. *Except the cases* $(a,b) \neq (0,0)$, $(n,q) = (7,3),(7,7)$, $N_{q,n}(a,b) > 0$.

We use the Cohen Sieve Method [2] for $(a,b) \neq (0,0)$, $(n,q) = (7,3)$, $(7,7)$. Let $e \mid q^n - 1$. Define

$$\mathbf{T}(e) = \{\xi \in \mathbf{F}_{q^n} \mid \xi \text{ is a solution of } (1)$$
$$\text{and } \xi \text{ is not any kind of } e\text{-th power in } \mathbf{F}_{q^n}\}.$$

It is obvious that $|\mathbf{T}(q-1)| = N_{q,n}(a,b)$. We have

$$\mathbf{T}(e_1) \bigcap \mathbf{T}(e_2) = \mathbf{T}([e_1, e_2])$$
$$\mathbf{T}(e_1) \bigcup \mathbf{T}(e_2) = \mathbf{T}((e_1, e_2)).$$

In the above, $e_1|q^n - 1, e_2|q^n - 1, [e_1, e_2]$ and (e_1, e_2) denote separately the least common multiple and the greatest common factor of e_1 and e_2. If $[e_1, e_2] = q - 1$, then

$$\begin{aligned} N_{q,n}(a,b) &= |\mathbf{T}(q-1)| \\ &= |\mathbf{T}(e_1)| + |\mathbf{T}(e_2)| - |\mathbf{T}((e_1, e_2))|. \end{aligned} \qquad (2)$$

Now we can prove

Theorem 3.6. *If $n \geq 7$, $N_{q,n}(a,b) > 0$ for any $a, b \in \mathbf{F}_q$.*

Proof. Suppose $(a,b) \neq (0,0)$, we consider the case $(n,q) = (7,7)$. Let $e_1 = 174, e_2 = 9466$, then $[e_1, e_2] = q^n - 1, (e_1, e_2) = 2$. Using Lemma 2.2 instead of Lemma 2.1 in the proof of Theorem 3.2, we obtain

$$\begin{aligned} |\mathbf{T}(174)| &\geq 3367; \\ |\mathbf{T}(9466)| &\geq 6895; \\ |\mathbf{T}(2)| &\leq 9079. \end{aligned}$$

By (2), we obtain $N_{q,n}(a,b) > 0$. For $(n,q) = (7,3)$, we take $e_1 = 2, e_2 = 1093$. A similar computation gives $N_{q,n}(a,b) > 0$. Hence we finish the proof of Theorem 3.6. $\qquad \Box$

By Lemma 3.1 and Theorem 3.6, we can easily give the following corollaries.

Corollary 3.7. *Suppose $n \geq 7$. Then there exists a primitive polynomial in $\mathbf{F}_q[x]$ of degree n with the first and second coefficients prescribed in advance.*

Corollary 3.8. *Suppose $n \geq 7$. There are at least q primitive polynomials in $\mathbf{F}_q[x]$ of degree n with the first or second coefficient prescribed in advance.*

Corollary 3.8 shows that Hansen-Mullen's conjecture holds for $i = 2$ if $n \geq 7$. To give a complete list of the possible exceptions for the cases $n = 4, 5, 6$, we suggest Cohen Sieve Method [2],[15] as a means of attack. The analysis of these cases is a future work. But for $n = 4$, $a_2 = 0$, our method seems to be inaccessible.

3.2. Even Characteristic

In this section, we obtain estimates for the number of primitive polynomials with prescribed second coefficient alone. Since the work of the previous section generally yields stronger results whenever q is odd but is inapplicable when q is even, we shall throughout assume that q is even.

Let

$$a(x) = \sum_{0 \leq i < j < n} x^{q^i + q^j}. \qquad (3)$$

We have the following simple lemma.

Lemma 3.9. *Let $a \in \mathbf{F}_q$. Then there exists a primitive polynomial of degree n with the second coefficient a over $\mathbf{F}_q$ if and only if there exists a primitive element $\xi \in \mathbf{F}_{q^n}$ such that $a(\xi) = a$.*

Proof. Firstly, let $f(x)$ be a primitive polynomial over $\mathbf{F}_q$ of degree n with the second coefficient a and let ξ be a root of $f(x)$ in $\mathbf{F}_{q^n}$. Then

$$f(x) = (x - \xi)(x - \xi^q) \cdots (x - \xi^{q^{n-1}}).$$

Therefore

$$a = \sum_{0 \le i < j < n} \xi^{q^i + q^j}.$$

That is, ξ is a primitive element such that $a(\xi) = a$. Conversely, let ξ be a primitive element such that $a(\xi) = a$. Then the minimal polynomial of ξ over $\mathbf{F}_q$ is a primitive polynomial of degree n with the second coefficient a. $\square$

By Lemma 3.9, Hansen-Mullen's conjecture for the second coefficient reduces to the existence of a primitive element solution x_a of the equation

$$a(x) = a, \text{ for each } a \in \mathbf{F}_q. \tag{4}$$

Let $N_{q,n}(a)$ denote the number of the primitive element solutions of (4). To prove $N_{q,n}(a) > 0$, we use Cohen Sieve Method. Let $e | q^n - 1$, define

$$M(e) = \{\xi \in \mathbf{F}_{q^n} | \xi \text{ is a solution of (4)}$$
$$\text{and } \xi \text{ is not any kind of } e\text{-th power in } \mathbf{F}_{q^n}\}.$$

It is obvious that $|M(q-1)| = N_{q,n}(a)$. Let $e_1, e_2 | q - 1$, then

$$M(e_1) \bigcap M(e_2) = M([e_1, e_2]);$$
$$M(e_1) \bigcup M(e_2) = M((e_1, e_2)).$$

where $[e_1, e_2]$ and (e_1, e_2) denote respectively the least common multiple and the greatest common factor of e_1 and e_2. If $[e_1, e_2] = q - 1$, then

$$N_{q,n}(a) = |M(q-1)|$$
$$= |M(e_1)| + |M(e_2)| - |M((e_1, e_2))|. \tag{5}$$

Remark 3.10. *(5) is a special case of Cohen Sieve Method which had 5 terms (see [15] Theorem 2). We also observe that this expression remains unaffected upon replacing e by e^*, the squarefree part of e.*

Now we can use Lemma 2.2 and Lemma 2.3 to construct a characteristic function for $M(e)$ and obtain the formula of $N_{q,n}(a)$ with exponential sums.

Theorem 3.11. *Let q be even and e a divisor of $q^n - 1$. Then*

$$|M(e)| = \frac{\theta(e)}{q}(q^n - (\delta(a)q - 1)\hat{S}(A, 1) - \delta(a)q$$
$$+ \sum_{1 < d | e} \frac{\mu(d)}{\varphi(d)} \sum_{\chi_d} G(\chi_d^{-1}, \psi_{-a})\hat{S}(A, \chi_d))$$

where

$$\theta(e) = \frac{\varphi(e)}{e} \qquad and \qquad \delta(a) = \begin{cases} 0 & \text{if } a \neq 0; \\ 1 & \text{if } a = 0. \end{cases}$$

and $G(\chi_d^{-1}, \psi_{-a})$ is the Gauss sum of χ_d^{-1} and ψ_{-a} over $\mathbf{F}_q$.

Proof. By Lemma 2.2 and Lemma 2.3, we have

$$
\begin{aligned}
|M(e)| &= \frac{\theta(e)}{q} \sum_{\xi \in \mathbf{F}_{q^n}^*} \sum_{c \in \mathbf{F}_q} \psi_c(a(\xi) - a) \sum_{d|e} \frac{\mu(d)}{\varphi(d)} \sum_{\chi_d} \chi_d(\xi) \\
&= \frac{\theta(e)}{q} \sum_{c \in \mathbf{F}_q} \sum_{d|e} \frac{\mu(d)}{\varphi(d)} \sum_{\chi_d} \sum_{\xi \in \mathbf{F}_{q^n}^*} \chi_d(\xi) \psi_c(a(\xi) - a) \\
&= \frac{\theta(e)}{q} \Big(q^n - 1 + \sum_{c \in \mathbf{F}_q^*} \sum_{\xi \in \mathbf{F}_{q^n}^*} \psi_c(a(\xi) - a) \\
&\quad + \sum_{c \in \mathbf{F}_q^*} \sum_{d|e} \frac{\mu(d)}{\varphi(d)} \sum_{\chi_d} \sum_{\xi \in \mathbf{F}_{q^n}^*} \chi_d(\xi) \psi_c(a(\xi) - a) \Big) \\
&= \frac{\theta(e)}{q} \Big(q^n - 1 + \sum_{c \in \mathbf{F}_q^*} \psi_c(a) \sum_{\xi \in \mathbf{F}_{q^n}} \psi(ca(\xi)) \\
&\quad + \sum_{1 < d|e} \frac{\mu(d)}{\varphi(d)} \sum_{\chi_d} \sum_{c \in \mathbf{F}_q^*} \chi_d^{-1}(c) \psi(-ca) \sum_{\chi_d} \chi_d(\xi) \psi(a(\xi)) \Big) \\
&= \frac{\theta(e)}{q} \Big(q^n - (\delta(a)q - 1)\hat{S}(A,\ 1) - \delta(a)q \\
&\quad + \sum_{1 < d|e} \frac{\mu(d)}{\varphi(d)} \sum_{\chi_d} G(\chi_d^{-1},\ \psi_{-c})\hat{S}(A,\ \chi_d) \Big).
\end{aligned}
$$

This finishes the proof of the theorem.   □

Theorem 3.12. *Let q be even with $e_1, e_2|q^n - 1$, and $[e_1,\ e_2] = q^n - 1$. Let $e_3 = (e_1,\ e_2)$ and set $e_i' = (\frac{q^n-1}{q-1}, e_i)$, $W_i = 2^{w(e_i)} - 1$, $W_i' = 2^{w(e_i')} - 1$, for $i = 1, 2, 3$; where $w(d)$ is the number of the prime factors of d. With these notations, the following holds. (i).*

$$
\begin{aligned}
q N_{q,n}(0) \geq\ & (\theta(e_1) + \theta(e_2) - \theta(e_3)) \Big(q^n - q - (q-1)\sqrt{q^{n+\varepsilon}} \\
& - \{\theta(e_1)W_1' + \theta(e_2)W_2' \\
& - \theta(e_3)W_3'(q-1)\} \cdot \sqrt{q - \sigma q^\varepsilon + 2(q^n - q^\varepsilon)\sqrt{q^n}} \Big).
\end{aligned}
$$

(ii). If $a \neq 0$, then

$$
\begin{aligned}
qN_{q,n}(a) \geq \quad & (\theta(e_1) + \theta(e_2) - \theta(e_3)) \Big(q^n - \sqrt{q^{n+\varepsilon}} \\
& - (\theta(e_1)W_1' + \theta(e_2)W_2' + \theta(e_3)W_3' \\
& - \{\theta(e_1)(W_1 - W_1') + \theta(e_2)(W_2 - W_2') \\
& - \theta(e_3)(W_3 - W_3')\} \sqrt{q}) \sqrt{q^n - \sigma q^\varepsilon + 2(q^n - q^\varepsilon)\sqrt{q^n}} \Big)
\end{aligned}
$$

where σ is the same as in Proposition 2.5 and where

$$
\varepsilon = \begin{cases} 1 & \text{if } n \text{ is odd;} \\ 0 & \text{if } n \text{ is even.} \end{cases}
$$

Proof. Let

$$
\Gamma_d = \frac{\mu(d)}{\varphi(d)} \sum_{\chi_d} G(\chi_d^{-1}, \ \psi_{-a}) \hat{S}(A(x), \chi_d).
$$

From (5) and Theorem 3.11,

$$
\begin{aligned}
qN_{q,n}(a) = \quad & (\theta(e_1) + \theta(e_2) - \theta(e_3))\{q^n - \delta(a)q - (\delta(a)q - 1)\hat{S}(A, \ 1) \\
& - \theta(e_1)\Big(\sum_{\substack{1<d|e_1'}} \Gamma_d + \sum_{\substack{d|e_1 \\ d\nmid e_1'}} \Gamma_d \Big) \\
& - \theta(e_2)\Big(\sum_{\substack{d|e_2' \\ d\nmid e_3'}} \Gamma_d + \sum_{\substack{d|e_2 \\ d\nmid e_2'}} \Gamma_d \Big) \\
& - (\theta(e_2) - \theta(e_3)) \sum_{\substack{1<d|e_3'}} \Gamma_d + \theta(e_3) \sum_{\substack{d|e_3 \\ d\nmid e_3'}} \Gamma_d \}.
\end{aligned}
$$

Since χ_d induces a trivial character on $\mathbf{F}_q$ if and only if $d|\frac{q^n-1}{q-1}$, we have

$$
|G(\chi_d, \ \psi_{-a})| = \begin{cases} q-1 & \text{if } d|\frac{q^n-1}{q-1} \text{ and } a = 0; \\ 0 & \text{if } d\nmid\frac{q^n-1}{q-1} \text{ and } a = 0; \\ 1 & \text{if } d|\frac{q^n-1}{q-1} \text{ and } a \neq 0; \\ \sqrt{q} & \text{if } d\nmid\frac{q^n-1}{q-1} \text{ and } a \neq 0; \end{cases} \tag{6}
$$

Using the estimates for $\hat{S}(A, \ \chi_d)$ from Theorem 2.6 and 2.7, together with (6) to bound Γ_d finishes the proof of Theorem 3.12. $\qquad\square$

Proposition 3.13. *With the notations of Theorem 3.12, we obtain the following conditions for $N_{q,n}(a) > 0$. (i) Let $n \geq 5$. If $\theta(e_1) + \theta(e_2) - \theta(e_3) > 0$ and*

$$
q^{\frac{n}{4}-1} \geq \frac{\sqrt{3}}{\theta(e_1) + \theta(e_2) - \theta(e_3)}(\theta(e_1)W_1' + \theta(e_2)W_2' - \theta(e_3)W_2'), \tag{7}
$$

then $N_{q,n}(0) > 0$. *(ii) Let $n \geq 4$ and $a \neq 0$. If $\theta(e_1) + \theta(e_2) - \theta(e_3) > 0$ and*

$$q^{\frac{n}{4} - \frac{1}{2}} \geq \frac{3}{\theta(e_1) + \theta(e_2) - \theta(e_3)} (\theta(e_1)W_1 + \theta(e_2)W_2 - \theta(e_3)W_3), \qquad (8)$$

then $N_{q,n}(a) > 0$.

To obtain a first lower bound on q which ensures $N_{q,n}(a) > 0$, we proceed as in Section 3 of [7]. To this end, we let $u_0 = (\frac{n}{4} - 1)^{-1}$, $u_1 = (\frac{n}{4} - \frac{1}{\varepsilon})^{-1}$. We suppress the details and produce only the tables, the reader can then easily reconstruct the complete proofs by consulting [7] for a guide.

Proposition 3.14. *(i). Let $n \geq 7$. Then $N_{q,n}(0) > 0$ if*

$$w(\frac{q^n - 1}{q - 1}) \geq A_n.$$

(ii). Let $n \geq 4$ and $a \neq 0$. $N_{q,n}(a) > 0$ if

$$w(\frac{q^n - 1}{q - 1}) \geq B_n.$$

In the above, A_n and B_n are taken from Table 3.1.

Table   3.1

n	u_0	u_1	A_n	B_n
4		8		145
5	20	$\frac{20}{3}$	*	66
6	12	6	*	46
7	$\frac{28}{3}$	$\frac{28}{5}$	42	37
8	8	$\frac{16}{3}$	145	31
9	$\frac{36}{5}$	$\frac{36}{7}$	46	28
≥ 10	$\frac{20}{3}$	5	66	26

By choosing e_1 and e_2 suitably, we can obtain better lower bounds for $q^{\frac{n}{4} - \frac{1}{2}}$ which ensures $N_{q,n}(a) > 0$. We proceed as in [2]. Let

$$q^n - 1 = \gamma_1^{\beta_1} \cdots \gamma_k^{\beta_k} p_1^{\alpha_1} \cdots p_s^{\alpha_s}$$

where $\gamma_i s$, $p_i s$ are distinct prime numbers with $p_1 < p_2 < \cdots < p_s$ and γ_i such that

$$\gamma_i | q - 1, \quad \gamma_i \nmid \frac{q^n - 1}{q - 1}.$$

Take

$$
\begin{aligned}
e_1 &= \gamma_1^{\beta_1} \cdots \gamma_k^{\beta_k} p_1^{\alpha_1} \cdots p_t^{\alpha_t} p_{t+1}^{\alpha_{t+1}} \cdots p_l^{\alpha_l}; \\
e_2 &= \gamma_1^{\beta_1} \cdots \gamma_k^{\beta_k} \cdot p_1^{\alpha_1} \cdots p_t^{\alpha_t} p_{l+1}^{\alpha_{l+1}} \cdots p_s^{\alpha_s};
\end{aligned}
$$

where $l = \left[\frac{s-t}{2}\right] + t$. Then using the earlier notations we find that

$$
\begin{aligned}
e_3 &= \gamma_1^{\beta_1} \cdots \gamma_k^{\beta_k} \cdot p_1^{\alpha_1} \cdots p_t^{\alpha_t}; \\
e_1' &= p_1^{\alpha_1} \cdots p_l^{\alpha_l}; \\
e_2' &= p_1^{\alpha_1} \cdots p_t^{\alpha_t} \cdot p_{l+1}^{\alpha_{l+1}} \cdots p_s^{\alpha_s}; \\
e_3' &= p_1^{\alpha_1} \cdots p_t^{\alpha_t}.
\end{aligned}
$$

Since $\theta(e_1) = \frac{\theta(e_3)\theta(e_1')}{\theta(e_3)'}, \theta(e_2) = \frac{\theta(e_3)\theta(e_2')}{\theta(e_3)'}$. Condition (6) simplifies to

$$
q^{\frac{n}{4}-1} \geq \frac{\sqrt{3}}{\theta(e_1') + \theta(e_2') - 1}(\theta(e_1')W_1' + \theta(e_2')W_2') \tag{9}
$$

when $\theta(e_1') + \theta(e_2') - 1 > 0$. We consider the case $n = 7$ and take $t = 0$, then $l \leq 21$, which gives $\theta(e_1') \geq \frac{6}{7} \times \frac{28}{29} \times \frac{42}{43} \times \frac{70}{71} \times \left(\frac{112}{113}\right)^{17} \doteq 0.685$ and $\theta(e_2) \geq 0.685$. Hence if $q^{\frac{n}{4}-1} > 3 \cdot 2^{22}$, then $N_{q,n}(0) > 0$. Similarly we can get the lower bounds appearing in the following table

Table   3.2

n	t	$\theta(e_1)$	$\theta(e_2)$	C_n
7	0	0.685	0.685	$3 \cdot 2^{22}$
8	23	0.531	0.531	2^{88}
9	0	0.512	0.512	$21 \cdot 2^{25}$
10	4	0.532	0.532	2^{39}

Proposition 3.15. . Let $n \geq 7$ and C_n be taken from Table 3.2. If

$$
q^{\frac{n}{4}-1} \geq C_n, \tag{10}
$$

then $N_{q,n}(0) > 0$.

For $a \neq 0$, we take a similar process for which $N_{q,n}(0) > 0$ to get the lower bounds. Let

$$
\begin{aligned}
q^n - 1 &= p_1^{\alpha_1} \cdots p_s^{\alpha_s}, \text{ with } p_1 < p_2 < \cdots < p_s, \\
e_1 &= p_1^{\alpha_1} \cdots p_t^{\alpha_t} \cdot p_{t+1}^{\alpha_{t+1}} \cdots p_l^{\alpha_l}, \\
e_2 &= p_1^{\alpha_1} \cdots p_t^{\alpha_t} p_{l+1}^{\alpha_{l+1}} \cdots p_s^{\alpha_s},
\end{aligned}
$$

where $l = \left[\frac{s-t}{2}\right] + t$. Then $e_3 = (e_1,\ e_2) = p_1^{\alpha_1} \cdots p_t^{\alpha_t}$. Let $E_1 = \frac{e_1}{e_3}$, $E_2 = \frac{e_2}{e_3}$. The condition (8) is changed into

$$q^{\frac{n}{4}-\frac{1}{2}} \geq \frac{3}{\theta(E_1) + \theta(E_2) - 1}(\theta(E_1)W_1 + \theta(E_2)W_2). \tag{11}$$

In a similar manner to Proposition 3.15, we obtain the following table:

Table 3.3

n	t	$\theta(e_1) \geq$	$\theta(e_2) \geq$	D_n
4	23	0.531	0.531	2^{88}
5	4	0.532	0.532	2^{39}
6	3	0.537	0.537	21×2^{25}
7	3	0.575	0.575	7×2^{21}
8	3	0.606	0.606	5×2^{17}
9	3	0.620	0.620	13×2^{15}
10	3	0.634	0.634	3×2^{16}

Proposition 3.16. *Let $n \geq 4$ and $a \neq 0$. If*

$$q^{\frac{n}{4}-\frac{1}{2}} \geq D_n, \tag{12}$$

where D_n is taken from Table 3.3, then $N_{q,n}(a) > 0$.

To conclude, we check all q^n not satisfying (10) or (12). We factor completely $q^n - 1$ when q^n is small and find small prime factors of $q^n - 1$ when q^n is large. Then we use one of the inequalities (7) or (8) or Theorem 3.12 to check when $N_{q,n}(a) > 0$. Doing this we obtain

Proposition 3.17. *(i). Let*

$$S_0 = \{(7, 2^4), (7, 2^3), (7, 2^2), (8, 2^3), (8, 2^2), (9, 2^2), (10, 2^2), (12, 2^2)\}.$$

If $n \geq 7$ and $(n, q) \notin S_0$, then $N_{q,n}(0) > 0$. (ii). Let

$$S_1 = \{(4, 2^k)|2 \leq k \leq 12\} \cup \quad \{(4, 2^{15}), (5, 2^8), (5, 2^6), (5, 2^4), (5, 2^3),$$
$$(5, 2^2), (6, 2^6), (6, 2^5), (6, 2^4), (6, 2^3), (6, 2^2),$$
$$(7, 2^4), (7, 2^2), (8, 2^3), (8, 2^2), (9, 2^2), (10, 2^2),$$
$$(12, 2^2), (14, 2^2)\}.$$

If $a \neq 0$ and $(n, q) \notin S_1$, then $N_{q,n}(a) > 0$.

For every pair $(n, q) \in S_0 \cup S_1$, we have designed a program with Mathematica to search for a primitive polynomial with the given second coefficient a by using a computer and found that $N_{q,n}(a) > 0$. So we finished the proof of

Theorem 3.18. *Let $a \in \mathbf{F}_q$, $(n,a) \neq (4,0),(5,0),(6,0)$, then there exists a primitive polynomial of degree n with the second coefficient a.*

At last we observe that by Proposition 3.13, Hansen-Mullen's conjecture holds for the second coefficient if $(n,a) \neq (0,0)$ and q is large enough. This will be discussed further in another case.

Acknowledgement. The work was done while the author visited National University of Singapore, he is indebted to Chaoping Xing for his inviting, to Keqin Feng for his encouragement and continuous support, to Igor Shparlinski and the referee for their comments.

References

[1] S. D. Cohen, *Primitive elements and polynomials with arbitrary traces,* Discrete Math., **83** (1990), 1–7.

[2] S. D. Cohen, *Primitive elements and polynomials: existence results,* Lect. Notes in Pure and Applied Math. 141, edited by G. L. Mullen and P. J. Shiue, Dekker, New York, (1992), 43–55.

[3] L. Carlitz, *Evaluation of exponential sums over a finite field,* Math. Nachr., **139** (1980), 319–339.

[4] H. Davenport, *Bases for finite fields,* J. London Math. Soc., **43** (1968), 21–39.

[5] A. Garcia and H. Stichtenoth, *A class of polynomials over finite fields,* Finite Fields and Their Applications, **5** (1999), 424–435.

[6] W-B. Han, *Primitive roots and linearized polynomials,* Advance in Math. (China), **22** (1994), 460–462.

[7] W-B. Han, *The coefficients of primitive polynomials over finite fields,* Math. of Comp., **65** (1996), 331–340.

[8] W-B. Han, *On two exponential sums and their applications,* Finite Fields and Their Applications, **3** (1997), 115–130.

[9] T. Hansen and G. L. Mullen, *Primitive polynomials over finite fields,* Math. of Comp., **59** (1992), 639–643.

[10] D. Jungnickel and S. A. Vanstone, *On primitive polynomials over finite fields,* J. of Algebra, **124** (1989), 337–353.

[11] R. Lidl and H. Niederreiter, *Finite fields,* Addison-Wesley, Reading, MA, (1983).

[12] H. W. Lenstra and R. J. Schoof, *Primitive normal bases for finite fields,* Math. of Comp., **48** (1987), 217–232.

[13] O. Moreno, *On the existence of a primitive quadratic trace over $GF(p^m)$,* J. of Combin. Theory Ser. A, **51** (1989), 104–110.

[14] G. L. Mullen and Igor. Shparlinski, *Open problems and conjectures in finite fields,* Finite Fields and Application, edited by S. D. Cohen and H. Niederreiter, Lect. Notes Ser. 233, London Math.Soc., (1996), 243–268.

[15] G. McNay, *Cohen's sieve with quadratic conditions,* to appear in Utilitas Math.

[16] W. M. Schmidt, *Equations over Finite Fields: an elementary approach*, Lect. Notes in Math. 536, Springer-Verlag, Berlin/Heidelberg/New York, (1976).

[17] I. Shparlinski, *Coefficients of primitive polynomials*, Matem. Zametki, **38** (1985), 810–815 (in Russian).

[18] I. Shparlinski, *On primitive polynomials*, Problemy Peredachi Inform., **23** (1987), 100–103 (in Russian).

[19] Q. Sun and W-B. Han, *On absolute trace and primitive roots in a finite field (in Chninese)*, Chinese Annals of Math., **11A2** (1990), 202–205.

[20] A. Weil, *Sur certains groupes d'opérateurs unitaires*, Acta Math., **111** (1964), 143–211.

Department of Applied Mathematics,
Zhengzhou Polytechnic Institute,
P.O.Box 1001 No.40,
Zhengzhou 450002, P.R.China
E-mail address: wb.han@netease.com

Progress in Computer Science and Applied Logic, Vol. 20
© 2001 Birkhäuser Verlag Basel/Switzerland

The Distribution of the Quadratic Symbol in Function Fields and a Faster Mathematical Stream Cipher

Jeffrey Hoffstein and Daniel Lieman

Abstract. We present a stream cipher based on mathematical considerations which is much faster then many other mathematical ciphers. Its security is based on the uniformity of the distribution of the quadratic symbol in function fields.

1. Introduction

One of the basic building blocks of cryptographic systems is the stream cipher, which produces from a fixed input a sequence of ones and zeros which ideally have the property that an observer cannot predict the nth bit, even after observing the first $n - 1$ bits. In its most straightforward use, a cipher allows the simple construction of a secret key cryptosystem, where the key is the initial input to the cipher, and the message to be encoded is combined with the sequential output of the cipher by XOR operations.

Cipher systems have historically fallen into one of two camps: those based on mathematical foundations, which are sometimes provably secure, but slow; and those which are (often quite) fast, but whose security is based on a statistical analysis and the simple fact no one has announced an algorithm to compromise their security, rather than on any provable security.

The purpose of this paper is to describe a mathematically based cipher which is much faster than most existing mathematical ciphers. (We have run this cipher at 1–2 Mbits per second on a Sparc 1; we will discuss timing considerations later.) While this cipher is *not* provably secure, it is based on an extremely difficult mathematical problem, which is well beyond the range of current techniques. This cipher is thus of some interest, as it provides an alternative to either of the traditional types of ciphers mentioned above.

Both authors partially supported by various NSF grants. The authors wish to thank MSRI, Berkeley, for its hospitality during part of the period during which this research was conducted, and Dan Bump, Burt Kaliski, Michael Rosen, and Yiqun Lisa Yin for helpful conversations. We also wish to thank Dorian Goldfeld for stimulating our interest in cryptographic problems, and Igor Shparlinski and the referees for many helpful comments on the final version of this paper.

Anshel and Goldfeld [1, 2] have proposed and patented ciphers based on sequential coefficients of zeta functions over global fields. Although this includes the function field case, they have not explicitly developed algorithms in this setting. By working over function fields, we can realize marked increases in speed over the number field algorithms presented in Anshel and Goldfeld [1, 2]. In addition, we propose a cipher design which incorporates a pre-generation table construction followed by the cipher generation proper. This process, described more fully in Section 5, is the key to the speed of this cipher. It is well suited to applications where the length of the encrypted message is much longer than an initial setup time (video-on-demand, or archival backups of large amounts of data, for example).

We also remark that ciphers based on residue symbols have also been proposed by other authors (in some cases before this paper was originally circulated in 1994, in other cases afterwards), cf. [4, 6, 8, 10, 12].

We end the introduction by explicitly stating the basis of our cipher.

Let $f(T) = a_d T^d + a_{d-1} T^{d-1} + \cdots + a_1 T + a_0$. We assume that f is square-free and neither even nor odd. Fix an odd prime q which is not too large, say $q < 50,000$ (and for which $d < q$). Let α vary mod q, that is, $1 \leq \alpha \leq q$, and consider the collection of quadratic Legendre symbols: $\left\{ \left(\frac{f(\alpha)}{q} \right)_2 \right\}_{\alpha=1}^q$. Our cipher is based on the following claim:

Claim 1.1. *Let I denote any proper subset of the integers $\{1, 2, \ldots, q\}$. Then if given the values $\left\{ \left(\frac{f(\alpha)}{q} \right)_2 \right\}_{\alpha \in I}$, the only way to find the remaining values is to reconstruct the polynomial f. Even if given all the values: $\left\{ \left(\frac{f(\alpha)}{q} \right)_2 \right\}_{\alpha=1}^q$, the polynomial f can not be reconstructed in time less than a constant times $(q/2)^d$. Moreover, the distribution of values of $\left(\frac{f(\alpha)}{q} \right)_2$ is uniform in a sense that will be made explicit in Section 2 below.*

We have not been able to prove this claim in its entirety. However, in the following, we will prove part of this claim. We will also show how it is related to other problems which are generally accepted as being hard, and why attempts to determine f in time shorter than $(q/2)^d$ are not possible by any known technique.

Remark 1.2. *All of the methods in this paper work for n^{th} order residue symbols as well as the Legendre symbol. In this case, we obtain an increase of $\log_2 n$ in speed, but with the need of a corresponding increase in parameters in order to ensure the security of the cipher. We also note that the claim seems to be true for all f satisfying our hypotheses.*

2. Some Foundational Material

First, we'll give some background on function fields. Let f be a polynomial as above, and consider the reduction of f mod q, that is, of every coefficient of f mod q. We then say that $f \in \mathbf{F}_q[T]$. This simply means that f is a polynomial in T

with coefficients in $\mathbf{F}_q = \mathbf{Z}/q\mathbf{Z}$. Let $g \in \mathbf{F}_q[T]$ be another such polynomial. If g is irreducible then we define the function field quadratic residue symbol as follows:

$$\left(\frac{f}{g}\right)_2 = \begin{cases} 0 & \text{if } f \text{ and } g \text{ are not relatively prime} \\ 1 & \text{if } f \equiv \text{a square mod } g \\ -1 & \text{otherwise.} \end{cases}$$

This can be extended to abitrary polynomials g by requiring that it have the same basic properties as the usual quadratic symbol. It is multiplicative in the numerator:

$$\left(\frac{fg}{h}\right)_2 = \left(\frac{f}{h}\right)_2 \left(\frac{g}{h}\right)_2;$$

its value depends only on the congruence class of f mod h:

$$\left(\frac{f + hg}{h}\right)_2 = \left(\frac{f}{h}\right)_2;$$

and it satisfies the law of quadratic reciprocity:

$$\left(\frac{f}{g}\right)_2 \left(\frac{g}{f}\right)_2 = (-1)^{d(f)d(g)(q-1)/2} (\operatorname{sgn} f)^{d(g)} (\operatorname{sgn} g)^{d(f)}.$$

Here, $d(f)$ denotes the degree of f, and if $f(T) = a_d T^d + \ldots$ then $\operatorname{sgn} f = \left(\frac{a_d}{q}\right)_2$, where this is the usual Legendre symbol.

These properties make it easy to evaluate the symbol in the cases required for this paper (it can also be extended to arbitrary g, but we will not required that here). For example, we have:

Proposition 2.1. *Let $f \in \mathbf{F}_q[T]$ and let $\alpha \in \mathbf{F}_q$. Then*

$$\left(\frac{f}{T - \alpha}\right)_2 = \left(\frac{f(\alpha)}{q}\right)_2,$$

where the left-hand symbol is the function field Legendre symbol, and the right-hand symbol is the usual one.

Proof. To see that this is the case, divide $T - \alpha$ into f, obtaining $f = (T - \alpha)h + f(\alpha)$, for some polynomial h. Applying the second property we see that $\left(\frac{f}{T-\alpha}\right)_2 = \left(\frac{f(\alpha)}{T-\alpha}\right)_2$. Applying quadratic reciprocity, and using the fact that $d(f(\alpha)) = 0$, $d(T - \alpha) = 1$, and $\left(\frac{T-\alpha}{f(\alpha)}\right)_2 = 1$, the result follows. $\qquad\square$

We may now define an L-series associated to f:

$$L(u, f) = 1 + C_f(1)u + C_f(2)u^2 + \cdots + C_f(d-1)u^{d-1}.$$

Here

$$C_f(i) = \sum_{n \text{ monic, with } \deg n = i} \left(\frac{f}{n}\right)_2.$$

It is easily verified that for $i \geq d$, $C_f(i) = 0$. In fact, as we will see shortly, $C_f(d-1)$ does not vanish, so $L(u, f)$ is a polynomial of degree $d-1$ in u.

There is an extra term, $L_\infty(u, f)$, which needs to be included as part of the standard normalization of this L-series. This is defined as follows:

$$L_\infty(u, f) = (1 - \epsilon u)^{-1},$$

where

$$\epsilon = \begin{cases} 0 & \text{if } d \text{ odd} \\ \left(\dfrac{a_d}{q}\right)_2 & \text{if } d \text{ even.} \end{cases}$$

The symbol here is the usual Legendre symbol, and a_d is the leading coefficient of f. If we define

$$\tilde{L}(u, f) = L(u, f) L_\infty(u, f),$$

then we have

$$\tilde{L}(u, f) = \prod_{i=1}^{D} (1 - \alpha_i u),$$

where

$$D = \begin{cases} d - 1 & \text{if } d \text{ odd} \\ d - 2 & \text{if } d \text{ even.} \end{cases}$$

Here D is twice the genus of the curve $y^2 = f(x)$.

The Riemann Hypothesis for curves (proved by A. Weil [14]) says that $|\alpha_i| = \sqrt{q}$ for $i = 1, \ldots, D$. The α_i occur in pairs: $\alpha_i, \overline{\alpha_i}$, and there is a functional equation:

$$\tilde{L}(u, f) = q^{D/2} u^D \tilde{L}(q^{-1} u^{-1}, f).$$

Notice that it follows immediately from this that if d is odd then $C_f(d-1) = q^{D/2}$, while if d is even $C_f(d-1) = -\epsilon q^{D/2}$ (so in particular $C_f(d-1)$ does not vanish).

We will now prove a function field analog of the main result of [9].

Proposition 2.2. *Given $f \in \mathbf{F}_q[T]$, square-free, with degree of $f = d(f)$, let $g \in \mathbf{F}_q[T]$ be any other polynomial, degree $g = d(g)$, with $d(g) \leq d(f)$. Suppose that $\left(\dfrac{f(\alpha)}{q}\right)_2 = \left(\dfrac{g(\alpha)}{q}\right)_2$ for every $\alpha \in \mathbf{F}_q$. Then if $q > (d(f) + d(g))^2$ it follows that $f = g$. In particular, if $q > 4d(f)^2$ then there are no $g \neq f$, $d(g) \leq d(f)$, with this property.*

Proof. Let f, g be as above. Let h equal the square free part of fg. The assumption that f and g are not equal implies that h is a polynomial of degree at least 1. If $d =$ the degree of h, $d \leq d(f) + d(g)$. Also, for every $\alpha \in \mathbf{F}_q$,

$$\left(\frac{h(\alpha)}{q}\right)_2 = \left(\frac{f(\alpha)g(\alpha)}{q}\right)_2 = \left(\frac{f(\alpha)}{q}\right)_2 \left(\frac{g(\alpha)}{q}\right)_2 = 1$$

(or 0, of course, but this happens sufficiently rarely that we will ignore it).

Thus by Proposition 1, for any monic n of degree 1, $\left(\frac{h}{n}\right)_2 = 1$, so $C_h(1) = q$ and

$$L(u, h) = 1 + qu + C_h(2)u^2 + \cdots + C_h(d-1)u^{d-1}.$$

On the other hand, moving the L_∞ factor to the other side of the equation,

$$L(u, h) = (1 - \epsilon u) \prod_{i=1}^{D}(1 - \alpha_i u),$$

so

$$C_h(1) = -\epsilon - \sum_{i=1}^{D} \alpha_i.$$

By the Riemann Hypothesis for curves

$$q = C_h(1) \leq 1 + (d-1)\sqrt{q} \leq d\sqrt{q},$$

so $q \leq d^2 \leq (d(f) + d(g))^2$. Thus there is a contradiction as soon as $q \geq (d(f) + d(g))^2$. $\qquad\square$

3. The Distribution of Values of $\left(\frac{f(\alpha)}{q}\right)_2$

In fact, one expects the α_i to be uniformly distributed about the circle of radius $\sqrt{q}$ (except that below the real axis is a mirror image of above the real axis). The next two relations follow from two conjectures generally believed to be true: a certain generalization of the Shimura-Taniyama-Weil conjecture, and the generalized Riemann Hypothesis. A large part of the original Shimura-Taniyama-Weil conjecture has recently been proved by Wiles [15], Taylor and Wiles [13], and Diamond [7].

These conjectures predict that in reality $C_f(1) = C_f(1, q) = -\epsilon - \sum_{i=1}^{D} \alpha_i$ is usually far smaller than $d\sqrt{q}$. More precisely, if one varies q and fixes f one expects

$$\sum_{q < x} \frac{\log q}{\sqrt{q}} C_f(1, q) \ll \log(\mathrm{disc}(f))(\log x)\sqrt{x}$$

and

$$\sum_{q < x} \frac{\log q}{q} C_f(1, q)^2 \sim x.$$

These relations begin to hold when x is of size roughly $\log(\mathrm{disc}(f))$. Note that by the prime number theorem $\sum_{q < x} \log q \sim x$. Thus the second relation means that on average $|C_f(1, q)|$ has size $\sqrt{q}$, and the first means that the sign of $C_f(1, q)$ varies in a uniform way.

Similar results are presumed true (and follow from standard conjectures) when q is fixed and f varies, but there are more ways of varying f. For example, if one fixes α and varies f by adding all multiples νg, where $\nu \in \mathbf{F}_q$ and g is a polynomial with $g(\alpha) \neq 0$ then

$$\sum_{\nu \in \mathbf{F}_q} \left(\frac{f + \nu g}{T - \alpha} \right)_2 = 0,$$

while if one only added terms corresponding to $1 \leq \nu \leq i$ for $i < q$, the sum would always be $\ll \sqrt{q}$. (Again, this is conjectural; the best known result is $\ll \sqrt{q} \log q$.) A situation closer to that used in our cipher stream would be to let f vary over prime polynomials of fixed degree. This is because the Riemann Hypothesis essentially says that within the density constraints of the prime number theorem, the distribution of primes is random. This applies equally well to ordinary rational primes and to prime polynomials in $\mathbf{F}_q[T]$. Thus the distribution of values of $f(\alpha)$ for fixed α and randomly chosen f of degree d should be approximated by the behavior of $f(\alpha)$ as f varies over prime polynomials of degree d. The following proposition shows that this behavior is exactly what we would expect.

Proposition 3.1. *Let the notation be as above and let $1 \leq \alpha \leq q$ be fixed. Then*

$$\sum_{p \text{ prime, with } \deg p = d} \left(\frac{p(\alpha)}{q} \right)_2 \leq q^{d/2}.$$

Proof. Consider the L-series $L(u, T - \alpha)$. The degree of this L-series is 0, so in fact $L(u, T - \alpha) = 1$. Thus its logarithmic derivative is 0. On the other hand $L(u, T - \alpha)$ also has an Euler product expansion:

$$L(u, T - \alpha) = \prod_p \left(1 - \left(\frac{p}{T - \alpha} \right)_2 u^{\deg p} \right)^{-1}.$$

Taking the logarithmic derivative, collecting the coefficient of u^{d-1}, and setting it equal to zero we obtain

$$\sum_{k | d} \frac{d}{k} \sum_{\deg p = d/k} \left(\frac{p}{T - \alpha} \right)_2^k = 0.$$

The entire contribution of this sum from $k \geq 2$ can be no larger than $dq^{d/2}$. Thus the $k = 1$ contribution is also bounded by $dq^{d/2}$ and the proposition follows immediately. □

Note that as the number of prime polynomials p of degree d is approximately q^d/d, the total sum above is less than $\sqrt{d}$ times the square root of the length.

4. On Recovering f

Proposition 2 established that if one tried to find a shorter polynomial that generated the same sequence of values $\left\{ \left(\frac{f(\alpha)}{q} \right)_2 \right\}_{\alpha=1}^q$, one could not possibly succeed.

The only other method known to the authors of determining the rest of the sequence from part of it is to determine f itself. So let us consider the question of determining f. Even though this might not be true in practice, we'll assume that one is given the entire sequence $\left\{\left(\frac{f(\alpha)}{q}\right)_2\right\}_{\alpha=1}^q$, and the knowledge of q and d (indeed, this is more information than one would obtain from the cipher as implemented in Section 5).

For each α there is an equation

$$a_d\alpha^d + a_{d-1}\alpha^{d-1} + \cdots + a_1\alpha + a_0 \equiv f(\alpha) \mod q.$$

There are q equations in $d+1$ unknowns, a highly overdetermined system. All an attacker knows about the values $f(\alpha)$, however, is what square class they fall into. Thus there are $(q-1)/2$ choices for each $f(\alpha)$. To solve for the unknown coefficients a_i, an attacked would need to take a subset of $d+1$ equations, and there would be then $((q-1)/2)^{d+1}$ possibilities to investigate. For each case the attacker would have to solve the subset of selected equations, and then check consistency with the remaining equations. If $(q/2)^d$ is large, this becomes impossible.

Is there another way to determine f? The only approach known is via L-series associated to f, for example, $L(u, f)$. However, there are two things blocking this approach. The main obstacle is that to extract any information from an L-series one needs to be able to compute its value very precisely. However a general principle operating in the study of L-series is that in order to compute a value with any precision, one must have a knowledge of the coefficients up to a certain point. In the case of $L(u, f)$, that point is the quantity $q^{D/2}$, which appears on the right-hand side of the functional equation. For any related L-series the point would be even higher. This means that one would need a knowledge of the coefficients up to $C_f(D/2)$. However, one is only given the knowledge of $C_f(1)$. This principle is one which cannot be proved. It is always possible that tomorrow someone could come up with an entirely new method of extracting information from L-series. However, it has been shown recently by Bombieri and Friedlander [3], in an analogous situation, that if an L-series is approximated by a Dirichlet polynomial, that polynomial will begin to give a good approximation to the L-series only after the length of the polynomial passes the previously mentioned point.

A second obstruction is that even if one could compute the L-series values, it is not at all clear how that would help to compute f. The coefficients $C_f(i)$ are symmetric functions of the roots α_i. It is conceivable that if the values $C_f(i)$ were known for all i one could determine f, but once again, one does not have that information. Only $C_f(1)$ is given.

5. Some More on the Cipher

We now present an implementation of the cipher which is extremely fast. As mentioned in the introduction, we have coded a modification of the cipher (using 16th order residue symbols) which runs at 1–2 megabits/second on a Sparc 1 (this is the speed of the cipher, and does *not* include table generation, a distinction which will be made clear below). This particular implementation is thus one to two orders of magnitude faster than any previous "mathematical" pseudorandom number generator.

Remark 5.1. *Other methods (such as the Naor-Reingold pseudorandom function, see [11]) have been developed since the original circulation of this paper which are also quite fast. It is not clear whether our method is faster than these methods since we last implemented a version of this cipher in 1994.*

Our implementation here has a non-trivial setup cost (to build tables) and a non-trivial memory cost (to hold tables). Nonetheless, it is well suited to applications where security is important, setup time is not important, and memory costs is not important (for example, a good application would be in a cable box on top of a TV, and a bad application would be in a smartcard), and where one needs to generate a very large stream of pseudorandom bits.

The idea of our table-driven cipher is as follows. We fix d and q with $d < q$, write down a key polynomial of degree d (the secret key of the system), and use this to generate a moderately large number l of polynomials of degree d. (This generation process may come from the Legendre symbols of the values of the key polynomial, or any other one-way function. In our time trials, we chose to generate 4000 polynomials recursively from this one key polynomial.) For each of these l polynomials (call them, say, f_j), we store in our table the values $f_j(\alpha)$ mod q, where α varies over some range (see below). We then set our starting polynomial f to be the sum of exactly r of these l polynomials. The choice of r itself and which r polynomials are to be initially used can be variable and part of the key, or fixed forever. The key idea is that now to compute the values $f(\alpha)$, one merely sums over the corresponding table entries, instead of doing any new polynomial evaluation. The cipher stream itself consists then of the bitstream $\left\{ \left(\frac{f(\alpha)}{q} \right)_2 \right\}_\alpha$. The process of computing a residue symbol of two integers is quite fast [5]; since we need only execute $r - 1$ additions and one residue symbol computation for each bit, this is *extremely* efficient. When it is time for a new polynomial, we do not generate (from f) a new random table, instead we generate some new random choices of indices and we set the new f', say, to be the sum of those f_j corresponding to those newly generated indices.

Remark 5.2. *If we were using n^{th} order residue symbols in our cipher, we would obtain $\log_2(n)$ bits at this point from each computation.*

The first issue that arises is "what if f is not square-free?" We assumed in the above that our f was square-free. In fact, everything remains true if f is replaced by the square-free part of f. The only danger is that the degree of the square-free part is then reduced. In particular, it could happen that the degree of the square-free term were so low that, given a sequence of Legendre symbols, it could be determined by an exhaustive search.

It is easily checked that if f has degree 32 and q is about 4000, then the chances that the square-free part of f has degree ≤ 16 is less than 1 in 10^{28}. This probability can be made as close to zero as desired by increasing d and q. If the degree of f is 16 then the time involved in an exhaustive search, given the Legendre symbol sequence, would be proportional to 10^{50}.

Recall that we used f to generate a stream of Legendre symbols $\left(\frac{f(\alpha)}{q}\right)_2$ as α varies only over a reasonably short range. In our timing tests, we have let α vary from 1 to q, but when q is large it may prove to save setup time and table storage space to have α vary in a smaller range, for example 1 to 500.

One concern we must then consider is what the probability would then be that a polynomial of low degree chosen at random could duplicate these first 500 values. In fact, the chances of one of the approximately 4000^{32} polynomials of degree ≤ 32 coresponding to one of the 2^{500} possible sequences of Legendre symbols is less than 1 in 10^{34}.

After completing this stream of residue symbols, we then use the end of the stream, or new values (not the placed into the the cipher stream) from continuing the process of computing symbols, to generate the next choice of r polynomials that get added to create the new f. Again, the computation of the symbols of the values of the new polynomial requires only sums of table entries. Thus from one starting polynomial, we get a recursive list of polynomials, each of which generates a series of residue symbols, and then the next polynomial.

The resulting cipher has a moderate startup cost, but extremely quick stream generation, as claimed.

References

[1] M. Anshel and D. Goldfeld, *Zeta functions, one-way functions, and pseudorandom number generators*, Duke Mathematical Journal **88** (1997), 371–390.

[2] M. Anshel and D. Goldfeld, *Multi-purpose high speed cryptographically secure sequence generator based on zeta-one-way functions*, U.S. Patent 5,577,124 (1996).

[3] E. Bombieri and J.B. Friedlander *Dirichlet polynomial approximations to zeta functions*, Ann. Scuola norm. Sup. Pisa. Cl. Sci. **22** (1995), 517–544.

[4] D. Boneh and R. Lipton, *Algorithms for black-box fields and their applications to cryptography*, Lect. Notes in Comp. Sci., Springer-Verlag, Berlin, **1109** (1996), 283–297.

[5] H. Cohen, *A Course in Computational Number Theory*, Grad. Texts in Math., Graduate Texts in Mathematics, **138** (1993), Springer-Verlag, New York.

[6] T. W. Cusic, C. Ding and A. Renvall, *Stream ciphers and number theory*, Elsevier, Amsterdam, 1998.

[7] F. Diamond, *On deformation rings and Hecke rings*, Annals of Math., **144** (1996), 137–166.

[8] I. B. Damgård, *On the randomness of Legendre and Jacobi sequences*, Lect. Notes in Comp. Sci., Springer-Verlag, Berlin, **403** (1990), 163–172.

[9] D. Goldfeld and J. Hoffstein, *On the number of Fourier coefficients that deter-mine a modular form,* A Tribute to Emil Grosswald: Number Theory and Related Analysis, Comtemp. Math **143** (1993) 385–393.

[10] C. Mauduit and A. Sárközy, *On finite pseudorandom binary sequences 1: Measure of pseudorandomness, the Legendre symbol,* Acta Arithm., **82** (1997), 365–377.

[11] M. Naor and O. Reingold, *Number-theoretic constructions of efficient pseudo-random functions,* Proc. 38th IEEE Symp. on Foundations of Comp. Sci. (FOCS'97), Miami Beach, IEEE (1997), 458–467.

[12] R. Peralta, *On the distribution of quadratic residues and nonresidues modulo a prime number,* Math. Comp., **58** (1992), 433–440.

[13] R. Taylor and A. Wiles, *Ring-theoretic properties of certain Hecke algebras,* Annals of Math. **141** (1995), 553–572.

[14] A. Weil, *On the Riemann hypothesis in function fields,* Proc. Nat. Acad. Sci. U.S.A. **27** (1941), 45–347.

[15] A. Wiles, *Modular elliptic curves and Fermat's last theorem,* Annals of Math. **141** (1995), 443–551.

Department of Mathematics
Brown University
Box 1917, Providence, RI 02912 USA
E-mail address: jhoff@math.brown.edu

Department of Mathematics
University of Georgia
Athens, GA 30602 USA
E-mail address: dlieman@math.uga.edu

Progress in Computer Science and Applied Logic, Vol. 20
© 2001 Birkhäuser Verlag Basel/Switzerland

Rational Groups of Elliptic Curves Suitable for Cryptography

David R. Kohel

Abstract. We give an overview of methods of construction of elliptic curves which contain a subgroup of large prime order. Variations on the standard random curve selection and complex multiplication methods are presented for constructing elliptic curves containing a subgroup of large prime order. The results of random curve selection and the CM method are qualitatively contrasted in terms of the randomness of the resulting curves; in particular we note that the CM method fails any reasonable measure of randomness if applied over a base field of predetermined characteristic. We analyze both practical and theoretical considerations in the choice of the group of points used for cryptographic applications.

1. Introduction

An elliptic curve E over a finite field k is a nonsingular projective plane curve defined by a Weierstrass equation

$$Y^2 Z + (a_1 X + a_3 Z) Y Z = X^3 + a_2 X^2 Z + a_4 X Z^2 + a_6 Z^3,$$

with all a_i in k. There is a unique point $O = (0 : 1 : 0)$ on the line $Z = 0$, which is specified as a distinguished point of the elliptic curve. It is thus standard to set $x = X/Z$ and $y = Y/Z$ and defined E to be defined by the affine model

$$y^2 + (a_1 x + a_3) y = x^3 + a_2 x^2 + a_4 x + a_6,$$

of the curve, together with the distinguished point O at infinity.

The elliptic curve E over of field k can be identified with the set of points

$$E(\overline{k}) = \{O\} \cup \{(\alpha, \beta) \mid \alpha, \beta \in \overline{k}, \ \beta^2 + (a_1 \alpha + a_3)\beta = \alpha^3 + a_2\alpha^2 + a_4\alpha + a_6\},$$

where $\overline{k}$ is an algebraic closure of k. Conversely the equation of the curve E is the unique equation interpolating the points in the set $E(\overline{k})$.

The set $E(\overline{k})$ has the structure of an abelian group with identity O, under the rule which says that three collinear points sum to O. More generally for any field extension K/k we have a finite subgroup $E(K)$ of K-rational points, and for any k-algebra homomorphism $K \to L$ we have a homomorphism of groups $E(K) \to E(L)$. We thus, in particular, distinguish the groups $E(K)$, from the geometric object E/k, which is a map associating a group to each field extension K/k and defining compatible systems of maps between them.

The theory of elliptic curves and its connections to modular forms and class field theory has a long and active history. This theory is treated, from a modern viewpoint, in books of Silverman [26, 27], Hüsemoller [9], and Knapp [10]. Groups of points on elliptic curves were suggested for use in cryptography independently by Koblitz [11] and Miller [21]. The acceptance of elliptic curve cryptography is attested in the new generation of books emphasizing elliptic curves over finite fields from a cryptographic point of view, including the books of Menezes [19]; Blake, Seroussi and Smart [1]; and Enge [5].

A general group $E(K)$ has the advantage over the corresponding multiplicative group K^* of the field, because the discrete logarithm problem appears to be harder in general for the former. Moreover, for any given field K there exists a large choice of curves. The multiplicative group K^* can, in fact, be interpreted as the group of points on a degenerate elliptic curve, or singular cubic curve (see Chapter 3, §7, Hüsemoler [9]). It is therefore not unreasonable to expect that a generic discrete logarithm algorithm for elliptic curves would apply as well to the multiplicative group of a field.

In this paper we describe methods for constructing point groups of elliptic curves of use for cryptography, and present some variants on the standard random curve and complex multiplication (CM) constructions. Because the structure of the groups $E(K)$ is governed by the structure of the endomorphism ring of E, we begin with background on endomorphisms of elliptic curves. In the following sections we describe variants of the random curve and CM methods for construction of curves with known numbers of points. Examples are chosen for pedagogic purposes. In particular the bit size of the point groups (> 350) are in excess of the current recommendations of 160–250 bits for commercial applications (see Lenstra and Verheul [16]), but demonstrate the effectiveness of a construction where point counting becomes nontrivial. On the other hand, this bit size is not unreasonably large for military applications in which a 50 year life span of confidentiality is insufficient or in cases where an additional security margin is sought. We take the conservative position that to ensure a sufficiently general construction, the elliptic curve should be chosen at random from a pool large enough to be effectively innumerable. This contrasts in particular with suggestions for use of curves over small fields, curves generated by the complex multiplication method over any field of prespecified characteristic, or, in the extreme, of one of the two ordinary elliptic curves over $\mathbb{F}_2$. In this view, the given example of the CM method over a field of characteristic 2 is taken purely for the sake of comparison of the CM and random curve methods, since the use of the CM method over a field of fixed characteristic violates this principle. In the final section we discuss this randomness criterion, make qualitative contrasts of "random" and "CM" curves, and discuss the implications of cryptographic use of point groups over proper extensions of the base field, in particular in light of the work of Gaudry, Hess and Smart [7].

2. Endomorphism Structure of Elliptic Curves

The structure of the abelian groups $E(K)$ is intimately related to the endomorphism ring structure of E and, in particular, to the distinguished Frobenius element. We therefore recall some background material on endomorphisms of elliptic curves as a means of constructing and analyzing rational point groups appropriate for cryptographic use.

An endomorphism $\phi : E \to E$ is a rational polynomial map

$$(x, y) \mapsto (f(x), g(x, y)),$$

where $g(x, y) = g_1(x)y + g_0(x)$ with $f(x)$, $g_1(x)$, and $g_0(x)$ in $k(x)$, such that

$$g(x, y)^2 + (a_1 f(x) + a_3)g(x, y) = f(x)^3 + a_2 f(x)^2 + a_4 f(x) + a_6,$$

and which takes O to O. As a consequence of the definition, an endomorphism ϕ induces a homomorphism $E(K) \to E(K)$ for any field extension K/k. Thus the addition law on the curve gives a well-defined addition of endomorphisms, and composition defines a compatible associative multiplication operation. This gives a ring structure to the set $\mathrm{End}_k(E)$ of endomorphisms, in which the multiplication-by-n maps $[n]$ define a subring isomorphic to $\mathbb{Z}$.

For an elliptic curve E/k, where $|k| = q$, we define the distinguished *Frobenius endomorphism* $\phi : E \to E$ by

$$(x, y) \mapsto (x^q, y^q).$$

Clearly $\phi(P) = P$ if and only if the point P is in $E(k)$. By a standard result of Hasse (see Silverman [26]), the Frobenius endomorphism satisfies a characteristic equation $X^2 - tX + q = 0$, where t is an integer satisfying $|t| \le 2\sqrt{q}$.

If the trace of Frobenius t is congruent to 0 mod p, we say that the curve is *supersingular*, otherwise we call it *ordinary*. Menezes, Okamoto and Vanstone [20] have proved that the discrete logarithm on supersingular elliptic curves can be reduced to a discrete logarithm in the multiplicative group of a finite extension field of degree generally 2, and at most 6, over the base field. This reduction to finite fields holds in general. However, in the ordinary case, Koblitz [12] has proved that the degree of the extension is generically large. We thus consider only the ordinary case for the purpose of cryptography. The following theorem, however, on the structure of endomorphisms, holds in general.

Theorem 1. *Let ψ be an endomorphism of E not contained in $\mathbb{Z}$. Then ψ has an irreducible characteristic polynomial $X^2 + aX + b$ and generates a ring isomorphic to the imaginary quadratic order $\mathcal{O} = \mathbb{Z}[X]/(X^2 + aX + b)$. The map*

$$\rho = [n] + [m]\psi \longmapsto \widehat{\rho} = [n - am] - [m]\psi$$

defines an automorphism of $\mathbb{Z}[\psi]$ and $\mathrm{Tr}(\rho) = \rho + \widehat{\rho}$ and $\mathrm{N}(\rho) = \rho\widehat{\rho}$ agree with the trace and norm from $\mathcal{O}$ to $\mathbb{Z}$. In particular we have $\mathrm{Tr}(\psi) = -[a]$ and $\mathrm{N}(\psi) = [b]$.

Proof. This follows from the standard properties of the dual isogeny, for which we refer to Chapter III §6 of Silverman [26], noting that $\hat{\rho}$ is the dual of ρ and $N(\rho)$ is its degree. ∎

For an ordinary elliptic curve E, the endomorphism ring $\text{End}_k(E)$ is a commutative ring of rank 2 over $\mathbb{Z}$, and the full endomorphism ring is generated by an element ψ satisfying $\phi = [n] + [m]\psi$ for integers n and m. If K is a field extension of degree r over k, then ϕ^r acts as the identity on the points in $E(K)$, so that $\phi^r - 1$ is in the kernel of the action of $\text{End}_k(E)$ on $E(K)$. In fact the following stronger result holds.

Theorem 2. *There exists a noncanonical isomorphism of* $\text{End}_k(E)$-*modules*

$$E(K) \cong \text{End}_k(E)/(\phi^r - 1).$$

In particular, the number $|E(K)|$ *of* K-*rational points is given by* $N(\phi^r - 1)$.

Proof. The isomorphism appears in Theorem 1 of Lenstra [15]. An analysis of this isomorphism shows that

$$\text{End}_k(E)/(\phi^r - 1) \cong \mathbb{Z}/m_1\mathbb{Z} \times \mathbb{Z}/(N(\phi^r - 1)/m_1)\mathbb{Z},$$

where m_1 is the largest divisor of $[\text{End}_k(E) : \mathbb{Z}[\phi]]$ such that $\phi^r \equiv 1 \bmod m_1$. ∎

This isomorphism will be the main tool used for the construction and analysis of groups of rational points on elliptic curves. The group order $N(\phi^r - 1)$ can be rapidly computed from Theorem 1. Let t be the trace of of Frobenius, and write

$$X^r - 1 \equiv uX + v \bmod (X^2 - tX + q).$$

Then we find the explicit form

$$N(\phi^r - 1) = u^2 q + t\,uv + v^2,$$

for the number of points in $E(K)$. We note that $m = [\text{End}_k(E) : \mathbb{Z}[\phi]]$ is a well-defined invariant of the curve E/k, and together with the trace t, suffices to determine the group structure of $E(K)$ for all extensions K/k. For further information in this direction we refer to the thesis of the author [13].

An elliptic curve over a finite field can be considered as the reduction of an elliptic curve over a number ring. As such it is an element of a two parameter family. The first parameter is the j-invariant of the curve, and the second is the prime of reduction. There are two approaches to the problem of choosing suitable elliptic curves for cryptographic use, which correspond to fixing one of the parameters and choosing the second at random. In the *complex multiplication* (CM) method we choose a suitable j-invariant of a curve, and the prime of reduction is determined subsequently. The *random curve* method first fixes the base field, essentially fixing a prime of reduction, then selects random curves to find one with good properties. We explore aspects of these two methods in the sections which follow.

3. Random Curve Selection

We let k be fixed, and choose E/k at random. The order of $E(k)$ can be computed in polynomial time using the SEA method of Schoof [24], with improvements of Elkies [4] and Atkin. We continue until we find a curve whose order is divisible by a large prime.

When the construction time is critical or a sophisticated point counting algorithm unavailable, we can consider the following variant of the random curve method. Let K be a fixed finite field, and let k be a proper subfield. We choose E/k at random, compute the group order $E(k)$ over the smaller field. Then by Theorem 2, the order of $E(K)$ is given by $N(\phi^r - 1)$, where ϕ is the Frobenius endomorphism relative to k. Since $N(\phi - 1)$ is a divisor of relatively small order, the objective is to find a large prime factor in

$$|E(K)/E(k)| = N(\phi^r - 1)/N(\phi - 1).$$

Example 3. Choosing random elliptic curves over $k = \mathbb{F}_{2^{31}}$, we obtain a particular example:

$$y^2 + xy = x^3 + \gamma,$$

having $\mathrm{Tr}(\phi) = 77689$, where $\gamma = w^{217980880}$ and $w^{31} + w^3 + 1 = 0$. Over an extension $K = \mathbb{F}_{2^{403}}$ of degree 13, we find the order of the group $E(K)/E(k)$ to be:

$$N(\phi^{13} - 1)/N(\phi - 1) = 79 \cdot p_{366}$$

containing a 366-bit prime factor p_{366}.

We note that the endomorphism ring has discriminant $D = -2554353871$ and class number 42966. The size of the class number makes this curve impractical to construct by the complex multiplication method which follows. We return to the implications of this in the final section.

4. CM Constructions

For any negative integer D congruent to 0 or 1 mod 4, there exists, up to isomorphism, a unique imaginary quadratic order $\mathcal{O} = \mathbb{Z}[(D + \sqrt{D})/2]$. In the complex multiplication method we choose a discriminant D of an imaginary quadratic order and find a finite field of q elements such that $m^2 D = t^2 - 4q$ for integers t and m and such that $q - t + 1$ contains a large prime factor. We denote by (D/p) the Legendre symbol for the prime p. The existence and construction of an elliptic curve with this discriminant is given by the following theorem.

Theorem 4. *For $\mathcal{O}$ to be the endomorphism ring of an ordinary elliptic curve E over k of characteristic p, it is necessary and sufficient that*

1. $D = \mathrm{disc}(\mathcal{O})$ *satisfies* $(D/p) = 1$; *and*
2. *The order in the class group of a prime over p divides* $[k : \mathbb{F}_p]$.

Proof. The results follow from the classical class field theory for complex multiplication, going back to Deuring [3]. The first condition ensures that the elliptic
curve is ordinary, and follows from Chapter 10 or Theorem 12 of Chapter 13 in
Lang [14]. The second condition defines the minimal field of definition of E, and
is a consequence of Theorem 7 of Chapter 12 of the same volume.   □

4.1. Class Polynomial Construction

In order to construct an elliptic curve with complex multiplication we use some
classical constructions of class field theory. The following theorem is the effective
variant of Theorem 4 used to produce the j-invariant of a particular elliptic curve.

Theorem 5. *Let $\mathcal{O}$ be an imaginary quadratic order of discriminant D and class
number $h(\mathcal{O})$. There exists a unique monic irreducible polynomial $H_D(X)$ of degree
$h(\mathcal{O})$ such that E is an ordinary elliptic curve over a field k of characteristic p
with endomorphism ring $\mathcal{O}$ if and only if $(D/p) = 1$ and the j-invariant of E is a
root of $H_D(X)$ mod p in k.*

The theorem is effective, since the roots of the class polynomial $H_D(X)$ can
be computed over $\mathbb{C}$ as special values of the modular function $j(\tau)$ on points τ
representing the R-ideal classes (see Section 7.6.2 of Cohen [2]). By computing the
roots to sufficient precision, the polynomial $H_D(X)$ is determined as an element
of $\mathbb{Z}[X]$. For the discriminant -47, we find that $H_{-47}(X)$ equals

$$X^5 + 2257834125X^4 - 9987963828125X^3 + 5115161850595703125X^2$$
$$- 149824728508286132281250X + 160429296006238708496093750375$$

On the other hand the size of the coefficients rapidly becomes an obstacle, as
suggested by this small example. For D in restricted congruence classes we obtain
alternative class polynomials over $\mathbb{Z}$ with smaller coefficients. For instance Yui
and Zagier [31] prove the existence of a class polynomial $W_D(X)$ defined for $D \equiv
1 \bmod 8$ and $D \neq 0 \bmod 3$ using Weber functions. The class polynomial $W_{-47}(X)$
takes the more compact form:

$$X^5 - 2X^4 + 3X^3 - 3X^2 + X + 1.$$

A root α of this equation corresponds to the j-invariant $j = (\alpha^{24} - 16)^3/\alpha^{24}$, from
which we can construct a curve with the desired endomorphism ring j-invariant j
is by Theorem 6 which follows.

Since one is interested only in the roots of the class polynomials modulo a
prime, the size of the coefficients of $W_D(X)$, computed using special values of analytic functions in $\mathbb{C}$, is only relevant to prevent coefficient explosion over $\mathbb{Z}$. Similar
class polynomials, of reduced coefficient size, can be defined for discriminants in
other congruence classes (see Gee [8]).

4.2. Isomorphism Classes of Elliptic Curves

In order to pass from a class polynomial to an elliptic curve with known number
of points, we require a construction for curves with given j-invariant. An elliptic
curve, however, may have several nonisomorphic *twists* with the same j-invariant.

The following theorem classifies all elliptic curves over a finite field k with given j-invariant.

Theorem 6. *Let j be an element of a finite field k, and denote by k^{*n} the subgroup of n-th residues in k^*.*

If k has characteristic 2, then all isomorphism classes of curves E with j-invariant j are given by the following equations:

(1) $y^2 + a_3 y = x^3 + a_4 x + a_6$, *if $j = 0$,*
(2) $y^2 + xy = x^3 + a_2 x^2 - 1/j$, *if $j \neq 0$.*

In the first, supersingular, case, the coefficient a_3 is a unit whose class in k^*/k^{*3} is determined by the isomorphism class of the curve. Two curves with coefficients a_3, a_4, a_6 and a_3', a_4', a_6' are isomorphic exactly when $a_3' = u^3 a_3$; when a_4' and $u^4 a_4$ are in the same additive class mod $\ker(\mathrm{Tr}_\ell^k)$, where $\ell = k \cap \mathbb{F}_4$; and when a_6' and $u^6 a_6$ are in the same additive class mod $\ker(\mathrm{Tr}_{\mathbb{F}_2}^k)$. In the second, ordinary, case, the isomorphism class is uniquely determined by the class of a_2 mod $\ker(\mathrm{Tr}_{\mathbb{F}_2}^k)$.

If k has characteristic 3, then all isomorphism classes of curves with j-invariant j are given by the following equations:

(1) $y^2 = x^3 + a_4 x + a_6$, *if $j = 0$,*
(2) $y^2 = x^3 + a_2 x^2 - a_2^3/j$, *if $j \neq 0$.*

In the first, supersingular, case, the coefficient a_4 is a unit whose class in k^*/k^{*4} is determined by the isomorphism class of the curve. Two curves with coefficients a_4, a_6 and a_4' a_6' are isomorphic exactly when $a_4 = u^4 a_4'$ and the equation $r^3 + a_4 r + a_6 = u^6 a_6'$ has a solution r in k. In the second, ordinary, case, the isomorphism class is uniquely determined by the class of a_2 in k^*/k^{*2}.

If k has characteristic ≥ 5, then all isomorphism classes of curves with j-invariant j are given by the following equations with t in k^*:

(1) $y^2 = x^3 + a_6$, *if $j = 0$,*
(2) $y^2 = x^3 + a_4 x$, *if $j = 12^3$,*
(3) $y^2 = x^3 - a_2^2 j x/48(j - 1728) + a_2^3 j/864(j - 1728)$ *otherwise.*

The isomorphism class is uniquely determined by the class of a_n in k^*/k^{*n}.

Proof. This follows by explicit verification, starting from the form of an isomorphism given in Appendix A of Silverman [26]. For the conditions in characteristic 2, we observe that $a \equiv b \mod \ker(\mathrm{Tr}_{\mathbb{F}_2}^k)$ is equivalent to $\mathrm{Tr}_{\mathbb{F}_2}^k(a) = \mathrm{Tr}_{\mathbb{F}_2}^k(b)$ and also to the existence of a solution r to the equation $r^2 + r = a + b$ over k.   □

We note that Morain [22] finds a more refined expression for the supersingular elliptic curves in characteristic 3, which also classifies the corresponding trace. This builds on Schoof [25], who does a complete enumeration of the abstract isomorphism classes of elliptic curves in terms of endomorphism rings structure. A more refined analysis of the supersingular case in characteristic 2 would provide a similar classification of the trace in terms of explicit equations.

4.3. CM Example

By Theorem 4 and Theorem 5 the degree of the field extension $k/\mathbb{F}_p$ must divide the degree of the class polynomial $H_D(X)$ or of an alternate class polynomial. For large degree extensions over $\mathbb{F}_p$ of small characteristic, the computation of the class polynomial becomes computationally expensive. To compensate, we indicate how to employ an intermediate degree extension to useful effect.

By sieving over small discriminants, we choose a discriminant $D = -8647$ with class number 31, the class group being generated by a prime $\mathfrak{p}_2$ over 2. Therefore there exists a curve E over $k = \mathbb{F}_{2^{31}}$ with endomorphism ring of this discriminant. Prior to doing any computations with curves, we find that the ring $\mathbb{Z}[\phi]$ generated by Frobenius is isomorphic to $\mathbb{Z}[X]/(X^2 - tX + 2^{31})$, where $t = \pm 35875$.

Over a degree 13 extension $K = \mathbb{F}_{2^{403}}/k$ we find that the group order of $E(K)/E(k)$ is either equal to a large composite number with small prime factors if $t = -35875$, or equals

$$\mathrm{N}(\phi^{13} - 1)/\mathrm{N}(\phi - 1) = 157 \cdot 7333 \cdot p_{352},$$

for a 352-bit prime p_{352} when $t = 35875$.

In order to construct an elliptic curve with this trace, we compute the Weber class polynomial $W_{-8647}(X)$, which takes the form:

$$
\begin{aligned}
X^{31} &- 33X^{30} + 135X^{29} + 1585X^{28} + 16905X^{27} + 77577X^{26} \\
&+ 261396X^{25} + 677142X^{24} + 1406953X^{23} + 2509293X^{22} \\
&+ 4044270X^{21} + 6101029X^{20} + 8852701X^{19} + 12285213X^{18} \\
&+ 15808518X^{17} + 18153439X^{16} + 17693230X^{15} + 13330467X^{14} \\
&+ 5493408X^{13} - 3612428X^{12} - 10816811X^{11} - 13646625X^{10} \\
&- 11600862X^{9} - 6330185X^{8} - 678696X^{7} + 3083034X^{6} \\
&+ 4212540X^{5} + 3382143X^{4} + 1882711X^{3} + 683247X^{2} \\
&+ 136725X + 1
\end{aligned}
$$

Then $W_{-8647}(X)$ mod 2 has a root $\alpha = w^{1023401681}$ over $\mathbb{F}_2[w] = \mathbb{F}_{2^{31}}$, where w has minimal polynomial $X^{31} + X^3 + 1$. From the associated j-invariant, we first construct the curve $y^2 + xy = x^3 + \gamma$ by Theorem 6, where $\gamma = w^{1878640454}$ and find that the trace of the Frobenius endomorphism is -35875. Passing to the quadratic twist we find the curve

$$y^2 + xy = x^3 + x^2 + \gamma,$$

with the desired trace 35875.

4.4. Sieving for Discriminants

For a fixed field k of cardinality $q = p^s$ one can sieve for discriminants D up to a particular bound which satisfy the conditions of Theorem 4. The condition $(D/p) = 1$ implies that there exists a prime $\mathfrak{p}$ over p, and one can quickly test if

$\mathfrak{p}^s$ is a principal ideal. The following corollary of Theorem 4 gives a more direct approach to finding a suitable D.

Corollary 7. *Let k be a field of q elements and let m be an integer coprime to q. For any integer t such that $t^2 \equiv 4q \bmod m^2$ with $|t| \le 2\sqrt{q}$, the integer $D = (t^2 - 4q)/m^2$ is the discriminant of the endomorphism ring of an ordinary elliptic curve over k.*

While this approach is constructive, using factorization modulo the prime divisors of m and Hensel lifting, the problem remains that the computation of class polynomials appears to have exponential complexity, which effectively implies an absolute bound on the discriminant. Thus, in practice, it is necessary to choose m sufficiently large such that $|D|$ is of a prescribed size.

5. Cryptographic Considerations

5.1. Enumerability of Curves

In the examples, we have specified the exact field $\mathbb{F}_{2^{31}}$ over which we are to choose an elliptic curve. The size of the field is sufficiently large that it is meaningful to speak of a randomly selected curve as being generic. However, if the CM method is used, then the number of curves available for use becomes severely constrained. The discriminants with $|D| < 10^4$, $D \equiv 1 \bmod 8$, and in which a prime over 2 has order 31 in the class group are limited to the 11 values

$$-719, -911, -2471, -2927, -3727, -4159, -4247, -4951, -6439, -7639, -8647.$$

With $|D| < 10^5$, $|D| < 10^6$, and $|D| < 10^7$ these numbers are 50, 191, and 623, respectively. The expected size of the class number of D and the size of the coefficients of a class polynomial place an effective absolute upper bound on the size of $|D|$. Thus for fixed field k there are a very constrained set of discriminants of small absolute value which can be the endomorphism ring of a curve over k. Thus, as the above example shows, the CM method, when used over fixed base ring or characteristic, is tantamount to prescribing a fixed enumerable set of curves of use for cryptography. In order to regain a reasonable concept of randomness, the CM method should only be applied in a context where the characteristic of the finite field is not prespecified.

5.2. Qualitative Distinction of CM and Random Curves

The methods presented here for computing class polynomials severely limit the size of a CM discriminant. As indicated above this means that the number of constructible curves over any particular field will be be contained in an enumerable set of curves. These curves will have the property that the discriminant of the endomorphism ring is exceptionally small compared to the general case. Indeed the method by which a curve was produced can be effectively determined by the size of the discriminant. While defined over the same field, the random curve method gave rise to a curve with $D = -2554353871$ and class number 42966, while in the CM method we constructed a curve of discriminant $D = -8647$ with class number 31. While no specific attacks profit exceptionally from the special form of elliptic curves constructed by the CM method, the potential remains.

5.3. Working over Field Extensions

In both examples given above we chose to define a curve over a field k and work in the group of rational points over a proper extension K. By working over an extension field of degree r, one loses r bits due to the small subgroup $E(k)$ of order $N(\phi - 1)$. The benefit is a more time efficient construction of the elliptic curve.

One advantage of these sort of composite degree extensions is the use of the Frobenius endomorphism with respect to k on the group $E(K)$. The Frobenius endomorphism can be rapidly computed without divisions in the field K. The following corollary of Theorem 2 describes the action of Frobenius on the cyclic subgroups of interest in cryptography.

Corollary 8. *Let E be an elliptic curve over k and let K be an extension of degree r over k. Suppose that n is a prime divisor of $|E(K)|$ and is coprime to $|E(K)|/n$. Then $E(K)$ contains a subgroup H of order n and*

$$\operatorname{End}_k(E)/(\phi^r - 1, n) \cong H \cong \mathbb{Z}/n\mathbb{Z},$$

where $(\phi^r - 1, n)$ is the ideal generated by $\phi^r - 1$ and n. In particular ϕ acts as $[a]$ on H for some r-th root of unity modulo n.

This permits scalar multiplication to be computed in base ϕ or base 2. Indeed, the work of Müller [23] in characteristic 2 and Smart [28] in odd characteristic, based on [18] and [29], exploit base ϕ representations for efficiency of scalar multiplication. The work of Wiener and Zuccherato [30] and Gallant, Lambert, and Vanstone [6], shows that an attacker also benefits by a factor of $r^{1/2}$, where $r = [K : k]$, in the discrete logarithm on such a curve. Since the algorithm remains exponential, this work is relevant when applied to a cryptosystem of critical security margin.

The recent work of Gaudry, Hess, and Smart [7] shows how the process of Weil decent in characteristic 2 can reduce a discrete logarithm on an elliptic curve over and extension k of $\mathbb{F}_2$ to a discrete logarithm in the Jacobian J of a hyperelliptic curve C over a proper subfield ℓ. This method maps the discrete logarithm in $E(k)$ to that in $J(\ell)$, with an explicit criterion for testing whether the map is injective. This, however, does not apply to the discrete logarithm problem in the point group $E(K)$ when K is a proper extension K/k. Thus an interesting open question is whether a discrete logarithm in $E(K)$ can be mapped injectively into a discrete logarithm problem in $J(L)$ for a proper subfield L of K. The method of Gaudry et al. was found to be effective when the extension degree was < 5. In the event of an affirmative answer to this question, the choice of base field $\mathbb{F}_{2^{31}}$ of the examples taken for this exposition should be of sufficiently large prime degree over $\mathbb{F}_2$ that the size of the genus of the curve C found by Weil descent would make this reduction an impractical means of attack. Moreover, the same construction, when applied in odd characteristic, even $p = 3$, fails to give a reduction to a discrete logarithm on a curve of particularly simple form.

We note that the use of a proper subgroup of $E(K)$, as presented in the examples, is potentially susceptible to the protocol attack of Lim and Lee [17]. The proper design of an encryption protocol, while not treated here, is of equal importance to the proper construction and choice of the group. In this instance, to prevent the leakage of bits, the design of a cryptoscheme on such a group must incorporate a verification of the order of an input message, or include a premultiplication by the cofactor, to eliminate the telltale "witness" to the secret key.

Acknowledgements. The author is grateful to the referees for helpful comments.

References

[1] I. Blake, G. Seroussi, and N. Smart, *Elliptic Curves in Cryptography*, London Mathematical Society, Lecture Notes Series, **265**, Cambridge University Press, 1999.

[2] H. Cohen, *A Course in Computational Algebraic Number Theory*, Graduate Texts in Mathematics, **138**, Springer-Verlag, Berlin, 1993.

[3] M. Deuring, *Die Typen der Multiplikatorenringe elliptischer Funktionenkörper*, Abh. Math. Sem. Hamburg, **14**, (1941), 197–272.

[4] N. Elkies, *Elliptic and modular curves over finite fields and related computational issues*, in *Computational perspectives on number theory (Chicago, IL, 1995)*, 21–76, AMS/IP Stud. Adv. Math., **7**, Amer. Math. Soc., Providence, RI, 1998.

[5] A. Enge, *Elliptic Curves and their Application to Cryptography: An Introduction*, Kluwer Academic Publishers, Boston, 1999.

[6] R. Gallant, R. Lambert, and S. Vanstone, *Improving the parallelized Pollard lambda search on binary anomalous curves*, Math. Comp., to appear.

[7] P. Gaudry, F. Hess, and N. P. Smart, *Constructive and destructive facets of Weil descent on elliptic curves*, HP Labs Technical Report, 2000.

[8] A. Gee, *Class invariants by Shimura's reciprocity law*, Les XXèmes Journées Arithmétiques (Limoges 1997), J. Théor. Nombres Bordeaux **11** (1999), no. 1, 45–72

[9] D. Husemöller, *Elliptic Curves*, Springer-Verlag, New York, 1987.

[10] A. Knapp, *Elliptic curves*, Mathematical Notes, **40**, Princeton University Press, Princeton, NJ, 1992.

[11] N. Koblitz, *Elliptic curve cryptosystems*, Math. Comp., **48** (1987), 203–209.

[12] N. Koblitz, *The improbability that an elliptic curve has subexponential discrete log problem under the Menezes-Okamoto-Vanstone algorithm*, J. Cryptography, **11** (1998), no. 2, 141–145.

[13] D. Kohel, *Endomorphism rings of elliptic curves over finite fields*, Thesis, U.C. Berkeley, 1996.

[14] S. Lang, *Elliptic functions*, Springer–Verlag, New York, 1987.

[15] H. W. Lenstra, Jr., *Complex multiplication structure of elliptic curves*, J. of Number Theory, **56** (1996), no. 2, 227–241.

[16] A. K. Lenstra and E. R. Verheul, *Selecting Cryptographic Key Sizes*, In H. Imai and Y. Zheng, eds., *Proceedings of the Third International Workshop on Practice and Theory in Public Key Cryptosystems, PCK 2000*, Lecture Notes in Computer Science, **1751**, 2000, 446–465.

[17] C. H. Lim and P. J. Lee, *A key recovery attack on discrete log-based schemes using a prime order subgroup*, In B. S. Kaliski, Jr., ed., *Advances in cryptology — CRYPTO '97 (Santa Barbara, CA, 1997)*, Lecture Notes in Computer Science, **1294**, Springer, Berlin, 1997, 249–263.

[18] W. Meier and O. Shaffelbach, *Efficient multiplication on certain nonsupersingular elliptic curves*, in *Advances in Cryptology — CRYPTO '92 (Santa Barbara, CA, 1992)*, 333–344, Lecture Notes in Comp. Sci., **740**, Springer, Berlin, 1993.

[19] A. Menezes, *Elliptic Curve Public Key Cryptosystems*, Kluwer Academic Publishers, Boston, MA, 1993.

[20] A. J. Menezes, T. Okamoto, and S. Vanstone, *Reducing elliptic curve logarithms to logarithms in a finite field*, IEEE Trans. Inf. Theory, **39** (1993), no. 5, 1639–1646.

[21] V. Miller, *Uses of elliptic curves in cryptography*, Advances in Cryptology — CRYPTO'85, Springer, 1986, 417-426.

[22] F. Morain, *Classes d'isomorphismes des courbes elliptiques supersingulières en caractéristique ≥ 3*, Util. Math., **52** (1997), 241–253.

[23] V. Müller, *Fast multiplication on elliptic curves over small fields of characteristic two* J. Cryptology, **11** (1998), no. 4, 219–234.

[24] R. Schoof, *Elliptic curves over finite fields and the computation of square roots mod p*, Mathematics of Computation, **44**, (1985), 483–494.

[25] R. Schoof, *Nonsingular plane cubic curves over finite fields*, J. Combin. Theory Ser. A, **46**, no. 2, (1987), 183–211.

[26] J. Silverman, *The Arithmetic of Elliptic Curves*, Graduate Texts in Mathematics, **106**, Springer-Verlag, New York, 1986.

[27] J. Silverman, *Advanced topics in the arithmetic of elliptic curves*, Graduate Texts in Mathematics, **151**, Springer-Verlag, New York, 1994.

[28] N. Smart, *Elliptic curve cryptosystems over small fields of odd characteristic*, J. Cryptology, **12** (1999), no. 2, 141–151.

[29] J. Solinas, *An Improved Algorithm for Arithmetic on a Family of Elliptic Curves*, in *Advances in Cryptology — CRYPTO'97 (Santa Barbara 1997)*, 357–371, Lecture Notes in Comp. Sci., **1294**, Springer, Berlin, 1997.

[30] M. Wiener and R. Zuccherato, *Faster attacks on elliptic curve cryptosystems*, in *Selected areas in cryptography (Kingston, ON, 1998)*, 190–200, Lecture Notes in Comp. Sci., **1556**, Springer-Verlag, Berlin, 1999.

[31] N. Yui and D. Zagier, *On the singular values of Weber modular functions*, Math. Comp., **66** (1997), no. 220, 1645–1662.

David R. Kohel
School of Mathematics and Statistics, F07
University of Sydney, NSW 2006,
Sydney, Australia
E-mail address: kohel@maths.usyd.edu.au

Progress in Computer Science and Applied Logic, Vol. 20
© 2001 Birkhäuser Verlag Basel/Switzerland

Effective Determination of the Proportion of Split Primes in Number Fields

Kwok Yan Lam and Francesco Sica

Abstract. Let K be a number field, d_K the absolute value of its discriminant. The number $N_1(x)$ of prime ideals of K of residual degree one predominate among all prime ideals of K of norm less than x. Indeed, $N_1(x) \sim x/\log x$ as x tends to infinity, which is also the asymptotic behaviour of the number of all prime ideals. However, for small values of x, there are some discrepancies depending on the field. We will study these irregularities in two cases.

Under the Generalized Riemann Hypothesis for number fields (GRH) we show that on average $|N_1(x)\log x/x - 1| \leq \epsilon$ as soon as $x > c\log^2 d_K$ for some constant $c > 0$ depending only on ϵ. A slightly weaker estimate follows from [8] and is applied to an irreducibility algorithm by Weinberger [13].

In the second part, we try to lift the GRH. We assume that there are no Siegel zeros, as well as some zero-density estimates and zero-free regions corresponding to the large sieve for number fields which generalize known results in the case of cyclotomic fields. We find that much of the same analysis can be carried out under those new assumptions. For example, the same inequality as above holds for $x > c\log^A d_K$, where c depends on ϵ and A is absolute.

1. Introduction and Notations

The problem of factorising a polynomial is an old one and has been studied extensively. It is known that given a polynomial with integer coefficients, it is highly probable that it will be irreducible. The method to ascertain irreducibility that we present here is based on the effective determination of the proportion of split primes in the extension generated by the polynomial. This proportion is directly related to the number of irreducible factors of the polynomial f. After this work was on its way, we discovered that the identical idea was already presented in [13]. However, Weinberger assumes the general Riemann hypothesis (GRH).

In this work we will relax the GRH to a set of two conjectures, both implied by the GRH. We shall see that almost the same results hold under those weaker hypothesis. For the sake of completeness, we recall Weinberger's method, with some simplifications, arising from our use of mean estimates whereas Weinberger

The research of the second author was supported by the Strategic Research Programme on Computer Security (RP 960668/M).

uses the Lagarias-Odlyzko results in [8] (cf. also [12] for a different formulation of these results). We shall not compute absolute constants involved in O-terms, although we stress that they are effectively computable, at least assuming GRH.

We will discuss at the end some possible improvements.

Definition 1.1. *Let K be a number field (finite extension of $\mathbb{Q}$). For $n \in \mathbb{N}$, denote $\omega_K(n)$ the number of different prime ideals $\mathfrak{p}$ dividing the principal ideal (n). The Von Mangoldt function of K is the arithmetical function defined as*

$$\Lambda_K(n) = \begin{cases} \omega_K(n) \log N\mathfrak{p} & \text{if } n = N\mathfrak{p}^k \text{ for some } k \geq 1 \text{ and } \mathfrak{p} \text{ prime ideal of } K, \\ 0 & \text{otherwise.} \end{cases}$$

This function is important in the study of the zeta function of K, which is defined by the formula

$$\zeta_K(s) = \sum_{\mathfrak{a} \text{ ideal of } K} \frac{1}{N\mathfrak{a}^s} = \prod_{\mathfrak{p} \text{ prime ideal of } K} \frac{1}{1 - N\mathfrak{p}^{-s}}.$$

This formula is valid for $\Re s > 1$ and defines $\zeta_K(s)$ as a non vanishing analytic function there. This function has been studied extensively and we summarize here the properties which we will use.

Theorem 1.2 (Hecke).

1. $\zeta_K(s)$ *can be continued to the whole plane. The continuation is analytic except for a simple pole at $s = 1$.*
2. *Let d_K be the absolute value of the discriminant of K, and write*

$$n_K = [K : \mathbb{Q}] = r_1 + 2r_2,$$

where r_1 is the number of real embeddings of K and r_2 the number of complex conjugate embeddings. Define

$$A = 2^{-r_2} d_K^{1/2} \pi^{-n_K/2} \tag{1}$$

$$\xi_K(s) = A^s \Gamma\left(\frac{s}{2}\right)^{r_1} \Gamma(s)^{r_2} \zeta_K(s). \tag{2}$$

Then we have the functional equation $\xi_K(s) = \xi_K(1 - s)$.

The function ξ_K has infinitely many zeros ρ, which all satisfy $0 < \Re\rho < 1$. The generalized Riemann hypothesis (GRH) predicts that in fact $\Re\rho = 1/2$. Until otherwise specified, we will suppose this hypothesis true in order to derive sharp bounds in our estimations.

In order to study the split prime ideals of K, we will introduce the following functions

$$\pi_K(x) = \sum_{N\mathfrak{p} \leq x} 1$$

and the easier to study

$$\psi_K(x) = \sum_{n \leq x} \Lambda_K(n)$$

$$\theta_K(x) = \sum_{n = \mathrm{N}\mathfrak{p} \leq x} \Lambda_K(n).$$

Also we introduce, after Hadamard and de la Vallée Poussin, the regularized

$$\pi_K^{(1)}(x) = \int_0^x \pi_K(t)\, dt = \sum_{\mathrm{N}\mathfrak{p} \leq x} (x - \mathrm{N}\mathfrak{p})$$

and

$$\psi_K^{(1)}(x) = \int_0^x \psi_K(t)\, dt = \sum_{n \leq x} (x - n)\Lambda_K(n),$$

$$\theta_K^{(1)}(x) = \int_0^x \theta_K(t)\, dt = \sum_{\mathrm{N}\mathfrak{p} \leq x} (x - \mathrm{N}\mathfrak{p}) \log \mathrm{N}\mathfrak{p}.$$

These functions generalize the corresponding ones for the ζ function (that is, when $K = \mathbb{Q}$).

We now state some easy lemmas relating those functions and then introduce the explicit formulas which will enable us to estimate their size. We should note that the corresponding work was done in [8] for $\pi_K(x)$ and $\psi_K(x)$ although not for the same purpose.

2. Relations between π, ψ and θ

Lemma 2.1 (Partial summation or Abel's lemma). *Let $(a_n)_{n \geq 1}$ be a sequence of complex numbers and h a continuously differentiable function $h \colon [1, x] \to \mathbb{C}$. Then the following formula holds:*

$$\sum_{n \leq x} a_n h(n) = \left(\sum_{n \leq x} a_n \right) h(x) - \int_1^x \left(\sum_{n \leq t} a_n \right) h'(t)\, dt.$$

Proof. This is nothing else than integration by parts. In this case this can be proved simply by writing $S(x) = \sum_{n \leq x} a_n$ so that $a_n = S(n) - S(n-1)$ and changing order of summation. $\qquad\square$

The functions θ_K and ψ_K are clearly related. In fact

$$\psi_K(x) = \theta_K(x) + \sum_{\mathrm{N}\mathfrak{p}^2 \leq x} \log \mathrm{N}\mathfrak{p} + \sum_{\substack{k > 2 \\ \mathrm{N}\mathfrak{p}^k \leq x}} \log \mathrm{N}\mathfrak{p}$$

$$= \theta_K(x) + O(n_K \sqrt{x} + n_K \sqrt[3]{x} \log x) = \theta_K(x) + O(n_K \sqrt{x}), \qquad (3)$$

because for each $p \in \mathbb{Z}$ prime and $k \geq 1$ there are at most n_K prime ideals $\mathfrak{p} \mid p$ satisfying $N\mathfrak{p} = p^k$ (note that we have an effective bound in view of a famous theorem of Tchebychev, cf. [5, p. 341]). Integrating, we obtain

$$\psi_K^{(1)}(x) = \theta_K^{(1)}(x) + O(n_K x^{3/2}). \tag{4}$$

We use partial summation to relate $\pi_K(x)$ to $\psi_K(x)$. Indeed

$$\pi_K(x) = \sum_{N\mathfrak{p} \leq x} 1 = \sum_{N\mathfrak{p} \leq x} \frac{\log N\mathfrak{p}}{\log N\mathfrak{p}}$$

$$= \frac{\theta_K(x)}{\log x} + \int_2^x \frac{\theta_K(t)}{t \log^2 t} \, dt = \frac{\psi_K(x)}{\log x} + \int_2^x \frac{\psi_K(t)}{t \log^2 t} \, dt + O\left(n_K \frac{\sqrt{x}}{\log x}\right)$$

by (3). We shall see that $\psi_K(x) \sim x$ and since we shall be concerned mainly with the error term, we might as well rewrite the last formula substituting $\psi_K(x) = x + E_K(x)$ to get

$$\pi_K(x) = \int_2^x \frac{1}{\log t} \, dt + \frac{E_K(x)}{\log x} + \int_2^x \frac{E_K(t)}{t \log^2 t} \, dt + O\left(n_K \frac{\sqrt{x}}{\log x}\right). \tag{5}$$

It is customary to call $\int_2^x \frac{1}{\log t} \, dt$ the logarithmic integral and denote it li x. Define also $\mathrm{li}^{(1)}(x) = \int_2^x \mathrm{li}\, t \, dt$. Then, after integrating (5) and performing integration by parts, we get

$$\pi_K^{(1)}(x) = \mathrm{li}^{(1)} x + \frac{E_K^{(1)}(x)}{\log x} + 2 \int_2^x \frac{E_K^{(1)}(t)}{t \log^2 t} \, dt$$
$$+ \int_2^u \int_2^x E_K^{(1)}(t) \frac{\log t + 2}{t^2 \log^3 t} \, dt \, dx + O\left(n_K \frac{x^{3/2}}{\log x}\right), \tag{6}$$

where we put

$$E_K^{(1)}(x) = \int_2^x E_K(t) \, dt = \psi_K^{(1)}(x) - \frac{x^2}{2}.$$

We will work mainly with $\psi_K^{(1)}(x)$ and then use (6) to relate to $\pi_K^{(1)}(x)$.

3. Explicit Formulas

Explicit formulas are the way to relate the behaviour of prime ideals of K to the zeros of ζ_K. We refer to [10] for details. We give here the statement in our case.

Theorem 3.1 (Explicit formulas). *Let $F \colon \mathbb{R} \to \mathbb{R}$ be an even function satisfying:*
- *there is an $\epsilon > 0$ such that $F(x) \exp((1/2+\epsilon)x)$ is integrable and of bounded variation,*
- *the function $(F(x) - F(0))/x$ is of bounded variation.*

Let $\phi = \hat{F}$ be the Fourier transform of F.
Then

$$\sum_{\zeta_K\left(\frac{1}{2}+i\gamma\right)=0} \phi(\gamma) = 2F(0)\log A + 2\phi\left(\frac{i}{2}\right) + \frac{r_1}{\pi}\int_{-\infty}^{\infty}\frac{\Gamma'}{\Gamma}\left(\frac{1}{4}+i\frac{t}{2}\right)\phi(t)dt$$

$$+ \frac{r_2}{\pi}\int_{-\infty}^{\infty}\frac{\Gamma'}{\Gamma}\left(\frac{1}{2}+it\right)\phi(t)dt - 2\sum_{n=1}^{\infty}\frac{\Lambda_K(n)}{\sqrt{n}}F(\log n)$$

where the sum on the left-hand side is over the zeroes of $\zeta_K(s)$ in the critical strip $0 < \Re\left(\frac{1}{2}+i\gamma\right) < 1$.

Corollary 3.2. *We have unconditionally*

$$\sum_{\zeta_K\left(\frac{1}{2}+i\gamma\right)=0}\frac{2}{1+\gamma^2} \le \log d_K + O(n_K) = O(\log d_K) \qquad (7)$$

Proof. We take the function $F(u) = e^{-|u|}$, whose Fourier transform is $\phi(t) = 2(1+t^2)^{-1}$. Then it suffices to notice that $\Lambda_K(n) \ge 0$ hence we can discard the last sum. The last equality follows from a famous theorem of Minkowski (cf. [9, p. 121]). $\qquad\square$

The next result is classical and follows from Stirling's formula (cf. [6, p. 57,footnote]).

Theorem 3.3. *The following asymptotic expansion holds for Γ'/Γ:*

$$\frac{\Gamma'(s)}{\Gamma(s)} = \log s + O(|s|^{-1}),$$

uniformly for $-\pi + \epsilon < \arg s < \pi - \epsilon$. The error term is effective (say) for $|s| > \epsilon$ in this region.

4. The Error Term under GRH

We now turn our attention to the estimation of $\psi_K^{(1)}(x)$. Our aim is to prove the following

Theorem 4.1. *Under the GRH we have*

$$\psi_K^{(1)}(x) = \frac{x^2}{2} + O(x^{3/2}\log d_K)$$

and

$$\pi_K^{(1)}(x) = \mathrm{li}^{(1)}\,x + O\left(\frac{x^{3/2}}{\log x}\log d_K\right),$$

with effective constants.

Remark 4.2. *The averaging process has the effect of cancelling* $\log^2 x$ *from the estimate in the lemma of* [13, p. 181].

Proof. We use the explicit formula theorem with test function

$$
F(u) = \begin{cases} e^{|u|/2} & \text{if } |u| < \log x, \\ \frac{\sqrt{x}}{2} & \text{if } |u| = \log x, \\ 0 & \text{otherwise.} \end{cases}
$$

Then an easy computation proves the following

Lemma 4.3. *Let F be as above. Then its Fourier transform ϕ is equal to*

$$
\phi(t) = \int_{-\infty}^{+\infty} e^{itu} F(u)\, du = \frac{x^{1/2+it}}{1/2+it} + \frac{x^{1/2-it}}{1/2-it} - \left(\frac{1}{1/2+it} + \frac{1}{1/2-it} \right),
$$

and $\phi(-i/2) = x - 1 + \log x$.

Hence by the explicit formula theorem

$$
2\sqrt{x} \sum_{\zeta_K(1/2+i\gamma)=0} \frac{x^{i\gamma}}{1/2+i\gamma} - \sum_{\zeta_K(1/2+i\gamma)=0} \frac{1}{1/4+\gamma^2}
$$

$$
= 2\log A + O(r_1 + r_2) + \frac{r_1\sqrt{x}}{\pi} \int_{-\infty}^{\infty} \frac{\Gamma'}{\Gamma}\left(\frac{1}{4} + i\frac{t}{2} \right) \left(\frac{x^{it}}{1/2+it} + \frac{x^{-it}}{1/2-it} \right) dt
$$

$$
+ \frac{r_2\sqrt{x}}{\pi} \int_{-\infty}^{\infty} \frac{\Gamma'}{\Gamma}\left(\frac{1}{2} + it \right) \left(\frac{x^{it}}{1/2+it} + \frac{x^{-it}}{1/2-it} \right) dt
$$

$$
+ 2x + 2\log x - 2\psi_K(x) + \Lambda_K(x), \quad (8)
$$

where $\Lambda_K(x) = \Lambda_K(n)$ if x equals an integer n and is zero otherwise. Notice that since Γ'/Γ is differentiable on $\Re s = 1/2$ and $\Re s = 1/4$ and in view of Theorem 3.3 the two integrals will be continuous in x and tend to zero as $x \to \infty$. Hence they are bounded and the whole formula can be rewritten under GRH as

$$
\psi_K(x) = x - \sum_{\zeta_K(1/2+i\gamma)=0} \frac{x^{1/2+i\gamma}}{1/2+i\gamma} + O\big(\log d_K + \sqrt{x}(r_1 + r_2)\big),
$$

where we have used (7). We can now invoke the fact (cf. [2, p. 109]) that the sum converges uniformly on closed intervals of x not containing any prime power and that at prime powers there is only a jump discontinuity to deduce that we can integrate the series termwise in x to get

$$
\psi_K^{(1)}(x) = \frac{x^2}{2} - \sum_{\zeta_K(1/2+i\gamma)=0} \frac{x^{3/2+i\gamma}}{(1/2+i\gamma)(3/2+i\gamma)} + O\big(x\log d_K + x^{3/2}(r_1 + r_2)\big).
$$

$$
(9)
$$

It remains to estimate the sum, which is done easily under the GRH, because

$$\left| \sum_{\zeta_K(1/2+i\gamma)=0} \frac{x^{3/2+i\gamma}}{(1/2+i\gamma)(3/2+i\gamma)} \right| \leq x^{3/2} \sum_{\zeta_K(1/2+i\gamma)=0} \frac{1}{|(1/2+i\gamma)(3/2+i\gamma)|}$$

$$\leq x^{3/2} \sum_{\zeta_K(1/2+i\gamma)=0} \frac{1}{(1/4+\gamma^2)} = O(x^{3/2}\log d_K)$$

This completes the proof of the first statement of the theorem. To prove the second, use (6) together with the estimate found on $E_K^{(1)}(x)$. □

5. An Application

Let f be an irreducible polynomial of degree δ with integer coefficients, $\bar{f}$ be the reduced polynomial modulo the prime p, and let $\rho_f(p)$ be the number of roots of $\bar{f}$ in $\mathbb{F}_p$. Then we can define the function

$$\pi_f^{(1)}(x) = \sum_{p \leq x} \rho_f(p)(x-p).$$

We are now in the position to prove the following theorem.

Theorem 5.1. *Under the GRH for the function ζ_K we have*

$$\pi_f^{(1)}(x) = \frac{x^2}{2\log x} + O\left(\frac{x^{3/2}}{\log x} \log d_f + \log^2 d_f \right),$$

where the constant involved in the O-term is effective.

Proof. Define K to be the field $\mathbb{Q}(\alpha)$ where $f(\alpha) = 0$. It is known that $d_K \mid d_f$. Furthermore, if $p \nmid d_f/d_K$, then $\rho_f(p)$ is the number of prime ideals $\mathfrak{p} \mid p$ of K of residual degree 1. Hence, if $\nu(n) = \sum_{p|n} 1$, we have

$$\pi_f^{(1)}(x) = \sum_{\substack{N\mathfrak{p} \text{ prime} \\ N\mathfrak{p} \leq x}} (x - N\mathfrak{p}) + O(n_K \nu(d_f/d_K)) = \pi_K^{(1)}(x) + O(n_K(x^{3/2} + \log d_f)),$$

by the same kind of reasoning as in (3) and the estimate stemming from Tchebychev's theorem that $\nu(n) = O(\log n)$. Hence Theorem 5.1 follows from Theorem 4.1. □

Corollary 5.2. *Let f be a polynomial in $\mathbb{Z}[x]$ of degree δ and let d_f be the absolute value of its discriminant. Assume the GRH. Then there exists an effective absolute constant $c > 0$ such that*

- *f is irreducible iff*

$$\pi_f^{(1)}(x) \leq \frac{3}{2} \operatorname{li}^{(1)} x \quad \text{for } x \geq c\log^2 d_f.$$

- *The number of irreducible factors of f is equal to the nearest integer to*

$$\frac{\pi_f^{(1)}(x)}{\operatorname{li}^{(1)} x} \quad \text{for } x \geq c\delta^2 \log^2 d_f.$$

Remark 5.3. *The corresponding lower bound on x in [13] is $c\delta^2 \log^2 d_f (\log \log d_f)^2$.*

Proof. Suppose $f = f_1 \cdots f_r$ where the f_i are irreducible polynomials and let $d_i = d_{f_i}$. Then $(d_1 \cdots d_r) \mid d_f$ and if we call $\pi_i^{(1)}(x) = \pi_{f_i}^{(1)}(x)$, we have

$$\pi_f^{(1)}(x) = \sum_{i=1}^{r} \pi_i^{(1)}(x) + O\left(\delta\nu\left(\frac{d_f}{d_1 \cdots d_r}\right)\right) = \sum_{i=1}^{r} \pi_i^{(1)}(x) + O(\delta \log d_f)$$

by Tchebychev's theorem again, hence

$$\pi_f^{(1)}(x) = r\frac{x^2}{2\log x} + O\left(r\frac{x^{3/2}}{\log x}\log d_f + \log^2 d_f\right).$$

Now if $r > 1$ the equality

$$r\frac{x^2}{2\log x} + O\left(r\frac{x^{3/2}}{\log x}\log d_f + \log^2 d_f\right) = \frac{x^2}{2\log x} + O\left(\frac{x^{3/2}}{\log x}\log d_f + \log^2 d_f\right)$$

is possible only when $x = O(\log^2 d_f)$. Hence there exists an absolute constant $c > 0$ such that the knowledge of $\pi_f^{(1)}(c \log^2 d_f)$ determines whether f is irreducible.

Similarly, if $\lambda \neq \mu$ and $\max(\lambda, \mu) \leq r$ the equality

$$\lambda\frac{x^2}{2\log x} + O\left(\lambda\frac{x^{3/2}}{\log x}\log d_f + \log^2 d_f\right) = \mu\frac{x^2}{2\log x} + O\left(\mu\frac{x^{3/2}}{\log x}\log d_f + \log^2 d_f\right)$$

can only happen when $x = O(r^2 \log^2 d_f)$. Hence there exists an absolute constant $c > 0$ such that the knowledge of $\pi_f^{(1)}(c\delta^2 \log^2 d_f)$ determines the number of factors of f. $\qquad\square$

6. The Error Term: Zero-free Regions and Zero Density Estimates

In this section we do not assume the validity of GRH. We will suppose that we have the following zero-free region for $\zeta_K(s)$.

Conjecture 6.1. *Let K be a number field and d_K the absolute value of its discriminant. Denote $\sigma = \Re s$ and $t = \Im s$. Then there exists an absolute constant $c_1 > 0$ such that $\zeta_K(\sigma + it) \neq 0$ in the region*

$$\sigma > 1 - \frac{c_1}{\log\left((|t| + 2)\log d_K\right)},$$

with at most one real and simple zero β_K. Furthermore, there exists $c_2 > 0$ such that

$$1 - \beta_K > \frac{c_2}{d_K^{1/2}\log d_K}.$$

With d_K instead of $\log d_K$, this is known as the Landau-Page lemma. The number β_K, if it exists, is called the Siegel zero for ζ_K.

The existence of a zero very close to 1 is the main obstruction to obtaining fast (with respect to d_K) asymptotics in the prime ideal theorem. To get down to polynomial-time estimates, we still need another conjectural (if K is not cyclotomic) ingredient in the form of a zero-density estimate.

Conjecture 6.2. *Let $N(\alpha, T, K)$ denote the number of zeros $\rho \neq \beta_K$ of ζ_K in the box*

$$\alpha \le \sigma \le 1, \quad 0 \le |t| \le T.$$

Then there exists absolute constants $c_3, c_4 > 0$ such that

$$N(\alpha, T, K) \ll T^{c_3(1-\alpha)},$$

whenever $T > \log^{c_4} d_K$.

Remark 6.3. *It is known that $N(\frac{1}{2}, T, K) \sim \frac{T}{2\pi} \log \frac{T d_K}{2\pi}$, but the important statement in this conjecture is that the zeros thin out near $\Re s = 1$.*

We now arrive to the main theorem of this section.

Theorem 6.4. *Assume Conjectures 6.1 and 6.2, and that there is no Siegel zero for K. We then have*

$$\psi_K^{(1)}(x) = \frac{x^2}{2} + O(x^2 B^{-A}) \quad \text{for } x \ge \log^A d_K,$$

and for some $A' = A'(A) > 0$

$$\pi_K^{(1)}(x) = \mathrm{li}^{(1)} x + O\left(\frac{x^2}{\log x} B^{-A}\right),$$

for $x \ge \log^{A'} d_K$.

Proof. The second part follows easily from the first by partial summation using (6). We focus on proving the first equation.

Without using GRH, we can rewrite the left-hand side of (8) as

$$2\left(\sum_{\xi_K(\rho)=0} \frac{x^\rho}{\rho} - \frac{1}{\rho} \right)$$

$$= 2\left(\sum_{|\rho| \le \frac{1}{3}} \frac{e^{\rho \log x} - 1}{\rho} + \sum_{\substack{|\rho| > \frac{1}{3} \\ \Re\rho < \frac{1}{2}}} \frac{x^\rho}{\rho} + \sum_{\Re\rho \ge \frac{1}{2}} \frac{x^\rho}{\rho} - \sum_{|\rho| > \frac{1}{3}} \frac{1}{\rho} \right)$$

$$= O(x^{1/3} \log d_K) + 2 \sum_{\substack{|\rho| > \frac{1}{3} \\ \Re\rho < \frac{1}{2}}} \frac{x^\rho}{\rho} + 2 \sum_{\Re\rho \ge \frac{1}{2}} \frac{x^\rho}{\rho} + O(\log d_K).$$

The two O-estimates follow from (7) (with the application of the maximum modulus principle for the first one). Also, after integration of the two middle series, it is easy to see that

$$\sum_{\substack{|\rho|>\frac{1}{3} \\ \Re\rho<\frac{1}{2}}} \frac{x^{\rho+1}}{\rho(\rho+1)} \ll x^{3/2}\log d_K.$$

Thus the explicit formula for $\psi_K^{(1)}$ can be written in this case as

$$\psi_K^{(1)}(x) = \frac{x^2}{2} - \sum_{\Re\rho\geq\frac{1}{2}} \frac{x^{\rho+1}}{\rho(\rho+1)} + O(x^{3/2}\log d_K).$$

We then need to examine the contribution of

$$\sum_{\Re\rho\geq\frac{1}{2}} \frac{x^{\rho+1}}{\rho(\rho+1)},$$

under the assumption that ξ_K has no Siegel zeros.

Note that

$$\sum_{\Re\rho\geq\frac{1}{2}} \frac{x^{\rho+1}}{\rho(\rho+1)} = \sum_{\substack{\Re\rho\geq\frac{1}{2} \\ |\Im\rho|\leq T}} \frac{x^{\rho+1}}{\rho(\rho+1)} + O\left(\frac{\log(Td_K)}{T}\right) \tag{10}$$

by the remark following Conjecture 6.2.

We have

$$\sum_{\substack{\Re\rho\geq\frac{1}{2} \\ |\Im\rho|\leq T}} \frac{x^{\rho+1}}{\rho(\rho+1)} \leq x \sum_{\substack{\Re\rho\geq\frac{1}{2} \\ |\Im\rho|\leq T}} x^\rho = -x \int_{\frac{1}{2}}^1 x^\alpha \, dN(\alpha,T,K)$$

$$= N(\tfrac{1}{2},T,K)x^{3/2} + x\log x \int_{\frac{1}{2}}^1 x^\alpha N(\alpha,T,K)\,d\alpha$$

$$\ll x^{3/2}T^2\log d_K + x\log x \int_{\frac{1}{2}}^{1-\frac{c_1}{\log(T\log d_K)}} x^\alpha \, T^{c_3(1-\alpha)}\,d\alpha$$

Let $T = \log^{c_4} d_K$. The last expression is bounded by

$$x^{3/2}\log^{c_6} d_K + x\log x\,T^{c_3}\int_{\frac{1}{2}}^{1-\frac{c_7}{\log\log d_K}} \left(\frac{x}{T^{c_3}}\right)^\alpha d\alpha$$

$$\ll x^{3/2}\log^{c_6} d_K + x^2\frac{\log x}{\log(x/T^{c_3})}x^{-\frac{c_7}{\log\log d_K}}.$$

We now let $x = \log^{c_8} d_K$ for a large enough constant c_8. It is then clear that the last expression becomes $O(x^2 e^{-c_7 c_8})$ thus proving our claim with $B = e^{c_7}$, $A = c_8$ and hence our main theorem. $\qquad\qquad\square$

The following corollary follows formally from the theorem in the same way as Corollary 5.2 follows from Theorem 5.1.

Corollary 6.5. *Suppose that there are no Siegel zeros for any K, as well as Conjectures 6.1 and 6.2. Then the algorithm of the previous section still determines with 100% certainty the irreducibility of f in polynomial time.*

Furthermore, if $n \leq \delta = \deg f$, then up to n factors of f can be detected in polynomial time, if n is fixed.

More precisely, there exists an absolute constant c_5 such that

- *f is irreducible iff*

$$\pi_f^{(1)}(x) \leq \frac{3}{2} \mathrm{li}^{(1)} x \quad \text{for } x \geq \log^{c_5} d_f.$$

- *$n \leq \delta$ irreducible factors of f can be detected by finding the nearest integer to*

$$\frac{\pi_f^{(1)}(x)}{\mathrm{li}^{(1)} x} \quad \text{for } x \geq \log^{c_5 \log n} d_f.$$

7. Final Remarks

Conjectures 6.1 and 6.2 are known to be true for cyclotomic extensions (generated by a primitive m-th root of unity), thanks in particular to the large sieve. It is possible also to remove the obstruction induced by the Siegel zero by considering the difference $\psi_K^{(1)}(x) - \psi_K^{(1)}(y)$ for a suitable $y = y(x)$.

The method can also be generalized to deal with polynomials over number fields. It is then known that Siegel zeros occur very rarely at most (cf. [11]).

We should also mention that the same hypotheses relaxing on the GRH can be applied in a variety of problems where one needs to be assured of the existence of a small prime p such that a given irreducible polynomial has a root in the finite field with p elements. For instance, we have the following refinement of [4]:

Theorem 7.1. *Assuming Conjectures 6.1 and 6.2 and the nonexistence of Siegel zeros for any number field, then nondivisibility of sparse polynomials belongs to NP.*

Finally, we would like to thank Igor Shparlinski for his suggestions concerning the presentation of this material.

References

[1] L.M. Adleman and A.M. Odlyzko, *Irreducibility testing and factorization of polynomials*, Math. Comp. **41** (1983), n. 164, 699–709.

[2] H. Davenport, *Multiplicative number theory*, GTM **74**, Springer-Verlag (1980).

[3] E. Dehn, *Algebraic equations*, Dover reprint (1960).

[4] D.Grigoriev, M. Karpinski and A. M. Odlyzko, *Short proof of nondivisibility of sparse polynomials under the Extended Riemann Hypothesis*, Fundamenta Informaticae, **28** (1996) no. 3–4, 297–301.

[5] G.H. Hardy and E.M. Wright, *An introduction to the theory of numbers*, OUP (1989).

[6] A.E. Ingham, *The distribution of prime numbers*, CUP re-issued in 1990.

[7] J.C. Lagarias and A.M. Odlyzko, *Computing $\pi(x)$: an analytic method*, Journal of Algorithms 8 (1987), 173–191.

[8] J.C. Lagarias and A.M. Odlyzko, *Effective versions of the Chebotarev density theorem* in Algebraic number fields: L-functions and Galois properties (Proc. Sympos. Univ. Durham, Durham, 1975), (1977), 409–464, Academic Press, London.

[9] S. Lang, *Algebraic number theory* GTM **110**, Springer-Verlag (1986).

[10] J.F. Mestre, *Formules explicites et minorations de conducteurs de variétés algébriques*, Comp. Math. **58** (1986), 209–232.

[11] A.M. Odlyzko and C.M. Skinner, *Nonexistence of Siegel zeros in towers of radical extensions*, Contemporary Math. **143** (1993), 499–511.

[12] J.-P. Serre, *Quelques applications du théorème de densité de Chebotarev*, Publ. Math. IHES, **54** (1982), 123-201.

[13] P.J. Weinberger, *Finding the number of factors of a polynomial*, Journal of Algorithms **5** (1984), 180–186.

School of Computing,
National University of Singapore,
Lower Kent Ridge Road,
Singapore 119260
E-mail address: lamky@comp.nus.edu.sg

E-mail address: sica@comp.nus.edu.sg

Progress in Computer Science and Applied Logic, Vol. 20

Algorithms for Generating, Testing and Proving Primes: A Survey

Preda Mihăilescu

Abstract. We survey methods of testing and proving primality and their implementation for generation of cryptographic primes. While discussing a wider variety of primality tests of theoretical or practical relevance, the focus is on criteria for practical use.

We give a new model for sources producing prime numbers with biased distributions and use it for measuring the security of biases against unknown attacks (adapted solutions to the discrete logarithm or integer factoring problems) which could make use of knowledge of the bias. Some results can be proved based solely upon the bias distribution, without prior knowledge of the attacks. Thus an important class of sources with *polynomially bounded* bias are secure in the sense that algorithms which can use the bias with a performance gain, can be turned into improvements to state of the art attacks in presence of uniform distributed sources.

The paper concludes with an overview of some outstanding schemes for generation of cryptographic primes. These are compared according to their performance and confidence of decision.

1. Introduction

The generation of primes is a cryptographical problem, given the multitude of public key algorithms, which rely in different ways upon knowledge of some large primes. Prime generation requires a search method that yields well distributed expected primes of a given length, and a method for actually proving primality of these candidates. Since primality proving is a classical problem in number theory, it is no surprise that it has a very wide literature. A good overview may be found in [65], [11], [26]. Search strategies have received comparatively less attention.

In this paper we give a survey that should be of interest for the use of cryptographers and general readers, alike. The exposition of primality proving will cover several topics which may be considered as theoretical background of the practically relevant information. Due to the focus on practical applications, some topics of interest in primality testing will not be mentioned or exposed in detail.

Key words and phrases. Primes, Lucas-Lehmer, Cyclotomy, ECPP.

The survey is structured as follows. In Section 2 we give an overview of the properties of primes used for testing and proving primality. We also discuss sufficient conditions for primality, which lead to conditional, deterministic primality proving methods. Section 3 covers some of the research on pseudoprimes and discusses the binding of pseudoprime tests in probabilistic proving methods. In Section 4 we describe most of the search methods which have been analysed in the literature. We then consider some deeper results concerning the average probability that iterated pseudoprime tests yield wrong answers, when repeated in the context of a prime search. We describe also in this context methods for producing provable primes for cryptographic use. In Section 5 we provide some detail about the currently implemented algorithms for general primality proving. Section 6 discusses some security issues connected with search distributions and proves a general security result for primes generated by sources with biased random distributions.

In the final section we reconsider some of the algorithms exposed from the perspective of the *implementer* and compare the most common approaches for prime number generation with respect to *performance* and *confidence*.

A complete reference list is too long to be included in this frame. Instead, we provided at least one important recent reference which refers to further useful literature for the respective topic.

2. Conditions for Primality

The naive approach to proving that a given positive number $n > 1$ is a prime, consists in verifying that it is not divisible by any prime $p \leq \sqrt{n}$. Although this method has exponential complexity and is thus impracticable for larger numbers, it suffices for numbers beneath some small bound B, say $B \sim 10^7$. An equally classical refinement of this idea, the *Eratosthenes sieve*, distributes the weight of the same computations over the task of simultaneously finding all the primes in a given interval $a < n < b$. In this case, all the multiples of primes $p < \sqrt{b}$ are successively eliminated from the interval, thus eventually leaving out only those numbers of the interval which are primes. Provided $b - a \ll a$, at the cost of a polynomial increase of the number of operations compared to the naive method above, one proves simultaneously $\approx \frac{b-a}{\log(\frac{a+b}{2})}$ primes. Eratosthenes sieves are efficient in the context of prime search methods in sequences of consecutive integers or arithmetic progressions. Optimized sieve methods are analysed in [59], [60].

Those properties of primes, which can be efficiently verified are ideal for "testing" for primality. The idea of a primality test, in the sense of some *quick* verification of the fact that a number *may* be prime has been applied for many decades. Such tests have high chances for eliminating composite numbers, yet do not, in general, provide a proof that the given number is indeed prime. We shall next give the example of an important property used for testing primes and use this example in order to distinguish between different types of tests.

Theorem 2.1. *Let* $n = 2^k \cdot m + 1$ *with* $k > 0$ *be an integer and* $1 < a < n$ *an other random integer. Consider the sequence:*

$$a_i \equiv a^{2^i \cdot m} \mod n, \quad for \ i = 0, 1, \dots, k. \tag{1}$$

If n *is prime, then either* $a_i \equiv 1 \mod n, \forall i$ *or* $\exists i \geq 0$ *such that* $a_i \equiv -1 \mod n$ *and* $a_j \equiv 1 \mod n$, *for* $k \geq j > i$.

Proof. If n is prime, then $a^{n-1} \equiv 1 \mod n$ by little Fermat. Since $(\mathbb{Z}/(n \cdot \mathbb{Z}))$ is a field in this case and the sequence a_i is produced by successive squarings, it follows that $a_i \equiv \pm 1 \mod n$ if $a_{i+1} \equiv 1 \mod n$. The assertion is proved by induction on i, for i decreasing from k to 0 or the first value for which $a_i \equiv -1 \mod n$.

An alternative argument is given by Dubois [53] who considers the algebraic factorization $a^{2^k \cdot m} - 1 = (a^m - 1) \cdot \prod_{j=0}^{k-1} (a^{2^j \cdot m} + 1)$. $\square$

Theorem 2.1 had been used as a primality test in the sense given above apparently long before the term of "probabilistic primality test" was introduced by Solovay and Strassen in [70]. In their paper, for a and n as above, they use the property

$$a^{\frac{n-1}{2}} = (a/n), \tag{2}$$

with (a/n) the Jacobi symbol. This property holds when n is a prime. We shall give more detail on probabilistic primality tests and also the use of Theorem 2.1 in Section 4.

It will be very useful to define here the complexity unit of 1 *selfridge*, which is the time required by one exponentiation, or equivalently, by $\log_2 n$ multiplications modulo n. Note the dependence of 1 selfridge upon a given size of integers $|n|$ and the independence from underlying methods for multiplications of long integers. We shall assume, that for fixed sizes, the same multiplication will be used.

The verification of conditions like Theorem 2.1 and condition (2) is both very efficient – 1 selfridge – and independent of any additional information about n. The verifications are likely to fail when applied to almost any composite number, in a sense that shall be made more strict later. It is for this reason that current terminology shifts towards the term of "compositeness tests", which is a better description of what the tests achieve. The term "primality test" catches better the intended information one is actually searching.

It is natural to require an actual proof about the fact that a given number is indeed a prime. An older line of thought ([65], [11]) was expressed in the following:

Lemma 2.2 (Pocklington). *Let* $n > 1$ *be an integer and* $q | n - 1$ *a prime. Suppose there is an integer* $a_q \in \mathbb{Z}/(n \cdot \mathbb{Z})^*$ *such that*

$$a_q^{n-1} = 1 \quad and \tag{3}$$

$$(a_q^{\frac{n-1}{q}} - 1, n) = 1.$$

Then, each prime $p | n$ *is of the shape* $p = k \cdot q + 1$, *for some integer* $k > 1$.

As a consequence, we have the following:

Corollary 2.3. *Let $n > 1$ be an integer and $n - 1 = F \cdot R$ with integers F, R and such that F is completely factored. If for each prime $q|F$ there is an integer $a_q \in \mathbb{Z}/(n \cdot \mathbb{Z})^*$ verifying (3), then all prime divisors of n verify $p = k \cdot F + 1$. In particular, if $F > \sqrt{n}$, then n is a prime.*

Proof. For proving the theorem, assume that $p|n$ is a prime. We set $r_q \equiv a_q^{\frac{n-1}{q}}$ mod p, $r_q \in \mathbb{Z}/(p \cdot \mathbb{Z})$ and by (3) $r_q \neq 1$. Since $r_q^q = 1$, it follows that r_q is an q-th primitive root of unity modulo p and in particular $q|(p-1)$, the order of the multiplicative group $\mathbb{Z}/(p \cdot \mathbb{Z})^*$ in which the cycle generated by r_q is a subgroup.

The corollary follows by noting that the condition (3) being verified for all primes $q|F$ implies the existence of primitive F-th roots of unity modulo any prime $p|n$; this means the order of any $p|n$ is divisible by F and implies the first assertion. If n is composite, it has at least one prime factor $< \sqrt{n}$, which cannot be the case when $F > \sqrt{n}$ and $p \equiv 1 \mod F$.   □

The following simple observation was written down by Quisquater and Couvreuer [20]. It allows choosing $F > \sqrt[3]{n}$ in Corollary 2.3.

Proposition 2.4. *Suppose the conditions of Corollary 2.3 are met for $F > \sqrt[3]{n}$ and let $R = u \cdot F + v$, $0 \le v < F$. If, additionally, the quadratic discriminant $\Delta = v^2 - 4 \cdot u$ is not the square of an integer, then n is a prime.*

Proof. By the corollary, possible prime factors of n are $\equiv 1 \mod F$. Thus n is built up of at most two prime factors $p_i = c_i \cdot F + 1$, $i = 1, 2$. After expanding $n = p_1 \cdot p_2$ in powers of F and comparing, we get

$$c_1 \cdot c_2 \cdot F + (c_1 + c_2) = u \cdot F + v.$$

Since $c_1 + c_2 < F$, it follows that the integers c_i are roots of the quadratic equation $z^2 - v \cdot z + u = 0$. This only has integer zeroes if Δ is a square, which completes the proof.   □

While the proofs of primality using the last propositions are still efficient in terms of complexity, they rely upon important additional knowledge about the prime candidate n. Indeed, finding the factored part $F|(n-1)$ would require in the general case factorization of $n - 1$ itself. These theorems are used in cases when either $n - 1$ has large factored parts due to a special shape of n, or at least algebraic factors of $n - 1$ are easy to find; for example, $n = k \cdot p^m + 1$, for some prime p with $p^m > k$.

The core idea of the Pocklington lemma, is to prove the existence of an element of "large" order in some group modulo n, so that the premise of n being composite together with the natural reduction of this group modulo the possible factors of n leads to a contradiction. It was used in many contexts different from the initial group $\mathbb{Z}/(n \cdot \mathbb{Z})^*$. We shall give here a formal generalization of this idea and discuss subsequently some known concrete instances.

Theorem 2.5. *Let* $\mathbf{A}(n), \forall n \in \mathbb{N}, n > 1$ *a family of rings and* $\mathbf{G}(n) \subset \mathbf{A}(n)$ *subsets provided with a composition rule* $\circ : \mathbf{G}(n) \times \mathbf{G}(n) \longrightarrow \mathbf{A}(n)$ *which makes* $\mathbf{G}(n), \circ)$ *into a group;* $\mathbf{G}(n)$ *is the domain of the relation* $\circ$. *Suppose that there is a natural reduction map* $\pi_{n,m} : \mathbf{A}(n) \longrightarrow \mathbf{A}(m), \forall m | n$, *which preserves the group law.*

Assume $f : \mathbb{N} \longrightarrow \mathbb{R}$ *is a monotonically increasing function upper bounding the group sizes:* $o(n) = \#\mathbf{G}(n) < f(n)$, $\forall n > 1$. *Let* n *be an integer, such that* $o(n) = F \cdot R$ *and* F *is completely factored. Suppose that for all primes* $q|F$ *there is some element* $\alpha_q \in \mathbf{A}(n)$, *such that:*

$$o(n) \odot \alpha_q = 0_n \in \mathbf{A}(n) \quad and \tag{4}$$

$$\pi_{n,m}\left(\frac{o(n)}{q} \odot \alpha_q\right) \neq 0_m, \quad \forall m|n,$$

where 0_m *is the neutral element in the group* $\mathbf{G}(m)$. *For an integer* k *and* $\beta \in \mathbf{G}(m)$, *we write* $k \odot \beta$ *for the* k-*fold composition* $\underbrace{\beta \circ \beta \ldots \circ \beta}_{k \; times}$. *Then for each prime* $p|n$, *the group* $\mathbf{G}(p)$ *contains an element of order* F. *In particular, if* $F > f(\lceil\sqrt{n}\rceil)$, *then* n *is prime.*

Proof. Consider the element $\alpha = \bigcirc_{q|F} \pi_{n,p}(\alpha_q) \in \mathbf{G}(p)$. Under the conditions (4), one shows that α has the order F in $\mathbf{G}(p), \forall p|n$, and thus $\mathbf{G}(p)$ has a cyclic subgroup with F elements. For $F > f(\lceil\sqrt{n}\rceil)$, this condition cannot be achieved by any prime $p < \sqrt{n}$, since f is increasing. This shows that n cannot be composite. $\square$

If condition (4) is replaced by the simpler condition

$$o(n) \odot \alpha_q = 0_n \in \mathbf{A}(n) \quad and \tag{5}$$

$$\pi_{n,m}(o(n) \odot \alpha_q) \neq 0_m, \quad \forall m|n,$$

no additional knowledge about n is required, and one has the simplest primality test, in a general form. In particular, for $\mathbf{A}(n) = \mathbb{Z}/(n \cdot \mathbb{Z})^*$, this is the Fermat test which consists in verifying that the small theorem of Fermat holds for a given n and some base $\alpha \in \mathbb{Z}/(n \cdot \mathbb{Z})^*$. It is consistent to refer to (5) as the *generalized Fermat* test. This variation of the Fermat test was intensively used in research on pseudoprimes. We shall discuss aspects about the surprising strength of this test in Section 4.

Generalizations of the Pocklington lemma to different group structures allow more variation to the functions defining the number of elements $o(n) = \#\mathbf{G}(n)$. Thus they increase the chance of finding factorizations of $o(n)$ and one finds primality proofs for larger classes of integers. We now give an overview of primality proof methods based on the generalized Pocklington lemma.

2.1. Lucas-Lehmer Sequences and Tests

Let $\mathbf{A}(n) = \mathbb{Z}/(n \cdot \mathbb{Z})[X]/(g(X))$, where

$$g(X) = \sum_{j=0}^{k} g_j \cdot X^j \in \mathbb{Z}/(n \cdot \mathbb{Z})[X]$$

is some, not necessarily irreducible polynomial of degree k. These rings have been considered earlier in connection with the recursive sequences

$$a_{i+k} = - \sum_{j=0}^{k-1} g_j \cdot a_{i+j}. \tag{6}$$

These recursive sequences are uniquely determined by a set of k initial values and have $g(X)$ as characteristic polynomials. The theory of finite differences shows that they are linear combinations of the sequences of successive powers of the single roots of $g(X)$:

$$(a)_{n \in \mathbb{N}} = \sum_{i=1}^{k} \lambda_i \cdot (\alpha_i^n)_{n \in \mathbb{N}},$$

with α_i being the roots of $g(x)$ in some extension of $\mathbb{Z}/(n \cdot \mathbb{Z})^*$. Note that for irreducible g, $\#\mathbf{A}(n) = n^k - 1$ and the proper factors of $n^k - 1$ – that is, those who do not divide $n^h - 1$ for any $h < k$ are divisors of $\Phi_k(n)$ with Φ_k the k-th cyclotomic polynomial. This shows that factors of $\Phi_k(n)$ are exactly the gain one may expect by raising the degree of the polynomial g.

For $k = 2$, sequences (6) bear the name of *Lucas sequences*. For $k = 3$, Perrin studied some special instances, while an extensive study may be found in [4]. From Lucas' fundamental paper [41] until the eighties, a vast number of variations on this theme have been analysed, with k ranging within a set of small values $k = 2, 3, 4, 6, 8$. The larger even values in this set are reached by taking g with coefficients in a quadratic extension of $\mathbb{Z}/(n \cdot \mathbb{Z})^*$. We shall use the generic name of *Lucas-Lehmer sequences* and Lucas-Lehmer tests to denote any instance belonging to this family. An important detailed analysis of the possible combination of Lucas-Lehmer tests for $k = 1$ and $k = 2$ is [14]. See also [4], [73], [75], [76] and their secondary literature, for further special cases. A beautiful and extensive presentation of Lucas-Lehmer tests can be found in [74].

We expose next the classical Lucas-Lehmer test for Mersenne primes, that is, primes of the shape $M(p) = 2^p - 1$, with p itself prime. Using quadratic reciprocity, Lucas noted that $3 \bmod M(p)$ is a quadratic nonresidue, whenever $M(p)$ is a prime. Thus, the polynomial $g(X) = X^2 - 2X - 2$ with reduced discriminant $\Delta = 3$ is irreducible and the sequences

$$
\begin{aligned}
U_m &= \frac{\alpha^m - \overline{\alpha}^m}{\alpha - \overline{\alpha}} \quad \text{and} \\
V_m &= \alpha^m + \overline{\alpha}^m,
\end{aligned}
\tag{7}
$$

with $\alpha, \overline{\alpha}$ being the zeroes of $g(X)$ in a quadratic extension of $\mathbb{Z}/(M(p) \cdot \mathbb{Z})^*$, will vanish modulo $M(p)$ with a period dividing $M(p) + 1 = 2^p$. This is because $V_m | U_{2m}$ and $\alpha^P = \overline{\alpha}^P = \alpha \cdot \overline{\alpha}$, for $P = M(p) + 1$ and $M(p)$ prime. Finally, $-2 = \alpha \cdot \overline{\alpha}$ being a quadratic nonresidue for prime Mersennes, the primality condition

$$
\begin{aligned}
V_{2^{p-1}} &\equiv 0 \bmod M(p) \\
V_{2^{p-2}} &\not\equiv 0 \bmod M(p),
\end{aligned}
$$

for Mersenne primes $M(p)$ follows from the above and Theorem 2.5. This condition is particularly simple to verify, using the recursion

$$V_{2m} = V_m^2 - 2 \cdot g_0^m,$$

where g_0 is the free term of the polynomial g. A further improvement was brought by Lehmer 50 years later, who observed that the condition $g_0 = 1$ would cut the computations for a Mersenne test in half. This condition can be achieved by the substitution $\alpha \mapsto \alpha + 1$ which corresponds to $g(X) \mapsto \tilde{g}(X) = X^2 - 4 \cdot X + 1$. With this, the Lucas-Lehmer test for Mersenne primes is:

Proposition 2.6. *Let p be a prime and $M(p) = 2^p - 1$. Let $r_n \in \mathbb{Z}/(M(p) \cdot \mathbb{Z})$ be a recursive sequence with $r_1 = 4$ and induction law:*

$$r_{i+1} = r_i^2 - 2, \ i > 0.$$

The number $M(p)$ is prime iff $r_i \neq 0$ for $i < p-1$ and $r_{p-1} = 0$.

This very version of the Lucas-Lehmer test was used for producing the largest currently known primes. The major contributions of recent days to this old times task consisted on the one hand in the implementation of a dedicated variant of the fast Fourier multiplication for long integers, which takes advantage of the special shape of Mersenne numbers [21]. On the other hand, computations were distributed over the Internet to participants willing to offer computer time for finding the next largest Mersenne prime [16]. The current largest Mersenne prime was thus found by the GIMPS effort; it is $2^{6972593} - 1$.

Deterministic tests similar to proposition 2.6, may be easily designed for classes of numbers of the shape $\prod p_i^{e_i} - 1$, for small primes p_i, by using explicit laws of reciprocity, for example, see [48]. Since every fraction of performance is important for the large scale computations involved in finding the largest primes, one can by this extend the range of candidates within which the search is performed. However, the performance of single tests will necessarily be slightly slower then the very clean Lucas-Lehmer test for Mersenne primes. An alternate class of primes, among which records are currently sought are such of the form $n = k2^m \pm 1$, also called Proth primes, since Proth first proved Theorem 2.5 for this case. Note that the sequences U_m, V_m defined in (7) allow restating the generalized Fermat test for n over quadratic extensions of $\mathbb{Z}/(n \cdot \mathbb{Z})^*$ by the condition:

$$V_{\frac{n+1}{2}} \equiv 0 \mod n.$$

A noteworthy problem of this generation of tests is the difficulty to draw common information from the known factors of $F_k | \Phi_k(n)$ for different values of k, after application of the Pocklington lemma for those k. Thus for instance, while tests for $k = 2$ require a factored part $F > \sqrt{n}$ of $n + 1$, when combining with tests for $k = 3$, not less then $F_2 \cdot F_3 > n^{2/3}$ was known to provide a final primality proof. Here $F_k | \Phi_k(n), k = 2, 3$ are the known factored parts. This problem was solved in 1981 by Hendrik Lenstra, who showed in [36] that by adding some condition related to Frobenius actions, one can use factored parts $F | \Phi_k(n)$, $F > \sqrt{n}$ for arbitrary large k. We shall discuss Lenstra's theorem in more detail in the context of general primality proving.

Theorem 2.7 (Lenstra,1981). *Let $s \in \mathbb{Z}_{>0}$. Let* $\mathbf{A}$ *be a commutative ring which contains* $\mathbb{Z}/(n \cdot \mathbb{Z})$ *as a subring. Suppose that there exists* $\alpha \in \mathbf{A}$ *satisfying the following conditions:*

$$\alpha^s = 1,$$
$$\alpha^{s/q} - 1 \in \mathbf{A}^*, \text{ for every prime } q|s, \tag{8}$$
$$\Psi_\alpha(X) = \prod_{i=0}^{t-1} \left(X - \alpha^{n^i} \right) \in \mathbb{Z}/(n \cdot \mathbb{Z})[X], \text{ for some } t \in \mathbb{Z}_{>0}$$

Then for every divisor

$$r|n, \ \exists \ i(r), \ 1 <= i(r) < t : r \equiv n^{i(r)} \mod s. \tag{9}$$

If $s > \sqrt{n}$ and $r|n$ is a prime $< \sqrt{n}$, it is equal to the minimal positive representative of $n^{i(r)} \mod s$.

Proof. Suppose that (8) holds and let $r|n$ be a prime and $\mathcal{R} \subset \mathbf{A}$ a maximal ideal containing r. Let $\nu : \mathbf{A} \to \mathbb{L}$ be the natural projection, with $\mathbb{L} = \mathbf{A}/\mathcal{R}$ being a field. If $\tilde{\alpha} = \nu(\alpha)$ and $\Psi_{\tilde{\alpha}} = \nu(\Psi_\alpha)$, then

$$\Psi_{\tilde{\alpha}}(X) \in (\mathbb{Z}/(r \cdot \mathbb{Z}))[X],$$

since $\nu(\mathbb{Z}/(n \cdot \mathbb{Z})) = \mathbb{Z}/(r \cdot \mathbb{Z})$. It follows that $\Psi_{\tilde{\alpha}}$ is divided by the minimal polynomial of $\tilde{\alpha}$ and, in particular, has also $\tilde{\alpha}^r$ as a zero. The ring $(\mathbb{Z}/(r \cdot \mathbb{Z}))[X]$ being factorial, we must have $\tilde{\alpha}^r = \tilde{\alpha}^{n^i}$ for some $i < t$ and since $\tilde{\alpha}^s = 1$, it follows that $r \equiv n^i \mod s$. $\qquad\square$

If $\mathbf{A} \subset \mathcal{O}(\mathbb{K})/(n \cdot \mathcal{O}(\mathbb{K}))$, where $\mathcal{O}(\mathbb{K})$ are the integers in a number field $\mathbb{K}$, we have equivalence in Theorem 2.7. This theorem was written in the context of spreading the new unconditional algorithm for proving primality, proposed by Adleman, Pomerance and Rumely [5]. That test is currently referred to as "Jacobi sum" test, while Lenstra's [36], [37] and ulterior contributions [15], [46], which pointed out to the intimate relationship it has to classical Lucas-Lehmer tests, lead to the more general notion of *cyclotomy primality tests*. These include the Jacobi sum test as a partial aspect. Cyclotomy is the only currently known efficient *unconditional* deterministic primality proving method; it is, though, not polynomial, having an asymptotic complexity of $\mathcal{O}\left((\log n)^{c \log \log \log n}\right)$, for an effectively computable absolute constant $c > 0$. Note that *conditional* deterministic polynomial time primality proving algorithm had been known, for example, depending on the extended Riemann conjecture, see [11].

2.2. Elliptic Curves

Let p be a prime; an elliptic curve over a ring $\mathbf{R}$ is given by a Weierstrass equation in the form:

$$\mathcal{E} : Y^2 = X^3 + aX + b, \quad \text{for } (a, b) \in \mathbf{R} \times \mathbf{R} \setminus \{(0, 0)\}. \tag{10}$$

Points $P = (X, Y) \in \mathcal{E}$ can be added according to an addition deduced from the "chord or tangent law". The sum $R = P \oplus Q$ of two points P and Q is the reflection of the third intersection with $\mathcal{E}$ of the chord or tangent (if $P = Q$) through P and Q, for example, see [69], **III**, §2 for the explicit algebraic definition.

If $\mathbf{R} = \mathbb{F}_p$ is a finite field, by Deuring's theory [23], [34], there is a curve $\mathcal{E}_{\mathbb{C}}$ over $\mathbb{C}$ which has complex multiplication in a complex imaginary quadratic extension $\mathbb{K} = \mathbb{Q}[\sqrt{-d}]$ and good reduction at p, such that $\mathcal{E}_{\mathbb{C}}$ reduces to $\mathcal{E}$ modulo p. If p is inert in $\mathbb{K}$, then the number of points on $\mathcal{E}$ is $o(p) = p + 1$, otherwise, $p = \pi \cdot \bar{\pi}$ is its splitting in $\mathbb{K}$ and

$$\sharp\mathcal{E} = o(p) = p + 1 \pm \mathbf{Tr}(\pi); \tag{11}$$

one of the sign choices corresponds to $\mathcal{E}$, the other to its twist.

The curve structure may be defined in general over rings modulo integers $\mathbf{R} = \mathbb{Z}/(n \cdot \mathbb{Z})^*$ and has a reduction $\pi_{n,m}$ to curves over $\mathbb{Z}/(m \cdot \mathbb{Z})^*$. Define $f(x) = (\sqrt[4]{x} + 1)^2$, $\mathbf{A}(n) = \mathbb{Z}/(n \cdot \mathbb{Z}) \times \mathbb{Z}/(n \cdot \mathbb{Z})$ and $\mathbf{G}(n) = \mathcal{E} \subset \mathbf{A}(n)$ with the curve addition as a composition law. The premises of Theorem 2.5 are herewith fulfilled.

Let n be an integer, $\mathcal{E}$ be a curve with complex multiplication over a field $\mathbb{K}$ and suppose that n splits in two principal ideals in $\mathcal{O}(\mathbb{K})$, the ring of the integers in $\mathbb{K}$: $n = \nu \cdot \bar{\nu}$. The Fermat test (5) in this setting consists in checking that a random point $P \in \mathcal{E}$ verifies $[n + 1 \pm \mathbf{Tr}(\nu)]P = \mathcal{O}$. Here $\mathcal{O}$ is the neutral element of $\mathbf{G}(n)$ (point at infinity of the curve) and $[a]P$ denotes the a-fold repeated addition on the curve $(a \odot P$, in our general notation $)$.

In [18], the Chudnowski brothers use deeper properties of complex multiplication and develop a series of rich and beautiful elliptic curve analogies to Lucas-Lehmer tests for integers with special forms. Some of their results do not directly fit in the scheme of generalized Fermat tests (5), but can very naturally be understood as Fermat analogues. We avoid excess of generality and refer, rather, directly to this beautiful piece of applied classical mathematics.

In the same year, Goldwasser and Kilian [28] remarked that the Pocklington Lemma 2.5 can be simply applied, if $o(n)$ has some large pseudoprime factor. They thus were the first to show that the elliptic curve group structure may be used for a random polynomial general primality test, which will be considered in more detail in Sections §4, 5.

2.3. Certificates

In the process of proving primality of a given number n by using, for example, Theorem 2.2, a lot of computing time may be spent in trial and error steps, which help gathering the information necessary for the proof. For instance, sufficient prime factors of $n - 1$ or some other function $o(n)$ need to be computed, then some element generating a cycle with those primes as order must be sought, etc. The idea of a *certificate* for primality responds to the wish to resume the essential information gathered during a primality proving process in a short list, allowing to reproduce the proof efficiently.

We can use our setting for the general Pocklington lemma for defining a large class of certificates for primality. Suppose that a family of rings $\mathbf{A}(n)$ together with their group structures $(\mathbf{G}(n), \circ)$ has been chosen for proving the primality of an integer P. Assume that the function f verifies the additional condition:

$$\exists c \in \mathbb{R}_+, c < 1, \text{ such that } f(\lceil \sqrt{n} \rceil) < c \cdot n, \ \forall n > B, \tag{12}$$

where B is a positive fixed given bound.

The certificate will be defined by downwards recursion as follows (see also [11]):

T. A prime p is defined to be terminal, if $p < B$ for a given bound B which allows proving primality of p by trial division. Its certificate consists of $\{p\}$ itself.

R. Suppose n is an integer such that $F|o(n)$, $F > f(\lceil \sqrt{n} \rceil)$ and all the prime factors of $o(n)$ are known. Suppose furthermore, that the elements $\alpha_q, \forall q|F$ verifying (4) are determined. A certificate for n consists of the set:

$$\mathcal{C}(n) = \{F\} \cup \{(q, \alpha_q; \mathcal{C}(q)) : q|F, \ q \text{ is prime }\},$$

where $\mathcal{C}(q)$ are the primality certificates for q. Note that (q, α_q) are pairs for which (4) must be verified for the node n, whereas $\mathcal{C}(q)$ are pointers to certificates for primes $q < c \cdot n$.

This construction makes a certificate into a directed tree with root P and terminal leaves at primes $p < B$. We show that this definition allows to build a certificate in finitely many steps and the information it contains allows verification of a primality proof. For the construction we use induction by the size of P. If $P < B$, the certificates trivially exist by condition T. Suppose that all primes $p < c \cdot P$ have a certificate. By (12), all the factors of $F|o(P)$ will have a certificate, and R. shows that the certificate for P itself can thus be constructed.

The verification of a primality proof using the certificate tree defined above is simple and consists in verifying (4) for all pairs $(\alpha_q, q) \in \mathcal{C}(n)$ and for all non terminal nodes n in the certificate tree $\mathcal{C}(P)$. For terminal leaves, the proof is done by a trial division.

We shall call a certificate *linear*, if the associated certificate tree is such that each parent node is terminal or attached to exactly one child node. The case when the group $(\mathbf{G}, \circ)$ is simply the multiplicative group of $\mathbb{Z}/(P \cdot \mathbb{Z})^*$ was naturally the first to be considered – [58], [57]. In this case we have the estimate $f(x) = x$ for the subgroup sizes and the elements α_q are primitive q-th roots of unity modulo their node n. Note that Proposition 2.4 allows to chose $F > \sqrt[3]{n}$ in this case.

Goldwasser and Kilian proposed in [28] to use elliptic curves $\mathcal{E}$ such that

$$\#\mathcal{E} = f \cdot q \quad \text{for some pseudoprime } q, \tag{13}$$

with f a small factored part. Actually in [28] they proposed $f = 2$. Since $\mathcal{E}(n) < n \cdot \left(1 + \frac{2}{\sqrt{n}}\right)$, it follows that this condition fits (12) with the constant $1 > c >$

$\frac{1}{2} + \frac{1}{\sqrt{B}}$. The way of counting the number of points on $\mathcal{E}$ which they proposed was the recently discovered algorithm of Schoof [67]. This approach was later improved by use use of complex multiplication, for example, see [7]. It was the important contribution of [28] to prove that curves verifying (13) can be found in random polynomial time for all but an exponentially thin set of integers of fixed size. More precisely, for a fraction of at least $1 - \mathcal{O}\left(2^{\frac{-n}{\log\log n}}\right)$ of all the inputs of length $\log n$, an elliptic curve certificate can be found in polynomial time. The authors thus prove that *almost all numbers can be quickly certified*. It is interesting to note that the initial paper suggested instead of T. a terminal condition in which the numbers to be certified were reduced in size bellow some limit, for which cyclotomy proofs could be completed in polynomial time with respect to the initial input.

Several years later, Adleman and Huang [3] used the same kind of *relative termination condition* while trying to eliminate the residue set of numbers, for which certificates could not be found in random polynomial time in the [28] setting. For this they used the Jacobian of hyperelliptic curves for the groups $\mathbf{G}(n)$, and showed that curves of genus $g > 1$ such that $o(n) = f \cdot q$, with f some small factors and q a large pseudoprime, can *unconditionally* be found in random polynomial time. They thus achieved the goal of an unconditional random polynomial primality proof method. The trouble of this approach is that the convergence condition (12) is not verified, and indeed the terminal condition is rather surprising. The sizes of primes may actually grow, while descending in an Adleman-Huang certificate tree. It is shown that the process stops by eventually finding a node for which Goldwasser-Kilian certificates are proved to exist.

3. Pseudoprimes and Probabilistic Tests

We already used the term of *pseudo-prime* numbers, short "pseudoprimes", in the definition of certificates, without giving a strict definition. In the context of certificates, the term is intended as *numbers which are expected to be prime* after having failed to be found composite by several compositeness tests. Their actual primality is subject to recursive proof in the process of verification of a certificate. Since a certificate yields full primality proofs of each single node, the pseudoprime character of its nodes is only temporary and indicative for the process of *generation* of the certificate. At the time a certificate is completed, all nodes can actually be proved to be primes.

In practice, one may quickly verify some of the many *necessary* conditions for primality listed in the previous section and would like to know more about the odds of a composite number n for passing these verification. It is in this context that pseudoprimes arise, as integers verifying some fixed necessary conditions for primality. Of course, pseudoprimes are related to the special conditions they verify, and the terminology tends to be more and more complex, while additional tests and combinations are taken into account. We refer to [31] for an attempt to order known notions of pseudoprimality, while we shall stick here, to the explicit statement of the properties which define a given number as pseudoprime.

The study of pseudoprime numbers aims to produce good estimations of the repetition of composites which verify given property that primes have. If this goal can be reached, bounds for the probability that a given composite number verifies one or more primality conditions can be deduced. This leads to the probabilistic primality tests.

In many cases however, all one knows is the fact that no proof has been found yet, that given set of conditions would imply primality. In such cases, not even existence of pseudoprimes is granted, although they can very well be defined. We shall start with a simple example of this kind. Let $n \equiv 2$ or 3 mod 5 be an integer and V_m be the Fibonacci sequence, that is, the sequence (7) with respect to the polynomial $g(X) = X^2 - X - 1$. Such a number n is a PSW (Pomerance, Selfridge, Wagstaff) pseudoprime, if it passes a Fermat test for base 2 and a generalized Fermat test (8) for the sequence V_m. There is a \$ 620 prize offered by the three initiators of the challenge for either an instance of such a pseudoprime or the proof that none exists. The expectations of the proposers as to which issue is more likely, appear to be split. See [29] for interesting details about this challenge. A similar test, in which $g(X)$ is a polynomial of third degree, was proposed by A.O.L. Atkin, [6], and is considered by him as infallible for *practical purposes*, that is, for eliminating composites in the range of magnitude which can be covered by present time computing power. Of course, no proof is known, whether pseudoprimes for the Atkin test exist or how large they may be.

We now return to better understood pseudoprimes. Carmichael investigated at the beginning of the century [17] composites which pass the simple Fermat test $a^{n-1} \equiv 1$ mod n, $\forall(a,n) = 1$; such integers bear his name. For instance 1729, the number on Hardy's cab, which was recognized by Ramanujan as being expressible as sum of two cubes in two different ways, is also a Carmichael number. It is interesting that both Ramanujan and Hardy do not mention the additional property of 1729 of being Carmichael. It is known that they are squarefree and built up of at least 3 prime factors. Both very large and very highly composite Carmichael numbers have been produced [29], since their definition. The fact that there exists an infinity of such pseudoprimes has been expected, but was only proved in 1994 by Alford, Granville and Pomerance, [2]. Grantham [30] generalized the methods used for proving existence of an infinity of Carmichael numbers with respect to some generalized Fermat tests (5) for some polynomials $g(X)$ of degree 3 (Perrin sequences).

By definition, Carmichael numbers pass the Fermat test for a fraction of $\delta_C = \frac{\varphi(n)}{n-1}$ out of the possible bases a. In general, the bases for which a composite verifies some primality condition are called *false witnesses* and δ will thus be their density bound. In 1977, Solovay and Strassen [70] set a landmark in primality testing by proving that if the Fermat test was replaced by (2), the density bound was provably $\delta_{SS} < \frac{1}{2}$. This allowed to use probabilistic considerations and argue that an integer which passes the test (2) for k independent random bases a is composite with probability $< \left(\frac{1}{2}\right)^k$. This is the beginning of probabilistic primality testing: one knows to measure the likelihood of making a false assumption of primality, after performing several simple and efficient compositeness tests. The request for a proof appears to be less stringent for practical purposes. At the same time, Rabin showed in [64] that the known condition of Theorem 2.1 had the better density bound $\delta_R = \frac{1}{4}$. It became current to verify that condition, now called strong pseudoprime test or Miller-Rabin test, for probabilistic verifications of primality.

As the instance of PSW pseudoprimes showed, little is known in the simplest cases about numbers passing *different* pseudoprime tests. Density bounds for false witnesses are thus not easy. Recently, Grantham [31, 32] proposed to add a condition on the Frobenius signature to Fermat tests in quadratic extensions and succeeded to actually prove a density bound $\delta_{Fr} = \frac{1}{7710}$ for the resulting algorithm. This suggests that the probability of failure granted by more than 6 strong pseudoprime tests may be achieved by one single Frobenius test, which is evaluated by the author to require 3 selfridges. We shall see however in the context of search methods, that much better average probability bounds are obtained for the strong pseudoprime test, when the whole process of generation of a prime of given length is considered. These bounds make the simple test (2.1) practically sufficient, if probabilistic tests are to be used.

We close the discussion on pseudoprimes by mentioning a type which has not been investigated and is of some theoretical relevance. Let s be a highly factored integer, so that $s > e^{q/\log q}$, $\forall q | s$, q prime. We define n to be an s-cyclotomic pseudoprime, if $n < s^2$ and

$$\forall r | n, \ \exists i(r) : r \equiv n^{i(r)} \mod s.$$

Without the condition $n < s^2$, s - cyclotomic pseudoprimes are easy to find for small s: for example, 91 is cyclotomic pseudoprime for $s = 3$ but not for $s = 9$. The question is theoretically important since, if it could be proved that no s-cyclotomic pseudoprimes exist beyond some bound on s, the cyclotomy primality proof could be completed in polynomial time, provided some precomputed tables.

4. Base Choice and Prime Search Strategies; Generation of Certifiable Primes

We have mentioned that the initial research on probabilistic primality tests assumes uniform random distributed bases for their tests. It was also assumed that the prime candidates one chooses in the process of generation are uniform distributed. We shall review here some alternatives to both choices.

If one considers random bits as not being for free, like some authors consequently did – E. Bach, V. Shoup, etc. – it is natural to use a sequence of consecutive integers as bases for strong pseudoprime tests, thus choosing only the initial value

k / t	1	2	3	4	5	6	7	8	9	10
100	5	14	20	25	29	33	36	39	41	44
150	8	20	28	34	39	43	47	51	54	57
200	11	25	34	41	47	52	57	61	65	69
250	14	29	39	47	54	60	65	70	75	79
300	19	33	44	53	60	67	73	78	83	88
350	28	38	48	58	66	73	80	86	91	97
400	37	46	55	63	72	80	87	93	99	105
450	46	54	62	70	78	85	93	100	106	112
500	56	63	70	78	85	92	99	106	113	119
550	65	72	79	86	93	100	107	113	119	126
600	75	82	88	95	102	108	115	121	127	133

TABLE 1. Lower bounds for $p_{k,t}$: from [24]

randomly. The probability of failure of the test, when bases are chosen this way, was analysed by Bach [8] and improved by Peralta and Shoup [62]. They show that on an average of $\log n$ bases, the error probability for the strong pseudoprime test is in this case $< 1/2$. The Solovay-Strassen case is also analysed, but the results there are weaker and depend also upon the number of factors of n.

The evaluation of $\delta_R = \frac{1}{4}$ made by Rabin is an absolute worst case bound. The question what error probability one should realistically expect when running t strong pseudoprime tests with random distributed bases on random candidates of fixed length $k = \log n$ takes into account the scarce distribution of the worst cases. The probability $p_{k,t}$ that this process outputs a composite number is evaluated in [24] and explicit bounds are found, some instances of which are given in the table below. The paper gives also good bounds for the distribution of the worst case candidates, for which the number of false witnesses approaches the Rabin estimate $\delta_R = 1/4$.

These results will be used in the final section, when discussing the confidence of different prime number generation strategies.

We now consider the process of generation of prime numbers. Given the size m in bits of the required prime, the process may be split in general in the following three stages:

S. Generate prime candidates n with m bits, by some given search method.
D. For all small primes $p < B$ for a given fixed bound B, check by trial division whether $p|n$. If this is the case go to S.
PT. Build the sequence (1) for a strong pseudoprime test on some randomly or sequentially chosen base a. If n fails this test, go to S. Repeat PT k

times, where the constant k is chosen in agreement with Table 1 and the wished bound of error. If n passes these tests, output n.

Obviously the most important amount of time will be spent in step PT., where a composite candidate is rejected at the cost of at least 1 selfridge. Since the probability that a random number of length m is prime is $\mathcal{O}(1/m)$, the complexity of a prime number generation process is well approximated by $\mathcal{O}(\log n)$ selfridges. Possible improvements are in the constants. The number of composites n which have no prime factor $q < B$ is evaluated by Mertens' theorem:

$$\sharp\{n : \log_2 n = m, q|n \Rightarrow q > B\} \sim \frac{e^{-\gamma}}{\log B} \cdot n.$$

This dependence allows good estimates of the number of exponentiations which are saved by trial division. Maurer suggested in [42] for the first time to use this fact for rigorous optimization of the trial division size B. The idea was adopted in later research and it is advisable for practical implementation to use some empirical values of the average times t_d and t_e for a trial division and an exponentiation, respectively. This leads to the cost function:

$$c(B) = \left(\frac{B}{\log B} \cdot t_d + \frac{e^{-\gamma}}{\log B} \cdot t_e\right) \cdot \log n.$$

The optimal trial division size is obtained by a simple extremum computation.

Like for the bases for probabilistic tests, the candidates in step S. were assumed to be produced by a uniform random distributed source. The search of primes in incremental search was first investigated in [12]. The advantages are here not only in saving random bits. The trial division may be far improved by the incremental search, since a complete trial division must only be performed for the starting candidate n_0. The remainders $r_p \equiv n_0 \mod p$, $\forall p < B$ are kept in a table, $n_i \equiv n_0 + 2i \equiv r_p + 2i \mod p$. The trial division for the further candidates amounts to mere addition of the increment 2. The procedure is equivalent to an Eratosthenes sieve and the term $\frac{B}{\log B}$ in the cost function is reduced by a factor $\frac{1}{\log n}$.

The drawback of incremental search is a bias in the distribution of the generated primes. Indeed, if one considers the start point n_0 to be randomly distributed, such that $p_- \leq n_0 \leq p_+$, with p_-, p_+ being its next primes, it will be more likely that a random choice of n_0 hits between two primes separated by a longer gap. The probability $P = \frac{p_+ - p_-}{2^m}$ of producing the prime p_+ by this search method is thus proportional to the length of the gap which separates it from its predecessor. Brandt and Damgård [12] give a beautiful proof that the gaps are Poisson distributed and the expected value $E(p_+ - p_-) = 2\log n$, being thus twice the average gap length between uniform distributed primes. They finally prove that the entropy of the primes produced is asymptotically equal to the entropy of uniform distributed primes.

Although probabilistic primality tests yield primes with high confidence, in the process of prime number generation one has enough degrees of freedom for

producing the required information even for certifiable primes. The simplest way
to do this consists in searching the primes p in sequences of candidates $n \equiv 1$
mod q, where q is a prime such that $q > \sqrt[3]{n}$. The primes q can be produced in
the same way, recursively, and a linear certificate of primality using Proposition
2.4 results. It is likely that this method has been known for a long time. It was
explicitly implemented and described in an internal research paper [20] in 1982,
and published for internal use of Philips Research, [63]. For reasons identical to the
ones discussed for incremental search, certifiable primes produced by this recursion
have a distribution bias, which was not analysed in [20].

Several years later, Maurer considered the problem of generating certifiable
primes under the assumption [43] that any deviation from uniform distribution of
the produced primes may generate unpredictable security leaks, when using the
method for cryptographic purposes. Maurer's method [42] thus produces primes
p with a complete, recursively defined, factorization of $p - 1$. The simulation of
uniform distribution is accomplished by means of some estimates of Trabb-Pardo
and Knuth for the average relative sizes of the three largest factors of a random
integer. Using a sampling algorithm by Bach [10], a factored part $F > \sqrt{p}$ is
recursively built, such that the relative sizes of the prime factors of $F|(p - 1)$
match the estimates [35]. This estimates being applied to $p - 1$, some additional
heuristics [44] are used for showing that the factorization of $p - 1$ is "close" to
the one of a random integer. The approach may be compared to the procedure for
generating certificates defined in the precedent section. Here, for each recursive
call requiring a prime p of m bits, the following steps are necessary:

1. Set up a list of sizes $m_1 > m_2 > \cdots$ of prime factors q_i, such that $p - 1 = F \cdot R = R \cdot q_1 \cdot q_2 \cdots$ is a factorization for which the relative sizes of the largest factors are distributed according to the distribution for random integers.

2. Make a recursive call to the procedure for all sizes m_i, returning the primes q_i, and build p.

3. For each $q_i | p - 1$ find a base a_i such that $a_i^{\frac{p-1}{q_i}}$ is a primitive q_i−th root of unity modulo p.

Maurer's algorithm provides an elegant solution for simulating uniform distribution when generating cryptographic primes. It has however the disadvantage
of losing performance compared to the simple strong pseudoprime tests, due to
extended recursive calls necessary for the simulation. In fact, the largest prime
factor $q_M | n$ of a composite n has, by [35], the expected size $\frac{\log q_M}{\log n} \sim 0.71$, so the
recursion is relatively slow, the performance detriment with respect to strong pseudoprime tests being around 60%, [42]. An earlier definition of the general method of
Maurer, which does not address however this author's central problem of uniform
distribution, nor run-time and implementation details, is found in Plaisted [55].

The question, whether the natural and simple approach described in [20] is
not more recommendable for practical use, still had to be addressed. The procedure of building certifiable primes by search in arithmetic progressions, that is,

by searching primes in sequences $n_i = 2 \cdot (t_0 + i) \cdot q + 1$, where $q > \sqrt[3]{n_i}$ is a recursively computed prime and t_0 is a random start point of adapted size, was analysed in [45]. It is shown that the method is superior to generation of probabilistically proved primes both from the point of view of performance and due to the property of producing a certificate of the generated primes. For the first, the possibility of using Eratosthenes sieves for incremental prime search is evaluated for the first time. Compared to Maurer's algorithm, further performance gain stems from the fast recursion. The recursion step requires $\sim 2\%$ of the overall run time for one prime generation. Of course, the possibility of proving primality by one single application of Pocklingotn's lemma reduces the time required for the actual proof to 1 selfridge. The sum of the time spent for the final proof plus the recursion time is still inferior to the time required by the minimal number of exponentiations one should iterate for sufficient confidence, when using probabilistic tests. Concerning the distribution bias of the primes generated in arithmetic progressions, it is shown in [46], in a similar way with [12], that the entropy of primes generated with this source is asymptotically equal to the one of random distributed primes. This, together with a new argument which will be exposed in Section 6, shows that the assumption [43] does not hold in this case. An example of a large (50000 bit) prime built by this method may be found at http://www.utm.edu/research/primes/lists/single_primes/50005bit.html.

5. Implemented General Primality Proofs

General primality proving algorithms can produce a proof of primality for *any* integers $n > 2$. This means in particular that no additional information about n or some function $o(n)$ is required, as is the case when applying variants of Theorem (2.5). Two families of algorithms based on different ideas have been studied in more depth and are implemented at this time.

The elliptic curve primality proving method (ECPP) was first proposed in [28]. The algorithm was implemented by Morain, based upon Atkin's idea of using complex multiplication for determining the number of points of elliptic curves, [7]. If there is a curve $\mathcal{E}_{\mathbb{C}}$ with complex multiplication in a quadratic field $\mathbb{K} = \mathbb{Q}[\sqrt{-d}]$ which has good reduction at a prime p, then the Hilbert class polynomial $h_{\mathbb{K}}(X)$ must split completely over $\mathbb{F}_p$. Furthermore, if $H_{\mathbb{K}}(\jmath) \equiv 0 \mod p$, an elliptic curve having $\jmath$ as invariant can be easily constructed and its number of points is given by (11): this explains Atkin's idea. By using this approach, the number of points on $\mathcal{E}$ can actually be determined directly by the choice of $\mathbb{K}$, before even computing the parameters of $\mathcal{E}$. This is particularly efficient in the ECPP context, where one requires curves such that (13) holds. Fields $\mathbb{K}$ for which this condition is not met will be discarded before doing all the computations required for determining $\mathcal{E}$. There is a drawback, however: the degree $\partial H_{\mathbb{K}}(X) = \sharp \mathcal{C}(\mathbb{K})$ is equal to the class number of $\mathbb{K}$. For polynomial time computations, one must thus restrict to the thin subset of imaginary quadratic extensions with polynomially bounded

P. Mihăilescu

(with respect to the length $\log n$ of the input) class number. In practice, the fields together with their discriminants, class numbers and Hilbert polynomials (or some related, simpler modular function) are tabulated. The practical experience did not bring up cases in which (13) could not be verified for at least one of the tabulated fields. Note however that this empirical fact is not granted by the proof in [28], the choice of curves among which the random choices are done being far more restricted than in Goldwasser and Kilian theorem.

The implementation [25] has been largely used for primality proving and been improved over the last decade. Its current complexity is $\mathcal{O}(\log^6 n)$ and the linear certificates may be verified in $\mathcal{O}(\log^4 n)$ operations. We refer to [52] for an update of the latest improvements. From this paper we gather the following timing for ECPP and the verification of the certificates it produces. The data are in seconds, measured on a DEC Alpha 125 MHz machine.

p	build	proof	check
$10^{499} + 153$	29559	298	156
$10^{599} + 2161$	46287	852	253
$10^{699} + 1279$	68758	4055	580

b	min	max	mean	s.d.
128	0.08	0.82	0.26	0.15
	0.05	0.48	0.17	0.09
	0.15	1.02	0.43	0.21
	7.00	14.00	9.82	1.88
	0.10	0.35	0.20	0.05
256	1.17	6.38	2.95	1.42
	0.38	2.10	0.89	0.31
	1.70	7.20	3.84	1.61
	9.00	23.00	16.54	2.75
	0.45	0.95	0.72	0.12
512	12.40	45.02	25.63	8.43
	5.75	27.00	10.18	4.16
	18.17	68.53	35.81	11.31
	24.00	39.00	31.08	3.56
	3.73	6.65	4.99	0.66

Efficient cyclotomy proofs are possible based upon the following theorem from analytic number theory:

Theorem 5.1 (Prachar, Odlyzko, Pomerance). *There exists a positive, effectively computable constant c such that $\forall\, n > e^e$, $\exists t > 0$ satisfying*

$$ t \;<\; (\log n)^{c \cdot \log\log\log n} \quad and \quad s^2 = \left(\prod_{\{q \in \mathcal{P},\, (q-1)|t\}} q \right)^2 > n. \tag{14} $$

With $f(n) = (\log n)^{c \cdot \log\log\log n}$, this shows that a factor $F | n^{f(n)} - 1$, $F > \sqrt{n}$ can be found unconditionally. Theorem 2.7 suggests that cyclotomy proofs are related to the factoring of cyclotomic polynomials $\Phi_s(X) \mod n$. Concepts for unifying Jacobi-sum tests and generalized Lucas-Lehmer were developed by H. W.

Lenstra, Jr. [36], [37] and first adapted for implementation by Bosma and van der Hulst, [15]. The relation of this unified cyclotomy test to the factoring of cyclotomic polynomials was put in evidence in [46], [47]. We shall give here a brief description of the ideas, overlooking variations which allow for computational improvements.

If n is an integer to be proved prime and s, t are defined as in (14), Theorem 5.1 implicitly defines a list of pairs,

$$\mathcal{P}(n) = \{\wp = (p^k, q) : \forall p^k | (q-1), \forall q | s\}.$$

Jacobi-sum tests are performed in some extension $\mathbf{T} \supset \mathbb{Z}/(n \cdot \mathbb{Z})$ which contains p^k-th roots of unity ζ_{p^k}, $\forall p^k | t$. For each pair $\wp \in \mathcal{P}$, elements $\alpha_\wp \in \mathbf{T}$ are computed by using Jacobi sums. They must verify:

$$\alpha_\wp^{\frac{n^{\varphi(p^k)}-1}{p^k}} \in < \zeta_{p^k} >, \tag{15}$$

which is a condition similar to (5). Explicit higher reciprocity laws are implied by (15). It can be shown that if the conditions are fulfilled for all $\wp \in \mathcal{P}$ and we set $\mathbf{A}(n) = \mathbf{T}$, then an element $\alpha \in \mathbf{A}(n)$ together with a polynomial $\Psi_s(X) \in (\mathbb{Z}/(n \cdot \mathbb{Z}))[X]$ can be deterministically constructed, so that the conditions of Theorem 2.7 hold. The Jacobi-sum test may thus be understood as a set of necessary conditions in order to complete in a deterministic way the construction of a factor $\Psi_s(X) | \Phi_s(X) \mod n$ with degree $\partial \Psi_s(X) = \mathrm{ord}_s(n)$. Of course, for smaller values of t, elements of given order $s_l | s$ can be sought by trial and error using (2.5) in some extension of degree t. In this case the generalized Lucas-Lehmer variant of cyclotomy is used. It can be shown that partial proofs by the two variants can be combined without additional conditions. This is particularly practical for cases when some large factored part $F | \Phi_k(n)$ is known for a small value of k, yet F is not sufficiently large in order to complete a generalized Lucas-Lehmer test. In such cases, the Jacobi-sum test provides for the missing information. The largest current proof of a general prime $P = \frac{2^{11279}+1}{3}$, [50] was completed this way.

It is assumed that a cyclotomy proof can use large tables of precomputed Jacobi-sums. Note that in order to complete a primality proof using Theorem 2.7, after having proved (8), one must verify that none of the minimal positive remainders $r_i \equiv n^i \mod s$ are divisors $r_i | n$. Since by Theorem 5.1 the remainders are in superpolynomial number $f(n)$, the complete proof is not polynomial. Assuming the Jacobi-sum tables, the proof of (8) may be completed in polynomial time $\mathcal{O}\left((\log n)^{4+\epsilon}\right)$. Furthermore, the exponent of $\log n$ in $f(n)$ being $\log \log \log n$, it is a very slowly diverging function and the overall complexity of a cyclotomy proof is bounded by $\mathcal{O}\left((\log n)^5\right)$ for numbers with less than 10^8 decimal digits. This makes cyclotomy into the practically best performing general method for primality proving, although it is not asymptotically polynomial. From the initial version [5] of the Jacobi-sum test to the current cyclotomy test, this family of algorithms has been repeatedly implemented, see [19], [15], [47]. Beside its practical efficiency it is also the only general primality proving method which has a *deterministic* variant.

nDec. Digits	Av. Time (sec)	Max Time (sec)	Min Time (sec)	Mod.Exp (sec)	Max Ext Expo: (sec)	Ratio	Max Deg
20	0.30	1.05	0.15	0.00	0.04	n.a.	3
50	0.61	0.91	0.36	0.01	0.15	60?	2
100	3.97	7.18	1.49	0.16	1.36	25	5
120	5.63	7.38	2.37	0.20	1.99	28	6
140	9.31	16.81	4.61	0.25	2.79	37	5
160	12.63	24.98	7.08	0.32	3.56	40	3
180	17.72	31.38	4.67	0.38	4.67	47	6
200	23.83	41.40	13.70	0.42	4.67	57	6
300	77.59	117.97	45.76	0.54	12.13	144	4
500	730.01	965.97	573.87	3.74	20.87	195	12
600	1100.76	1794.03	569.34	4.54	32.93	244	12
700	1890.92	3201.22	1087.71	14.02	48.25	135	9
800	3068.57	5117.85	1878.57	19.15	68.32	160	8
1000	8321.05	11216.48	5045.94	23.61	119.12	353	6

TABLE 2. Performance **CYCLOPROVE**

A drawback of cyclotomy was until recently the absence of some efficiently verifiable certificate. Still assuming Jacobi-sum tables and not considering the trial division verification of (9) as certifiable, it has been shown that the proof of (8) allows certificates which may be verified in $\mathcal{O}\left((\log n)^4\right)$ operations, [51].

It is a problem of current research, whether the use of Shimura reciprocity can – in a certain analogy to the Jacobi-sum test – improve the use that elliptic curve tests can make of intermediate information from tests on different curves. The existence of a polynomial time deterministic general primality proving method is an open question.

We give in Table 2 some performance data as stated in [47]. The times are all measured on a DEC Alpha 500 MHz workstation.

Considering the sizes of primes with cryptographical relevance to vary currently between 512 and 2048 bits, the table shows that these can be proved – also a posteriori – within seconds up to less than an hour, on a modern desktop workstation. Even the proof of 4096 bit primes – a still very uncommon size for cryptographic applications – can be completed within a few hours.

6. Security Aspects

Public key cryptosystems which use prime numbers rely their security on the generally assumed intractability of one of the following number theoretical problems:

IF The problem of factoring rational integers.

DL The discrete logarithm in the multiplicative group of a finite field.

EL The discrete logarithm in the group structure of an elliptic curve over a finite field.

The last two decades brought important developments of the known solutions for the first two problems. With the number field sieve, see [40] for example, both problems can be solved for an input n in random $\mathcal{O}\left(\exp\left(c\sqrt[3]{\log n \cdot (\log\log n)^2}\right)\right)$ time. The notion of *smoothness* of an integer is fundamental to most solutions of the two problems. An integer n is y-smooth if all of its prime factors are below y:

$$n \text{ is } y \text{ - smooth iff } \quad q|n \Rightarrow q < y.$$

Early subexponential factorization algorithms were based upon the odds that $p \pm 1$ is smooth. In order to avoid the generation of vulnerable keys for the RSA algorithm, Gordon proposed in [27] to impose for cryptographic primes that $p \pm 1$ be divisible by some large prime q. Primes respecting this constraint are called *Gordon strong primes*. They are still required by different cryptographic standards. However, the elliptic curve factoring method discovered in 1986 by H. W. Lenstra, Jr. [39] makes this caution superfluous. This algorithm takes advantage of any elliptic curve $\mathcal{E}$ mod n for which the number of points $\sharp\mathcal{E}$ mod p is smooth, for some prime $p|n$. Since the number of points is randomly distributed in an interval, when curve parameters are random, it is not feasible to generate primes which make elliptic curve factoring harder then average. The elliptic curve factoring method makes thus the Gordon condition irrelevant for the factoring problem. Note that using Gordon primes, which are known to be strong *at least* against Pollard $p - 1$ factoring, is surely not a disadvantage. The probability that $p - 1$ be smooth being far lower then the probability then the one that some random elliptic curve has smooth order, the *security gain* is however unconvincing. These facts are commented in depth in [66], showing " ... that, contrary to common belief, it is unnecessary to use strong primes in the RSA cryptosystem".

For the DL problem, some larger factor of $p - 1$ is still necessary in order to avoid Chinese remainder theorem attacks (Pohlig-Hellman, [54]). Some schemes require such factors explicitly, for example, DSA.

6.1. Security of Primes with Biased Distribution

We mentioned the argument according to which biased prime generation sources may create a security breach for cryptographic schemes. Such a breach would allow yet undiscovered solutions to the relevant number theoretical problems IF and DL to take advantage of the given biases and reduce in this way the time for breaking a scheme, under the presence of biases.

On the other hand, mathematical uniform distribution is hard to conceive in practice and some error of the random sources should be realistically accepted as part of the analysis. Also, as Bach argues [8], random numbers are not for free. Finally, as described in previous sections, some sources with low bias may have other interesting properties, which one would want to take advantage of.

The conception that "bias equals potential insecurity" does not allow a differentiated balance of these two aspects. It is important to have some measurable conditions in order to evaluate, how much bias can be allowed to a source without the risk of a more efficient attack then those against uniform distributed sources. One approach for answering this question consists in comparing the entropy of a biased source to the one of uniform distributed ones, based on the idea that no risk is taken, if the entropies are close or asymptotically equal, [12].

We shall describe here the ideas of [49], where some classes of biases are defined. It can be proved in that context that there is no possible improved strategy against certain "well behaved biases", which does not lead to improvements of the general, uniform distribution based strategy.

Consider for simplicity the problem of generating an m-bit prime, that is,

$$p \in I_m = \{n \in (2^{m-1}, 2^m) : n \text{ prime}\}.$$

The set I_m has $|I_m| \sim \cdot \frac{2^{m-1}}{m \log 2}$ elements and a uniform distributed source $\mathcal{S}$ will produce each one with equal probability $P_u = \ell = 1/|I_m|$. A biased source $\mathcal{S}_b$ on the contrary, will produce each prime $p \in I_m$ with probability

$$P_b(p) = \lambda(p) \cdot \ell, \quad 0 \le \lambda \le |I_m|.$$

This defines the bias as a function $\lambda : I_m \to [0, |I_m|]$. Let $c \in \lambda(I_m)$ be some value in the image of λ. We define the following sets and measures, which quantify the bias:

$$
\begin{aligned}
a(c) &= \{p \in I_m : P_b(p) < c\ell\}, & m(c) &= |a(c)| \\
A(c) &= \{p \in I_m : P_b(p) \ge c\ell\}, & M(c) &= |A(c)|
\end{aligned}
\tag{16}
$$

In order to model unknown algorithms for breaking DL or IF, we shall assume that they have individual strategies for every single prime or prime pair and that out of these strategies one can compute the average run time $\Sigma_b(c)$ for any bias $c \in \lambda(I_m)$. We denote both the strategies as algorithms and their average run time by Σ. The run time for the state of the art algorithm for uniform distribution will be Σ_u and it is assumed that this will always be preferred, when no better solution is known for some prime p. Thus:

$$\Sigma_b(c) \le \Sigma_u, \quad \forall c \in \lambda(I_m).$$

Note that an adapted strategy must take advantage of the large biases. An improvement for the primes produced by $\mathcal{S}_b$ with higher probability is the most likely to produce an overall improvement for any output of the source. Of course an adapted strategy is most efficient if it knows the exact bias of the primes involved in any instance of a DL or IF problem which it tries to break. This condition is possible but not necessary in our analysis. A gain of Σ_b over Σ_u which is polynomial in m should be accepted as irrelevant, in consistency with the general complexity theory applied in security analysis. We may thus make the following:

Definition 6.1. *An order $\preceq$ on the strategies Σ for solving either IF pr DL is defined by:*

$$\Sigma_b \preceq \Sigma'_b \Leftrightarrow \sum_{c \in \lambda(I_m)} < g(1/m) \cdot \sum_{c \in \lambda(I_m)} \Sigma'_b(c), \tag{17}$$

where $g(x) \in \mathbb{Q}[x]$ is a polynomial which does not depend upon m. The equivalence relation $\Sigma_b \sim \Sigma'_b$ is defined naturally by $\Sigma_b \preceq \Sigma'_b$ and $\Sigma'_b \preceq \Sigma_b$. A biased source S_b has **tolerable bias**, *if for each strategy Σ_b for the biased source there is a strategy Σ'_u with uniform distribution, such that $\Sigma'_u \preceq \Sigma_b$.*

This definition is meaningful, since it ascertains, that for a tolerable bias, any gaining strategy Σ_b can be converted into a uniform distributed strategy Σ'_u with comparable runtime (in the sense of complexity theory), thus improving upon the state of the art Σ_u. It is natural to consider in such cases that the gain is not due to the bias, but to some real improvement of the state of the art.

We are now able to introduce a class of biases which can be proved to be tolerable with respect to Definition 6.1.

Definition 6.2. *Let S_b be a biased source and $A(c) = \{p \in I_P : P_b(p) \geq \ell \cdot c\}$. If there is a polynomial $g(x) \in \mathbb{Q}[x]$ such that*

$$M(g(m)) < g(m),$$

the distribution has a **polynomially bounded bias**.

With this definitions, one proves [49] the following:

Theorem 6.3. *Polynomially bounded biases are* tolerable *in the sense of definition 6.1.*

The proof uses simple measure theory and shows how to convert biased strategies Σ_b into equivalent uniform distributed ones, if the bias is polynomially bounded.

Since the sources of primes in arithmetic progression or incremental search are Poisson distributed, it can be easily shown that they have polynomially bounded bias. The above theorem shows basically that a bias which is centered around polynomially bounded values is tolerable. Large values should be in excess in order to break the limits of tolerance as defined here.

A typical example of a source which has no tolerable bias is the following. Let p be a given prime and consider sources S which produce elliptic curves $\mathcal{E}(a, b)$ mod p, together with $N = \sharp\mathcal{E}$. The uniform distributed source will produce uniform distributed parameters a, b and then compute N with some recent variant of the Schoof algorithm, for example, see [68]. The CM source S_{CM} chooses a complex multiplication field $\mathbb{K}$ from a list of fields with polynomial discriminant and easily finds $N = p + 1 \pm \mathbf{Tr}(\pi)$, provided that $p = \pi \cdot \overline{\pi}$ splits in $\mathbb{K}$. Since the curves produced by $\mathcal{S}_{CM}$ are an exponentially thin subset of all the elliptic curves over $\mathbb{F}_p$, this source is a typical example of a massively biased source. Although no EL algorithm dedicated for curves with CM in fields with small discriminant is known or made plausible, the source $\mathcal{S}_{CM}$ is obviously incompatible with our general notion of tolerable biases.

7. Conclusions and Practical Comments

We have reviewed a series of topics, both of theoretical and of practical interest, connected to the generation and testing of primes. We shall summarize here some simple conclusions for practical use.

The choices to be made when selecting a prime number generation scheme concern:

1. Confidence of the primality testing algorithm chosen.
2. Performance.
3. Search strategies.
4. Possible additional requirements.

The **confidence** of probabilistic tests is highest when evaluated by using the worst case bounds δ_R, δ_F. With this, an error probability $p < 2^{-64}$ requires as much as 32 selfridges, when using the strong pseudoprime test and roughly the half, with the Frobenius test. In both cases, this overhead is often larger then the the amount of computation required by the whole prime search. The striking results of Table 1 show that a good cryptographic error bound of, say, $p < 2^{-64}$, can *in average* be achieved by very few – between 2 and 6 – strong pseudoprime tests. This is a judicious indicator of the number of tests that should be performed for a chosen average confidence. Given the simplicity and performance of the test 2.1, it is thus not very likely that other tests should be used with the probabilistic approach for generation of cryptographic primes. While the results of [24] are best indicators for the realistic power of pseudoprime tests, it should also be noted that the evaluations are made under assumption of uniform distributed search and choice of witnesses. Rigorously, some additional research would be required in order to use pseudo prime tests together with such practical "nice-to-have" as incremental search [12] and sequential witnesses [62].

The confidence is naturally highest for certifiable primes – in that case the generated numbers have an actual proof of primality. Also, search in arithmetical progressions is a natural ingredient and performance is best among all strategies for generation of primes. This fact is noteworthy, since the conception that a greater amount of information (a certificate versus a probabilistic statement) should come for more amount of computation spent, is sometimes encountered. A possible explanation of this apparent paradox may be the fact that the expected additional work is in the proofs of correctness and security of the schemes and not in the computation at run time.

Search /	uniform	incremental
Test	distrib.	arith. prog.
strong	3c ; 3p	4c ; 2p
psp.-test	[24]	[12]
certi-	1c ; 4p	1c ; 1p
fiable	[42]	[45]

TABLE 3. Overview of prime generation strategies, their performance and confidence

In terms of **performance**, the run time of the most extreme combinations (except for general primality proving algorithms) of prime generation schemes, when measured with the same base arithmetic, only *differs by a small factor, or even percentages*. In terms of complexity, the run time is naively $\mathcal{O}(\log n)^4$), or $\mathcal{O}(\log n)$ selfridges in a measure which is independent of the implementation of long modular multiplication. The time required for the generation of a prime of cryptographic size on a modern workstation is in the order of milliseconds, while a general primality proof for the same sizes is carried out in the order of seconds or minutes.

The **search** of primes in incremental sequences or general arithmetic progressions has, beside the lower use of random bytes, the advantage of reducing the time for trial divisions. In the later case, provable primes are produced and the performance is best, due to the savings both in the trial division step as in the final proof, where only 1 selfridge is required for the proof – which is less then one needs for a confident probabilistic test. The depth of recursion is negligible ($< 2\%$) when primes with linear certificates are produced, whereas in the uniform distributed case, it may account for a performance loss of $\sim 60\%$ ([42]). See also [13] for a list of practical implementation tricks, concerning also the arithmetic.

The following table gives an overview of some outstanding combinations of prime generation strategies. They are listed with one paper in which they have been investigated and are given some relative ranks based on the criteria discussed above. The ranks for performance are 1p (best) to 4p and those for confidence, 1c–4c. While the performance ranks are obvious and based both on analysis and practical experience, the confidence ranks require some remarks. Certifiable methods have total confidence independently of the source distribution, and thus have equal rank. Although the evaluations of $p_{k,t}$ in Table 1 are made for uniform distributed search, the methods suggest that the probability will increase in presence of incremental search. This explains the rank difference of the two methods.

In some cases, the primes required may need to fulfill **additional requirements**, such as having a fixed size prime $q|(p-1)$ or being a Sophie Germain prime (that is, $\frac{p-1}{2}$ is prime too), etc. In [71], Vanstone and Zuccherato study the requirement to produce two primes p, q such that their product $n = p \cdot q$ has an amount of predefined bits, while not compromising the security of the product against factoring. This allows fitting some identification data directly into the public keys n, thus saving bandwidth or data space in settings where this is important. After some simple arithmetic considerations, the requirement is reduced to producing some primes in short intervals – compared to the size of the primes themselves. This can be achieved both with certifiable primes and with probabilistic tests.

8. Conclusions

We have reviewed a subject about which very much is known. Without hoping to achieve an exhaustive overview, we intended to structure the major topics of research by the common ideas and point out some open questions which deserve investigation. For the practice oriented implementer we resumed in the last section the various possible combinations of prime generation strategies according to few criteria. Despite of some expectation that certifiable primes should be harder to obtain, it turns out that they are recommendable both from the point of view of confidence and of performance. We also presented a new model allowing to evaluate the security of biased distributed sources of cryptographic keys. It is proved by using this model, that the large class of polynomially bounded biases are tolerable from the point of view of security.

Acknowledgments I owe to discussions with numerous people along the years the development of the views presented here. For the theoretical parts, special thanks are due to H. W. Lenstra, Jr. and F. Morain. Stimulating exchanges with U. Maurer helped to improve the view upon the cryptographical applications of primality proving. I thank M. Hallett, L. Jaschke, U. Maurer and F. Pappalardi for careful reading of early versions of the paper and their valuable remarks. I thank Richard Brent for his careful and generous reading of the manuscript and the valuable remarks which it brought.

References

[1] L. M. Adleman and H. W. Lenstra, Jr.: *Finding irreducible polynomials over finite fields*, Proc. 18-th Ann. ACM Symp. on Theory of Computing (STOC) (1986), pp. 350–355.

[2] W. R. Alford, A. Granville and C. Pomerance: *There are infinitelym many Carmichael numbers*, Ann. Math., **140** (1994), pp. 703–722.

[3] L. M. Adleman and M.A. Huang: *Primes in random polynomial time*, Proc. 19-th Ann ACM Symp. on Theory of Computing **STOC** (1987), pp. 462–469.

[4] W. Adams and D. Shanks: *Strong primality tests that are not sufficient*, Math. Comp., **vol. 39**, Nr. 159 (July 1982), pp. 255–300.

[5] L. M. Adleman and C. Pomerance, R.S. Rumely: *On Distinguishing Prime Numbers from Composite Numbers*, Ann. Math., **117** (1983), pp. 173–206.

[6] A. O. L. Atkin: *Intelligent primality test offer*, Computational Perspectives on Number Theory, Proceedings of a Conference in Honor of A.O.L. Atkin, International Press, 1998, pp. 1–11.

[7] A. O. L. Atkin and F.Morain: *Elliptic curves and primality proving.*, Math. Comp., **vol. 61** (1993), pp. 29–68.

[8] E. Bach: *Realistic analysis of some randomized algorithms*, J. Comput. Sys. Sci., **42** (1992), pp. 30–53.

[9] E. Bach: *Explicit bounds for primality testing and related problems*, Math. Comp., **55** (1990) pp. 355–380.

[10] E. Bach: *Exact analysis of a priority queue algorithm for random variate generation*, Proc. ACM – SIAM Simp. on Discrete Algorithms (SODA) (1994), pp. 48–56.

[11] E. Bach and J. Shallit: *Algorithmic number theory*, MIT Press, 1996.

[12] J. Brandt and I. Damgård: *On generation of probable primes by incremental search.* In Ernest F. Brickell, editor, Advances in Cryptology – CRYPTO'92, LNCS **740**, pp. 358–370. Springer-Verlag, 1993.

[13] J. Brandt, I. Damgård and P. Landrock: *Speeding up prime number generation* Proc. of Asiacrypt 91, Springer Verlag Lecture Notes.

[14] J. Brillhart, D. H. Lehmer and J.L. Selfridge: *New primality criteria and factorization of* $2^m \pm 1$, Math. of Comp., **vol. 29**, Number 130 (April 1975), pp. 620–647.

[15] W. Bosma and M. van der Hulst: *Primality proving with cyclotomy*, Doctoral Thesis, Universiteit van Amsterdam 1990.

[16] C. Caldwell: *The Prime Pages*, http://www.utm.edu/research/primes

[17] R. D. Carmichael: *On Composite numbers P which satisfy the Fermat congruence* $a^{Pn-1} \equiv 1 \mod P$, Amer. Amth. Monthly, **19** (1912), pp. 22–27.

[18] D. V. Chudnowsky and G. V.Chudnowski: *Sequences of numbers generated by addition in formal groups and new primality and factorization tests*, Advances in Applied Math., **7**(1986), pp. 385–434.

[19] H. Cohen and H. W. Lenstra Jr.: *Primality testing and Jacobi sums*, Math. Comp., **vol. 48** (1984), pp. 297–330.

[20] C. Couvreur and J.-J. Quisquater: *An introduction to fast generation of large primes*, Philips Journal of Research, **vol. 37**, (1982) pp. 231–264, Plenum Press, New York and London 1990.

[21] C. Crandall and B. Fagin: *Discrete weighted transforms and large-integer arithmetic*, Math. Comp., **62** (1994) pp. 305–324.

[22] http://www.inf.ethz.ch/~mihailes, Homepage of Cyclotomy, Preda Mihăilescu.

[23] M. Deuring: *Die Typen der Multiplikatorenringe elliptischer Funktionenkörper*, Abh. Math. Sem. Hamburg, **14** (1941), pp. 197–272.

[24] I. Damgård, P. Landrock and C. Pomerance: *Average case bounds for the strong probable prime test*, Math. Comp., **61**, no.203, pp. 177–194.

[25] http://lix.polytechnique.fr/~morain/Prgms/ecpp.english.html, Site for downloading the elliptic curve primality test software of F. Morain.

[26] J. von zur Gathen and J. Gerhard: *Modern computer algebra*, Cambridge University Press, (1999).

[27] J. Gordon: *Strong primes are easy to find*, Advances in Cryptology – EUROCRYPT '84, LNCS, **209**, (1984), pp. 216–223.

[28] S. Goldwasser and J. Kilian: *Almost all primes can be quickly certified*, Proc. 18-th Annual ACM Symp. on Theory of Computing (1986), 316–329.

[29] J. Grantham's homepage: http://www.clark.net/~grantham.

[30] J. Grantham: *There are infinitely many Perrin pseudoprimes*, preprint (1997).

[31] J. Grantham: *A probable prime test with high confidence*, J. Number Theory, **72**(1998), pp. 32–47.

[32] J. Grantham: *Frobenius Pseudoprimes*, Math. Comp. (2000), to appear, see [29].

[33] S. Gurak: *Pseudoprimes for higher-order linear recurrence sequences*, Math. Comp., **55** (1990) 783–813.

[34] D. Husemöller: *Elliptic curves*, Springer Verlag, 1987.

[35] D. Knuth and L. Trabb Pardo: *Analysis of a simple factorization algorithm*, Theoretical Computer Science, **3** (1976), pp. 157–165.

[36] H. W. Lenstra, Jr.: *Primality testing algorithms (after Adleman, Rumely and Williams)*, Seminaire Bourbaki # 576, Lectures Notes in Mathematics, **vol. 901**, pp. 243–258.

[37] H. W. Lenstra, Jr.: *Galois theory and primality testing*, in "Orders and Their Applications", Lecture Notes in Mathematics, **vol. 1142**, (1985) Springer Verlag.

[38] H. W. Lenstra, Jr.: *Divisors in residue classes*, Math. Comp., **vol. 48** (1984), pp. 331–334.

[39] H. W. Lenstra, Jr.: *Factoring integers with elliptic curves*, Annals of Mathematics, **126**, (1987), pp. 649-673.

[40] A. K. Lenstra and H. W. Lenstra, Jr. (eds.): *The development of the number field sieve*, Lecture Notes in Mathematics **1554** (1993).

[41] E. Lucas: *Théorie des fonctions numériques simplement périodiques*, Amer. J. of Math., **1** (1878), pp. 184–240 and 289–321.

[42] U. Maurer: *Fast generation of prime numbers and secure public-key cryptographic parameters*, Journal of Cryptology., **8** (1995), Pages: 123–156.

[43] U. Maurer: private communication.

[44] U. Maurer: *Some number theoretic conjectures and their relation to generation of cryptographic primes*, Cryptography and Coding II, C. Mitchell (ed.), Oxford University Press (1992), pp. 173–191.

[45] P. Mihăilescu: *Fast generation of provable primes using search in arithmetic progressions*, Proceedings CRYPTO94, pp. 282–293.

[46] P. Mihăilescu: *Cyclotomy of rings & primality testing*, dissertation 12278, ETH Zürich, 1997.

[47] P. Mihăilescu: *Cyclotomy primality proving – recent developments*, Proceedings of the Third International Symposium ANTS III, Portland, Oregon, Lecture Notes in Computer Science **vol. 1423** (1998), pp. 95–111.

[48] P. Mihăilescu: *Recent developments in primality proving*, Mathematics and Computers in Simulation **49** (1999), pp. 193–204.

[49] P. Mihăilescu: *Measuring the cryptographic relevance of biased public key distributions*, manuscript (1998).

[50] P. M. Mihăilescu: *New Wagstaff prime proved* , EMail to the NMBRTHRY mailing list; available on http://listserv.nodak.edu/archives/nbrthry.html, January 1998.

[51] P. Mihăilescu and F. Morain: *Cyclotomy primality proofs. part II: Certification*, submitted Math.Comp.

[52] F. Morain: *Primality proving using elliptic curves: an update*, J.P.Buhler (Ed.) Proceedings of the Third International Symposium ANTS III, Portland, Oregon, Lecture Notes in Computer Science **vol.** 1423 (1998), pp. 111–128.

[53] Number Theory List NMBRTHRY@LISTSERV.NODAK.EDU, F. Morain, *Nomenclature*, post from Tue, 18 Feb 1997.

[54] S. Pohlig and M. Hellman: *An imporved algorithm for computing discrete logarithms in $GF(p)$ and its cryptographic significance*, IEEE Transactions on Information Theory, **24, nr. 1**, (Jan. 1978) pp. 106–111.

[55] D. A. Plaisted: *Fast verification, testing and generation of large primes*, Theoretical Computer Science, **vol. 9** (1979), pp. 1–17.

[56] H. C. Pocklington: *Determination of the prime or composite nature of large numbers by Fermat's theorem*, Proc. Cambridge Philos. Soc., **18** (1914–16), pp. 29–30.

[57] C. Pomerance: *Very short primality proofs*, Math. Comp., **48** (1987), pp. 315–322.

[58] V. R. Pratt: *Every prime has a succinct certificate*, SIAM J. Comput., **4** (1975) 214–220.

[59] P. Pritchard: *A sublinear additive sieve for finding prime numbers*, Comm. ACM, **24** (1981), pp. 18–23.

[60] P. Pritchard: *Explaining the wheel sieve*, Acta Informatica, **17**, (1982), pp. 477–85.

[61] F. Proth: *Théorèmes sur les nombres premiers*, C.R. Acad. Sci. Paris, **87**, (1878), p. 926.

[62] R. Peralta and V. Shoup: *Primality testing with fewer random bits*, Computational Complexity, **3** (1993), pp. 355–367.

[63] J.-J. Quisquater: *private communication*, The second part of [20] had internal circulation at Philips reserch and described the implementation of generation of linear certifiable primes.

[64] M. O. Rabin: *Probabilistic algorithms in finite fields* , SIAM J.Comput. vol. 9, May 1980, pp. 273–280.

[65] H. Riesel: *Prime numbers and computer methods for factorization*, Birkhäuser, 1994.

[66] R. Rivest and B. Silverman: *Are 'strong' primes needed for RSA?*, Preprint, 1999.

[67] R. Schoof: *Elliptic curves over finite fields and the computation of square roots mod p*, Math. Comp., **44** (1985), pp. 483–494.

[68] R. Schoof: *Counting points on elliptic curves over finite fields*, J. de Théorie des Nombres, Bordeaux, **7** (1995), 219–254.

[69] J. Silverman: *The arithmetic of elliptic curves*, Springer, Graduate texts in Math., **106**.

[70] R. Solovay and V. Strassen: *A fast Monte Carlo test for primality*, SIAM J. Comput., **6** (1977), pp. 64–85.

[71] S. Vanstone and R. Zuccherato: *Short RSA keys and their generation*, J. of Cryptology, **8** (1995), pp. 101–114.

[72] A. E. Western: *On Lucas and Pepin's test for primeness of Mersenne numbers*, J. of the London Math. Society, **vol. 7/I** (1932).

[73] H. C. Williams: *Primality testing on a computer*, Ars Combin., **vol. 5** (1978), pp. 127–185.

[74] H. C. Williams: *Edouard Lucas and primality testing*, Canadian Society Series of Monographs and Advanced Texts, **vol. 22**, John Wiley and Sons (1998).

[75] H. C. Williams and J. S. Judd: *Some algorithms for prime testing, using generalized Lehmer functions*, Math. Comp., **vol. 30** (1976), 867–886.

[76] H. C. Williams and C. R. Zarnke: *Some prime numbers of the forms* $2A3^n + 1$ *and* $2A3^n - 1$, Math. Comp., **vol. 26** (October 1972), pp. 995–998.

Institut für Wissenschaftliches Rechnen, ETH,
8090 Zürich, Switzerland
E-mail address: mihailes@inf.ethz.ch

Progress in Computer Science and Applied Logic, Vol. 20
© 2001 Birkhäuser Verlag Basel/Switzerland

Elliptic Curve Factorization Using a "Partially Oblivious" Function

René Peralta

Abstract. Let $N = P \cdot R$ where P is a prime not dividing R. We show how a special class of functions $f : Z_N \to Z$ can be used to help obtain P given N. The requirements of f are that it be non-trivial and that $f(x) = f(x \bmod P)$. Such a function does not "see" R. Hence the name *partially oblivious*.

1. Introduction

It is not known how to efficiently factor a large integer N. Currently, the algorithm with best asymptotic complexity is the Number Field Sieve (see [6]). For numbers below a certain size (currently believed to be about 100 decimal digits), either the Quadratic Sieve [12] or Lenstra's Elliptic Curve Method (ECM) [7] are faster. Which of these algorithms to use depends on the size of N and of the smallest prime factor of N. When the size of the smallest factor is sufficiently smaller than $\sqrt{N}$, ECM is the fastest of the three. This note describes a speedup of ECM under special conditions.

Suppose $N = P \cdot R$, where P is a prime not dividing R. We assume the size, in bits, of P is known. That is, we assume knowledge of $\hat{P}$ such that $\frac{\hat{P}}{2} \leq P \leq \hat{P}$. The algorithm makes use of a special class of functions $f : Z_N \to Z$ to speedup ECM. The requirements of f are:

- $f(x) = f(x \bmod P)$;
- for uniformly chosen $x \in Z_N$, the probability that $f(x) = 1$ is in the range $[0 + \epsilon, 1 - \epsilon]$ for some positive ϵ independent of N.

We call such functions *partially oblivious* since they are dependent on some prime factors of N and oblivious of the others. The method is most useful when P is much smaller than R. A function that is oblivious to a factor R of N will be referred to as an *R-oblivious function*. The algorithm is a two-phased variation of ECM. Access to an R-oblivious function is assumed in the second phase.

There is at least one class of numbers for which polynomial-time computable oblivious functions exist. These are numbers of the form $N = P \cdot R$, where R is a perfect square. In this case, the Jacobi function $\left(\frac{x}{N}\right) = \left(\frac{x}{P}\right)$ and thus is oblivious to R. Numbers of this form have been used in various cryptographic applications (see [5, 10, 13]). The algorithm presented in this paper is not fast enough to threaten the security of these applications. However, secure parametrization of such applications must take into account the speed of our algorithm. Should partially oblivious functions be discovered for general composite numbers, our methods would allow for the factorization of numbers which are larger than are possible to factor with currently known techniques.

2. The algorithm: Phase I

Let $w = (x, y)$ be a point in an elliptic curve $y^2 = x^3 + ax + b$. We will use additive notation for the elliptic curve operation. Let ξ be the cardinality of the elliptic curve modulo P. Provided some minimal conditions are met when choosing the parameters (w, a, b), we can consider ξ a uniformly random number of order P (see [7]). Let the prime factorization of ξ be $\xi = q_1 \ldots q_{m-1} q_m$ with $q_i \leq q_{i+1}$. If $q_m \leq U$, then ξ is said to be U-smooth. If $q_{m-1} \leq U$ and $q_m \leq V$, then ξ is said to be (U, V)-semismooth [1].

Let p_i be the ith prime. For each prime p_i, let e_i be maximum such that $p_i^{e_i} \leq U$. Finally, let $M = \prod_{i=1}^{k} p_i^{e_i}$, where p_k is the largest prime not exceeding U. If ξ is (U, V)-semismooth then, with overwhelming probability, $\tau = M \cdot w$ is either the identity element or has order q_m in the elliptic curve modulo P.[1] If τ is the identity element then, with overwhelming probability, computing τ on the curve modulo N will require a modular inversion of a multiple of P. One iteration of ECM consists of choosing (w, a, b) randomly and computing τ. If a multiple of P is encountered, then ECM calculates and returns P after a GCD operation with N. If τ is not the identity element, the ECM iteration fails. Thus, the number of iterations in ECM is expected to be the inverse of the probability that ξ is U-smooth. This probability is heuristically assumed (and experimentally verified) to be closely approximated by Dickman's ρ-function $\rho(u)$ where u is such that $U = P^{1/u}$ (see [4] or [1]). Phase I of the method described in this note is exactly one iteration of ECM.[2] The result of this phase is the element τ in the elliptic curve. Phase II will find P in $O(\sqrt{V})$ elliptic curve operations provided ξ is (U, V)-semismooth.

3. The algorithm: Phase II

This phase uses τ, a, b as generated in phase I. Let $\langle \tau \rangle$ be the group generated by τ on the elliptic curve $y^2 = x^3 + ax + b$ modulo P. If ξ is (U, V)-semismooth then the order of τ is no more than V. By the birthday paradox, a random walk on

---

[1] There is a small chance that ξ is (U, V)-semismooth but divisible by a prime power p^k such that $p \leq U$ but $p^k > U$. We have not yet developed a closed form or numerical approximation to this small probability.
[2] We are referring here to Lenstra's original algorithm. Other "two-phase" variants of ECM have been proposed (see [8, 2], also [3]). The speed of these variants, when implemented on special-purpose hardware, may be competitive with the results presented here. Furthermore, the above variants do not require access to a partially oblivious function.

$\langle \tau \rangle$ will repeat an element after $O(\sqrt{V})$ steps. As in Pollard's rho method [11], we make the heuristic assumption that the same is true of a pseudo-random walk on $\langle \tau \rangle$. We implement such a walk using an R-oblivious function f. If ψ is a point in $\langle \tau \rangle$ then a "step" in the pseudo-random walk consists of adding either $\psi = \psi + \tau$ or $\psi = \psi + \psi$. Which of the two paths to take is decided by computing f on the x coordinate of ψ. As in Pollard's rho method, we use two points (ψ_0, ψ_1) in $\langle \tau \rangle$, where ψ_1 advances at twice the speed of ψ_0.

$$\begin{array}{ll} \text{if } (\ f(\psi.x) == 1\ ) & \psi = \psi + \tau \\ & \text{else } \psi = \psi + \psi; \\[2mm] \text{return}(\psi); & \end{array}$$

FIGURE 1. Function step(ψ)

$$\begin{array}{lll} \text{step} & 1: & \psi_0 = \tau; \quad \psi_1 = \tau; \quad counter = 0; \\ \text{step} & 2: & \psi_0 = step(\psi_0); \\ \text{step} & 3: & \psi_1 = step(step(\psi_1)); \\ \text{step} & 4: & q = GCD((\psi_0.x - \psi_1.x), N); \\ \text{step} & 5: & \text{if } (1 < q < N) \text{ return}(q); \\ \text{step} & 6: & counter + +; \\ \text{step} & 7: & \text{if } (counter > 2\sqrt{V}) \text{ return}(\text{"failure"}); \\ \text{step} & 8: & \text{go to Step 2;} \end{array}$$

FIGURE 2. Algorithm: phase II

The optimal number of iterations in each phase of our algorithm is dependent on implementation. It is not hard to show that a random walk of length m on a space of size V, has probability greater than $1 - e^{-\frac{m(m-1)}{2V}}$ of repeating itself. In the following analysis we will make the simplifying assumption that a repetition will occur before $\sqrt{V}$ steps. In practice, however, our pseudo-random walk is of length $2\sqrt{V}$. Figures 1 and 2 describe our implementation of phase II.

4. Analysis

We will assume the bias

$$\frac{\text{Prob}\,(f(x) = 1) - \text{Prob}\,(f(x) = 0)}{2}$$

of the function f on randomly chosen x is $0.^3$ The purpose of the following analysis
is to get a ball-park figure of the optimal values of (U, V); to show the method
by which these values can be found for any given implementation; and to measure
the speedup obtained over ECM.

The number C_I of elliptic curve operations required for one exponentiation
during phase I of the algorithm is $\Theta(\ln U)$. By the Prime Number Theorem, the
number of exponentiations is $\pi(U) \sim \frac{U}{\ln U}$. Thus, phase I of the algorithm performs
$\Theta(U)$ elliptic curve operations.

Phase II of the algorithm is easily seen to perform $C_{II} = \Theta(\sqrt{V})$ elliptic
curve operations (the cost of the GCD calculation can be subsumed under this
term).

Letting $U = P^{1/u}$ and $P^{1/v}$, the total expected cost of the algorithm is

$$\frac{C_I + C_{II}}{\sigma(u, v)} = \frac{\alpha P^{1/u} + \beta P^{0.5/v}}{\sigma(u, v)}$$

where $\sigma(u, v)$ is the probability that an integer chosen uniformly at random in the
range $[1..P]$ is $(P^{1/u}, P^{1/v})$-semismooth and α, β are implementation-dependent
parameters. This cost expression is proportional to

$$c(u, v) = \frac{P^{1/u} + \gamma P^{0.5/v}}{\sigma(u, v)}$$

where $\gamma = \frac{\beta}{\alpha}$. Therefore the optimal values of the parameters U, V are determined
solely by the parameter γ. This also gives us a formula for the speedup of our algorithm over ECM. Since ECM is essentially phase I of our algorithm, the speedup
is given by

$$speedup_\gamma(\log_{10} P) = \frac{\min_u \left\{ P^{1/u}/\rho(u) \right\}}{\min_{u,v} \left\{ (P^{1/u} + \gamma P^{0.5/v})/\sigma(u, v) \right\}}.$$

For numbers of the form $N = PQ^2$, we can take $f(x)$ to be the Jacobi function
$f(x) = \left(\frac{x}{N} \right)$. On arithmetic modulo 100-digit numbers, implementation of elliptic
curve computation using projective coordinates and the computational number
theory library package "ln 3" [4] we measured $\gamma \approx 3.5$.

Numerical methods for calculating $\sigma(u, v)$ can be found in [1]. A plot of
$speedup_\gamma(\log_{10} P)$ for $\gamma = 3.5$ and $\log_{10} P$ between 20 and 40 yields a slightly
upward bending curve (the speedups at $20, 30, 40$ digits are $8.15, 13.35, 19.0$, respectively.). A $\Theta(\log P)$ speedup would yield a straight line. Thus, it would appear
the actual speedup is slightly larger than $\Theta(\log P)$. No attempt was made to study
this phenomenon analytically.

As an example, we consider the problem of extracting a prime P of size
30 digits from a composite $N = P \cdot Q^2$ of size 100 digits. Our analysis yields
approximately optimal values of $670,000$ for U and $58,000$ for V. With these

---

[3]Standard techniques can be used to construct an unbiased coin from a coin with an unknown
bias. The reader can verify that these same techniques can be used in the context of this paper.
[4]This package is available online at http://www.entropy.cs.uwm.edu/~ ln 3/

parameters, it is expected that P would be found after examining approximately 270 elliptic curves. We ran our implementation once on a workstation and found P after generating 156 elliptic curves. The experiment took 54 hours of computation.

5. Quadratic residuosity

It is an open question whether integer factorization is reducible to the problem of deciding quadratic residuosity. Thus it is worth noting that a quadratic residuosity oracle yields a partially oblivious function. Suppose N is the product of two primes P and Q. Let $r(x)$ return 1 if x is a quadratic residue modulo N and 0 otherwise. We may use the function r to find a quadratic non-residue z such that $\left(\frac{z}{N}\right) = 1$. Now factor z into $z = \mu_1 \cdot \mu_2 \bmod N$ where both μ_1 and μ_2 have Jacobi symbol -1. The reader can verify that one of the following two functions is Q-oblivious and the other P-oblivious:

$$f_1(x) = \left\{ \begin{array}{ll} r(x) & \text{if } \left(\frac{x}{N}\right) = 1, \\ r(x\mu_1) & \text{otherwise,} \end{array} \right.$$

$$f_2(x) = \left\{ \begin{array}{ll} r(x) & \text{if } \left(\frac{x}{N}\right) = 1, \\ r(x\mu_2) & \text{otherwise.} \end{array} \right.$$

Acknowledgements

A slower version of this algorithm appeared in a paper by E. Okamoto and this author [9]. J. Pollard and later D. Bleichenbacher suggested improvements leading to the algorithm in this note. The particular random walk in Figure 1. was suggested by E. Teske. F. Akbulut coded several implementations of this algorithm. The package ln 3 was designed in conjunction with D. Meilecke.

References

[1] E. Bach and R. Peralta, *Asymptotic semismoothness probabilities*, Mathematics of Computation, **65** (1996), 1701–1715.

[2] R. Brent, *Some integer factorization algorithms using elliptic curves*, Computer Science Laboratory, Australian National University, Report **TR-CS-96-02**, (September 1995).

[3] R. Brent, R. Crandall, K. Dilcher and C. van Halewyn, *Three new factors of Fermat numbers*, Preprint, 1999, (available online at http://www.mscs.dal.ca/ dilcher/Preprints/BCDH.ps).

[4] N. G. de Bruijn, *On the number of positive integers $\leq x$ and free of prime factors $> y$*, Indag. Math. **13** (1951), 50–60.

[5] A. Fujioka, T. Okamoto and S. Miyaguchi, *ESIGN: An efficient digital signature implementation for smart cards*, Advances in Cryptology – Proceedings of EUROCRYPT 91, Lecture Notes in Computer Science, **547**, Springer-Verlag (1991), 446–457.

[6] A. Lenstra and H. W. Lenstra, Eds., *The development of the number field sieve*, Lecture Notes in Mathematics, **1554**, Springer-Verlag (1993).

[7] H. W. Lenstra, *Factoring integers with elliptic curves*, Annals of Mathematics, **126** (1987), 649–673.

[8] P. Montgomery, *An FFT extension of the elliptic curve method of factorization*, Ph.D. dissertation, Mathematics Department, UCLA, (1992).

[9] E. Okamoto and R. Peralta, *Faster factoring of integers of a special form*, IEICE Transactions on Fundamentals of Electronics, Communications, and Computer Sciences, **E79-A** (1996), 489–493.

[10] T. Okamoto and S. Uchiyama, *A new public-key cryptosystem as secure as factoring*, Lecture Notes in Computer Science, Advances in Cryptology – Proceedings of EUROCRYPT 98, **1403**, Springer-Verlag (1998), 308–318.

[11] J. M. Pollard, *A Monte Carlo method for factorization*, BIT, **15** (1975), 331–334.

[12] C. Pomerance, *The quadratic sieve factoring algorithm*, Advances in Cryptology – Proceedings of EUROCRYPT 84, Lecture Notes in Computer Science, **209**, Springer-Verlag (1985), 169–182.

[13] T. Takagi, *Fast RSA-type cryptosystem modulo $p^k q$*, Lecture Notes in Computer Science, Advances in Cryptology - Proceedings of CRYPTO 98, **1462**, Springer-Verlag (1998), 318–326.

Department of Computer Science,
Yale University,
P.O. Box 208285,
New Haven, CT. 06520-8285, USA
E-mail address: peralta-rene@cs.yale.edu

Progress in Computer Science and Applied Logic, Vol. 20
© 2001 Birkhäuser Verlag Basel/Switzerland

The Hermite-Serret Algorithm and $12^2 + 33^2$

Alf van der Poorten

Abstract. Musing on the cute observation that $12^2 + 33^2 = 1233$ led me to remind myself of well-known techniques for writing a given integer n as a sum of two squares, given (or having already found) a square root z, say, of -1 modulo n. In brief, one applies the Euclidean algorithm to n and z, stopping at the first pair x and y of remainders that are smaller than $\sqrt{n}$. Then, lo! it happens that $n = x^2 + y^2$. Naturally, square roots of -1 properly different from z lead to different representations of n as sum of two squares. Obviously, so simple an algorithm must have an elegant and near trivial explanation, yet the literature contains some rather turgid proofs. I briefly point out that, in general, a representation of n by a reduced definite binary quadratic form can readily be found by symmetric decomposition of a symmetric matrix, a process well known as *reduction*; and that this does give insight into why certain remainders in the Euclidean algorithm applied to n and some square root modulo n yield the representation. My story is for our mild amusement, and provides a nice and easily comprehended story to tell our students.

1. 1233 and All That

The pleasing observation that $1233 = 12^2 + 33^2$ apparently emanates from Hendrik Lenstra. It came to me by way of the text [1]. Happily, it is more than just a numerical curiosity, being the augur of infinitely many such cute decompositions.

Lemma 1.1. *The identity $a^2 + b^2 = 10^u a + b$ is equivalent to the representation $10^{2u} + 1 = (10^u - 2a)^2 + (2b - 1)^2$.*

Proof. Multiplying $a^2 + b^2 - 10^u a - b = 0$ by 4 and completing squares readily yields $(10^{2u} - 4 \cdot 10^u a + 4a^2) + (4b^2 - 4b + 1) = 10^{2u} + 1$. □

It's enough, therefore, to look for representations of $10^{2u} + 1$ as a sum $x^2 + y^2$ with $x = 10^u - 2a$ even, and $y = 2b - 1$ odd.

Example 1.2. Because 101 is prime, its only representation as sum of squares is $10^2 + 1^2$, yielding $a = 0$ and $b = 1$ and so the boring decomposition $1 = 1^2$. However, $10^4 + 1 = 73 \cdot 137$ and so has a second potentially interesting decomposition given by $(8^2 + 3^2)(4^2 + 11^2) = (8 \cdot 11 - 3 \cdot 4)^2 + (8 \cdot 4 + 3 \cdot 11)^2 = 76^2 + 65^2$. That is

$a = 12$ and $b = 33$, as in our motivating example. Similarly, $10^6 + 1 = 101 \cdot 9901$ yields the possibly interesting decomposition $10^6 + 1 = 980^2 + 199^2$. But that is $a = 10$ and $b = 100$ and provides only the quite dull $10100 = 10^2 + 100^2$.

It is easy to see that there are plenty of u so that $10^{2u} + 1$ has lots of prime factors, but it's not entirely obvious that seemingly interesting sums of squares yield amusing decompositions. Indeed, the example $10^{10} + 1 = 101 \cdot 3541 \cdot 27961$ has an encouraging three prime factors but the $4 - 1 = 3$ possibly amusing decompositions are $2584043776 = 25840^2 + 43776^2$, $1765038125 = 17650^2 + 38125^2$, which are pretty interesting, but also the unacceptable $99009901 = 990^2 + 09901^2$. In all, the claim there are infinitely many 'cute' decompositions is not totally convincing.

Example 1.3. The following brief selection[1] possibly is more compelling

$$116788^2 + 321168^2 = 116788321168 \qquad 768180^2 + 2663025^2 = 7681802663025$$
$$1675455088^2 + 3734621953^2 = 16754550883734621953.$$

I take it as known that 2, and each prime p congruent to 1 modulo 4 has an essentially unique presentation as a sum of two positive integer squares. Then the identity $(a^2 + b^2)(c^2 + d^2) = (ad \pm bc)^2 + (ac \mp bd)^2$ readily shows by induction that a product of s distinct odd primes $\equiv 1$ modulo 4 has 2^{s-1} essentially different representations as a sum of squares.

However, we had best first discuss just how one finds the decomposition of an integer as a sum of two squares.

2. Representation of Integers in the Form $ax^2 + 2bxy + cy^2$

It seems more convincing to discuss the more general problem of representation by arbitrary quadratic forms of negative discriminant. The matter of representation of integers by binary quadratic forms is one of the oldest problems of number theory. Of course, the real issue is to explain just which integers are represented, and why, but it certainly is also of interest actually to find representations.

2.1. Symmetric Decomposition

My following remarks are a minor variant on known algorithms for determining representations by definite forms. The idea is conveniently illustrated by a toy example.

Consider the problem of finding nonnegative integers x and y so that

$$173 = 2x^2 + 3y^2.$$

We first solve the congruence $z^2 \equiv -3 \cdot 2 \pmod{173}$. Indeed $72^2 = 30 \cdot 173 - 6$. Accordingly we study the matrix

$$M = \begin{pmatrix} 173 & 72 \\ 72 & 30 \end{pmatrix} = \begin{pmatrix} 2 & 1 \\ 1 & 0 \end{pmatrix} \begin{pmatrix} 2 & 1 \\ 1 & 0 \end{pmatrix} \begin{pmatrix} 2 & 1 \\ 1 & 0 \end{pmatrix} \begin{pmatrix} 14 & 1 \\ 1 & 0 \end{pmatrix} \begin{pmatrix} 1 & 0 \\ 0 & 6 \end{pmatrix}.$$

---
[1]From a complete list up to $u = 10$ for which I thank Michael Volpato.

Here I have effected the decomposition by the Euclidean algorithm *on the rows* of M, with the details given by the array

	173	72
2	72	30
2	29	12
2	14	6
14	1	0
	0	6

Dually, we might have performed the Euclidean algorithm *on the columns* of M, obtaining

	2	2	2	14		
173	72	29	14	1	0
72	30	12	6	0	6

This dual decomposition is particularly friendly to left-handed mathematicians. But it yields the transpose of the previous decomposition, namely

$$M = \begin{pmatrix} 173 & 72 \\ 72 & 30 \end{pmatrix} = \begin{pmatrix} 1 & 0 \\ 0 & 6 \end{pmatrix} \begin{pmatrix} 14 & 1 \\ 1 & 0 \end{pmatrix} \begin{pmatrix} 2 & 1 \\ 1 & 0 \end{pmatrix} \begin{pmatrix} 2 & 1 \\ 1 & 0 \end{pmatrix} \begin{pmatrix} 2 & 1 \\ 1 & 0 \end{pmatrix}.$$

Something is clearly wrong here. M is a symmetric matrix, yet our methods of decomposition destroy that symmetry. So we try again, working symmetrically both by row and by column. Our working begins with the two steps, one *row operation*, the one *column operation*,

		2	
	173	72	
2	72	30	12
	29	12	5

reporting that

$$M = \begin{pmatrix} 173 & 72 \\ 72 & 30 \end{pmatrix} = \begin{pmatrix} 2 & 1 \\ 1 & 0 \end{pmatrix} \begin{pmatrix} 30 & 12 \\ 12 & 5 \end{pmatrix} \begin{pmatrix} 2 & 1 \\ 1 & 0 \end{pmatrix}.$$

Ultimately, we have

		2	2	1		
	173	72				
2	72	30	12			
2	29	12	5	2		
1		6	2	2	0	
			3	0	3	

showing that

$$M = \begin{pmatrix} 173 & 72 \\ 72 & 30 \end{pmatrix} = \begin{pmatrix} 2 & 1 \\ 1 & 0 \end{pmatrix} \begin{pmatrix} 2 & 1 \\ 1 & 0 \end{pmatrix} \begin{pmatrix} 1 & 1 \\ 1 & 0 \end{pmatrix} \begin{pmatrix} 2 & 0 \\ 0 & 3 \end{pmatrix} \begin{pmatrix} 1 & 1 \\ 1 & 0 \end{pmatrix} \begin{pmatrix} 2 & 1 \\ 1 & 0 \end{pmatrix} \begin{pmatrix} 2 & 1 \\ 1 & 0 \end{pmatrix}.$$

The array

$$\begin{matrix} & & 2 & 2 & 1 \\ \\ 0 & 1 & 2 & 5 & 7 \\ 1 & 0 & 1 & 2 & 3 \end{matrix}$$

details the computation

$$\begin{pmatrix} 2 & 1 \\ 1 & 0 \end{pmatrix} \begin{pmatrix} 2 & 1 \\ 1 & 0 \end{pmatrix} \begin{pmatrix} 1 & 1 \\ 1 & 0 \end{pmatrix} = \begin{pmatrix} 7 & 5 \\ 3 & 2 \end{pmatrix}.$$

So we have

$$M = \begin{pmatrix} 173 & 72 \\ 72 & 30 \end{pmatrix} = \begin{pmatrix} 7 & 5 \\ 3 & 2 \end{pmatrix} \begin{pmatrix} 2 & 0 \\ 0 & 3 \end{pmatrix} \begin{pmatrix} 7 & 3 \\ 5 & 2 \end{pmatrix}$$

and, indeed,

$$2 \cdot 7^2 + 3 \cdot 5^2 = 173.$$

Principal Remark. *Let m be a positive integer, and suppose $z^2 \equiv -m \pmod{n}$. Set $k = (z^2 + m)/n$. Then symmetric decomposition of the symmetric matrix $\begin{pmatrix} n & z \\ z & k \end{pmatrix}$ yields a unimodular nonnegative integer matrix $\begin{pmatrix} x & x' \\ y & y' \end{pmatrix}$ and integers a, b and c satisfying $0 \le b < a$, $0 \le b < c$ and $ac - b^2 = m$, so that*

$$M = \begin{pmatrix} n & z \\ z & k \end{pmatrix} = \begin{pmatrix} x & y \\ x' & y' \end{pmatrix} \begin{pmatrix} a & b \\ b & c \end{pmatrix} \begin{pmatrix} x & x' \\ y & y' \end{pmatrix},$$

and plainly

$$n = ax^2 + 2bxy + cy^2.$$

Conversely, every such representation of n is obtained by symmetric decomposition from a solution of $z^2 \equiv -m \pmod{n}$.

Proof. We perform the Euclidean algorithm on the rows of the matrix. That is, there is a greatest integer d, possibly zero, so that both $n - dz$ and $z - dk$ are nonnegative. Repeating the operation on the columns it suffices to see we already know that $z - dk$ is positive, and that, because $k \cdot \big((n - dz) - d(z - dk)\big) - (z - dk)^2$ is $m > 0$ it follows that also $(n - dz) - d(z - dk)$ must be positive. This leaves us with a nonnegative symmetric integer matrix of positive determinant m on which we may repeat the symmetric decomposition. The process terminates if d is zero for two consecutive steps or, equivalently, if z is less than both n and k.

Conversely, a representation of n yields both $an = (ax + by)^2 + my^2$ and $cn = (bx + cy)^2 + mx^2$, vindicating the final claim.   □

To add conviction, let me detail the case $83 = 2x^2 + 2xy + 3y^2$. Here $m = 5$ and $59^2 + 5 = 42 \cdot 83$. Indeed

		1	2	1	
	83	59			
1	59	42	17		
2	24	17	7	3	
1		8	3	2	1
		4	1	3	

and the continued fraction $[1\,,\,2\,,\,1]$ is $4/3$ whilst $[1\,,\,2] = 3/2$. Hence

$$2 \cdot 4^2 + 2 \cdot 4 \cdot 3 + 3 \cdot 3^2 = 83\,.$$

2.2. Algorithmic Considerations

My apparent innovation, though I do not claim any innovation at all, is the idea of symmetric decomposition of symmetric integer matrices. Otherwise, I hint at the well-known algorithm sometimes attributed to Cornacchia; for example in the fine text [1]; and more conveniently, at [2], §1.5.2. My recollection is that an allusion to the contribution of Serret and Hermite might be more appropriate, at any rate in the case $m = 1$ of particular interest to us here.

In that case, thus the case of representation as a sum of two squares, the symmetric matrix M is unimodular and its symmetry is precisely the symmetry of the continued fraction expansion of n/z. In [8], H. J. S. Smith gives a nice proof that there is indeed a z so that p/z has a symmetric continued fraction expansion whenever p is a prime $\equiv 1 \pmod 4$; we repeat that delightful story in [3]. The general algorithm is analysed in [4]; this is also a good source for earlier references, including those I do not detail above.

However, the reader will have noticed that the method sketched, though fine for toy examples done by hand, does not seem especially efficient. In particular, rather than doing everything twice it is plainly preferable in the case $m = 1$ just to perform the Euclidean algorithm on n and z. Then the pair of remainders x, y first both less than $\sqrt{n}$ yields $x^2 + y^2 = n$. Theorem 2.1 shows that. For $m > 1$ it is less obvious just when to stop, and exactly how to obtain x and y in terms of remainders appearing in the Euclidean algorithm. The cases $b = 0$ are detailed in [4] (and that readily generalizes, see for example [5]). By the way, although I suppose the central coefficient is even, that's no loss of generality. If that coefficient is odd, just consider representations of $2n$ by twice the given form.

Knowledgeable readers will complain that my remarks miss the point. The game is precisely to formulate the stopping rules and the like that allow one to apply the Euclidean algorithm on n and z to find a representation. For example, in [4] that's in effect done for the case $m = fg$, and representations $fx^2 + gy^2$, by first obtaining z so that $z^2 = -m \pmod{fn}$. One now applies the Euclidean algorithm to the pair of integers n and $w = z/f$; and x is the first remainder less than $\sqrt{n/f}$.

Whatever, my present remarks were in the first instance only my personal attempt to recall why such techniques work at all. I was also amused to be reminded that, given a square root of $-m$ (mod n), one obtains a representation of n of discriminant $-4m$ without needing to know, in advance, just which representation of that discriminant it would turn out to be.

3. A Curious Property of 17

Of course, I hide the real problem, namely that of finding a square root of $-m$ modulo n. Happily, any factor n of $10^{2u} + 1$ obviously satisfies $z^2 \equiv -1$ (mod n) with $z \equiv 10^u$ (mod n). Thus $10^8 + 1 = 17 \cdot 5882353$ leads us to compute

$$
\begin{array}{ccccc}
 & & 588 & 4 & \\
 & 5882353 & 10^4 & & \\
588 & 10^4 & 17 & 4 & \\
4 & 2353 & 4 & 1 & 0 \\
 & & 1 & 0 & 1
\end{array}
$$

That is,

$$
\begin{pmatrix} 5882353 & 10^4 \\ 10^4 & 17 \end{pmatrix} = \begin{pmatrix} 5882 & 1 \\ 1 & 0 \end{pmatrix} \begin{pmatrix} 4 & 1 \\ 1 & 0 \end{pmatrix} \begin{pmatrix} 4 & 1 \\ 1 & 0 \end{pmatrix} \begin{pmatrix} 5882 & 1 \\ 1 & 0 \end{pmatrix}
$$
$$
= \begin{pmatrix} 2353 & 588 \\ 4 & 1 \end{pmatrix} \begin{pmatrix} 2353 & 4 \\ 588 & 1 \end{pmatrix}.
$$

So $5882353 = 588^2 + 2353^2$, and we have *chanced* upon a cute decomposition!

It is now easy to confirm that each integer $(10^{8(4u+1)} + 1)/17$ has, as it were, an automatic decomposition. The next is

$$5882352941176470588235294117647 05882353$$
$$= 5882352941176470588^2 + 23529411764705882353^2.$$

However, the exponents $8(4u - 1)$ are a little less automatic. The case $u = 1$ yields

$$5882352941223529411 7648 = 5882352941^2 + 235294117648^2.$$

Proposition 3.1. *There are infinitely many amusing decompositions.*

Proof. Each integer $(10^{8(4u+1)} + 1)/17$ is represented by the sum of the squares of $(10^{4(4u+1)} - 4)/17$ and of $(4 \cdot 10^{4(4u+1)} + 1)/17$. The integers $(10^{4(4u-1)} + 4)^2/17$ decompose as the sum of $10^{4(4u-1)}$ times $a = (10^{4(4u-1)} + 4)/17$ and of $b = 4a$. The attentive reader will already have been hunting for other curious factors of $10^{2u} + 1$. Those attentive readers will see readily that the said representations are decompositions, and the said decompositions yield representations as sums of squares. Is it that factors $m^2 + 1$ are singular? Try $(10^{128} + 1)/257$. But I shouldn't say more, lest I spoil the fun for readers with too much time on their hands. □

4. Remarks

The educated reader will already have recognised that 'symmetric decomposition' is no more than reduction of a quadratic form. Set (a, b, c) to denote the form $aX^2 + 2bXY + cY^2$. Then the opening example of §2 amounts to the reduction

$$(173, 72, 30) \xrightarrow{2} (30, 12, 5) \xrightarrow{2} (5, 2, 2) \xrightarrow{1} (2, 0, 3).$$

Note that these forms correspond precisely to the residual matrices left after each double step of the symmetric decomposition. Further, we have kept track of the translations involved and can thus readily compute

$$\begin{pmatrix} 2 & 1 \\ 1 & 0 \end{pmatrix} \begin{pmatrix} 2 & 1 \\ 1 & 0 \end{pmatrix} \begin{pmatrix} 1 & 1 \\ 1 & 0 \end{pmatrix} = \begin{pmatrix} 7 & 5 \\ 3 & 2 \end{pmatrix},$$

obtaining $173 = 2 \cdot 7^2 + 3 \cdot 5^2$. In all, the algorithm I point to of course is well known and is reasonably efficient. By the way, although I was not obliged to say anything at all about representation by indefinite forms, the present reformulation of symmetric decomposition in terms of reduction of quadratic forms makes it clear how one will proceed in that case. A version of §2 appears as a section in [7].

4.1. Cornacchia's Algorithm

Detailed proof of the algorithm is given variously in [4] and [6]. Let me briefly hint why my remarks can lead to a proof of a generalized result. First, suppose that $n/z = [a_0, a_1, \ldots, a_s]$ with $\gcd(n, z) = 1$. Then it's not dead obvious, but not too difficult to see, that the remainders appearing in the Euclidean algorithm applied to n and z are the numerators of the quantities $[a_i, \ldots, a_s]$. Now let me illustrate what the Principal Remark says on an example from [6]. With $n = 1938758870912466947228$, take $z = 1838519813993681402789$ so that $z^2 \equiv -4755 = -15 \cdot 317 \pmod{n}$. Here $k = 1743463386375897354317$, and my symmetric decomposition yields $\left(\begin{smallmatrix} n & z \\ z & k \end{smallmatrix} \right) = N^t \left(\begin{smallmatrix} 15 & 0 \\ 0 & 317 \end{smallmatrix} \right) N$, where the first half of the continued fraction expansion of n/z is

$$[1, 18, 2, 1, 13, 5, 3, 1, 5, 1, 1, 6, 2, 40, 1, 91] \longleftrightarrow N^t = \begin{pmatrix} 11354668973 & 123452677 \\ 10767601996 & 117069841 \end{pmatrix}$$

and provides the transpose of N. By the Principal Remark we have

$$n = 15 \cdot 11354668973^2 + 317 \cdot 123452677^2.$$

But, seemingly, we've done that without even looking at the second half of the expansion and meeting remainders from the Euclidean algorithm on n and z.

However, first note that if the matrix N^t corresponds to a certain continued fraction expansion, then N corresponds to that same expansion in reverse order. Set $M = \left(\begin{smallmatrix} 15 & 0 \\ 0 & 317 \end{smallmatrix} \right)$, and note that, of course, MN sort of corresponds to the second half of the continued fraction expansion of n/z. Thus $N^t M$, recall that M is symmetric, similarly corresponds to that second half in reverse order. Precisely when M is of the shape $\left(\begin{smallmatrix} 1 & 0 \\ 0 & m \end{smallmatrix} \right)$, which is the case of representations $x^2 + my^2$ with which Cornacchia's algorithm concerns itself in the first instance, one finds that the entry in the $(1,1)$ position of $N^t M$, which is a remainder, coincides with that entry of N^t, to wit x. In our example we get $15x$ as that entry in $N^t M$.

There's a number of not altogether trivial details I gloss over, signalled in part by such words as 'half' and 'sort of', or hidden by particularities of the example. So these remarks just hint at an explanation, with a justification for Cornacchia's algorithm inter alia requiring a considerably more detailed introduction to the correspondence between continued fraction expansions and two by two matrices.

5. Acknowledgements

I am grateful to Frits Beukers for treating me as an honorary Dutchman and sending me [1], among other things provoking the remarks reported here. I also acknowledge with thanks useful advice from the referee inducing me to sketch a relation between my first principles remarks and methods directly relying on the Euclidean algorithm.

This work was supported in part by a grant from the Australian Research Council.

References

[1] Frits Beukers, *Getaltheorie voor Beginners*, Epsilon Uitgaven, Utrecht, 1999.

[2] Henri Cohen, *A course in computational algebraic number theory*, Graduate Texts in Mathematics, 138, Springer-Verlag, Berlin, 1993, xii+534 pp.

[3] Michel Dekking, Michel Mendès France and Alfred J. van der Poorten, *FOLDS! II: Symmetry disturbed*, The Mathematical Intelligencer **4** (1982), 173–181.

[4] Kenneth Hardy, Joseph B. Muskat and Kenneth S. Williams, *A deterministic algorithm for solving $n = fu^2 + gv^2$ in coprime integers u and v*, Math. Comp. **55** (1990), 327–343.

[5] Kenneth Hardy, Joseph B. Muskat and Kenneth S. Williams, *Solving $n = au^2 + buv + cv^2$ using the Euclidean algorithm*, Util. Math. **38** (1990), 225-236.

[6] Abderrahmane Nitaj, *L'algorithme de Cornacchia*, Exposition. Math. **13** (1995), 358–365.

[7] A. J. van der Poorten, *On Number Theory and Kustaa Inkeri*, to appear in Proc. Turku Symposium on Number Theory in Memory of Kustaa Inkeri (Turku, May 31–June 4, 1999), Matti Jutila and Tauno Metsänkylä eds., Walter de Gruyter, Berlin 2000, 13pp.

[8] H. J. S. Smith, *De compositione numerorum primorum formae $4\lambda + 1$ ex duobus quadratis*, J. für Math. (Crelle), **50** (1855), 91–92.

ceNTRe for Number Theory Research
Macquarie University, Sydney,
NSW 2109, Australia
E-mail address: alf@math.mq.edu.au

Progress in Computer Science and Applied Logic, Vol. 20
© 2001 Birkhäuser Verlag Basel/Switzerland

Applications of Algebraic Curves to Constructions of Sequences

Chaoping Xing

Abstract. We survey some recent constructions of various sequences based on algebraic curves over finite fields.

1. Introduction

Algebraic curves over finite fields have found many applications in various areas in recent years.

In the last few years, several new constructions of various sequences were proposed based on algebraic curves over finite fields [3, 17, 18, 19, 20, 21, 22]. In this paper we give a survey of these constructions. In Section 2 we review the necessary background on algebraic curves and their function fields. Section 3 presents the new construction of almost perfect sequences. A construction of almost perfect multi-sequences is included in Section 3. Section 4 is devoted to constructions of sequences with both low correlation and large linear complexity.

2. Background on algebraic curves

For the finite field $\mathbf{F}_q$, let $\mathcal{X}$ be a smooth, absolutely irreducible, projective algebraic curve defined over $\mathbf{F}_q$. We express this fact by simply saying that $\mathcal{X}/\mathbf{F}_q$ is an algebraic curve. The genus of $\mathcal{X}/\mathbf{F}_q$ is denoted by $g := g(\mathcal{X})$. A point on $\mathcal{X}$ is called *rational* if all its homogeneous coordinates belong to $\mathbf{F}_q$.

A divisor G of $\mathcal{X}$ is called *rational* if $G^\sigma = G$ for any automorphism $\sigma \in \mathrm{Gal}(\overline{\mathbf{F}}_q/\mathbf{F}_q)$. In this paper we always mean a rational divisor whenever a divisor is mentioned.

We denote by $\mathbf{F}_q(\mathcal{X})$ the function field of $\mathcal{X}$. A prime rational divisor of $\mathcal{X}$ is called a closed point of $\mathcal{X}$ or a place of $\mathbf{F}_q(\mathcal{X})$.

An element of $\mathbf{F}_q(\mathcal{X})$ is called a function. We write ν_P for the normalized discrete valuation corresponding to the close point P of $\mathcal{X}/\mathbf{F}_q$. Let $x \in \mathbf{F}_q(\mathcal{X})\backslash\{0\}$

Research supported by the NUS grant RP3991621.

and denote by $Z(x)$, respectively $N(x)$, the set of zeros, respectively poles, of x. We define the *zero divisor* of x by

$$(x)_0 = \sum_{P \in Z(x)} \nu_P(x) P$$

and the *pole divisor* of x by

$$(x)_\infty = \sum_{P \in N(x)} (-\nu_P(x)) P.$$

Then $(x)_0$ and $(x)_\infty$ are both rational divisors. Furthermore, the *principal divisor* of x is given by

$$\operatorname{div}(x) = (x)_0 - (x)_\infty.$$

The degree of $\operatorname{div}(x)$ is equal to zero, that is,

$$\deg((x)_0) = \sum_{P \in Z(x)} \nu_P(x) = \sum_{P \in N(x)} (-\nu_P(x)) = \deg((x)_\infty).$$

For an arbitrary divisor $G = \sum m_P P$ of $\mathcal{X}$, we denote by $\nu_P(G)$ the coefficient m_P of P. Then

$$G = \sum \nu_P(G) P.$$

For such a divisor G we form the vector space

$$\mathcal{L}(G) = \{x \in \mathbf{F}_q(\mathcal{X}) \setminus \{0\} : \operatorname{div}(x) + G \geq 0\} \cup \{0\}.$$

Then $\mathcal{L}(G)$ is a finite-dimensional vector space over $\mathbf{F}_q$, and we denote its dimension by $\ell(G)$. By the Riemann-Roch theorem (see [15, Chapter 1]), we have

$$\ell(G) \geq \deg(G) + 1 - g,$$

and equality holds if $\deg(G) \geq 2g - 1$.

Let P be a close point of $\mathcal{X}$. The integral ring of P is

$$\mathcal{O}_P = \{z \in F | \nu_P(z) \geq 0\}.$$

This is a local ring with the maximal ideal

$$\mathcal{P}_P = \{z \in \mathcal{O}_P | \nu_P(z) > 0\}.$$

The *residue class field* $\mathcal{O}_P/\mathcal{P}_P$, denoted by F_P, is a finite extension of $\mathbf{F}_q$. Thus F_P is also a finite field. The degree $[F_P : \mathbf{F}_q]$ is called the *degree* of P. It is denoted by $\deg(P)$. For an element $z \in \mathcal{O}_P$, we denote by $z(P)$ the residue class $\bar{z}$ of z in $F_P = \mathcal{O}_P/\mathcal{P}_P$.

Now we assume that P is a close of $\mathcal{X}$ of degree m, and t is a *local parameter* of P, that is, $\nu_p(t) = 1$ (such a local parameter always exists). Choose m elements $x_1, x_2, \ldots, x_m \in \mathcal{O}_P$ such that $x_1(P), x_2(P), \ldots, x_m(P)$ form an $\mathbf{F}_q$-basis of F_P.

For an element $y \in \mathcal{O}_P$, $y(P)$ can be represented as an $\mathbf{F}_q$-linear combination of $x_1(P), x_2(P), \ldots, x_m(P)$. Let $a_{10}, a_{20}, \ldots, a_{m0} \in \mathbf{F}_q$ satisfy

$$y(P) = \sum_{i=1}^{m} a_{i0} x_i(P).$$

The above equality is equivalent to

$$\nu_P(y - \sum_{i=1}^{m} a_{i0}x_i) \geq 1.$$

Hence $(y - \sum_{i=1}^{m} a_{i0}x_i)/t \in \mathcal{O}_P$. Let $a_{11}, a_{21}, \ldots, a_{m1} \in \mathbf{F}_q$ satisfy

$$\left(\frac{y - \sum_{i=1}^{m} a_{i0}x_i}{t} \right)(P) = \sum_{i=1}^{m} a_{i1}x_i(P),$$

that is,

$$\nu_P \left(\frac{y - \sum_{i=1}^{m} a_{i0}x_i}{t} - \sum_{i=1}^{m} a_{i1}x_i \right) \geq 1.$$

This is equivalent to

$$\nu_P \left(y - \sum_{i=1}^{m} a_{i0}x_i - (\sum_{i=1}^{m} a_{i1}x_i)t \right) \geq 2.$$

Hence
$(y - \sum_{i=1}^{m} a_{i0}x_i - (\sum_{i=1}^{m} a_{i1}x_i)t)/t^2 \in \mathcal{O}_P$.
By induction, we obtain a sequence of vectors $\{(a_{1j}, a_{2j}, \ldots, a_{mj})\}_{j=0}^{\infty}$ such that

$$\nu_P \left(y - \sum_{j=0}^{n} \left(\sum_{i=1}^{m} a_{ij}x_i \right) t^j \right) \geq n+1$$

for all $n \geq 0$. We express this fact by the formal series

$$y = \sum_{j=0}^{\infty} \left(\sum_{i=1}^{m} a_{ij}x_i \right) t^j. \tag{1}$$

The above series is called a *local expansion* of y at P. In particular, if $\deg(P) = 1$, that is, P is a rational point, then we get the local expansion

$$y = \sum_{j=0}^{\infty} a_j t^j, \tag{2}$$

where all a_j are elements of $\mathbf{F}_q$.

The coefficients of local expansions (1) and (2) will be used to construct our almost perfect sequences in Sections 3 and 4.

An $\mathbf{F}_q$-automorphism σ of $\mathcal{X}$ is an automorphism of $\mathbf{F}_q(\mathcal{X})$ keeping all elements of $\mathbf{F}_q$ fixed. $\mathrm{Aut}(\mathcal{X}/\mathbf{F}_q)$ denotes the group of all $\mathbf{F}_q$-automorphisms of $\mathcal{X}$. The following results can be easily proved.

Lemma 2.1 (see [1, 15]). *Let* $\sigma \in \mathrm{Aut}(\mathcal{X}/\mathbf{F}_q)$, *P a close point of* $\mathcal{X}$ *and* $f \in \mathbf{F}_q(\mathcal{X})$, *then*
(1) $\sigma(P)$ *is also a close points of* $\mathcal{X}$ *with* $\deg(\sigma(P)) = \deg(P)$;
(2) $\nu_{\sigma(P)}(\sigma(f)) = \nu_P(f)$;
(3) $\sigma(f)(\sigma(P)) = f(P)$ *if* $\nu_P(f) \geq 0$.

3. Constructions of almost perfect sequences

In this section we apply algebraic curves over finite fields to construct sequences that are relevant in the theory of stream ciphers. The security of a stream cipher is based on the quality of the keystream. A good keystream must possess satisfactory statistical randomness and complexity properties. The assessment of the quality of keystreams is a crucial task in the practice of stream ciphers. We refer to [13] for background on stream ciphers.

An important measure of the complexity of keystreams is based on the extent to which the keystream can be simulated by linear recurring sequences in $\mathbf{F}_q$. In the following, we view the zero sequence as a linear recurring sequence of order 0.

If n is a positive integer and S a sequence $s_1, s_2, \ldots$ of elements of $\mathbf{F}_q$, then the *nth linear complexity* $L_n(S)$ is defined to be the least k such that $s_1, s_2, \ldots, s_n$ form the first n terms of a kth-order linear recurring sequences in $\mathbf{F}_q$. We always have $0 \leq L_n(S) \leq n$. The sequence $L_1(S), L_2(S), \ldots$ is called the *linear complexity profile* of S. Since $L_n(S) \leq L_{n+1}(S)$, the linear complexity profile is a nondecreasing sequences of nonnegative integers.

A detailed investigation of the linear complexity profile of random sequences was carried out by Niederreiter [6].

There has been a strong interest in sequences S for which the deviation of $L_n(S)$ from $n/2$ is bounded. Such sequences were already considered by Rueppel [12, Chapter 4] and Niederreiter [4]. We follow the terminology of Niederreiter and Vielhaber [10] and call such a sequence *almost perfect*. More precisely, if d is a positive integer, then a sequence S of elements of $\mathbf{F}_q$ is called *d-perfect* if

$$|2L_n(S) - n| \leq d \quad \text{for all } n \geq 1.$$

In this section we present a method of constructing almost perfect sequences from algebraic curves over finite fields. We need the following ingredients for our construction.

$\mathcal{X}/\mathbf{F}_q$ – an algebraic curve over $\mathbf{F}_q$;
P – a rational point on $\mathcal{X}$;
t – a local parameter at P with $\deg((t)_\infty) = 2$;
f – a function in $\mathbf{F}_q(\mathcal{X})\backslash\mathbf{F}_q(t)$ with $\nu_P(f) \geq 0$.
The local expansion of f at P has the form

$$f = \sum_{i=0}^{\infty} s_i t^i$$

with all $s_i \in \mathbf{F}_q$. From this expansion we read off the sequence S_f of coefficients $s_1, s_2, \ldots$. The following result was shown in [19].

Theorem 3.1. *If $\nu_P(f) \geq 0$ and the integer d is such that $d \geq \deg((f)_\infty)$, then the sequence S_f constructed above is d-perfect.*

A systematic investigation of examples of almost perfect sequences that can be obtained from Theorem (3.1) was carried out by Kohel, Ling, and Xing [3]. This paper also discusses the efficient computation of local expansions by means of Hensel's lemma.

Example 3.1. Let $\mathcal{X}$ be the projective line over $\mathbf{F}_2$. Its function field is the rational function field $\mathbf{F}_2(x)$. We choose $P = 0, t = x^2 + x$, and $f = x/(x+1)$. Then we have the expansion

$$f = \frac{x^2}{t} = \sum_{h=1}^{\infty} t^{2^h - 1}.$$

The sequence S_f of coefficients $1, 0, 1, 0, 0, 0, 1, \ldots$ is 1-perfect by Theorem (3.1). This is the same sequence as the generalized Rueppel sequence.

Example 3.2. Consider the elliptic curve over $\mathbf{F}_2$ defined by

$$y^2 + y = x^3 + x^2 + x.$$

There are three points on this curve, that is, $(0,0), (0,1)$ and ∞. We only consider expansions of functions at $(0,0)$. x is a local parameter at $(0,0)$ and $\mathrm{div}(x) = (0,0) + (0,1) - 2\infty$.

The local expansion of y/x at $(0,0)$ is

$$\frac{y}{x} = 1 + \sum_{m=1}^{\infty} x^{3 \cdot 2^{m-1} - 1}.$$

The sequence S_f of coefficients $(0, 1, 0, 0, 1, 0, 0, 0, 0, 0, 1, \ldots)$ is 2-perfect.

4. Construction of almost perfect multi-sequences

In this section, we generalize almost perfect sequences to almost perfect multi-sequences. Furthermore, based on function fields over finite fields, we present a construction of perfect multi-sequences. This construction is a natural, but not obvious generalization of our construction for almost perfect sequences of Theorem (3.1) [17].

Consider a set A of multi-sequences of dimension $m \geq 1$

$$\mathbf{a}_1 = (a_{11}, a_{12}, a_{13}, \ldots) \in \mathbf{F}_q^\infty$$
$$\mathbf{a}_2 = (a_{21}, a_{22}, a_{23}, \ldots) \in \mathbf{F}_q^\infty$$

$$\cdot$$
$$\cdot$$
$$\cdot$$

$$\mathbf{a}_m = (a_{m1}, a_{m2}, a_{m3}, \ldots) \in \mathbf{F}_q^\infty$$

If n is a positive integer, then the *nth linear complexity* $L_n(A)$ is defined to be the least k such that $a_{i1}, a_{i2}, \ldots, a_{in}$ is generated by a fixed kth-order linear recurring shift register in $\mathbf{F}_q$ for all $i = 1, 2, .., m$. We always have $0 \le L_n(A) \le n$.

The sequence $L_1(A), L_2(A), \ldots$ is called the *linear complexity profile* of A. This linear complexity profile is also a nondecreasing sequences of nonnegative integers.

If d is a positive integer, then a multi-sequence A of dimension m of elements of $\mathbf{F}_q$ is called *d-perfect* if

$$L_n(A) \ge \lceil \frac{m(n+1) - d}{m+1} \rceil \quad \text{for all } n \ge 1.$$

If A is a d-perfect multi-sequence set of dimension m, then d is at least m. Furthermore, we have the following result [17].

Theorem 4.1. *A multi-sequence set $A = \{\mathbf{a}_1, \mathbf{a}_2, \ldots, \mathbf{a}_m\}$ of dimension m is m-perfect if and only if*

$$\mathcal{L}_n(A) = \lceil \frac{mn}{m+1} \rceil$$

for all $n \ge 1$.

Throughout this section, we have the following notations and assumptions.

$\mathcal{X}/\mathbf{F}_q$ – an algebraic curve over $\mathbf{F}_q$;
Q – a close point of degree m of $\mathcal{X}$;
$x_1, x_2, \ldots, x_m$ – m elements of $\mathcal{O}_Q$ satisfying the condition that $x_1(Q), x_2(Q), \ldots, x_m(Q)$ form an $\mathbf{F}_q$-basis of F_Q; t – a local parameter of Q with $\deg(t)_\infty = m+1$;
y – an element of $\mathcal{O}_Q$ satisfying $y \notin \bigoplus_{i=1}^m \mathbf{F}_q(t)x_i$.

Consider the local expansion of y at Q as in (1)

$$y = \sum_{j=0}^\infty \left(\sum_{i=1}^m a_{ij} x_i \right) t^j.$$

Put

$$\mathbf{a}_i(y) = (a_{i1}, a_{12}, a_{i3}, \ldots) \in \mathbf{F}_q^\infty$$

for any $1 \le i \le m$ and the multi-sequence set

$$A(y) = \{\mathbf{a}_i(y)\}_{i=1}^m. \tag{3}$$

Theorem 4.2 (see [17]). *Let $A(y) = \{\mathbf{a}_i(y)\}_{i=1}^m$ be constructed as in (3). Then $A(y)$ is d-perfect, where $d = \deg((y)_\infty \vee (x_1)_\infty \vee (x_2)_\infty \vee \cdots \vee (x_m)_\infty)$. In particular, $\{\mathbf{a}_i(y)\}_{i=1}^m$ are m-perfect multi-sequences if $d = m$.*

For any $m \ge 1$, m-perfect multi-sequences of dimension m can be constructed using the above theorem. The following example shows a 2-perfect binary sequence of dimension 2.

Example 4.1. $\mathcal{X}/\mathbf{F}_2$ – the projective line; Q – the unique zero of $x^2 + x + 1$; $t = x(x^2 + x + 1)$ – a local parameter of Q. Put $y = x^2$ and $x_1 = 1, x_2 = x$. Then $x_1(Q), x_2(Q)$ are an $\mathbf{F}_2$-basis of $\mathbf{F}_Q$ and $d = \deg((y)_\infty \vee (x_1)_\infty \vee (x_2)_\infty) = 2 = \deg(Q)$. The local expansion of y at Q is

$$y = (x_1 + x_2) + (x_1 + x_2)t + x_2 t^2 + (x_1 + x_2)t^3 + x_1 t^4 + x_1 t^6 + (x_1 + x_2)t^8 + \cdots.$$

Put

$$\mathbf{a}_1(y) = (1, 0, 1, 1, 0, 1, 0, 1, \cdots),$$
$$\mathbf{a}_2(y) = (1, 1, 1, 0, 0, 0, 0, 1, \cdots).$$

By Theorem (4.2), $\mathbf{a}_1(y), \mathbf{a}_2(y)$ are 2-perfect multi-sequences of dimension 2.

5. Constructions of sequences with low autocorrelation and large complexity

At the beginning of Section 3 we mentioned that a good keystream must possess both satisfactory statistical randomness and complexity properties. In Sections 3 and 4, we constructed sequences having large linear complexity. However, those sequences usually possess poor statistical randomness. In this section, we present a construction of periodic sequences possessing both satisfactory statistical randomness and complexity properties based on algebraic curves over finite fields.

A common method of measuring randomness of a sequence is based on autocorrelation of the sequence [2]. From now on, we will only consider binary periodic sequences. The autocorrelation at shift w of a binary periodic sequence $\mathbf{a} = (a_1, a_2, \dots)$ of period n is given by

$$\theta_{\mathbf{a}}(w) = \sum_{i=1}^{n} (-1)^{a_i + a_{i+w}}.$$

We denote by $\theta_{\mathbf{a}}$ the maximum of absolute value of $\theta_{\mathbf{a}}(w)$ for all w not congruent to 0 (mod n).

The *linear complexity* $L(\mathbf{a})$ of a sequence $\mathbf{a}$ of period n is defined to be the least k such that $a_1, a_2, \dots, a_{2n}$ form the first $2n$ terms of a kth-order linear recurring sequences in $\mathbf{F}_q$.

We will always assume that the characteristic p of $\mathbf{F}_q$ is equal to 2 in this section. We also fix some notation for this section.

$\mathcal{X}/\mathbf{F}_q$ – an algebraic curve over $\mathbf{F}_q$;
P – a rational point of $\mathcal{X}/\mathbf{F}_q$;
σ – an automorphism in $\mathrm{Aut}(\mathcal{X}/\mathbf{F}_q)$;
n – the least positive integer satisfying $\sigma^n(P) = P$, that is, $\sigma^n(P) = P$ and $\sigma^i(P) \neq P$ for all $1 \leq i \leq n$.

Theorem 5.1 (see [18]). *Put* $P_i = \sigma^i(P)$ *for all* $i \geq 1$. *Let* $z \in \mathcal{X}(\mathbf{F}_q)$ *satisfy* $\nu_{P_i}(z) \geq 0$. *Suppose that* Q *is the unique pole of* z *with* $(\nu_Q(z), 2) = 1$, *and* $Q, \sigma(Q), \sigma^2(Q), \ldots, \sigma^{n-1}(Q)$ *are pairwise distinct. Put*

$$\mathbf{a}_z := (Tr(z(P_1)), Tr(z(P_2)), \ldots, Tr(z(P_n)))$$

where Tr *is the trace map from* $\mathbf{F}_q$ *to* $\mathbf{F}_2$. *Then the linear complexity of* $\mathbf{a}_z$ *satisfies*

$$\mathcal{L}(\mathbf{a}_z) \geq \frac{2n - q - 1 - 2(2g(F) + d - 1)\sqrt{q}}{2d\sqrt{q}}$$

and the autocorrelation of $\mathbf{a}_z$ *satisfies*

$$\theta_{\mathbf{a}_z} \leq 2(2g(F) + 2d - 1)\sqrt{q} + |q + 1 - n| + 2(N(F) - n),$$

where d *is the degree of the pole divisor* $(z)_\infty$ *of* z.

By applying the above theorem to projective lines and elliptic curves, we obtain some interesting results in the following two examples.

Example 5.1. Let $m \geq 6$ and $q = 2^m$. Consider the projective line over $\mathbf{F}_q$. Then $\mathbf{F}_q(\mathcal{X})$ is the rational function field $\mathbf{F}_2(x)$. Let P be the unique zero of x and σ be the automorphism of $\mathcal{X}/\mathbf{F}_q$ sending x to ζx, where ζ is a primitive element of $\mathbf{F}_q$. Let $z(x) \in \mathbf{F}_2(x)$ with the unique pole being an irreducible polynomial of degree 2 in $\mathbf{F}_q[x]$.

Then by Theorem (5.1) the binary sequence $\mathbf{a}_z$ is of period $2^m - 1$. Moreover,

$$\mathcal{L}(\mathbf{a}_z) \geq 2^{m/2-2} - \frac{1}{2} - \frac{3}{2^{m/2+2}},$$

and

$$\theta_{\mathbf{a}_z} \leq 6(2^{m/2} + 1).$$

Example 5.2. Let $m \geq 8$ and let t satisfy one of the following conditions
(i) $t = 0$, or
(ii) $t = \sqrt{q} = 2^{m/2}$ if m is even, and $t = \sqrt{2q} = 2^{(m+1)/2}$ if m is odd.
Then there exists an elliptic curve $\mathcal{X}$ over $\mathbf{F}_q$ such that its rational points form a cyclic group of order $n := 1 + q + t$ (see [11]). Fix a rational point O as the zero point and let P be a generator of $\mathcal{X}(\mathbf{F}_q)$. Consider the automorphism σ of $\mathcal{X}/\mathbf{F}_q$ fixing O and sending P to $2P$ (such an automorphism exists (see [1]). Let Q be a close point of $\mathcal{X}$ of degree 2 (such a point also exists by the Weil theorem). By the Riemann-Roch theorem, there exists a function z with the unique pole Q. It can be proved that $Q, \sigma(Q), \ldots, \sigma^{n-1}(Q)$ are pairwise distinct.

Thus by Theorem (5.1) the binary sequence $\mathbf{a}_z$ is of period $2^m + 1 + t$. Moreover,

$$\mathcal{L}(\mathbf{a}_z) \geq 2^{m/2-2} - \frac{3}{2} - \frac{1 + 2t}{2^{m/2+2}}$$

and

$$\theta_{\mathbf{a}_z} \leq 10 \cdot 2^{m/2} + |t|.$$

References

[1] M. Eichler, *Introduction to the Theory of Algebraic Numbers and Functions*, Academic Press, New York, 1951.

[2] T. Helleseth and P.V. Kumar, *Sequences with Low Correlation*, a chapter in: Handbook of Coding Theory, edited by V. Pless and C. Huffman, Elsevier Science Publishers, 1998.

[3] D. Kohel, S. Ling, and C.P. Xing, *Explicit sequence expansions*, In: Sequences and Their Applications (C. Ding, T. Helleseth, and H. Niederreiter, eds.), pp. 308–317, Springer, London, 1999.

[4] H. Niederreiter, *Continued fractions for formal power series, numbers, and linear complexity of sequences*, In: Contributions to General Algebra 5, (Proc. Salzburg Conf., 1986), pp. 221–233, Teubner, Stuttgart, 1987.

[5] H. Niederreiter, *Sequences with almost perfect linear complexity profile*, In: Advances in Cryptology – EUROCRYPT '87(D. Chaum and W.L. Price, eds.), Lecture Notes in Computer Science, Vol. **304**, pp. 37–51, Springer, Berlin, 1988.

[6] H. Niederreiter, *The probabilistic theory of linear complexity*, In: Advances in Cryptology – EUROCRYPT '88 (C.G. Günther, ed.), Lecture Notes in Computer Science, Vol. **330**, pp. 191–209, Springer, Berlin, 1988.

[7] H. Niederreiter, *Keystream sequences with a good linear complexity profile for every starting point*, In: Advances in Cryptology – EUROCRYPT '89 (J.-J. Quisquater and J. Vandewalle, eds.), Lecture Notes in Computer Science, Vol. **434**, pp. 523–532, Springer, Berlin, 1990.

[8] H. Niederreiter, *Finite fields and cryptology*, In: Finite Fields, Coding Theory, and Advances in Communications and Computing (G. L. Mullen and P.J.-S. Shiue, eds.), pp. 359–373, Dekker, New York, 1993.

[9] H. Niederreiter, *Some computable complexity measures for binary sequences*, In: Sequences and Their Applications (C. Ding, T. Helleseth, and H. Niederreiter, eds.), pp. 67–78, Springer, London, 1999.

[10] H. Niederreiter and M. Vielhaber, *Linear complexity profiles: Hausdorff dimensions for almost perfect profiles and measures for general profiles*, J. Complexity, **13**(1997), 353–383.

[11] H.-G. Rück, *A Note on Elliptic Curves over Finite Fields*, Math. of Comp., **49**(1987), 301–304.

[12] R.A. Rueppel, *Analysis and Design of Stream Ciphers*, Springer, Berlin, 1986.

[13] R.A. Rueppel, Stream ciphers, *Contemporary Cryptology: The Science of Information Integrity* (G.J. Simmons, ed.), pp. 65–134, IEEE Press, New York, 1992.

[14] J.H. Silverman, *The Arithmetic of Elliptic Curves*, Springer-Verlag, New York, 1986.

[15] H. Stichtenoth, *Algebraic Function Fields and Codes*, Springer, Berlin, 1993.

[16] M.A. Tsfasman and S.G. Vlădut, *Algebraic-Geometric Codes*, Kluwer, Dordrecht, 1991.

[17] C.P. Xing, *Multi-sequences with almost perfect linear complexity profile and function fields over finite fields*, Journal of Complexity, to appear.

[18] C.P. Xing, V.J. Kumar and C.S. Ding, *Low correlation, large linear span sequences from function fields*, Preprint, 2000.

[19] C.P. Xing and K.Y. Lam, *Sequences with almost perfect linear complexity profiles and curves over finite fields*, IEEE Trans. Inform. Theory, **45**(1999), 1267–1270.

[20] C.P. Xing and N. Niederreiter, *Applications of algebraic curves to constructions of codes and almost perfect sequences*, to appear in Proceedings of the 5th International Conference on Finite Fields and Applications.

[21] C.P. Xing, H. Niederreiter, K.Y. Lam, and C.S. Ding, *Constructions of sequences with almost perfect linear complexity profile from curves over finite fields*, Finite Fields Appl. **5**, 301–313 (1999).

[22] C.P. Xing, K.Y. Lam and Z.H. Wei, *A class of explicit perfect multi-sequences*, In: Advanced in Cryptology – Asiacrypt'99 (K.Y. Lam, E. Okamoto, C.P. Xing, eds.), Lecture Notes in Computer Science, Vol. 1716, pp. 299–305, Springer-Verlag, Berlin, 1999.

Department of Mathematics,
National University of Singapore,
2 Science Drive 2,
S117543, Singapore
E-mail address: matxcp@nus.edu.sg

Cryptography

Progress in Computer Science and Applied Logic, Vol. 20

Designated 2-Verifier Proofs and their Application to Electronic Commerce*

Nikos Alexandris, Mike Burmester, Vassilis Chrissikopoulos, and Yvo Desmedt

Abstract. Designated verifier interactive proofs have been used as non-malleable mechanisms and to prevent man-in-the-middle attacks. In this paper we use designated 2-verifier proofs to design simple and secure electronic payment systems. Two on-line protocols are presented. These link the Customer (the prover) with the Merchant and Bank (the two verifiers) inextricably to their transactions. In the first protocol the identity of the Customer is traceable. This can be used for general electronic payment systems. The second protocol can be used for anonymous electronic cash payments. Both protocols have a simple structure and are provably secure.

1. Introduction

The recent wide utilization of international interconnected computer networks such as the internet and web applications, and the rapid expansion of electronic commerce have stimulated the demand for electronic payment systems. In particular, electronic commerce is recognized as a leading application in the information society. The internet is perceived as a virtual market place which offers significant advantages over the traditional market. The business potential of such applications is the major driving force for electronic payment systems. Payment systems must be simple and flexible, and must also offer a high level of security to protect all the parties involved.

The first electronic cash payment systems which emulate the properties of physical cash were proposed by D. Chaum [11, 12, 14]. Subsequently many other authors contributed to this area, for example, see [9, 16, 17, 37, 6]), which rapidly expanded to include general financial and payment systems, for example, see [23, 33, 34, 32]).

Some of the basic desired properties of electronic cash are *functionality, security, acceptability* and, for some applications, *anonymity*. Functionality requires that the basic provisions of physical money are satisfied. Security deals with forgery and double spending. Electronic payment systems should prevent these. Acceptability deals with the usefulness of payment mechanisms. Payment instruments must be accepted widely by different financial institutions to facilitate reconciliation.

---

* The present paper is an extension of [1].

In many financial transactions it is desirable that the identity of the customer is traceable, to prevent several frauds such as money laundering, blackmailing and illegal purchases. However there are transactions in which the identity of the customer must be protected. With these it should be impossible to monitor the spending patterns of the customer, or to determine the customer's source of income. There are many arguments for, as well as against, anonymous electronic payments. Depending on the application and the general social and legal constraints, an appropriate system needs to be used.

Electronic payments can be performed on-line, where a financial center is involved directly in each payment, or off-line, without direct contact. On-line systems are in general more secure but require more communication. Off-line systems usually require tamper-resistant hardware to prevent double spending.

There are essentially three types of electronic payment systems: *credit card* systems, *cheque* systems and *cash* systems. Credit card systems provide credit through a card infrastructure. Cheque systems involve payment instruments which are linked to the issuer. Cash systems involve tokens which are valid independently of the issuer. While for credit card and cheque systems the identity of the customer is traceable, for cash systems the customer may be anonymous.

Several protocols for electronic payment systems have been proposed in the literature, for example, iKP [3], SET [32], CyberCash [19], NetCheque [34], FSTC [27], Ecash [18], CAFE [5], NetCash [33]. These offer varying degrees of cryptographic security, which is based on the security of the underlying cryptographic mechanisms. However these are not proven secure against the most general type of attacks, for example, malleable attacks [22].

Jakobsson et al [31] introduced the notion of designated verifier proofs in which the prover designates a specific verifier who is the only party that can obtain a conviction of the correctness of the proof. These proofs can be used as non-malleable mechanisms and prevent man-in-the-middle attacks [4].

In this paper we use the notion of designated 2-verifier interactive proofs to design two simple and secure on-line electronic payment systems. The first one can be used for electronic cheque and credit card systems, while the second is for electronic cash systems. Our motivation is to design protocols which are provably secure in a strong sense, that is non-malleable. The notion of non-malleability is an extension of semantical security [22]. It guarantees that given encryptions, digitally signed or communicated data, it is not possible to generate encryptions, digital signatures or communications of different related data. For our first protocol we use two settings, an RSA setting and a DL (discrete logarithm) setting. Depending on the setting, the security of this protocol reduces to either the factoring problem or the DL problem. In the RSA setting we use a novel trapdoor commitment function which is of interest in itself. The security of the second protocol reduces to the DL problem.

The organization of this paper is as follows. In the rest of this section we consider some general threats for electronic payment systems. In Section 2 we discuss our approach and in Section 3 we present designated 2-verifier protocols for an RSA setting and a DL setting, and a protocol for secure electronic payments. In Section 4 we present a protocol for electronic cash systems. We conclude in Section 5.

1.1. Threats in Electronic Payments

Forgery and double spending are major problems for electronic cash systems. With off-line systems double spending can be controlled to some extend by using tamper-resistant devices. With on-line systems this problem is controlled by the instantaneous nature of communication, but other threats are possible, such as man-in-the-middle attacks and their variants. These attacks can be controlled by using designated verifier proofs. Below we illustrate some of these attacks.

1. *Double spending.* Customer A' attempts to spend the same electronic coin at Merchants M_1 and M_2.

$$A' < \begin{array}{l} M_1 \longrightarrow B \\ M_2 \longrightarrow B \end{array}$$

2. *Man-in-the-middle or "mafia" attack* [4]. Merchant M_2' (an active eavesdropper) attempts to get credited with the payment of Customer A to Merchant M_1 by (simply) relaying data.

$$A \longrightarrow M_1 \longrightarrow M_2' \longrightarrow B$$

3. *Impersonation attack.* Merchants M_1', M_2' conspire to fool Customer A and Bank B, so that A is debited for $\$y$ while purchasing goods for $\$x$.

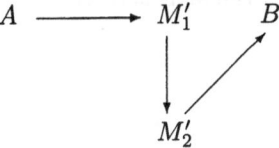

An example of such an attack is illustrated in the EPOS (Electronic Point-of-Sale) transaction in Figure 1, below. Observe that such attacks are possible with EPOS transactions because EPOS terminals only verify "who you are" and not "to whom the payment should be made" or "how much you spend". It is essential therefore that with EPOS transactions the Merchant is designated by the Customer.

4. *General man-in-the-middle attacks on interactive proof systems.* With interactive proof systems the on-line confidence which is conveyed to the verifier cannot necessarily be transferred off-line to third parties. In particular, with zero-knowledge interactive proof systems the communicated data can be simulated off-line by anybody (no extra knowledge is required

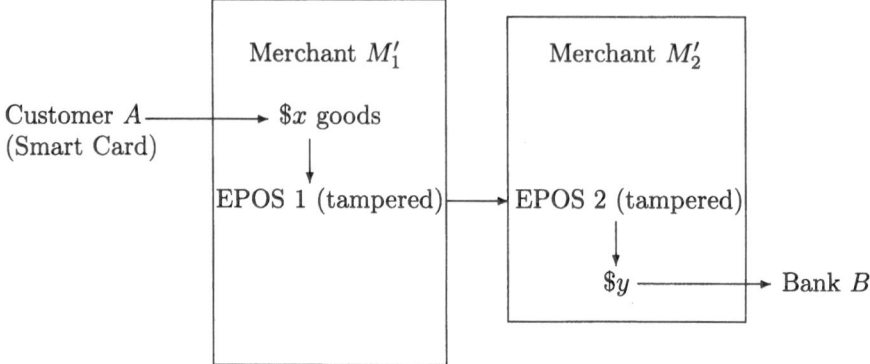

Figure 1. The Customer is debited $y while purchasing goods for $x.

for the simulation). Therefore no confidence can be converted off-line. However this is not necessarily the case with on-line confidence. General man-in-the-middle attacks on interactive systems are on-line attacks in which the adversary intercepts the data sent by the prover and verifier, and either:

- *Relay attacks* [4]: relays the data claiming to be the prover, as in attack 1 above (such attacks can be used, for example, with interactive identification schemes such as those in [26, 30]), or
- *Divertible attacks* [36, 21, 8]: modifies the data to get a *different* interactive proof system claiming to be the prover of the modified proof (as for example with *divertible* proofs – an illustration of such an attack is given in Figure 1, with M_2' the attacker, when $x \neq y$ and $M_1' = M_1$ is untampered).

There seems to be no easy theoretical way to defeat such attacks in the general case [21, 7, 8].[1] Recently an approach has been proposed which restricts the scope (the on-line power) of interactive proof systems [31]. With this the prover *designates* specific verifiers, who are the only ones who will gain the intended on-line confidence. We shall adopt this approach and use designated 2-verifier proofs to control such attacks. These can be regarded as implicit confirmations. Our proofs will link the parties involved in the designated proofs inextricably.[2]

---

[1]With interactive proof systems. Good signatures will thwart such attacks, but they may be less efficient – *cf.* the discussion in the last part of Remark 3.3, Section 3.4.

[2]With many applications it is sufficient to prevent only the relay attacks, since the divertible attacks are controlled by other mechanisms. This observation applies to all our protocols.

2. Approach and Model

There are three parties in our model: the Customer, the Merchant and the Bank. The Customer wishes to purchase goods from the Merchant (but only from this Merchant) by using the Bank. We consider two scenarios. In the first the Customer uses his account in the Bank (a credit card or an electronic cheque), and in the second the Customer uses electronic cash issued by the Bank. For these we will present interactive protocols which link the three parties on-line through designated verifier proofs. Interactive proof systems for which the verifier is designated have been used for software copyright applications by Jakobsson et al. [31]. We shall use such proof systems to design efficient and secure electronic payment systems.

Our protocols are based on proof system in which the prover P designates *two* verifiers U *and* W *jointly*. P will prove that she knows $x_p \vee f(x_u, x_w)$, where x_p is a secret key of P, "$\vee$" is inclusive-or, f is an appropriate function and x_u, x_w are the secret keys of U, W respectively. The prover P can do this because she knows the secret x_p. The verifiers U and W will be convinced that P knows x_p unless $f(x_u, x_w)$ is compromised. The function f is such that $f(x_u, x_w)$ is not known to either U or W but can be used in the protocol by U and W without having to reveal their secret keys x_u, x_w. Examples of such functions are $f(x_u, x_w) = x_u \cdot x_w$, $f(x_u, x_w) = x_u + x_w$.[3] Since the interactive proof system is designated by the prover P to the verifiers U and W, a different verifier U' cannot convert this proof to a proof of knowledge of $x_p \vee f(x'_u, x_w)$, or even to a proof of knowledge of $x'_p \vee f(x'_u, x_w)$ for $x'_p \neq x_p$. In our first protocol we will use such a proof system two times. The first time the prover will be the Customer and the verifier will be the Bank, with designated verifiers the Bank and the Merchant. The second time the prover will be the Bank and the verifier the Merchant, with designated verifiers the Merchant and the Customer. The Merchant and the Customer are linked with the transaction data in such a way so as to get designated proofs which authenticate the transactions. With our protocols an arbitrary party is able to determine who is designated because the interactive proof links the secret keys x_p and x_u, x_w of the prover and the designated verifiers (which correspond to their public keys). It is this property which is used to link the parties involved.

We will consider two settings, an RSA setting and a DL setting, and present for each setting an electronic payment protocol. We will then use Chaum's confirmation scheme for undeniable signatures [15] as modified by Jakobsson et al. [31], but designated to *two* verifiers to get an electronic cash system. For this protocol the prover is the Bank and the verifier is the Merchant. The Bank will designate the Merchant and the Customer.

With all our schemes we shall assume that the Bank *is trusted*. In particular, the Bank will not conspire with the Merchant to defraud the Customer, or the Customer to defraud the Merchant. The Bank will keep a record (a database) of all transactions for a limited period to settle possible disputes.

---

[3] U and W do not learn anything additional about each other's secret key because the value of $f(x_u, x_w)$ is not known to either.

154   N. Alexandris, M. Burmester, V. Chrissikopoulos, and Y. Desmedt

3. A Secure Payment System

3.1. An Interactive Designated 2-Verifier Proof $(P,V)_{U,W}$ for an RSA Setting

We use an RSA setting with modulus n, a product of two large primes. Let $m = \log_2 n$ and k be the security parameter. The prover P has k private keys, $x_{p,i} \in_R Z_n^*$, $i = 1, 2, \ldots, k$. We adopt the notation $a \in_R A$ to indicate that the element a is selected randomly with uniform distribution from the set A. The verifiers U, W each have m private keys, $x_{u,j}, x_{w,j} \in_R Z_n^*$, $j = 1, 2, \ldots, m$. The corresponding public keys are $y_{p,i} = x_{p,i}^2 \bmod n$, and $y_{u,j} = x_{u,j}^2 \bmod n$, $y_{w,j} = x_{w,j}^2 \bmod n$, $i = 1, 2, \ldots, k$, $j = 1, 2, \ldots, m$. The proof system will use the following commitment function with trapdoor the *product* of the private keys of U, W: $\{x_{u,j}x_{w,j}\}_{j=1}^m$ (*cf.* [31]). Let $b = b_1 b_2 \cdots b_m$ be a bitstring and $r \in_R Z_n^*$. The commitment for b is,

$$c = (y_{u,1}y_{w,1})^{b_1}(y_{u,2}y_{w,2})^{b_2} \cdots (y_{u,m}y_{w,m})^{b_m} \cdot r^2 \bmod n.$$

The commitment is opened by revealing b, r. If the trapdoor is known then it is possible to open the commitment c as b', r', for any bitstring $b' = b_1' b_2' \cdots b_m'$, by taking $r' = r \cdot \prod_{j=1}^m (x_{u,j}x_{w,j})^{b_i - b_i'} \bmod n$. However if (all) the keys of the trapdoor are not known then the commitment can only be opened as b, r, provided factoring n is hard.

The following protocol is based on the Fiat-Shamir interactive proof system [26], which is modified so that it is designated to U *and* W. In this, the prover P proves to the verifier V that she either knows the keys $\{x_{p,i}\}_{i=1}^k$ or the trapdoor $\{x_{u,j}x_{w,j}\}_{j=1}^m$. Let $hash_k$ be a collision resistant hash function with values in $\{0,1\}^k$, and let msg be a message.

An Interactive 2-Verifier Proof $(P,V)_{U,W}$ which authenticates a message msg for an RSA setting

1. The prover P selects $t, r \in_R Z_n^*$, computes $z = t^2 \bmod n$ and the trapdoor commitment

$$c = (y_{u,1}y_{w,1})^{z_1}(y_{u,2}y_{w,2})^{z_2} \cdots (y_{u,m}y_{w,m})^{z_m} \cdot r^2 \bmod n, \qquad (1)$$

   where z_i is the i-th bit of z. P sends to V: P, U, W, msg, c.
2. The verifier V selects a bitstring $e \in_R \{0,1\}^k$ and sends it to P.
3. P computes $h = hash_k(P, U, W, msg, c, e)$, and $y = t \cdot \prod_{i=1}^k x_{p,i}^{h_i} \bmod n$, where h_i is the i-th bit of h. P sends to V: z, r, y.
4. V computes $h = hash_k(P, U, W, msg, c, e)$. Then the verifier V checks that (1) is satisfied, and that $z \cdot \prod_{i=1}^k y_{p,i}^{h_i} \equiv y^2 \pmod{n}$. If the check fails, V halts and rejects.

Theorem 3.1. $(P,V)_{U,W}$ *is an interactive proof system of knowledge* [24] *with designated verifiers U and W in which P authenticates the message msg, if factoring is hard.*

Proof. This is an extension of the proof in [24]. Completeness is straightforward. Soundness follows by observing that, if the trapdoor $\{x_{u,j}x_{w,j}\}_{j=1}^m$ is not known then the commitment c can be opened only as z, r. Let P^* be a program which succeeds with probability $\varepsilon > 2^{-k}$ (better than guessing) in getting V to accept (the proof), and let AL be a probabilistic algorithm which uses P^* as a subroutine in order to factor the modulus n. AL chooses private keys $x_{p,i}, x_{u,j}, x_{w,j} \in_R Z_n^*$, $i = 1, 2, \ldots, k$, $j = 1, 2, \ldots, m$ and computes the corresponding public keys $y_{p,i}, y_{u,j}, y_{w,j}$ which it gives to P^*. Then AL runs P^* for $1/\varepsilon$ times and checks the outputs of P^*. It is easy to see that AL will find a pair $y, y' \in Z_n$ such that, $y^2 \equiv z \cdot \prod y_{p,i}^{h_i}(\mathrm{mod}\, n)$ and $y'^2 \equiv z \cdot \prod y_{p,i}^{h_i'}(\mathrm{mod}\, n)$. Then $(y'/y)^2 \equiv \prod y_{p,i}^{h_i - h_i'}(\mathrm{mod}\, n)$, with $h_i - h_i' \in \{0, 1, -1\}$. This implies that AL can find a witness $y'/y \bmod n$ for $\prod y_{p,i}^{h_i - h_i'} \bmod n$, which with probability $1/2$ is different from $\pm \prod x_{p,i}^{h_i - h_i'} \bmod n$ (which AL can compute from the $x_{p,i}$). Then AL can factor n.

Verifier designation: U, W together can generate a proof for the verifier V by using the trapdoor $\{x_{u,j}x_{w,j}\}_{j=1}^m$. In step 1, U, W send P, U, W, msg, c to V, where c is computed in the same way as with P. Let e be the 'challenge' of V and h the corresponding hash value. In step 3, U, W open c as a commitment for z', r', where $z' = t \cdot \prod_{i=1}^k y_{p,i}^{-h_i} \bmod n$, by solving the congruence

$$(x_{u,1}x_{w,1})^{z_1} \cdots (x_{u,m}x_{w,m})^{z_m} \cdot r \equiv (x_{u,1}x_{w,1})^{z_1'} \cdots (x_{u,m}x_{w,m})^{z_m'} \cdot r' \quad (\mathrm{mod}\, n),$$

for r'. They give the verifier V: z', r' and $y = t$. It is easy to see that the checks of V will be satisfied. $\qquad\square$

SECURITY AND EFFICIENCY It is unlikely that the proof $(P, V)_{U,W}$ is zero-knowledge, that is, that the communication of P in $(P, V')_{U,W}$, where V' is any verifier, can be simulated in polynomial time by V'). However the proof is *witness indistinguishable* (V' cannot find out which particular witness is actually used) and *witness hiding* [25] (V' cannot compute a witness that he did not know before participating in the protocol. Witness indistinguishability is preserved under arbitrary (polynomial) compositions of such protocols. The compositions we will consider are witness hiding.

Compared to the Fiat-Shamir identification scheme, we have the extra cost of the commitment in the proof $(P, V)_{U,W}$. This is roughly one exponentiation. Furthermore we need m keys for the commitment, so the length of the key is m^2.

3.2. An Interactive Designated 2-Verifier Proof $(P, V)_{U,W}$ for a DL Setting

We next consider a designated interactive proof system which uses a discrete logarithm setting, with modulus p a large prime, q a large prime divisor of $p - 1$, and $g \in Z_p^*$ an element of order q. The private keys of the prover P and the verifiers U *and* W are $x_p, x_u, x_w \in Z_q$, respectively. The corresponding public keys are $y_p = g^{x_p} \bmod p$, $y_u = g^{x_u} \bmod p$, and $y_w = g^{x_w} \bmod p$. The proof is based on the Schnorr identification scheme [40], which we modify by using a commitment function with trapdoor $x_u + x_w$, so that it is designated to U *and* W. In this proof, the prover P proves to the verifier V that she either knows x_p or $x_u + x_w$. Let k be the *security parameter*, $hash_q$ a collision resistant hash function with values in Z_q, and msg a message.

An interactive 2-verifier proof $(P, V)_{U,W}$ which authenticates a message msg for a DL setting

1. The prover P selects $t, r \in_R Z_q$ and computes $z = g^t \bmod p$ and the trapdoor commitment $c = g^z (y_u y_w)^r \bmod p$. P sends to V: P, U, W, msg, c.
2. The verifier V selects a number $e \in_R Z_q$ and sends it to P.
3. P computes $h = hash_q(P, U, W, msg, c, e)$ and $y = t + h x_p \bmod q$, and sends to V: z, r, y.
4. V computes $h' = hash_q(P, U, W, msg, c, e)$, checks that $z \cdot y_p^{h'} \equiv g^y \ (\bmod \ p)$ and that $c \equiv g^z \cdot (y_u y_w)^r \ (\bmod \ p)$. If the check fails, V halts and rejects.

Theorem 3.2. *$(P, V)_{U,W}$ is an interactive proof system of knowledge with designated verifiers U and W, in which P authenticates the message msg, if the discrete logarithm problem is hard.*
Proof. This is an extension of the proof in [40], and is similar to the proof of Theorem 3.1. We only show the last part.

Verifier designation: U, W together can generate a proof by using the trapdoor $x_u + x_w$. In step 1, U, W send P, U, W, msg, c to V, where c is computed in the same way as with P. Let e be the challenge of V and h the corresponding hash value. In step 3, U, W open c as a commitment to $z' = g^t y_p^{-h} \bmod p$, by solving the congruence

$$z + (x_u + x_w)r \equiv z' + (x_u + x_w)r' \pmod{q},$$

for r'. They give V: z', r' and $y = t$. $\qquad\Box$
SECURITY AND EFFICIENCY   This proof system is also witness indistinguishable and witness hiding [25].

Compared to the Schnorr identification scheme, we have the additional cost of the commitment in the system $(P, V)_{U,W}$. This requires two exponentiations and one multiplication. For this system only one key is required for the commitment.

3.3. Digital Signatures vs Interactive Proof Systems

Interactive proofs convey on-line confidence which is not necessarily transferable off-line in any useful way (as for example with witness hiding proofs). On the other hand digital signatures convey confidence which is transferable both off-line and on-line. Therefore digital signatures can be used to certify *directly* financial transactions without any interaction. However they are less efficient than interactive proof systems [26]. Furthermore, they must involve some sort of timestamps, to prevent replay attacks. Since timestamps are inherently weak [39], we cannot get provable security. For these reasons we have not used this approach.

We can also use digital signatures interactively in a challenge-respond mode. This is similar to the approach we shall use. However instead of using signatures explicitly, we use interactive proofs which implicitly link all the parties involved with the transaction data through the mechanism of a designated 2-verifier proof. This makes it possible to thwart both on-line and off-line attacks in a provable way.

Of course one can always convert a designated interactive proof to a non-interactive proof (and hence a digital signature) by using pseudo-random functions as in [26]. However even this approach may not be provably secure [10].

3.4. An Electronic Payment System

We will now use the interactive proof system $(P, V)_{U,W}$ with designated verifiers U, W to design a secure electronic payment system. The parties involved in this protocol are the Customer A, the Merchant M and the Bank B.

Protocol 1. *An electronic payment system*

The Merchant first makes a payment request, *payment_request_M*, to the Customer.

Subroutine (a). *The Customer makes a payment request to the Bank, which is designated to the Merchant M and the Bank.*
A confirms (authenticates) a *payment_request_A_to_M* with a designated interactive proof system $(A, B)_{M,B}$. The communication is sent via the Merchant M.

Subroutine (b). *The Bank confirms the payment requested by the Customer and the Merchant, to the Merchant. This is designated to the Customer and the Merchant.*
B confirms (authenticates) a *payment_transaction_A_to_M* by using a designated interactive proof system $(B, M)_{A,M}$. The communication is forwarded to A.

SECURITY AND EFFICIENCY   The overall security of this Protocol depends on the security of the designated proofs in Subroutine (a) and Subroutine (b). If we use the designated proof in Section 3.1 then we get provable security when the trapdoor commitment is secure. This reduces to the factoring problem. The model used is that in [24]. In particular Subroutine (a) will thwart ordinary man-in-the-middle attacks (attack 2 in Section 1.1) and general man-in-the-middle attacks (attack 4, Section 1.1). A similar result holds for the designated proof system in Section 3.2. We note that it is not strictly necessary to designate the Customer.[4] This designation serves as an on-line confirmation for the Customer (there is little extra cost for the Bank).

To minimize the number of rounds we can merge the subroutines as follows. In step 1, A sends a commitment $c_{\text{subr(a)}}$ to M, and M forwards this to B. In step 2, B sends to M a challenge $e_{\text{subr(a)}}$ together with a commitment $c_{\text{subr(b)}}$. M

---

[4]A signature acknowledgement would of course be sufficient, but as remarked earlier it may be less efficient.

forwards these to A. In step 3, A sends a response $r_{\text{subr(a)}}$ to M, and M forwards this to B together with the challenge $e_{\text{subr(b)}}$. Finally in step 4, B sends its response $r_{\text{subr(b)}}$ to M which forwards it to A, together with the challenge $e_{\text{subr(b)}}$.

For a practical implementation we can take as hash function any collision resistant function such as SHA-1 [28].

Remark 3.3. *Challenge-response mechanisms*
Our protocols are based on a challenge-response mechanism (inherent to interactive proof systems) which involves three steps: the prover commits to a bitstring c, the verifier sends a challenge e, and finally the prover sends a response r. These steps link the parties concerned to the payment transactions. Designation guarantees that the on-line confidence which the verifier gains cannot be used by anybody else on-line in an effective way. Off-line protection comes from the fact that the interactive proof system does not leak any usable knowledge. We note that our designated proofs can easily be modified to get non-interactive proofs by using pseudo-random functions as in [26].

An alternative approach would be to use digital signatures interactively, with the prover signing a challenge of the verifier, etc. This is similar to what we have done. However our approach is indirect. Instead of using digital signatures explicitly, we use interactive proofs which implicitly link all the parties involved through the mechanism of a designated verifier proof system which leaks no knowledge of any practical use. This makes it possible to thwart both on-line and off-line attacks in a provable way.

4. A Protocol for Electronic Cash Systems

Our last protocol is for anonymous on-line electronic cash payments. The setting is the same as in the previous subsection. In this case the prover is the Bank which designates as verifiers the Merchant and the Customer. We use Chaum's confirmation proof for undeniable signatures [15] as modified by Jakobsson *et al.* [31] and designate it to two verifiers, the Merchant and the Customer. To prevent the Merchant from cheating the Customer, and man-in-the-middle attacks, the data exchanged are initially encrypted using the Bank's public key. The Bank has two private keys $x_b, x_b' \in_R Z_q$ while the Merchant has only one private key $x_m \in_R Z_q$. The corresponding public keys are $y_b = g^{x_b} \bmod p$, $y_b' = g^{x_b'} \bmod p$ and $y_m = g^{x_m} \bmod p$. The public key y_b is used to confirm the coins whereas y_b' is used for encryptions. As mentioned earlier (Section 2) the Bank is assumed to be trusted.

Let (z, s), $z \in Z_p$, $s = z^{x_b} \bmod p$, be the coin of the Customer, and let $\tilde{A}$ be a pseudonym of the Customer. We assume that z has appropriate redundancy. The Customer chooses a private key x_a for $\tilde{A}$ and computes the public key $y_a = g^{x_a} \bmod p$. In the following protocol the Bank B proves to the Merchant M that either (z, s) is a valid coin or that it knows the key $x_m x_a$, by using an interactive proof system with is designated to M and $\tilde{A}$.

Protocol 2. Chaum's designated confirmation scheme $(B, M)_{M, \tilde{A}}$ **for a purchase by $\tilde{A}$ from M**

1. a. The Merchant M selects $a, b \in_R Z_q$ and gives these to $\tilde{A}$.

   b. The Customer $\tilde{A}$ computes the commitment $v = z^a g^b \bmod p$ and gives the encrypted string $E_1 = E_{y'_b}(M, y_a, v)$ to M.

   c. M forwards E_1 to B.

2. The Bank B decrypts E_1 to get M, y_a, v and computes $w = v^{x_b} \bmod p$. Then B computes the commitment $c = g^w (y_m y_a)^r \bmod p$, $r \in_R Z_q$, which it sends to M. M forwards the commitment to $\tilde{A}$.

3. $\tilde{A}$ sends the encryption $E_2 = E_{y'_b}(z, s, a, b)$ to M, and M forwards this to B.

4. B decrypts E_2 to get z, s, a, b, and checks these and that v is of the right form. If so, and if the coin (z, s) has not been spent, B de-commits c by sending w, r, v to M. Then B invalidates the coin (z, s) and credits the account of M with this coin.

5. The Merchant M gives w, r to the Customer $\tilde{A}$, who checks that c and w were properly constructed. If this is the case, $\tilde{A}$ gives the M the coin (z, s). M does the same checks, and also checks the encryptions E_1, E_2. If these are all correct, M gives $\tilde{A}$ the goods purchased.

Protocol 2 extends the designated confirmation scheme of Chaum as modified by Jakobsson *et al.* [31] in such a way that, the Customer $\tilde{A}$ can realize an on-line purchase from the Merchant M securely.

SECURITY AND EFFICIENCY   Protocol 2 will thwart all the attacks in Section 1.1. It links the transactions of the Customer, the Merchant and the Bank inextricably through the mechanism of the designated verifier proof. Its security depends on the security of Chaum's modified confirmation scheme. This has been shown to be a zero-knowledge designated verifier proof [31]. Our modification requires two additional encryptions. So the protocol is secure if a pseudorandom encryption scheme is used.

Remark 4.1. *Pseudonyms and the traceability of coins.*
Even if it is not possible to trace the identity of a Customer A from a pseudonym $\tilde{A}$ directly, it is possible to link the spent coins by $\tilde{A}$. In fact the pseudonym $\tilde{A}$ is visible with all coins of A which are withdrawn from the Bank B. It follows that the coins can be linked through the pseudonym $\tilde{A}$. As pointed out by several authors in the literature, this may make it possible to trace the identity of A by using conventional means [35, 37, 38, 20]. It may therefore be desirable for A not to use a pseudonym. However in this case there will be no proof of payment for the purchase. A compromise is to take a 'variable' pseudonym with private key $x_a = hash(A, random)$ as an identifier. In case of a dispute the Customer can then reveal his/her true identity and claim the purchase.

5. Conclusion

One of the main threats with on-line payment systems comes from general man-in-the-middle attacks. These make it possible for the adversary to achieve an on-line advantage by manipulating the communicated data. Designated verifier proofs can be used to 'target' specific parties, who will be the only ones who gain the intended on-line confidence. This will thwart man-in-the-middle attacks. Off-line attacks exploit the knowledge which may be leaked during the execution of the protocol. These can be thwarted by using zero-knowledge proofs (which leak no knowledge other than a conviction) or witness hiding proofs (which leak no knowledge which is of any practical use to the attacker).

In this paper we have presented two protocols for on-line electronic payments. These use 2-designated verifier proofs which link the parties involved in a transaction to the transacted data inextricably. The resulting schemes are as secure as the trapdoor mechanisms used. We note that our first scheme uses a novel trapdoor commitment for an RSA setting which is of interest in itself.

One might reasonably comment that our goal could be achieved in a simpler way by using digital signatures. Two possible ways come to mind. The first one is non-interactive and involves timestamps. As pointed out earlier this approach cannot be used to design provably secure systems. The second is to use digital signatures interactively, in a challenge-response mode. This is similar to our technique. However instead of using digital signatures explicitly, we have used interactive proofs which certify the transaction implicitly through the mechanism of a designated verifier proof system which leaks no knowledge of any practical use. This makes it possible to thwart both on-line and off-line attacks in a provable way.

References

[1] N. Alexandris, M. Burmester, V. Chrissikopoulos and Yvo Desmedt, *Secure Linking of Customers, Merchants and Banks in Electronic Commerce*, Security on the Web, Future Generation Computer Systems, Elsevier Science B.V., to appear.

[2] M. Bellare, P. Rogaway, *Random oracles are practical: a paradigm for designing efficient protocols*, Proceedings of the First Annual Conference on Computer and Communications Security, ACM Press (1993), 62–73.

[3] M. Bellare, J. Garay, R. Hauser, A. Herzberg, H. Krawczyk, M. Steiner, G. Tsudik and M. Waidner, *iKP – A Family of Secure Electronic Payment Protocols*, Proceedings of the 1st Usenix Electronic Commerce Workshop, New York (1995), 1–20.

[4] S. Bengio, G. Brassard, Y. G. Desmedt, C. Goutier and J.-J. Quisquater, *Secure implementations of identification systems*, Journal of Cryptology, 4 (1991), 175–183.

[5] J.P. Boly et al, *The ESPRIT Project CAFE*, Computer Security, ESORICS '94, Third European Research in Computer Security, Proceedings, Lecture Notes in Computer Science 875 (1994), Springer-Verlag, 217–230.

[6] S. Brands, *Untraceable Off-line Cash in Wallet with Observers*, Advances in Cryptology – Crypto '93, Lecture Notes in Computer Science **773** (1994), Springer-Verlag, 302–318.

[7] M. Burmester, Y. Desmedt, T. Itoh, H. Shizuya and M. Yung, *A progress report on subliminal-free channels*, Information Hiding, First International Workshop, Cambridge, Lecture Notes in Computer Science **1174** (1996), Springer-Verlag, 157–168.

[8] M. Burmester, Y. Desmedt, T. Itoh, K. Sakurai and H. Shizuya, *Divertible and subliminal-free zero-knowledge proofs for languages*, Journal of Cryptology, **12** (1999), 197–223.

[9] J. Camenish, U. Maurer, M. Stadler, *Digital payment systems with passive anonymity-revoking trustees*, ESORICS '96, Proceedings, Lecture Notes in Computer Science (1996), Springer-Verlag.

[10] R. Canetti, O. Goldreich and S. Halevi, *The Random Oracle Methodology, Revisited*, 30th Annual STOC, ACM Press (1998), 209–218.

[11] D. Chaum, *Blind Signatures for Untraceable Payments*, in: D. Chaum, R.L. Rivest, A. T. Sherman, eds., Advances in Cryptology – Crypto'82 **1983**, Plenum Press N.Y., 199–203.

[12] D. Chaum, A. Fiat and M. Naor, *Untraceable electronic cash*, in: C. Günther, ed., Advances in Cryptology – Crypto '88, Lecture Notes in Computer Science **300** (1990), Springer-Verlag, 319–327.

[13] D. Chaum, *Privacy Protected Payments – Unconditional Payer and/or Payee Untraceability*, SMART CARD 2000: The future of IC Cards, Proc. IFIP WG 11.6, North Holland (1989), 69–93.

[14] D. Chaum, *On-line Cash checks*, Advances in Cryptology – Eurocrypt '89, Lecture Notes in Computer Science **434** (1989), Springer-Verlag, 288–293.

[15] D. Chaum, *Zero-knowledge undeniable signatures*, in: I. Damgård, ed., Advances in Cryptology – Eurocrypt '90, Lecture Notes in Computer Science **473** (1991), Springer-Verlag, 458–464.

[16] D. Chaum, T.P. Pedersen, *Transferred cash grows in size*, in: R. Rueppel, ed., Advances in Cryptology – Eurocrypt '92, Lecture Notes in Computer Science **658** (1993), Springer-Verlag, 390–407.

[17] D. Chaum, T.P. Pedersen, *Wallet databases with Observers*, in: E. Brickell, ed., Advances in Cryptology – Crypto '92, Lecture Notes in Computer Science **740** (1993), Springer-Verlag, 89–106.

[18] *DigiCash, Ecash Web Server*, 1996, http://www.digicash.com/cash

[19] *CyberCash Web Server*, Reston, VA 1996, http://www.cybercash.com

[20] G. Davida, Y. Frankel, Y. Tsiounis and M. Yung, *Anonymity Control in Electronic Cash*, Financial Cryptography FC'97, Lecture Notes in Computer Science, **1997**.

[21] Y. Desmedt, *Subliminal-free authentication and signature*, Advances in Cryptology – Eurocrypt '88, Lecture Notes in Computer Science **330** (1988), Springer-Verlag, 23–33.

[22] D. Dolev, C. Dwork and M. Naor, *Non-Malleable Cryptography*, 23rd Annual STOC, ACM Press (1991), 542–552.

[23] S. Even, O. Goldreich and Y. Yacobi, *Electronic wallet*, Advances in Cryptology, Proc. of Crypto '83, Plenum Press (1984), 383–386.

[24] U. Feige, A. Fiat and A. Shamir, *Zero-knowledge proofs of identity*, Journal of Cryptology, **1** (1998) 210–217.

[25] U. Feige and A. Shamir, *Witness indistinguishable and witness hiding protocols*, Proceedings of the twenty second annual ACM Symp. Theory of Computing, STOC, ACM Press (1990), 416–426.

[26] A. Fiat and A. Shamir, *How to prove yourself: Practical solutions to identification and signature problems*, Advances in Cryptology –Crypto '86, Lecture Notes in Computer Science **263** (1987), Springer-Verlag, 186–194.

[27] *Financial Services Technology*, FTSC Electronic Cheque Project Details, (1995), http://www.fstc.org/projects/echeck/index.html

[28] *FIPS 180-1, Secure hash standard*, Federal Information Processing Standards Publication 180-1, U.S. Department of Commerce/N.I.S.T., National Technical Information Service, Springfiel, Virginia, USA, **1994**.

[29] A.S. Glass, *Could the smart card be dumb*, Abstract of Papers, Eurocrypt 86, Lecture Notes in Computer Science **1440** (1998), Springer-Verlag, 8–10.

[30] L. Guillou and J.-J. Quisquater, *A practical zero-knowledge protocol fitted to security microprocessor minimizing both transmission and memory*, Advances in Cryptology – Eurocrypt '88, Lecture Notes in Computer Science **330** (1991), Springer-Verlag, 123–128.

[31] M. Jakobsson, K. Sako, and R. Impagliazzo, *Designated verifier proofs and their applications*, in: U. Maurer, ed., Advances in Cryptology – Eurocrypt '96, Lecture Notes in Computer Science **1070** (1996), Springer-Verlag, 143–155.

[32] MasterCard and VISA Corporations, *Secure Electronic Transactions (SET)*, Books 1,2,3. http://www.mastercard.com/set

[33] G. Medvinsky and B.C. Neuman, *NetCash: A design for practical electronic currency on the Internet*, Proceedings of 1st ACM Conf. Computer & Communication Security '93, ACM Press (1993), 102–106.

[34] B.C. Neuman and G. Medvinsky, *Requirements for Network Payment: The Net-Cheque Perspective*, Proceedings IEEE Compcon '95, San Francisco, **1995**.

[35] T. Okamoto, *An Efficient Divisible Electronic Cash System*, Advances in Cryptology – Crypto '95, Lecture Notes in Computer Science **963** (1995), Springer-Verlag, 438–451.

[36] T. Okamoto, K. Ohta, *Divertible zero knowledge interactive proofs and commutative random self-reducibility*, Advances in Cryptology, Eurocrypt '89, Lecture Notes in Computer Science **434** (1990), Springer-Verlag, 134–149.

[37] T. Okamoto, K. Ohta, *Universal Electronic Cash*, Advances in Cryptology – Crypto '91, Lecture Notes in Computer Science **576** (1992), Springer-Verlag, 324–337.

[38] B. Pfitzmann, M. Waidner, *How to break and repair a "Provably Secure" Untraceable electronic Payment System*, Advances in Cryptology – Crypto '91, Lecture Notes in Computer Science **576** (1992), Springer-Verlag, 338–350.

[39] N. F. Ramsey, *Precise Measurement of Time*, American Scientist, **76** (1988), 42–49.

[40] C. P. Schnorr, *Efficient Identification and Signatures for Smart Cards*, Advances in Cryptology – Crypto '89, Lecture Notes in Computer Science **435** (1990), Springer-Verlag, 239–252.

Department of Informatics, University of Piraeus, 80 Karaoli & Dimitriou Str. 185 34 Piraeus, Greece.

Information Security Group, Royal Holloway – University of London, Egham, Surrey TW20 OEX, UK.
E-mail address: m.burmester@rhbnc.ac.uk

Department of Informatics, University of Piraeus, 80 Karaoli & Dimitriou Str. 185 34 Piraeus, Greece.
E-mail address: chris@unipi.gr

Department of Computer Science, Florida State University, Tallahassee, FL 32306-4530, USA, and Information Security Group, Royal Holloway – University of London, Egham, Surrey TW20 OEX, UK.
E-mail address: desmedt@cs.fsu.edu

Progress in Computer Science and Applied Logic, Vol. 20
© 2001 Birkhäuser Verlag Basel/Switzerland

A Survey of Divide and Conquer Attacks on Certain Irregularly Clocked Stream Ciphers

Ed Dawson, Leonie Simpson, and Jovan Golić

Abstract. Recent proposals for keystream generators for stream ciphers based on linear feedback shift registers (LFSRs) incorporate irregular clocking to provide resistance against the conventional correlation attacks. This paper presents a survey of recent attacks, including correlation attacks, on several such keystream generators. These results demonstrate that despite the use of irregular clocking, such generators may still be vulnerable to certain divide and conquer attacks.

1. Introduction

Many designs for keystream generators for stream ciphers are based on shift registers, particularly linear feedback shift registers (LFSRs), because LFSRs are easily implemented in hardware and they allow for high encryption and decryption rates. Also, sequences produced by LFSRs with primitive characteristic polynomials possess certain important cryptographic properties: large periods and good long-term and short-term statistical properties. However, due to the the linearity, LFSR sequences are easily predicted from a short segment [14], which makes LFSRs unsuitable for use as keystream generators on their own. Instead they are used as components in keystream generators. Nonlinearity is introduced into the keystream either by using a nonlinear Boolean function to combine the outputs from several LFSRs or several stages of one LFSR or by using irregular clocking.

If an LFSR-based keystream generator is regularly clocked, then a keystream bit is produced every time the underlying LFSRs are clocked. For regularly clocked keystream generators, nonlinearity is generally introduced into the keystream by the explicit use of a nonlinear Boolean function. For irregularly clocked keystream generators, keystream bits are not produced every time the underlying data shift registers are clocked, but at irregular intervals. Irregular clocking is an implicit source of nonlinearity.

Many recent stream cipher proposals use irregularly clocked keystream generators because some regularly clocked keystream generators have been shown to be vulnerable to correlation attacks [18] and fast correlation attacks [15, 17] which use a correlation measure based on the Hamming distance. For irregularly clocked keystream generators, this correlation measure can not be applied. Also, currently there are no implemented fast correlation attacks on irregularly clocked keystream generators. However, these generators may be susceptible to other divide and conquer attacks, including correlation attacks. In this paper, we examine the susceptibility of several irregularly clocked keystream generators to divide and conquer attacks.

For irregularly clocked LFSR-based keystream generators, the output of one LFSR may be used to control the clock of another. The simplest keystream generator designs use only two binary LFSRs; one of these is a control register, $LFSR_C$, and the other is the data generating register, $LFSR_D$, as illustrated in Figure 1. Usually $LFSR_C$ is regularly clocked, and the output of this register, c, controls the clocking of $LFSR_D$. $LFSR_D$ is described as a clock-controlled LFSR (CCLFSR). The output of $LFSR_D$ then forms the keystream sequence, z. The clocking of $LFSR_D$ is termed constrained if there exists some fixed maximum number of times $LFSR_D$ may be clocked before a keystream bit is produced. Otherwise, it is termed unconstrained.

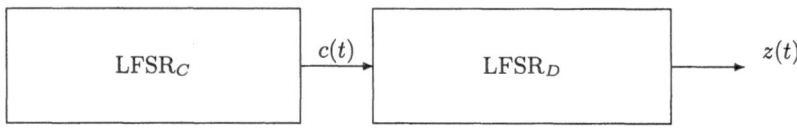

FIGURE 1. A simple clock-controlled LFSR scheme.

A simple example of a clock-controlled keystream generator for which the clocking is constrained is the step1-step2 generator, where $LFSR_C$ is regularly clocked, and controls the output of $LFSR_D$ in the following manner. At time instant t, $LFSR_C$ is stepped. If $c(t)$, the output of $LFSR_C$ at time t, is 0, then $LFSR_D$ is stepped once, and if $c(t)$ is 1, then $LFSR_D$ is stepped twice. The output of $LFSR_D$ forms the keystream bit, $z(t)$. The step1-step2 generator is vulnerable to a correlation attack: in [24], a constrained embedding attack on $LFSR_D$ is proposed, which can be used as a first stage in a divide and conquer attack on a step1-step2 generator. A more general attack, which can be applied to other constrained clock-controlled generators is proposed in [6]. An outline of this attack is given in Section 4.3.

The shrinking generator [2] is another example of a keystream generator with irregular output. The shrinking generator consists of two regularly clocked binary LFSRs. Denote these as $LFSR_A$ and $LFSR_S$, as shown in Figure 2, and denote the lengths of these LFSRs as L_A and L_S, respectively. The shrinking generator output is a "shrunken" version or subsequence of the output of $LFSR_A$, with the subsequence elements selected according to the position of 1's in the output sequence of $LFSR_S$: the keystream sequence z consists of those bits in the sequence a for which the corresponding bit in the sequence s is a 1. The other bits of a,

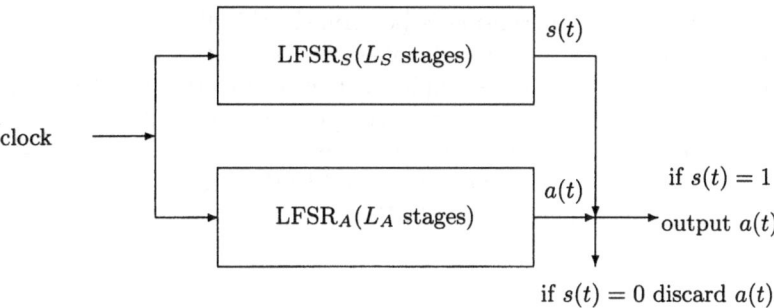

FIGURE 2. The shrinking generator.

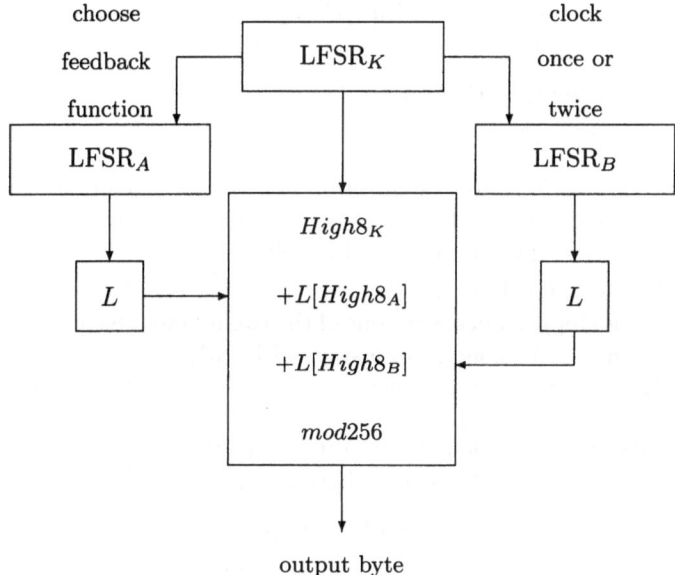

FIGURE 3. The ORYX keystream generator.

for which the corresponding bit of s is a 0, are deleted. The shrinking generator succumbs to a correlation attack [19], outlined in Section 5.

Another LFSR-based design using irregular clocking is the ORYX stream cipher, proposed for use in North American digital cellular systems to protect cellular data transmissions [21]. The ORYX cipher is based on three binary LFSRs.

However, the operation of the cipher differs from those described previously, as it operates on bytes rather than bits. The generator output is a random-looking sequence of bytes, and encryption is performed by XORing the keystream bytes with the data bytes to form bytes of ciphertext. Decryption is performed by XORing the keystream bytes with the ciphertext to recover the plaintext. The ORYX cipher, illustrated in Figure 3, has four components: three 32-bit LFSRs, denoted as $LFSR_A$, $LFSR_B$ and $LFSR_K$, and an S-box containing a known permutation L of the integer values 0 to 255, inclusive. The LFSR feedback functions are fixed as follows. The feedback function for $LFSR_K$ is

$$f_K = x^{32} + x^{28} + x^{19} + x^{18} + x^{16} + x^{14} + x^{11} + x^{10} + x^9 + x^6 \\ + x^5 + x + 1.$$

The feedback functions for $LFSR_A$ are

$$f_{A_1} = x^{32} + x^{26} + x^{23} + x^{22} + x^{16} + x^{12} + x^{11} + x^{10} + x^8 + x^7 \\ + x^5 + x^4 + x^2 + x + 1$$

and

$$f_{A_2} = x^{32} + x^{27} + x^{26} + x^{25} + x^{24} + x^{23} + x^{22} + x^{17} + x^{13} + x^{11} \\ + x^{10} + x^9 + x^8 + x^7 + x^2 + x + 1.$$

The feedback function for $LFSR_B$ is

$$f_B = x^{32} + x^{31} + x^{21} + x^{20} + x^{16} + x^{15} + x^6 + x^3 + x + 1.$$

The permutation L is fixed for the duration of a call, and is formed from a known algorithm, initialised with a value which is transmitted in the clear during the call setup. Each keystream byte is generated as follows:

1. $LFSR_K$ is stepped once.
2. $LFSR_A$ is stepped once, with one of the two different feedback polynomials, depending on the content of a stage of $LFSR_K$.
3. $LFSR_B$ is stepped either once or twice, depending on the content of another stage in $LFSR_K$.
4. The high bytes of the current states of $LFSR_K$, $LFSR_A$ and $LFSR_B$ are combined to form a keystream byte using the combining function:

$$Keystream = \{High8_K + L[High8_A] + L[High8_B]\} \bmod 256,$$

where $High8_X$ denotes the high byte of $LFSR_X$.

Note that $LFSR_K$ is regularly clocked, and provides the clock control for the two other LFSRs. $LFSR_A$ is regularly clocked, although with one of two possible feedback functions, determined by $LFSR_K$. $LFSR_B$ is irregularly clocked, again under the control of $LFSR_K$. The known permutation L, applied to the output bytes from $LFSR_A$ and $LFSR_B$, provides explicit nonlinearity in the combining function. However, the ORYX cipher has a weakness which is exploited in a divide and conquer attack [22], outlined in Section 6.2. This attack can be derived from the inversion attack [9] on nonlinear filter generators and the internal state reversion attack [10] on the alleged A5 generator which appeared earlier.

2. Properties of Irregularly Clocked Keystream Generators

Several properties of pseudorandom binary sequences are considered basic security requirements: a sequence that does not possess these properties is generally considered unsuitable for cryptographic applications. These requirements include a large period and a high linear complexity. A large period prevents observation of repetition of a segment of the keystream, which could be used to predict the entire keystream. High linear complexity makes the Berlekamp-Massey attack [14] infeasible. This attack can reproduce the keystream from a known keystream segment of length equal to twice the linear complexity of the sequence. One further property which appears common to many irregularly clocked keystream generators is the equivalent keys property, where several sets of initial states for the component LFSRs produce the same keystream. These basic properties and their implications for the security of keystream generators are discussed in the remainder of this section.

2.1. Period

Suppose that the output of the keystream generator can be viewed as an irregular decimation of an underlying periodic sequence, d, with period P_d. The decimation is determined by the clock control or decimating sequence, c, with period P_c, whose elements specify the number of steps per each output symbol. Then an upper bound on the period of the keystream sequence is given in the following theorem, from [1].

Theorem 2.1. *An upper bound on the period of the keystream sequence is given by the product of P_d and P_c. This upper bound is obtained if the length of the segment of the periodic sequence d spanned by one period of the clock control sequence is relatively prime to P_d.*

Some generalizations of this theorem can be found in [11].

2.2. Linear Complexity

If the output of the keystream generator can be viewed as an irregular decimation of an underlying pseudonoise (maximum-length) sequence, d, with period $P_d = 2^{L_d} - 1$, then an upper bound on the linear complexity of the keystream sequence is given in the following theorem, from [5].

Theorem 2.2. *When a pseudonoise sequence of period P_d is nonuniformly decimated by means of a sequence of period P_c, if the sum modulo P_d of P_c successive values of the decimating sequence equals K, then the decimated sequence has a maximum linear complexity of $P_c L_d$ only if the multiplicative order of 2 modulo $P_d \div gcd(P_d, K)$ is equal to L_d.*

In [5] it is also shown that, if the decimating sequence is chosen randomly, then the probability that the maximum linear complexity is obtained can be made arbitrarily close to one for appropriately chosen L_d and P_c.

2.3. Equivalent Keys Property

Assuming fixed LFSR feedback polynomials, so that the secret key of the generator is only the initial states of the component LFSRs, for some keystream generators, several sets of LFSR initial states can produce the same keystream.

As a simple example, consider the generation of the output from a shrinking generator with $L_A = 3$ and $L_S = 4$, and sequences a and s as shown in Table 1. Suppose the initial states of two LFSRs are as shown in bold in the table. After stepping both LFSRs once, no keystream bit has yet been produced. Hence, a secret key which consists of the LFSR contents at time instant $t = 0$ will produce the same keystream as the LFSR contents at time $t = 1$, and also the contents at time $t = 2$. If the secret key to the keystream generator is only the initial states of the two LFSRs, then there are at least 3 pairs of LFSR initial states that generate the keystream shown in this example. Thus, for the keystream $z = 1, 1, 0, 0, 1, \ldots$, there are at least three equivalent keys.

t	0	1	2	3	4	5	6	7	8	9
a	**0**	**0**	1	1	1	0	1	0	0	1
s	**0**	**0**	1	0	0	0	1	1	1	1
z			1				1	0	0	1

TABLE 1. Equivalent keys example: shrinking generator.

Two implications of this property should be considered. Firstly, for a particular known keystream, the equivalent keys property can be used to reduce the number of effective keys to be tested in a brute force attack. Secondly, if the LFSR initial states are chosen at random, although all keys are equally likely, all keystreams may not be. Some keystreams may be more likely to be produced than others, thus the probability that a naive brute force attack will be successful can be greater for some keystreams than it is for others.

Note that this property arises through the use of irregular clocking: regularly clocked keystream generators generally do not possess this property. However, keystream generators which use irregular clocking may avoid the equivalent keys property through appropriate design and parameter selection [20].

3. Attacks on Irregularly Clocked LFSR-based Keystream Generators

Several methods for attacking LFSR-based stream ciphers are well known. In this section, these methods, as applied to irregularly clocked keystream generators, are outlined.

3.1. Divide and Conquer Attacks

Divide and conquer attacks on keystream generators work on each component of the keystream generator separately, and sequentially solve for individual subkeys. The attack is generally performed as a known plaintext attack, although sufficient redundancy in the plaintext may permit a ciphertext only attack. For LFSR-based keystream generators, the objective of the attack is to recover the initial contents of a subset of the component LFSRs from a known segment of the keystream sequence. These attacks require knowledge of the structure of the keystream generator. If the entire structure of the generator is known, and the secret key is only the LFSR initial states, for a keystream generator consisting of n LFSRs, the total number of keys to be searched in a brute force attack is $\prod_{i=1}^{n}(2^{L_i} - 1)$, where L_i is the length of the ith LFSR. Using a divide and conquer attack that determines individual LFSR states sequentially reduces the total number of keys to be searched to $\sum_{i=1}^{n}(2^{L_i} - 1)$. Although divide and conquer attacks were devised for regularly clocked keystream generators, the approach can be extended to generators which incorporate irregular clocking. The attacks presented in this paper are intended for use in divide and conquer attacks on irregularly clocked generators.

3.2. Correlation Attacks

Correlation between two binary segments is a measure of the extent to which they approximate each other. Correlation attacks on LFSR-based keystream generators are based on statistical dependencies between the observed keystream sequence and underlying shift register sequences. The objective of the attacks is to use the correlation between the known keystream segment and the underlying LFSR sequences to sequentially recover the initial contents of each of the component LFSRs.

The first correlation attack, proposed by Siegenthaler [18], is a divide and conquer attack on a regularly clocked LFSR-based keystream generator, the non-linear combination generator. For this attack the keystream is viewed as a noisy version of an underlying LFSR sequence, where the noise is additive and independent of underlying LFSR sequence. The problem can be viewed as one of decoding a linear block code over a binary symmetric channel. As the clocking is regular, both the observed keystream segment and segments of the underlying LFSR sequences have the same length and the Hamming distance is used as a measure of correlation.

To identify the actual initial state of each component LFSR, the correlation between the keystream and the LFSR sequence is calculated for each possible LFSR initial state. The correct initial state is assumed to be that for which the correlation is highest. Fast correlation attacks, based on iterative error-correction algorithms, which outperform exhaustive search over the initial states of the individual shift registers, have also been proposed for regularly clocked LFSR-based keystream generators [15, 17], and more recently in [13].

More recently, correlation attacks which can be applied to keystream generators based on irregularly clocked LFSRs have been proposed [24, 6, 7, 8, 12].

For irregularly clocked keystream generators, a correlation attack may be used as
the first stage in a divide and conquer attack. The attack targets the component
LFSRs involved in producing the keystream bits. Once the correct initial states of
these LFSRs are identified, the clock control component can be attacked.

The correlation attack on irregularly clocked keystream generators follows the
same procedure as the basic correlation attack on regularly clocked keystream generators. That is, for each subkey (set of possible LFSR initial states for a subset of
the LFSRs), a segment of the underlying regularly clocked sequence is produced,
and the correlation between the segment and the known keystream segment is
computed. The correct subkey is assumed to be that for which the correlation is
highest. The attack is considered successful if there are only a few such subkeys.
Note that this is a basic attack: as the computation must be performed for each
possible subkey, this is not a fast correlation attack. Indeed, no implemented fast
correlation attack on irregularly clocked LFSR-based keystream generators appears in the available literature. The correlation attack procedure is implemented
by the following algorithm.

Basic Correlation Attack Procedure:

- *Input:* structure of keystream generator and observed keystream sequence
  of length n, $z^n = \{z(t)\}_{t=1}^n$.
- Calculate length m required for the underlying regularly clocked sequence
  $a^m = \{a(t)\}_{t=1}^m$.
- For each possible subkey:
  1. generate the sequence $a_i^m = \{a_i(t)\}_{t=1}^m$,
  2. compute the correlation between a_i^m and z^n,
  3. if correlation between a_i^m and z^n is high, preserve the subkey as a
     possible candidate.
- *Output:* set of candidate subkeys (LFSR initial states).

The major point of difference between the basic correlation attacks on regularly clocked and irregularly clocked keystream generators is in the measure of
correlation. For an irregularly clocked keystream generator, the keystream may
be viewed as an irregularly decimated version of some longer underlying sequence
produced by a regularly clocked keystream generator. As the known keystream
segment and the underlying segment are of different lengths, the Hamming distance is no longer useful as a basis of the measure of correlation. Instead, a different
measure of correlation is required.

A commonly used measure for the distance between two strings of different
lengths is the Levenshtein distance.

Definition 1. *The Levenshtein distance, $LD(a,b)$, between two binary strings a
and b is defined to be the minimum number of edit operations (deletion, insertion,
substitution) needed to transform a, of length N_1, into b, of length N_2.*

Note that for $N_1 \geq N_2$, in obtaining the minimum distance, the insertion operation
is not required, and that the minimum distance can not be less than $N_1 - N_2$. Where

clocking is constrained, the distance measure may be based on the constrained Levenshtein distance [6], which is similar to the Levenshtein distance, but with a constraint on the maximum number of consecutive deletions which can occur in any edit sequence.

Definition 2. *The constrained Levenshtein distance, $CLD(a, b)$, between two binary strings a and b is defined to be the minimum number of edit operations (deletion, insertion, substitution) needed to transform a, of length N_1, into b, of length N_2, subject to the constraint that no more than d consecutive bits of a may be deleted.*

All correlation attacks involve a process of hypothesis testing. For a sequence a^m produced by the regularly clocked keystream generator, and the observed keystream sequence z^n, the hypothesis to be tested is

$$H_0 : \quad z^n \text{ and } a^m \text{ are independent, versus}$$
$$H_1 : \quad z^n \text{ and } a^m \text{ are correlated.}$$

As with any hypothesis testing, there are two types of errors which can occur. Let the error made by deciding that the two strings are independent when, in fact, they are correlated be described as "missing the event", and the error made by deciding that the two strings are correlated when, in fact, they are independent be described as "a false alarm". Denote the probabilities of these events by P_m and P_f, respectively. A relationship between these probabilities exists, and is dependent on the lengths of the sequences, n and m. For a given value of n, m should be chosen as a function of n and P_m. The recovery will be successful if, for fixed P_m, increasing n will result in a decrease of P_f.

In [8], several correlation attacks using measures of correlation which are similar to, or based on the Levenshtein distance are analysed, and the conditions under which these attacks can be successful are determined. The so-called embedding and probabilistic correlation attacks on a CCLFSR are examined, for the constrained and unconstrained cases. Constrained embedding attacks can be successful if enough keystream is available, whereas unconstrained embedding attacks are successful only if the probability of bit deletion is less than 0.5. In that case, a probabilistic correlation attack may be successful. These attacks are outlined in Section 4.3 and Section 4.4.

4. Divide and Conquer Attack on a Single CCLFSR

There are two approaches which can be taken in a divide and conquer attack on a single CCLFSR: an attack which targets the clock control first, or an attack which targets the LFSR producing the keystream bits. These two approaches are outlined in the following sections.

4.1. Attacking the Clock Control

This known plaintext attack, conducted under the assumption that the structure
of the generator is known, is an application of the linear consistency test [23]. If the
clock control is provided by another LFSR, as in the simple scheme illustrated in
Figure 1, then assume the LFSR feedback polynomials are known, and the secret
key is only the LFSR initial states. Now, for each possible initial state of $LFSR_C$,
generate a segment of the $LFSR_C$ sequence. Use that segment and the known
keystream segment to determine bits in the $LFSR_D$ sequence. The initial state of
$LFSR_D$ can then be recovered by solving a system of linear equations based on
the $LFSR_D$ feedback polynomial. If the system of equations is not consistent, then
the candidate $LFSR_C$ initial state is incorrect.

4.2. Attacking the CCLFSR

In performing a divide and conquer attack on a single CCLFSR, note firstly that
the CCLFSR output sequence is an irregular decimation of the binary sequence
which would be produced if the LFSR was clocked regularly, as illustrated in Figure 4. That is, the keystream segment $\{z(t)\}_{t=1}^{n}$ can be obtained from $\{a(j)\}_{j=1}^{m}$,
$m \geq n$, by deleting some of the bits of a. If the clocking is constrained, then there
will be a constraint on the maximum number of consecutive bits of a which can
be deleted.

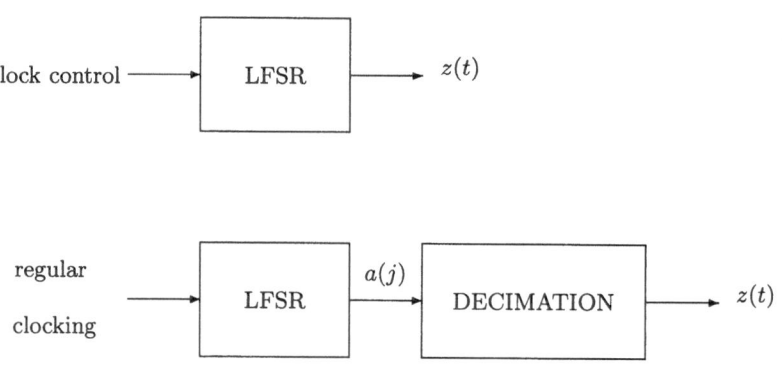

FIGURE 4. CCLFSR statistical model.

The objective of a correlation attack on a CCLFSR is to recover the initial
state of the LFSR from a given segment of the output sequence, without knowing
the decimation sequence. Correlation attack on a CCLFSR is conducted in the
same way as the basic correlation attack outlined in Section 3.2. That is, it is
a known plaintext attack, for which the length, L, and feedback polynomial of
the LFSR are known. Details of the clock control are not required, except for
the constraint on the maximum number of consecutive deletions permitted, if the
clocking is constrained.

The attack consists of the following steps. For every possible LFSR initial state, a sequence a^m is produced through regular clocking of the LFSR. The correlation between a^m and the known keystream segment z^n is computed, using some defined measure of correlation. The initial state of the LFSR is considered to be that for which the correlation is the highest. The correlation attack is considered successful if there are only a few such initial states, preferably only one.

For constrained clocking, a constrained embedding attack as outlined in Section 4.3 can be conducted. For unconstrained clocking, the probabilistic correlation attack outlined in Section 4.4 can be applied.

4.3. Constrained Embedding Attacks

A keystream sequence obtained through decimation is said to be embedded in the underlying regularly clocked sequence. Several embedding attacks have been proposed in the literature [24, 16]. The measure of correlation is based on the Levenshtein distance. The correlation is the highest between the keystream z^n the underlying string a^m for which the constrained Levenshtein distance $CLD(a^m, z^n)$ is minimal.

In the case where there is no additive noise, if the keystream z^n can be embedded in a^m, then only deletions are required, so that $CLD(a^m, z^n) = m - n$. In this case, the correct LFSR initial states produce a sequence a^m for which CLD is minimal possible, $CLD = m - n$. The constrained embedding attack is implemented by the following procedure:

Constrained Embedding Correlation Attack Procedure:

- Assume known: structure of the keystream generator, including LFSR lengths and feedback functions, the maximum number d of consecutive bits in the underlying regularly clocked LFSR sequence which can be deleted, and a segment of keystream $z = \{z_i\}_{i=1}^n$.
- Assume unknown: details of the clock control generator.
- Use the Basic Correlation Attack Procedure, with the CLD between a and z as a correlation measure.
- Correct LFSR initial state should have CLD $= m - n$.

It is important to note that the CLD can be computed efficiently by a recursive algorithm of complexity $O(n(m - n))$. The embedding attack is shown to be successful [24] for $d = 1$, whereas for $d > 1$, the theoretical approach from [8] shows that the attack can be successful, but the required keystream sequence length may be prohibitively large.

4.4. Probabilistic Correlation Attacks

Embedding attacks make no use of the decimation probability distribution, and so are not generally optimal. An optimal attack should use a measure of probability which incorporates the decimation probabilities in the computation of correlation between two sequences. Such attacks are referred to as probabilistic correlation attacks. For the constrained case, a generalized correlation attack based on a probabilistic constrained edit distance is given in [7]. For the unconstrained case, for a particular decimation probability distribution, a correlation attack using a measure of probability termed the joint probability is proposed in [8]. For practical applications, the joint probability too quickly approaches 0, so a normalized joint probability is used in [20]. The unconstrained probabilistic correlation attack procedure is implemented by the following algorithm.

Unconstrained Probabilistic Correlation Attack Procedure:

- Assume known: structure of the keystream generator, including LFSR lengths and feedback functions, p, the probability that a bit in the underlying regularly clocked sequence will be deleted, and a segment of keystream $z = \{z_i\}_{i=1}^n$.
- Assume unknown: details of the clock control generator.
- Use the Basic Correlation Attack Procedure, with the normalized joint probability of a and z as a correlation measure.
- Correct LFSR initial state should have high normalized joint probability.

5. Correlation Attack on the Shrinking Generator

In this section the security of the shrinking generator, illustrated in Figure 2 and described in Section 1, is examined. In a more precise description, let $a = \{a(t)\}_{t=1}^\infty$ denote the LFSR$_A$ sequence produced from a nonzero initial state $\{a(t)\}_{t=1}^{L_A}$, and let $s = \{s(t)\}_{t=1}^\infty$ denote the LFSR$_S$ sequence produced from a nonzero initial state $\{s(t)\}_{t=1}^{L_S}$. Let $z = \{z(k)\}_{k=1}^\infty$ denote the output sequence of the shrinking generator, that is, the keystream. Then $z(k) = a(t_k)$ where $\sum_{i=1}^{t_k} s(i) = k$. That is, t_k is the position of the kth 1 in the sequence s. In [19], the shrinking generator is shown to be vulnerable to a divide and conquer correlation attack.

In the attack, the output of shrinking generator is viewed as an irregularly decimated version of the LFSR$_A$ output sequence a, with the decimation controlled by the LFSR$_S$ sequence s. That is, the keystream can be viewed as the output of a clock-controlled LFSR$_A$, where the clocking is unconstrained.

As the first stage in a divide and conquer attack, the unconstrained probabilistic correlation attack outlined in Section 4.4 is applied to the shrinking generator to recover the initial state of LFSR$_A$ from a known segment of the keystream z, given the feedback polynomial of LFSR$_A$. At the end of the first stage of the attack, the LFSR$_A$ initial states are sorted based on their normalized joint probability values. States with a high normalized joint probability value are the most likely candidates for the actual LFSR$_A$ initial state.

Given a candidate initial state for LFSR$_A$, the attack proceeds to a second stage: reconstruction of the initial state of LFSR$_S$, using the candidate LFSR$_A$ sequence, the feedback function for LFSR$_S$ and the known segment of z. For a given LFSR$_A$ initial state candidate, the known LFSR$_A$ feedback function is used to generate a candidate LFSR$_A$ sequence $\hat{a}$. This sequence and the known keystream sequence can then be used to form candidate LFSR$_S$ initial states, using either an edit distance matrix or an iterative reconstructive process (see [19]).

5.1. Experimental Results

Experiments were conducted on a simulated shrinking generator, where LFSR$_A$ was a 15-bit LFSR and LFSR$_S$ was a 17-bit LFSR; thus the total key length was 32 bits. Experiments were performed for various keystream lengths, n. For each n, the attack was performed for fifty trials. Each trial required exhaustive search of the LFSR$_A$ keyspace: since $L_A = 15$, there are $2^{15} = 32767$ possible LFSR$_A$ initial states.

5.1.1. RECOVERY OF LFSR$_A$ INITIAL STATE   Experiments were performed for keystream lengths $n = 150, 225, 300$ and 600. The lengths of the underlying regularly clocked LFSR$_A$ sequences were calculated as a function of n, with $m(n) = n/(1-p) + 3\sqrt{n}$, with p, the bit deletion probability, set to $p = 0.5$.

Let N denote the number of LFSR$_A$ initial states with joint probability higher than that for the actual initial state. In each trial, the value of N was recorded. A summary of this data is presented in Table 2. Minimum and maximum are, respectively, the minimum and maximum values of N occurring in the fifty trials. Mean is the mean value of N. The median is also given as a measure of the centre of the distribution, as the data sets contain a few relatively large values which affect the mean. Similarly, the first and third quartiles, denoted as Q1 and Q3, respectively, are given as a robust measure of the spread of the data.

Statistics	Keystream length n			
	150	225	300	600
Minimum	4	0	0	0
Q1	83.5	17.75	10.75	10
Median	447.5	63	28	36
Mean	900.12	305.14	108.64	35.4
Q3	1091.25	301	82.5	58.25
Maximum	5378	4199	1278	98

TABLE 2. Number of LFSR$_A$ initial states with joint probability higher than that for the actual initial state.

The main observations are that the normalized joint probability of the correct initial state for LFSR$_A$ is among the highest normalized joint probability values, and that as the length of the known keystream increases, the number of initial states with normalized joint probability higher than the actual initial state decreases.

5.1.2. RECOVERY OF BOTH LFSR INITIAL STATES   For the keystream length $n = 300$, the complete attack was performed for fifty trials. Let N denote the number of LFSR$_A$ initial states tested before an initial state was found which could, with the appropriate LFSR$_S$ initial state, reproduce the keystream. In each trial the value of N was recorded. The minimum and maximum values of N occurring in the fifty trials were 0 and 815, respectively. The values for the first and third quartiles were 5 and 38.25, respectively. The median value of N was 18.5 and the mean was 74.76, heavily influenced by a few large values in the data set.

Comparing the values for the complete attack with the statistics for $n = 300$ in Table 2, a reduction in the values of all the statistics is observed, with the exception of the minimum value of N, already zero. This is due to the existence of equivalent pairs of LFSR initial states for some keystreams, with the normalized joint probability of the keystream and the sequence produced by the alternative LFSR$_A$ initial state being higher than that for the actual LFSR$_A$ initial state used to produce the keystream. An alternative pair of initial states, generating the required keystream, was produced in twenty-seven of the fifty trials.

5.2. Other Attacks

If the LFSR feedback polynomials are fixed, and assuming the structure of the generator is known to the cryptanalyst, then the secret key of the generator is only the initial states of the two LFSRs. If the all-zero initial states are avoided for either LFSR, the total number of secret keys for the shrinking generator is $(2^{L_A} - 1)(2^{L_S} - 1)$. In the worst case brute force attack, this is the number of trials required to recover the key. Moreover, the equivalent keys property can be applied to reduce the search space to the number of effective keys: $(2^{L_A} - 1)(2^{L_S - 1})$.

As an alternative to exhaustive search, one possible attack [2] is an application of the linear consistency test [23]. This divide and conquer attack is conducted under the same assumptions as the probabilistic correlation attack, but targets the LFSR$_S$ subkey, rather than the LFSR$_A$ subkey. The attack requires exhaustive search of the LFSR$_S$ keyspace. The complexity of the attack is $O(2^{L_S} L_A^3)$.

In [12], identifying the underlying state of LFSR$_A$ is treated as a decoding problem on the deletion channel, and an iterative algorithm to find the initial state of LFSR$_A$ is proposed. Additionally, it is noted that certain subsequences of the output sequence can be used to find the initial state of LFSR$_A$ with complexity less than exhaustive search.

5.3. Discussion

The shrinking generator output can be viewed as an irregularly decimated version of the output of an underlying regularly clocked LFSR$_A$, of length L_A. If the structure of the generator and a segment of the keystream are known, then the unconstrained probabilistic correlation attack can be applied to the shrinking generator as the first stage in a divide and conquer attack. Systematic computer simulations show that the unconstrained probabilistic correlation attack can be used to identify, with high probability the correct initial state of LFSR$_A$. The minimum keystream length required for this attack to be successful is about twenty times the length of LFSR$_A$ (see [19]).

The major limitation of the attack is that the first stage requires computation of the joint probability for all possible LFSR$_A$ initial states. Thus the attack is suitable only for generators where L_A is not large. Also, further computation is required to recover the initial state of LFSR$_S$. The computational complexity of the attack is $O(2^{L_A}L_S^2)$.

Divide and conquer attacks on the shrinking generator have complexity exponential in the length of one of the underlying LFSRs. In general, the generator may be attacked using the unconstrained probabilistic correlation attack if LFSR$_A$ is not large, or the linear consistency attack if L_S is not large. Therefore the shrinking generator can not be considered secure if either LFSR is short.

6. Attack on the ORYX Stream Cipher

The ORYX stream cipher depicted in Figure 3 and described in Section 1 is an LFSR-based keystream generator. In [22], the security of ORYX is examined with respect to a known plaintext attack, conducted under the assumption that the cryptanalyst knows the complete structure of the cipher, including the LFSR feedback functions. The secret key is only the initial states of the three 32 bit LFSRs: a total keysize of 96 bits. The generator weakness which makes it susceptible to attack, and the attack itself are explained in the remainder of this section.

6.1. Generator Weakness

The generator weakness is a combination of the following three properties, each of which can be a potential weakness if present in an LFSR-based keystream generator.

First, each time the LFSRs are stepped, the number of output bits, eight, used to form the keystream is bigger than one. In particular, it exceeds the number of LFSRs present. According to [9], the generalized inversion attack on an LFSR-based combiner with such a property may be modelled by a subcritical branching process which, together with a short input memory of a combiner, can make the attack feasible.

Second, the input memory of the combiner with respect to each LFSR is only seven bits, as only the high eight stages of each LFSR are used to form the output byte, with the two stages of LFSR$_K$ whose contents control the selection of the feedback polynomial for LFSR$_A$ and the number of times LFSR$_B$ is stepped both within the high eight stages of LFSR$_K$.

Third, the LFSR controlling the clocking, LFSR$_K$, is regularly clocked (that is, is stepped once for each byte of output produced) and its high byte output is linearly combined (modulo 256) to form the keystream byte. So, its high byte can be directly computed from the known keystream byte and the guessed high byte outputs of the other two LFSRs.

Rather than recovering the contents of the LFSRs sequentially, beginning with LFSR$_K$, the attack targets, simultaneously, one section of each of the three LFSRs. Since the keystream bytes are formed from the contents of the high eight stages of each of the three LFSR states, the keyspace is divided to focus on these 24 bits.

6.2. Attack Procedure

Denote the high eight bits of the three LFSRs at the time the tth byte of available keystream is produced by $High8_A(t)$, $High8_B(t)$ and $High8_K(t)$. The initial contents are $High8_A(0)$, $High8_B(0)$ and $High8_K(0)$, and all registers are stepped before the first byte of available keystream, denoted $Z(1)$, is produced. To produce a keystream byte $Z(t+1)$ at time instant $t+1$, $LFSR_K$ is stepped once, then $LFSR_A$ is stepped once, then $LFSR_B$ is stepped either once or twice. The contents of $High8_A(t+1)$, $High8_B(t+1)$ and $High8_K(t+1)$ are then combined to form the keystream byte $Z(t+1)$. Therefore, there is no need to guess all 24 bits: the contents of $High8_A(1)$ and $High8_B(1)$, the first byte of the known keystream $Z(1)$ and the combining function can be used to calculate the corresponding contents of $High8_K(1)$, which is essentially the idea used in [3] to attack the summation generator. Thus the attack requires exhaustive search of only a 16 bit subkey: the contents of $High8_A(1)$ and $High8_B(1)$.

For a particular 16-bit guess of $High8_A(1)$ and $High8_B(1)$, $Z(1)$ is used to calculate the corresponding contents of $High8_K(1)$. After this calculation, the attack proceeds iteratively to construct a path of guesses of $High8_A(t)$, $High8_B(t)$ and $High8_K(t)$ which are consistent with the known keystream. In each iteration a set of predictions for the next keystream byte is formed and the guess evaluated by comparing the known keystream byte with the predicted values. The attack exploits essentially the same ideas as the generalized inversion attack on nonlinear filter generators [9] and the internal state reversion attack on the alleged A5 generator [10].

In the tth iteration, after stepping the three LFSRs to produce the next output byte, $High8_K(t+1)$ and $High8_A(t+1)$ effectively have one unknown bit shifting into them and, depending on $High8_K(t+1)$, $High8_B(t+1)$ has either one or two unknown bits. As the unknown bit of $High8_K(t+1)$ is used for clock control, there are exactly 12 possible combinations of these new input bits, and there will be at most 12 distinct output bytes consistent with the guess of $High8_A(t)$, $High8_B(t)$ and $High8_K(t)$. The known keystream byte $Z(t+1)$ is compared with the predicted output bytes.

If $Z(t+1)$ is the same as one of the predicted output bytes, for the case where there are 12 distinct outputs, then a single possible set of values exists for $High8_K(t+1)$, $High8_A(t+1)$ and $High8_B(t+1)$. These values are used in the next iteration of the attack. Where there are less than 12 distinct outputs, with the keystream byte the same as the predicted output byte for more than one combination of new input bits, the path of consistent guesses branches. In this situation a depth-first search is conducted. If the keystream byte is not the same as any of the predicted output bytes, then the guessed contents of $High8_A(t)$ and $High8_B(t)$ were obviously incorrect. Another path must be taken from the last branching point. If each possible path is searched without finding a path of consistent guesses of length equal to the number of bytes of known keystream, then the guessed contents of $High8_A(1)$ and $High8_B(1)$ were obviously incorrect. A new 16-bit guess is made and the procedure repeated.

Once a sufficiently long path of consistent guesses is found, assume that the values for $High8_A(1)$, $High8_B(1)$ and $High8_K(1)$ were correct. This provides knowledge of the contents of the high eight states of each of the three LFSRs at the time that the first byte of keystream was produced. For the 24 consecutive guesses $High8_A(t)$, $High8_B(t)$ and $High8_K(t)$ for $2 \leq t \leq 25$, each set of values: $High8_K(t)$ and $High8_A(t)$ gives another bit in the state of LFSR$_K$ and LFSR$_A$, respectively, and $High8_B(t)$ gives either one or two bits in the state of LFSR$_B$. From the reconstructed 32-bit state of each LFSR at the time the first keystream byte was produced, the LFSR states can then be stepped back to recover the initial states of the three LFSRs: the secret key of the ORYX generator. Thus the entire key can be recovered using a minimum of 25 bytes of keystream, and at most 2^{16} guesses. More precisely, although the recovered initial states are consistent with given 25 keystream bytes, they need not be consistent with the entire keystream. So, a few more than 25 keystream bytes may be needed to resolve the residual ambiguities and identify the correct LFSR initial states, by comparing the candidate keystreams with the known keystream.

6.3. Implementation Issues for the Attack

It is possible that the keystream byte $Z(t+1)$ will be the same as one of the predicted output bytes, although the values for High8$_A(t)$ and High8$_B(t)$ are incorrect (a false alarm). For the attack to be effective, the probability of a false alarm occurring must be small. The probability of a false alarm at each step is upper-bounded by $1 - (255/266)^{12}$, which is less than five percent [22], with the corresponding probability of detecting the incorrect guess over ninety-five percent, at each step. On the other hand, the probability of failing to identify the actual initial states (missing the event) is equal to zero at each step.

The constructed search tree can be modelled by a subcritical branching process where the average number of branches leaving each node is 12/256, which is sufficiently small for an incorrect initial guess to be detected after a relatively small number of steps (see [10]). The theory shows that if the initial guess is correct, then there may exist multiple candidates for the LFSR initial states. The correct LFSR initial states are identified either by using a few more keystream bytes or by using the LFSR$_B$ recursion, because, due to the irregular stepping, an average of 44 bits of LFSR$_B$ are obtained in the process. Accordingly, the total computational complexity of the attack is $O(2^{16})$ in time units smaller than those for the brute force attack, which requires $O(2^{96})$ time.

The success of the attack critically depends on the average number of branches leaving each node in the produced search trees, which itself depends on the number of bits, m, in the output m-bit word (in ORYX $m = 8$). In general, the average number of branches equals $12/2^m$. For any $m \geq 4$, the associated branching process is subcritical and the attack will still be successful, and its complexity increases as m decreases. On the contrary, if $m = 1$, then the attack will not be successful.

The minimum length of keystream required for this attack to be successful is thus 25 bytes: one byte to obtain the required eight bit value for $High8_K(1)$, giving eight known bits in each of the three 32-bit LFSR initial states, and then one byte to recover each of the other 24 bits in the three LFSR initial states. The more keystream available, the more certain we are of successful reconstruction. However, for less than 25 bytes of known keystream, the attack can still be performed as outlined above to give a likely reconstruction of most of the LFSR states, with exhaustive search over the contents of the last few stages to find the actual key.

6.4. Experimental Results

The performance of the attack was experimentally analysed to find the the proportion of performed attacks for which the initial states can be successfully recovered, for the keystream lengths $n = 25$, 26, and 27. The experiments use the following procedure. Nonzero initial states are generated for LFSR$_A$,LFSR$_B$ and LFSR$_K$. A keystream segment of length n, $\{Z(t)\}_{t=1}^n$ is produced using the ORYX cipher. The attack, as described in Section 6.2, is launched on the produced segment of the keystream.

For each keystream length, the attack was performed one thousand times, using pseudorandomly generated LFSR initial states. The attack was considered successful if the reconstructed LFSR initial states were unique and hence the same as the actual LFSR initial states. Table 3 shows as the success rate the proportion of attacks conducted which were successful, for each value of n.

n	25	26	27
% Success	99.7	99.9	100.0

TABLE 3. Success rate (%) versus n.

The main observation from Table 3 is that, even for the minimum keystream length $n = 25$, the attack is usually successful. In a small number of cases, there exist multiple sets of LFSR initial states which produce the required keystream segment and the attack can not identify the actual states used. However, as noted in Section 6.2, only a small increase in known keystream length is required to eliminate these additional candidates. In the experiments, a known keystream length of 27 bytes was sufficient for the attack to be successful in every trial.

6.5. Discussion

ORYX is a simple stream cipher proposed for use as a keystream generator to protect cellular data transmissions. A known plaintext attack on ORYX, using a divide and conquer approach is conducted under the assumption that the cryptanalyst knows the complete structure of the cipher and the 96-bit secret key is only the initial states of the component LFSRs. The attack requires exhaustive search over 16 bits, and has over 99 percent probability of success if the cryptanalyst knows 25 bytes of the keystream. The probability of success is increased if the cryptanalyst has access to more than 25 bytes of the keystream. In experimental trials, a known keystream length of 27 bytes was sufficient for the attack to correctly recover the key in every trial. These results indicate that the ORYX algorithm offers a very low level of cryptographic security. The main property enabling the attack is the multiple bit output of ORYX, which rarely happens in the designs proposed in the open literature.

7. Conclusion

In this paper, examples are given to demonstrate that it may possible to conduct successful known-plaintext divide and conquer attacks on a number of stream ciphers which use irregularly clocked LFSR-based keystream generators, including the shrinking generator and the ORYX stream cipher.

Some of the features of irregularly clocked LFSR-based keystream generators which increase their susceptibility to attacks include using only implicit nonlinearity, that is, either constrained or unconstrained irregular clocking of a single LFSR; using weak combinations of implicit and explicit nonlinearity; using multiple bit outputs; and using short input memory in combining functions.

There are many areas for further research into attacks on irregularly clocked LFSR-based keystream generators. These include investigations into the possibility of fast correlation attacks on irregularly clocked keystream generators and developing attacks on other irregularly clocked keystream generators which exploit the equivalent keys property. Another area is the development of designs for keystream generators that minimise the opportunity for these attacks to be successful.

References

[1] G. R. Blakley and G. B. Purdy, *A necessary and sufficient condition for fundamental periods of cascade machines to be products of the fundamental periods of their constituent finite state machines*, Information Sciences, **24** (1981), 71–91.

[2] D. Coppersmith, H. Krawczyk and Y. Mansour, *The shrinking generator*, Advances in Cryptology – CRYPTO'93, Lecture Notes in Computer Science, **773** (1993), 22–39.

[3] E. Dawson and A. Clark, *Divide and conquer attacks on certain classes of stream ciphers*, Cryptologia, **18(1)** (1994), 25–40.

[4] C. Ding, G. Xiao and W. Shan, *The Stability Theory of Stream Ciphers*, Lecture Notes in Computer Science, **561** (1991).

[5] J. Dj. Golić and M. Živković, *On the linear complexity of nonuniformly decimated PN-sequences*, IEEE Transactions on Information Theory, **34** (1988), 1077–1079.

[6] J. Dj. Golić and M. J. Mihaljević, *A generalized correlation attack on a class of stream ciphers based on the Levenshtein distance*, Journal of Cryptology, **3(3)** (1991), 201–212.

[7] J. Dj. Golić and S. Petrović, *A generalized correlation attack with a probabilistic constrained edit distance*, Advances in Cryptology – EUROCRYPT'92, Lecture Notes in Computer Science, **658** (1993), 472–476.

[8] J. Dj. Golić and L. O'Connor, *Embedding and probabilistic correlation attacks on clock controlled shift registers*, Advances in Cryptology – EUROCRYPT'94, Lecture Notes in Computer Science, **950** (1994), 230–243.

[9] J. Dj. Golić, *On the security of nonlinear filter generators*, Fast Software Encryption – Cambridge'96, Lecture Notes in Computer Science, **1039** (1996), 173–188.

[10] J. Dj. Golić, *Cryptanalysis of alleged A5 stream cipher*, Advances in Cryptology – EUROCRYPT'97, Lecture Notes in Computer Science, **1233** (1997), 239–255.

[11] J. Dj. Golić, *Periods of interleaved and nonuniformly decimated sequences*, IEEE Transactions on Information Theory, **44** (1998), 1257–1260.

[12] T. Johansson, *Reduced complexity correlation attacks on two clock-controlled Generators*, Advances in Cryptology – ASIACRYPT'98, Lecture Notes in Computer Science, **1514** (1998), 342–356.

[13] T. Johansson and F. Jönsson, *Improved fast correlation attacks on stream ciphers via convolutional codes*, Advances in Cryptology – EUROCRYPT'99, Lecture Notes in Computer Science, **1592** (1999), 347–362.

[14] J. L. Massey, *Shift register synthesis and BCH decoding*, IEEE Trans. Inform. Theory, **15** (1969), 122–127.

[15] W. Meier and O. Staffelbach, *Fast correlation attacks on certain stream ciphers*, Journal of Cryptology, **1(3)** (1989), 159–167.

[16] M. J. Mihaljević, *An approach to the initial state reconstruction of a clock-controlled shift register based on a novel distance measure*, Advances in Cryptology – AUSCRYPT'92, Lecture Notes in Computer Science, **718** (1992), 349–356.

[17] M. Salmasizadeh, L. Simpson, J. Dj. Golić and E. Dawson, *Fast Correlation Attacks and Multiple Linear Approximations*, Information Security and Privacy – ACISP'97, Lecture Notes in Computer Science, **1270** (1997), 228–239.

[18] T. Siegenthaler, *Decrypting a class of stream ciphers using ciphertext only*, IEEE Transactions on Computers, **34(1)** (1985), 81–85.

[19] L. Simpson, J. Dj. Golić and E. Dawson, *A probabilistic correlation attack on the shrinking generator*, Information Security and Privacy – ACISP'98, Lecture Notes in Computer Science, **1438** (1998), 147–158.

[20] L. Simpson, *Divide and Conquer Attacks on Shift Register based Stream Ciphers*, PhD thesis, Information Security Research Centre, Queensland University of Technology, Brisbane, Australia, November 1999.

[21] T I A TR45.0.A, *Common Cryptographic Algorithms*, Telecommunications Industry Association, Vienna V A., USA June 1995, Rev B.

[22] D. Wagner, L. Simpson, E. Dawson, J. Kelsey, W. Millan and B. Schneier, *Cryptanalysis of ORYX*, Proceedings of the Fifth Annual Workshop on Selected Areas in Cryptography – SAC'98, Lecture Notes in Computer Science, **1556** (1998), 296–305.

[23] K. C. Zeng, C. H. Yang and T. R. N. Rao, *On the linear consistency test (LCT) in cryptanalysis with applications*, Advances in Cryptology – CRYPTO'89, Lecture Notes in Computer Science, **434** (1990), 164–174.

[24] M. Živković, *An algorithm for the initial state reconstruction of the clock-controlled shift register*, IEEE Transactions on Information Theory, **37** (1991), 1488–1490.

Information Security Research Centre, Queensland University of Technology, GPO Box 2434, Brisbane, Qld 4001, Australia
E-mail address: dawson@isrc.qut.edu.au

Information Security Research Centre, Queensland University of Technology, GPO Box 2434, Brisbane, Qld 4001, Australia
E-mail address: simpson@isrc.qut.edu.au

Faculty of Electrical Engineering, University of Belgrade, 11000 Belgrade, Yugoslavia
E-mail address: golic@galeb.etf.bg.ac.yu

Progress in Computer Science and Applied Logic, Vol. 20
© 2001 Birkhäuser Verlag Basel/Switzerland

New Results on the Randomness of Visual Cryptography Schemes

Annalisa De Bonis and Alfredo De Santis

Abstract. Visual cryptography schemes have been introduced in 1994 by Naor and Shamir as a mean to share images among a given group of participants. A visual cryptography scheme (VCS) for a set $\mathcal{P}$ of n participants encodes a secret image into n shadow images called shares each of which is given to a distinct participant. Certain *qualified* subsets of participants can recover the secret image, whereas *forbidden* subsets of participants have no information on the secret image. The shares given to participants in $X \subseteq \mathcal{P}$ are xeroxed onto transparencies. If X is *qualified* then the participants in X can visually recover the secret image by stacking their transparencies without any cryptography knowledge and without performing any cryptographic computation.

In this paper we provide a new technique to derive lower bounds on the randomness of VCSs for any access structure. We also analyse minimum randomness (k, k)-threshold VCSs thus giving a new insight into the structure of these schemes.

1. Introduction

Visual cryptography schemes have been introduced in 1994 by Naor and Shamir [13] as a mean to share images among a given group of participants. A visual cryptography scheme (VCS) for a set $\mathcal{P}$ of n participants encodes a secret image into n shadow images called shares each of which is given to a distinct participant. Certain *qualified* subsets of participants can recover the secret image, whereas *forbidden* subsets of participants have no information on the secret image. The specification of all qualified and forbidden subsets of participants constitutes an *access structure*. The shares given to participants in $X \subseteq \mathcal{P}$ are xeroxed onto transparencies. If X is *qualified* then the participants in X can visually recover the secret image by stacking their transparencies without any cryptography knowledge and without performing any cryptographic computation.

Naor and Shamir [13] analysed (k, n)-threshold visual cryptography schemes, that is schemes where any subset of k participants is qualified, whereas groups of less than k participants are forbidden. The model by Naor and Shamir has been extended in [1, 2] to general access structures.

Most of the previous papers on visual cryptography mainly focus on two parameters: the *pixel expansion*, which represents the number of subpixels in the encoding of the original image, and the *contrast* which measures the "difference" between a black and a white pixel in the reconstructed image. In particular, several results on the contrast and the pixel expansion of VCSs can be found in [2, 5, 6, 7, 11].

This paper analyses the amount of randomness needed to visually share a secret image. Random bits are a natural computational resource which must be taken into account when designing cryptographic algorithms. Considerable effort has been devoted to reduce the number of bits used by probabilistic algorithms (see for example [12]) and to analyse the amount of randomness required in order to achieve a given performance. Since "truly" random bits are hard to generate, it has also been investigated the possibility of using imperfect source of randomness in randomized algorithms [17].

The randomness of visual cryptography schemes has been analysed for the first time in [9]. That paper provides upper and lower bounds on the randomness of VCSs. Moreover, it provides a complete characterization of minimum randomness (k, k)-threshold VCSs with both minimum randomness and minimum pixel expansion.

In this paper we provide a new technique to derive lower bounds on the randomness of VCSs for any access structure. We also investigate the structure of minimum randomness (k, k)-threshold VCSs and provide a characterization of minimum randomness (k, k)-threshold VCSs verifying an hypothesis weaker than that of minimum pixel expansion.

2. The Model

Let $\mathcal{P} = \{1, \ldots, n\}$ be a set of *participants*, and let $2^{\mathcal{P}}$ denote the set of all subsets of $\mathcal{P}$. An *access structure* for $\mathcal{P}$ consists of a pair $(\Gamma_{\mathsf{Qual}}, \Gamma_{\mathsf{Forb}})$, with $\Gamma_{\mathsf{Qual}} \subseteq 2^{\mathcal{P}}$, $\Gamma_{\mathsf{Forb}} \subseteq 2^{\mathcal{P}}$, and $\Gamma_{\mathsf{Qual}} \cap \Gamma_{\mathsf{Forb}} = \emptyset$. The members of Γ_{Qual} are termed *qualified sets*, while those of Γ_{Forb} are termed *forbidden sets*.

A participant $P \in \mathcal{P}$ is an *essential* participant if there exists a set $X \subseteq \mathcal{P}$ such that $X \cup \{P\} \in \Gamma_{\mathsf{Qual}}$ but $X \notin \Gamma_{\mathsf{Qual}}$. A non-essential participant does not need to participate "actively" in the reconstruction of the image, since the information he has is not needed by any set in $\mathcal{P}$ in order to recover the shared image. In any VCS having non-essential participants, these participants do not require any information in their shares. If a participant P is not essential then we can construct a visual cryptography scheme giving him nothing as his share or a completely "white" share.

Let Γ_0 consist of all the minimal qualified sets: $\Gamma_0 = \{A \in \Gamma_{\mathsf{Qual}} : A' \notin \Gamma_{\mathsf{Qual}}$ for all $A' \subset A\}$. In the case where Γ_{Qual} is monotone increasing, Γ_{Forb} is monotone decreasing, and $\Gamma_{\mathsf{Qual}} \cup \Gamma_{\mathsf{Forb}} = 2^{\mathcal{P}}$, the access structure is said to be *strong*, and Γ_0 is termed the *basis* of the access structure. (This situation is the usual setting for traditional secret sharing.) In a strong access structure, $\Gamma_{\mathsf{Qual}} = \{C \subseteq \mathcal{P} : B \subseteq C \text{ for some } B \in \Gamma_0\}$, and we say that Γ_{Qual} is the *closure* of Γ_0.

Notice that if a set of participants X is a superset of a qualified set X', then they can recover the shared image by considering only the shares of the set X'. This does not in itself rule out the possibility that stacking all the transparencies of the participants in X does not reveal any information about the shared image.

We assume that the image consists of a collection of black and white pixels. The image is encoded pixel by pixel. Each pixel is encoded into n pixels called *shares*, one for each transparency. Each share is a collection of m black and white subpixels. The resulting structure can be described by an $n \times m$ boolean matrix $M = [m_{ij}]$ where $m_{ij} = 1$ iff the j-th subpixel in the i-th transparency is black. Therefore the grey level of the combined share, obtained by stacking the transparencies $i_1, \ldots, i_s$, is proportional to the Hamming weight $w(V)$ of the m-vector $V = OR(R_{i_1}, \ldots, R_{i_s})$ where $R_{i_1}, \ldots, R_{i_s}$ are the rows of M associated with the transparencies we stack. This grey level is interpreted by the visual system of the users as black or as white according with some rule of contrast.

Definition 2.1. *Let* $(\Gamma_{\mathsf{Qual}}, \Gamma_{\mathsf{Forb}})$ *be an access structure on a set of n participants. Two collections (multisets) of $n \times m$ boolean matrices $\mathcal{C}_0$ and $\mathcal{C}_1$ constitute a visual cryptography scheme* $(\Gamma_{\mathsf{Qual}}, \Gamma_{\mathsf{Forb}})$-*VCS if there exist a value $\alpha(m)$ and a collection* $\{(X, t_X)\}_{X \in \Gamma_{\mathsf{Qual}}}$ *satisfying:*

1. *Any (qualified) set* $X = \{i_1, i_2, \ldots, i_p\} \in \Gamma_{\mathsf{Qual}}$ *can recover the shared image by stacking their transparencies.*

   Formally, for any $M \in \mathcal{C}_0$, the "or" V of rows $i_1, i_2, \ldots, i_p$ satisfies $w(V) \le t_X - \alpha(m) \cdot m$; whereas, for any $M \in \mathcal{C}_1$ it results that $w(V) \ge t_X$.

2. *Any (forbidden) set* $X = \{i_1, i_2, \ldots, i_p\} \in \Gamma_{\mathsf{Forb}}$ *has no information on the shared image.*

   Formally, the two collections of $p \times m$ matrices $\mathcal{D}_b$, with $b \in \{0, 1\}$, obtained by restricting each $n \times m$ matrix in $\mathcal{C}_b$ to rows $i_1, i_2, \ldots, i_p$ are indistinguishable in the sense that they contain the same matrices with the same frequencies.

Notice that $\mathcal{C}_0$ ($\mathcal{C}_1$) is a multiset of $n \times m$ boolean matrices, therefore we allow a matrix to appear more than once in $\mathcal{C}_0$ ($\mathcal{C}_1$). Moreover, the size of the collections $\mathcal{C}_0$ and $\mathcal{C}_1$ does not need to be the same.

To share a white (black, respectively) pixel, the dealer randomly chooses one of the matrices in $\mathcal{C}_0$ ($\mathcal{C}_1$, resp.) and distributes row i to participant i, for $i = 1, \ldots, n$.

The first property of Definition 2.1 is related to the contrast of the image. It states that a qualified set of users who stack their transparencies can correctly recover the shared image. The value $\alpha(m)$ is called *relative difference* and the number $\gamma = \alpha(m) \cdot m$, which is assumed to be an integer, is referred to as the *contrast* of the image. We want the contrast to be as large as possible and at least one, that is, $\alpha(m) \ge 1/m$. The set $\{(X, t_X)\}_{X \in \Gamma_{\mathsf{Qual}}}$ is called the *set of thresholds* and t_X is the *threshold* associated to $X \in \Gamma_{\mathsf{Qual}}$. The second property is called *security*, since it implies that, even by inspecting all their shares, a forbidden set of participants cannot gain any information in deciding whether the shared pixel was white or black.

The model of visual cryptography we consider is the same as that described in [1, 2]. This model is a generalization of the one proposed in [13], since with each set $X \in \Gamma_{\mathsf{Qual}}$ we associate a (possibly) different threshold t_X. Further, the access structure is not required to be strong in our model.

Observe that any subset of a forbidden subset is forbidden, so Γ_{Forb} is necessarily monotone decreasing. Moreover, it is easy to see that no superset of a qualified subset is forbidden. Hence, a strong access structure is simply one in which Γ_{Qual} is monotone increasing and $\Gamma_{\mathsf{Qual}} \cup \Gamma_{\mathsf{Forb}} = 2^{\mathcal{P}}$.

3. The Randomness of Visual Cryptography Schemes

The *randomness* of a visual cryptography scheme represents the number of random bits per pixel used by the dealer to share an image among the participants. The randomness of a visual cryptography scheme has been defined in [9]. Since visual cryptography schemes are a special kind of secret sharing schemes, then the randomness of a VCS has been defined in accordance with the definition of randomness for general secret sharing schemes given in [8]. The randomness of a VCS realized by $\mathcal{C}_0$ and $\mathcal{C}_1$ is defined as

$$\mathcal{R}^{(\mathcal{C}_0,\mathcal{C}_1),p} = p\log|\mathcal{C}_0| + (1-p)\log|\mathcal{C}_1|,$$

where p denotes the probability (frequency) of the white pixels in the image to be encoded. Let $\Gamma = (\Gamma_{\mathsf{Qual}}, \Gamma_{\mathsf{Forb}})$ be a given access structure. Following [8], the dealer's randomness for the access structure Γ is defined as

$$\mathcal{R}_\Gamma = \inf_{\mathcal{A},\mathcal{I}} \mathcal{R}^{(\mathcal{C}_0,\mathcal{C}_1),p},$$

where $\mathcal{A}$ denotes the set of all pairs of collections $\mathcal{C}_0$ and $\mathcal{C}_1$ realizing a VCS for Γ, and $\mathcal{I} = (0,1)$ is the range of all values of the probability p. Each pixel is assumed to be white or black with nonzero probability. In [8] the above definition has been proved to be equivalent to the following

$$\mathcal{R}_\Gamma = \min_{\mathcal{A}} \log(\min\{|\mathcal{C}_0|, |\mathcal{C}_1|\}).$$

It follows that, given a pair of matrix collections $\mathcal{C}_0$ and $\mathcal{C}_1$ realizing a VCS for the access structure Γ, we are mainly concerned with the quantity $\log(\min\{|\mathcal{C}_0|, |\mathcal{C}_1|\})$. Hence, the randomness $\mathcal{R}(\mathcal{C}_0, \mathcal{C}_1)$ of a VCS realized by $\mathcal{C}_0$ and $\mathcal{C}_1$ is defined as

$$\mathcal{R}(\mathcal{C}_0, \mathcal{C}_1) = \log(\min\{|\mathcal{C}_0|, |\mathcal{C}_1|\}).$$

Notice that the randomness of any VCS is at least one in that the share assigned to any essential participant has to be chosen in a set of size at least two.

Most of the VCSs presented in literature [1, 2, 5, 13] can be represented by means of two $n \times m$ characteristic matrices, S^0 and S^1, called *basis matrices*. The collections $\mathcal{C}_0$ and $\mathcal{C}_1$ are obtained by permuting the columns of the corresponding basis matrix (S^0 for $\mathcal{C}_0$, and S^1 for $\mathcal{C}_1$) in all possible ways. This representation is advantageous when memory space is a major concern. Notice that Property 2. of 2.1 implies that for any $X = \{i_1, i_2, \dots, i_p\} \in \Gamma_{\mathsf{Forb}}$, the restrictions of S^0 and S^1 to rows $i_1, i_2, \dots, i_p$ are equal up to a column permutation. Since the collections $\mathcal{C}_0$ and $\mathcal{C}_1$ obtained from S^0 and S^1 have both size equal to $m!$, then the randomness of such schemes is $\log(m!)$.

3.1. Known Bounds on the Randomness of (k, n)-Threshold VCSs

In this section we briefly summarize the known upper and lower bounds on the randomness of (k, n)-threshold VCSs.

The following table presents some known upper bounds on the randomness of (k, n)-threshold VCSs. The first upper bound reported in the table is relative

Naor *et al.* [13]	$n^k \log(2^{k-1}!)$
Ateniese *et al.* [1]	$\log\left((O(k(2e)^k) \log n)!\right)$
Thm. 6 [9]	$\binom{n}{k-1} - 1$
Thm. 9 [9]	$(k-1)\binom{n}{k}$
Cor. 1 [9]	$O(k^2 e^k) \log n$
Cor. 2 [9]	$O(k^{2 \log^* n} \log n)$, for k constant

TABLE 1. Upper bounds on the randomness of (k, n)-threshold VCSs.

to the randomness of the very first (k, n)-threshold VCS described in literature [13]. This upper bound is larger than all other upper bounds given in the table.

The upper bound implied by Theorem 6 of [9] is always smaller than or equal to that implied by Theorem 9 of [9]. Notice that the upper bound of Corollary 1 of [9] is asymptotically smaller than that implied by Theorem 6 of [9] when k is a sublinear function of n, whereas the bound of Corollary 2 of [9] is asymptotically smaller than that implied by Theorem 6 of [9] for any fixed k with $2 \leq k \leq n$.

Most of the constructions for (k, n)-threshold VCSs, given so far in literature, are represented by basis matrices. Recall that the randomness of such a scheme with pixel expansion m is equal to $\log(m!)$. For that reason, it is interesting to compare the upper bounds of Corollary 1 and Corollary 2 of [9] with the (k, n)-threshold VCS having the smallest pixel expansion among those represented by basis matrices [1]. This (k, n)-threshold VCS has pixel expansion $O(k(2e)^k) \log n$.

By Stirling approximation formula, the asymptotic upper bound on the randomness of this scheme is larger than $O(k(2e)^k)(\log n)(\log(O(k2^k e^{k-1})\log n))$. As a consequence, this upper bound is larger than those of both corollaries.

We conclude this section by recalling the lower bound on the randomness of (k,n)-threshold VCSs given in [9]. That lower bound is equal to

$$(k-1)\log(n-k+2) \tag{1}$$

and is known to be tight for $k=2$ and $k=n$.

4. Deriving Lower Bounds on the Randomness of VCSs for General Access Structures

In this section we will present a new technique to derive lower bound on the randomness of VCSs. To this aim we need the following definition.

Definition 4.1. Let $(\Gamma_{\mathsf{Qual}}, \Gamma_{\mathsf{Forb}})$ be an access structure on a set of participants $\mathcal{P}$. A set $G \subseteq \Gamma_{\mathsf{Forb}}$ is a max-forb set if for any pair of distinct sets $A, B \in G$, there exists a set $C \in \Gamma_{\mathsf{Qual}}$ such that $C \subseteq A \cup B$.

Let $\mathcal{C}_0$ and $\mathcal{C}_1$ realize a VCS for the access structure $(\Gamma_{\mathsf{Qual}}, \Gamma_{\mathsf{Forb}})$ on a set of participants $\mathcal{P}$, and let G a max-forb set. For any $i \in \mathcal{P}$, let $G_i = \{A \in G : i \in A\}$ and let $d_i = |\{R : M[\{i\}] = R \text{ for some } M \in \mathcal{C}_0\}|$. Notice that the number d_i, $i \in \mathcal{P}$, is larger than or equal to two since it represents the number of distinct row vectors assigned as share to participant i. In [9] it has been proved that

$$|\mathcal{C}_b| \geq \max\{|G|, \max_{i \in \mathcal{P}}\{d_i \cdot |G_i|\}\}, \qquad b \in \{0,1\}. \tag{2}$$

Such a result is a useful tool to derive lower bounds on the randomness of VCSs for general access structures. As an application of this result, we derive a lower bound on the size of the matrix collections realizing a VCS for a (k,n)-threshold structure $(\Gamma_{Qual}, \Gamma_{Forb})$. A (k,n)-threshold structure $(\Gamma_{Qual}, \Gamma_{Forb})$ on a set $\mathcal{P}$ of n participants is any access structure in which

$$\Gamma_0 = \{B \subseteq \mathcal{P} : |B| = k\} \text{ and } \Gamma_{Forb} = \{B \subseteq \mathcal{P} : |B| \leq k\}.$$

A VCS for a (k,n)-threshold structure is called (k,n)-threshold VCS.

Let G denote the family of all subsets of $\mathcal{P}$ of size $k-1$. It is $G \subseteq \Gamma_{\mathsf{Forb}}$. Moreover, for any $A, B \in G$, with $A \neq B$, one has that $A \cup B$ contains at least a subset of $\mathcal{P}$ of size k. Hence, G is a max-forb set. Let $G_i = \{A \in G : i \in A\}$, for $i = 1, \ldots, n$. It is $|G_i| = \binom{n-1}{k-2}$. For $i = 1, \ldots, n$, let d_i denote the number of distinct row vectors assigned to participant i as share. Then, (2) implies that $\mathcal{C}_0$ and $\mathcal{C}_1$ have both size larger than or equal to both $\binom{n}{k-1}$ and $(\max_{i \in \{1,\ldots,n\}}\{d_i\})\binom{n-1}{k-2} \geq 2\binom{n-1}{k-2}$.

Then, one has that for any $2 \leq k \leq n$, the randomness of a (k, n)-threshold VCS is larger than or equal to

$$\max \left\{ 1 + \log \binom{n-1}{k-2}, \log \binom{n}{k-1} \right\}. \tag{3}$$

The above lower bound is tight for $k = 2$. We refer the interested reader to [9] for a construction of minimum randomness $(2, n)$-threshold VCSs. Notice that lower bound (1) outperforms lower bound (3) for any value of $2 < k \leq n$.

Given an access structure $(\Gamma_{\text{Qual}}, \Gamma_{\text{Forb}})$ on a set of participants $\mathcal{P}$, it is relatively easy to construct a max-forb set G, while it might be difficult to obtain good lower bounds on the number d_i, $i = 1, \ldots, n$, of shares assigned to participant i. In the following we will provide a new technique to derive a lower bound on the number of distinct shares assigned to some of the participants.

Theorem 4.2. Let $C_0 = \{M_0^0, \ldots, M_{c_0}^0\}$ and $C_1 = \{M_0^1, \ldots, M_{c_1}^1\}$ realize a VCS for the access structure $(\Gamma_{\text{Qual}}, \Gamma_{\text{Forb}})$ on a set of participants $\mathcal{P}$. Let $G \subseteq \Gamma_{\text{Forb}}$ be a max-forb set. For any $A \in G$, let S_A denote the set of $|A|$-tuple representing all possible share assignments for the members of A, that is, $S_A = \{M[A] : M \in C_b\}$, $b \in \{0, 1\}$. One has either that $|S_A| \geq |G|$, for all $A \in G$, or that there exists at least a $B \in G$ such that $|S_B| \geq |G| + 1$. Consequently, there exists an index $i \in \mathcal{P}$ such that the number of all possible share assignments for participant i is $d_i \geq \lceil |G|^{1/|A|} \rceil$, for any $A \in G$ such that $i \in A$.

Proof. Let $G = \{A_1, \ldots, A_{|G|}\}$. For $b \in \{0, 1\}$ and $i = 1, \ldots, c_b$, let $\tilde{M}_i^b$ denote the $|G|$-row matrix having as r-th row, $r = 1, \ldots, |G|$, the vector obtained by computing the bitwise or of the rows of $M_i^b[A_r]$. It is easy to verify that the collections $\{\tilde{M}_1^0, \ldots, \tilde{M}_{c_0}^0\}$ and $\{\tilde{M}_1^1, \ldots, \tilde{M}_{c_1}^1\}$ realize a $(2, |G|)$-threshold VCS. Theorem 3.4 of [10] states that in any $(2, |G|)$-threshold VCS, if no participant is assigned more than $|G|$ distinct row vectors as share, then all participants are assigned exactly $|G|$ distinct row vectors as share. Hence, it follows that either there exists an index $r \in \{1, \ldots, |G|\}$ such that $|S_{A_r}| \geq |G| + 1$, or it results $|S_{A_r}| \geq |G|$, for any $r = 1, \ldots, |G|$. $\square$

The following example illustrates how Theorem 4.2 can be applied to derive a lower bound on the randomness of any given access structures.

Example 4.3. Let $\Gamma = (\Gamma_{\text{Qual}}, \Gamma_{\text{Forb}})$ be the strong access structure on the set of participants $\{1, 2, 3, 4, 5\}$ with basis $\Gamma_0 = \{\{1, 2\}, \{1, 4\}, \{2, 3\}, \{2, 4\}, \{2, 5\}, \{3, 5\}\}$. We will prove that the randomness for the access structure Γ is at least $\log 6$. Let C_0 and C_1 be two matrix collections realizing a VCS for the access structure Γ, and let $G = \{\{1, 3\}, \{1, 5\}, \{3, 4\}, \{4, 5\}, \{2\}\}$. Notice that G is a max-forb set. By Theorem 4.2 one has that either $|S_A| \geq 5$, for any $A \in G$, or there exists a set $B \in G$ such that $|S_B| \geq 6$. In the former case one has $|S_{\{1,3\}}| \geq 5$ and as a consequence at least one of participants 1 and 3 is assigned more than two distinct row vectors as share. Since 1 and 3 are both contained in two sets of G, that is

$|G_1|, |G_3| = 2$, then (2) implies that C_0 and C_1 have size at least six. In the latter case, one has $|C_b| \geq |\{M[B] : M \in C_b\}| = |S_B| \geq 6$, $b \in \{0, 1\}$. Hence, in both cases C_0 and C_1 have size at least six and, consequently, the desired lower bound on the randomness for the access structure Γ is proved.

The above lower bound is tight. Indeed, the following matrix collections realize a minimum randomness VCS for Γ.

$$
C_0 = \left\{
\begin{array}{c}
\begin{bmatrix} 001111 \\ 010111 \\ 010111 \\ 011110 \\ 010111 \end{bmatrix},
\begin{bmatrix} 001111 \\ 101101 \\ 101101 \\ 101011 \\ 101101 \end{bmatrix},
\begin{bmatrix} 110011 \\ 010111 \\ 010111 \\ 110101 \\ 010111 \end{bmatrix}, \\[3em]
\begin{bmatrix} 110011 \\ 111010 \\ 111010 \\ 101011 \\ 111010 \end{bmatrix},
\begin{bmatrix} 111100 \\ 101101 \\ 101101 \\ 110101 \\ 101101 \end{bmatrix},
\begin{bmatrix} 111100 \\ 111010 \\ 111010 \\ 011110 \\ 111010 \end{bmatrix}
\end{array}
\right\}
$$

$$
C_1 = \left\{
\begin{array}{c}
\begin{bmatrix} 001111 \\ 111010 \\ 010111 \\ 110101 \\ 101101 \end{bmatrix},
\begin{bmatrix} 001111 \\ 111010 \\ 101101 \\ 110101 \\ 010111 \end{bmatrix},
\begin{bmatrix} 110011 \\ 101101 \\ 010111 \\ 011110 \\ 111010 \end{bmatrix}, \\[3em]
\begin{bmatrix} 110011 \\ 101101 \\ 111010 \\ 011110 \\ 010111 \end{bmatrix},
\begin{bmatrix} 111100 \\ 010111 \\ 101101 \\ 101011 \\ 111010 \end{bmatrix},
\begin{bmatrix} 111100 \\ 010111 \\ 111010 \\ 101011 \\ 101101 \end{bmatrix}
\end{array}
\right\}.
$$

5. Characterization of Minimum Randomness (k, k)-Threshold VCSs

In this section we will investigate the structure of the minimum randomness (k, k)-threshold VCSs. Notice that lower bound (1) implies that the randomness of a (k, k)-threshold VCS is at least $k - 1$.

In [9] it has been provided a complete characterization of (k, k)-threshold VCSs having both minimum randomness and minimum pixel expansion. At the end of this section we will present a characterization of minimum randomness (k, k)-threshold VCSs verifying an hypothesis weaker than that of minimum pixel expansion.

First we will briefly recall the results of [9] providing the characterization of (k, k)-threshold VCSs having both minimum randomness and minimum pixel expansion. In the following we will refer to columns having even weight as *even columns* and to those having odd weight as *odd columns*. Moreover, we will denote with $\mathcal{M}_{h,k}$, for $k \geq 2$ and $h \geq 1$, a $k \times h2^{k-1}$ matrix which contains all even columns with multiplicity h and no odd columns. For any $i = 1, \ldots, k$, let $\check{R}_i$ and $\tilde{R}_i$ denote two row vectors and let $\mathbf{v}$ be a k-entry binary vector. We will denote with $\mathbf{M}(\mathbf{v}, \check{R}_1, \ldots, \check{R}_k, \tilde{R}_1, \ldots, \tilde{R}_k)$ the k-row matrix whose i-th row, $i = 1, \ldots, k$, is equal to $\check{R}_i$ if the i-th entry of $\mathbf{v}$ is 0, and to $\tilde{R}_i$ otherwise.

For $i = 1, \ldots, k$, let R_i denote the i-th row of $\mathcal{M}_{h,k}$ and $\overline{R}_i$ denote the bitwise complement of R_i. The construction for minimum randomness (k, k)-threshold VCSs given in [9] works as follows. Every time a white pixel has to be shared among the k participants, the dealer randomly chooses

$$\begin{cases} \text{a} & k\text{-entry binary vector } \mathbf{v} \text{ of even weight if the pixel is white} \\ \text{a} & k\text{-entry binary vector } \mathbf{v} \text{ of odd weight if the pixel is black,} \end{cases}$$

and for $i = 1, \ldots, k$, selects the i-th row of $\mathbf{M}(\mathbf{v}, R_1, \ldots, R_k, \overline{R}_1, \ldots, \overline{R}_k)$ as share for participant i. The matrix collections realizing the scheme are

$$\mathcal{C}_0 = \{\mathbf{M}(\mathbf{v}, R_1, \ldots, R_k, \overline{R}_1, \ldots, \overline{R}_k) : w(\mathbf{v}) \text{ is even}\}$$

$$\mathcal{C}_1 = \{\mathbf{M}(\mathbf{v}, R_1, \ldots, R_k, \overline{R}_1, \ldots, \overline{R}_k) : w(\mathbf{v}) \text{ is odd}\}$$

The scheme has pixel expansion $h2^{k-1}$, relative difference $\alpha(m) = 1/2^{k-1}$ and minimum randomness $k - 1$.

An application of the above construction is given by the following example.

Example 5.1. *The following matrix collections are obtained by applying our construction for minimum randomness (k, k)-threshold VCSs with $k = 3$ and $h = 1$.*

$$\mathcal{C}_0 = \left\{ \begin{bmatrix} 0011 \\ 0101 \\ 0110 \end{bmatrix}, \begin{bmatrix} 0011 \\ 1010 \\ 1001 \end{bmatrix}, \begin{bmatrix} 1100 \\ 0101 \\ 1001 \end{bmatrix}, \begin{bmatrix} 1100 \\ 1010 \\ 0110 \end{bmatrix} \right\},$$

$$\mathcal{C}_1 = \left\{ \begin{bmatrix} 0011 \\ 0101 \\ 1001 \end{bmatrix}, \begin{bmatrix} 0011 \\ 1010 \\ 0110 \end{bmatrix}, \begin{bmatrix} 1100 \\ 0101 \\ 0110 \end{bmatrix}, \begin{bmatrix} 1100 \\ 1010 \\ 1001 \end{bmatrix} \right\}.$$

$\triangle$

In order to characterize the structure of a minimum randomness (k, k)-threshold VCS, we fix the value of the contrast $\gamma = \alpha(m) \cdot m$. Recall that such a quantity measures the "difference" between a black and a white pixel in the reconstructed image. For any $k \geq 2$ and $\gamma \geq 1$, $m^*(k, \gamma)$ will denote the pixel expansion of the (k, k)-threshold VCS with smallest pixel expansion among those having contrast γ.

In [9] it has been proved that in any (k, k)-threshold VCS with contrast $\gamma \geq 1$ and minimum pixel expansion $m^*(k, \gamma) = \gamma 2^{k-1}$ each matrix of $\mathcal{C}_0$ consists of all even columns each occurring with multiplicity γ, whereas each matrix of $\mathcal{C}_1$ consists of all odd columns each occurring with the same multiplicity γ. Moreover, if the (k, k)-threshold VCS has both minimum randomness and minimum pixel expansion, then each participant is assigned two distinct shares one being the bit-wise complement of the other. Consequently, for any given $\gamma \geq 1$, our construction is the only one providing a (k,k)-threshold VCS with minimum pixel expansion $m^*(k,\gamma) = \gamma 2^{k-1}$, minimum randomness $k-1$, and contrast γ.

In the remainder of this section we first investigate the structure of general minimum randomness (k,k)-threshold VCSs and then provide a characterization of a special kind of minimum randomness (k,k)-threshold VCSs.

In order to prove our results, we will resort on the following lemma stated in [9] (see [10] for a complete proof of the lemma).

Lemma 5.2. *Let C_0 and C_1 be two matrix collections realizing a minimum random-ness (k,k)-threshold VCS, $k \geq 2$. Then, for any $i = 1, \ldots, k$, the set $\{M[\{i\}] : M \in C_0 \cup C_1\}$ consists of only two row vectors.*
Let $M_t^0 \in C_0$ and let $\{M[\{i\}] : M \in C_0 \cup C_1\} = \{R_i, \tilde{R}_i\}$, with $R_i = M_t^0[\{i\}]$, for $i = 1, \ldots, k$. Then, it results $C_0 = \{\mathbf{M}(\mathbf{v}, R_1, \ldots, R_k, \tilde{R}_1, \ldots, \tilde{R}_k) : w(\mathbf{v})$ is even$\}$ and $C_1 = \{\mathbf{M}(\mathbf{v}, R_1, \ldots, R_k, \tilde{R}_1, \ldots, \tilde{R}_k) : w(\mathbf{v})$ is odd$\}$.

Lemma 5.2 implies that in any minimum randomness (k,k)-threshold VCS each participant is assigned exactly two row vectors as share. We will show that for any minimum randomness (k,k)-threshold VCS with contrast γ there exists a set of at least $\gamma 2^{k-1}$ positions such that the restrictions to those positions of the two shares assigned to participant i, for $i = 1, \ldots, k$, yields two row vectors each being the bitwise complement of the other.

First we need to present a few concepts. Let us consider a (k,k)-threshold VCS with contrast γ realized by two collections C_0 and C_1 consisting of the same number of $k \times m$ matrices $c = |C_0| = |C_1|$. Let M^0 and M^1 be the $k \times cm$ matrices obtained by concatenating the columns of $M_1^0, \ldots, M_c^0$ and the columns of $M_1^1, \ldots, M_c^1$, respectively. Observe that M^0 and M^1 are the basis matrices of a (k,k)-threshold VCS with pixel expansion $m' = c \cdot m$ and relative difference $\alpha'(m') = \gamma c/m' = \gamma/m = \alpha(m)$. M^0 and M^1 are termed the basis matrices *induced* by C_0 and C_1. Let S^0 and S^1 be the two matrices obtained by removing from M^0 and M^1, respectively, the columns which appear in both matrices. In [10] it has been proved that for S_0 and S_1 the following result holds.

Lemma 5.3. *Let C_0 and C_1 be two equally sized matrix collections realizing a (k,k)-threshold VCS with contrast γ and let M^0 and M^1 be the basis matrices induced by C_0 and C_1. Let S^0 and S^1 denote the two matrices obtained by removing from M^0 and M^1, respectively, the columns which appear in both matrices. Then, S^0 and S^1 are the basis matrices of a (k,k)-threshold VCS with pixel expansion $m'' = \gamma c 2^{k-1}$ and relative difference $\gamma c/m''$, and have the property that all even columns appear in S^0 with multiplicity γc and all odd columns appear in S^1 with the same multiplicity.*

We will refer to S^0 and S^1 as to the *reduced* basis matrices *induced* by C_0 and C_1. Now we are ready to prove the following theorem.

Theorem 5.4. *Let $C_0 = \{M_1^0, \ldots, M_{c_0}^0\}$ and $C_1 = \{M_1^1, \ldots, M_{c_1}^1\}$ be two collections of $k \times m$, $k \geq 2$, matrices realizing a minimum randomness (k, k)-threshold VCS with contrast γ. For all $i = 1, \ldots, k$, let R_i and $\tilde{R}_i$ denote the two row vectors assigned as share to participant i. Then, there exists a subset $I \subseteq \{1, \ldots, m\}$ of size at least $\gamma 2^{k-1}$ such that for any $j \in I$ and for any $i \in \{1, \ldots, k\}$ the j-th entry of R_i is the complement of the j-th entry of $\tilde{R}_i$.*

Proof. Let us assume for the moment that the collections C_0 and C_1 consist of the same number of matrices c. Let M^0 and M^1 be the basis matrices induced by C_0 and C_1, and let S^0 and S^1 be the reduced basis matrices induced by C_0 and C_1. Lemma 5.3 implies that S^0 and S^1 are the basis matrices of a (k, k)-threshold VCS with pixel expansion $m'' = \gamma c 2^{k-1}$ and relative difference $\gamma c / m''$, and have the property that all even columns appear in S^0 with multiplicity γc and all odd columns appear in S^1 with the same multiplicity γc.

Let $F_j = \{i : \text{the } j\text{-th entry of } R_i \text{ is equal to the } j\text{-th entry of } \tilde{R}_i\}$, and let f_j denote the size of F_j. Suppose by contradiction that the number of column indices j for which $f_j = 0$ is less than $\gamma 2^{k-1}$.

Assume without loss of generality that, for some matrix $M_t^0 \in C_0$, it is $M_t^0[\{i\}] = R_i$, for $i = 1, \ldots, k$. Then, from Lemma 5.2 one has

$$C_0 = \{\mathbf{M}(\mathbf{v}, R_1, \ldots, R_k, \tilde{R}_1, \ldots, \tilde{R}_k) : w(\mathbf{v}) \text{ is even}\}$$

and

$$C_1 = \{\mathbf{M}(\mathbf{v}, R_1, \ldots, R_k, \tilde{R}_1, \ldots, \tilde{R}_k) : w(\mathbf{v}) \text{ is odd}\}.$$

In the remainder of the proof we will use the shorter notation $\mathbf{M}(\mathbf{v})$ to denote $\mathbf{M}(\mathbf{v}, R_1, \ldots, R_k, \tilde{R}_1, \ldots, \tilde{R}_k)$.

Let $f_\ell > 0$, for some $\ell \in \{1, \ldots, m\}$. Let $\mathbf{c}$ denote the ℓ-th column of some matrix $\mathbf{M}(\mathbf{w})$, for some k-entry binary vector $\mathbf{w}$. Then, $\mathbf{c}$ is the ℓ-th column of $\mathbf{M}(\mathbf{v})$, for any k-entry binary vector $\mathbf{v}$ coinciding with $\mathbf{w}$ in all entries with index in $\{1, \ldots, k\} \setminus F_\ell$. The total number of distinct such vectors is 2^{f_ℓ}. Notice that $2^{f_\ell - 1}$ of these vectors have even weight, whereas the remaining $2^{f_\ell - 1}$ vectors have odd weight. Hence, the former vectors are associated with matrices in C_0, whereas the latter are associated with matrices in C_1. Consequently, $\mathbf{c}$ occurs $2^{f_\ell - 1}$ times as the ℓ-th column of a matrix in C_0 and $2^{f_\ell - 1}$ as the ℓ-th column of a matrix in C_1. For this reason, the ℓ-th column of each matrix of C_0 (C_1, resp.) is removed from M^0 (M^1, resp.) when we go from M^0 (M^1, resp.) to S^0 (S^1, resp.). Therefore, for any $j \in \{1, \ldots, m\}$ and any $M_v^0 \in C_0$, the j-th column of M_v^0 is in S^0 only if $f_j = 0$. Since the number of column indices j for which $f_j = 0$ is strictly less than $\gamma 2^{k-1}$, then the number of columns of S^0 is strictly less than $2^{k-1}\gamma c$. This is absurd since we know that S^0 and S^1 consist of $2^{k-1}\gamma c$ columns each. It follows that the number of column indices j for which $f_j = 0$ must be at least $\gamma 2^{k-1}$.

Now let us assume that C_0 and C_1 have different sizes c_0 and c_1, respectively. Following a technique introduced in [1], we construct two equally sized matrix collections C_0' and C_1' realizing a (k, k)-threshold VCS with the same pixel expansion and the same contrast as the (k, k)-threshold VCS realized by C_0 and C_1. The collection $\mathcal{C}'_0$ ($\mathcal{C}'_1$, resp.) is constructed by replicating c_1 (c_0, resp.) times all matrices of $\mathcal{C}_0$ (C_1, resp.). Then, we can apply to $\mathcal{C}'_0$ and $\mathcal{C}'_1$ the same argument previously applied to equally sized matrix collections to show that there exists a subset $I \subseteq \{1, \dots, m\}$ of size at least $\gamma 2^{k-1}$ such that for any $j \in I$ and for any $i \in \{1, \dots, k\}$ the j-th entry of R_i is the complement of the j-th entry of $\tilde{R}_i$. Since the row vectors assigned as share to participant i, for $i = 1, \dots, k$, in the (k, k)-threshold VCS realized by $\mathcal{C}_0$ and $\mathcal{C}_1$ are the same as those assigned in the (k, k)-threshold VCS realized by $\mathcal{C}'_0$ and $\mathcal{C}'_1$, then one has that $\mathcal{C}_0$ and $\mathcal{C}_1$ verify the statement of the theorem, too.   □

The following example illustrates Theorem 5.4.

Example 5.5. *The following matrix collections define a minimum randomness* $(3,3)$*-threshold VCS with* $m = 8$, $t_{\{1,2,3\}} = 7$ *and* $\alpha(m) = 1/8$. *In this example the set* I *of Theorem 5.4 is* $\{1, 2, 3, 4\}$.

$$\mathcal{C}_0 = \left\{ \begin{bmatrix} 00110011 \\ 00001001 \\ 00110100 \end{bmatrix}, \begin{bmatrix} 00110011 \\ 11110000 \\ 11000010 \end{bmatrix}, \begin{bmatrix} 11001100 \\ 11110000 \\ 00110100 \end{bmatrix}, \begin{bmatrix} 11001100 \\ 00001001 \\ 11000010 \end{bmatrix} \right\},$$

$$\mathcal{C}_1 = \left\{ \begin{bmatrix} 11001100 \\ 00001001 \\ 00110100 \end{bmatrix}, \begin{bmatrix} 00110011 \\ 11110000 \\ 00110100 \end{bmatrix}, \begin{bmatrix} 00110011 \\ 00001001 \\ 11000010 \end{bmatrix}, \begin{bmatrix} 11001100 \\ 11110000 \\ 11000010 \end{bmatrix} \right\}.$$

△

A characterization better than that provided by Theorem 5.4 can be provided for minimum randomness (k, k)-threshold VCSs which verify a particular assumption on the pixel expansion. This assumption is far weaker than that of minimum pixel expansion considered in [9]. Indeed, we consider the case when all matrices in the collections $\mathcal{C}_0$ and $\mathcal{C}_1$ contain only columns which contribute to the relative difference of the VCS, that is, columns whose removal would cause the contrast to decrease.

Definition 5.6. *Let* $\mathcal{C}_0$ *and* $\mathcal{C}_1$ *be two collections of* $n \times m$ *boolean matrices constituting a* (k, k)*-threshold VCS,* $k \geq 2$, *with relative difference* $\alpha(m)$. *The* j*-th column, with* $1 \leq j \leq m$, *is redundant for* $\mathcal{C}_0$ *and* $\mathcal{C}_1$ *if the matrix collections* $\mathcal{C}'_0$ *and* $\mathcal{C}'_1$ *obtained by removing the* j*-th column from all matrices in* $\mathcal{C}_0$ *and* $\mathcal{C}_1$, *realize a* (k, n)*-threshold VCS with pixel expansion* $m' = m - 1$ *and relative difference* $\alpha'(m') > \alpha(m)$.

Obviously, for any $\gamma \geq 1$, two matrix collections $\mathcal{C}_0$ and $\mathcal{C}_1$ constitute a (k, k)-threshold VCS with minimum pixel expansion $m^*(k, \gamma)$ only if the matrices of $\mathcal{C}_0$ and $\mathcal{C}_1$ contain no redundant column.

Example 5.7 shows a $(3, 3)$-threshold VCS realized by two collections $\mathcal{C}_0$ and $\mathcal{C}_1$ of 3×5 matrices for which the 5-th column is redundant.

In the following, we will simply denote the threshold $t_{\{1,\dots,k\}}$ of Definition 2.1 with t.

Example 5.7. *The following collections C_0 and C_1 realize a $(3,3)$-threshold VCS.*

$$C_0 = \left\{ \begin{bmatrix} 00110 \\ 01010 \\ 01100 \end{bmatrix}, \begin{bmatrix} 00110 \\ 10101 \\ 10011 \end{bmatrix}, \begin{bmatrix} 11001 \\ 01010 \\ 10011 \end{bmatrix}, \begin{bmatrix} 11001 \\ 10101 \\ 01100 \end{bmatrix} \right\},$$

$$C_1 = \left\{ \begin{bmatrix} 00110 \\ 01010 \\ 10011 \end{bmatrix}, \begin{bmatrix} 00110 \\ 10101 \\ 01100 \end{bmatrix}, \begin{bmatrix} 11001 \\ 01010 \\ 01100 \end{bmatrix}, \begin{bmatrix} 11001 \\ 10101 \\ 10011 \end{bmatrix} \right\}.$$

The matrices in C_0 and C_1 verify Property 1. of Definition 2.1 with $t = 5$ and $\alpha(5) = 1/5$. If we remove the 5-th column from each matrix of C_0 and C_1, then the matrices in the resulting collections verify Property 1. of Definition 2.1 with $t = 4$ and $\alpha(4) = 1/4$. Thus, the 5-th column is redundant for C_0 and C_1. $\triangle$

Theorem 5.8. *Let C_0 and C_1 be two collections of $k \times m$, $k \geq 2$, boolean matrices constituting a (k,k)-threshold VCS with relative difference $\alpha(m)$. A column with index j, for some $1 \leq j \leq m$, is redundant for C_0 and C_1 if and only if at least one of the following conditions holds:*

a) *The j-th columns of all matrices in C_0 and C_1 are identical.*

b) *The j-th columns of all matrices in C_0 have weight larger than zero.*

c) *For any matrix $M_0 \in C_0$ such that the j-th column of M_0 is the all-zero column, one has that the "or" V_0 of the rows of M_0 has weight $w(V_0) < t - \alpha(m)m$.*

d) *For any matrix $M_1 \in C_1$ such that the j-th column of M_1 has weight larger than zero, one has that the "or" V_1 of the rows of M_1 has weight $w(V_1) > t$.*

Proof. It is easy to see that if the j-th column verifies one of the conditions a), b), c) and d), then it is redundant. Let us prove that if the j-th column satisfy none of conditions a), b) c) and d), then the j-th column is not redundant. Let us assume that the j-th column does not satisfy any of conditions a), b), c) and d). Then, there is a matrix $M_0 \in C_0$ which has the j-th column equal to the all-zero column and such that the "or" V_0 of the rows of M_0 satisfies $w(V_0) = t - \alpha(m)m$, and a matrix $M_1 \in C_1$ which has the j-th column different from the all-zero column and such that the "or" V_1 of all rows of M_1 satisfies $w(V_1) = t$. Let M_0' and M_1' denote the matrices obtained by removing the j-th column from the matrices M_0 and M_1, respectively. The "or" V_1 of all rows in M_1' has weight $t - 1$, whereas the "or" V_0 of all rows in M_0' has weight equal to $t - \alpha(m)m = t - \alpha(m) \cdot (m-1) - \alpha(m) = t - 1 - (\alpha(m) - \frac{1-\alpha(m)}{m-1})(m-1)$. Since $\alpha(m) \leq 1$, then M_0' and M_1' verify Property 1. of Definition 2.1 with a value $\alpha'(m-1) = \alpha(m) - \frac{1-\alpha(m)}{m-1} \leq \alpha(m)$. $\square$

Let C_0 and C_1 realize a minimum randomness (k,k)-threshold VCS. If there is no redundant column for C_0 and C_1, then the following theorem holds.

Theorem 5.9. *Let C_0 and C_1 be two matrix collections realizing a minimum randomness (k,k)-threshold VCS, $k \geq 2$, with contrast $\gamma \geq 1$. If the matrices in C_0 and C_1 contain no redundant column, then each matrix in C_0 contains at least $\gamma 2^{k-1}$ even columns, whereas each matrix in C_1 contains at least $\gamma 2^{k-1}$ odd columns.*

Proof. Since the matrices in $\mathcal{C}_0$ and $\mathcal{C}_1$ do not contain any redundant column then one has that $\mathcal{C}_0$ does not verify condition *b)* of Theorem 5.8, in other words, for any column index j there exists a matrix in $\mathcal{C}_0$ having the j-th column equal to the all-zero column.

From Lemma 5.2 it follows that in any minimum randomness (k,k)-threshold VCS each participant is assigned two distinct row vectors as share. For any $i = 1, \ldots, k$, let R_i and $\tilde{R}_i$ denote the row vectors assigned as share to participant i. From Theorem 5.4 it follows that there exists a set I of size at least $\gamma 2^{k-1}$ such that for any $j \in I$ and for any $i \in \{1, \ldots, k\}$, the j-th entry of R_i is the complement of the j-th entry $\tilde{R}_i$. Suppose by contradiction that for some $j \in I$, there exists a matrix $M^1 \in \mathcal{C}_1$ whose j-th column has weight $2t$ for some t, $0 \le t \le \lfloor k/2 \rfloor$. Assume without loss of generality that the first $2t$ entries of this column are equal to 1. Then a matrix $M^0 \in \mathcal{C}_0$ has the j-th column equal to the all-zero column if and only if $M^0[\{i\}] \neq M^1[\{i\}]$ for $1 \le i \le 2t$, and $M^0[\{i\}] = M^1[\{i\}]$ for $2t + 1 \le i \le k$. Assume without loss of generality that $M^0[\{i\}] = R_i$, for $i = 1, \ldots, k$. Then, Lemma 5.2 implies

$$\mathcal{C}_0 = \{\mathbf{M}(\mathbf{v}, R_1, \ldots, R_k, \tilde{R}_1, \ldots, \tilde{R}_k) : w(\mathbf{v}) \text{ is even}\}$$

and

$$\mathcal{C}_1 = \{\mathbf{M}(\mathbf{v}, R_1, \ldots, R_k, \tilde{R}_1, \ldots, \tilde{R}_k) : w(\mathbf{v}) \text{ is odd}\}.$$

Since each of the first $2t$ rows of M^0 is different from the corresponding row of M^1, whereas each of the last $k - 2t$ rows of M^0 is equal to the corresponding row of M^1, then it results $M^1 = \mathbf{M}(\mathbf{v}, R_1, \ldots, R_k, \tilde{R}_1, \ldots, \tilde{R}_k)$ with $\mathbf{v}$ being the k-entry binary vector having the first $2t$ entries equal to 1 and the remaining $k - 2t$ entries equal to 0. This is absurd since $M^1 \in \mathcal{C}_1 = \{\mathbf{M}(\mathbf{v}, R_1, \ldots, R_k, \tilde{R}_1, \ldots, \tilde{R}_k) : w(\mathbf{v}) \text{ is odd}\}$. Hence, we proved that for any $j \in I$, the j-th columns of all matrices in $\mathcal{C}_1$ have odd weight. Moreover, it can be proved that if for some $j \in I$ there exists a matrix of $\mathcal{C}_0$ with an odd column in the j-th position, then there exists a matrix of $\mathcal{C}_1$ with an even column in the j-th position. Indeed, let $\bar{M}^0 \in \mathcal{C}_0$ and suppose by contradiction that for some $j \in I$ the j-th column of $\bar{M}^0$ is an odd column. Let $\mathbf{z}$ be the k-entry vector such that $\bar{M}^0 = M(\mathbf{z}, R_1, \ldots, R_k, \tilde{R}_1, \ldots, \tilde{R}_k)$ and let $\mathbf{w}$ be any k-entry vector which differs in exactly one entry from $\mathbf{z}$. By Lemma 5.2 one has that $\mathbf{z}$ has even weight and that the matrix $\bar{M}^1 = M(\mathbf{w}, R_1, \ldots, R_k, \tilde{R}_1, \ldots, \tilde{R}_k)$ belongs to $\mathcal{C}_1$. Since the j-th columns of $\bar{M}^0$ and $\bar{M}^1$ differ only in one entry then one has that the j-th column of $\bar{M}^1$ is even. This is absurd since we have just proved that for any $j \in I$, the j-th columns of all matrices in $\mathcal{C}_1$ have odd weight. Then it follows that, for any $j \in I$, the j-th column of all matrices in $\mathcal{C}_0$ has even weight. □

Notice that the minimum randomness $(3,3)$-threshold VCS of Example 5.5 is realized by two matrix collections containing no redundant column and has pixel expansion larger than $m^*(3,1)$.

References

[1] G. Ateniese, C. Blundo, A. De Santis, and D. R. Stinson, *Visual cryptography for general access structures*. Information and Computation **129-2**, 86–106 (1996).

[2] G. Ateniese, C. Blundo, A. De Santis, and D. R. Stinson, *Constructions and bounds for visual cryptography*. ICALP 1996, LNCS **1099**, 416–428.

[3] A. Beimel and B. Chor, *Universally ideal secret sharing schemes*. IEEE Trans. on Information Theory **40**(3), 786–794 (1994).

[4] C. Blundo, A. De Santis, G. Persiano and U. Vaccaro, *Randomness complexity of private computation*. Computational Complexity **8**, 145-168 (1999).

[5] C. Blundo, P. D'Arco, A. De Santis, and D. R. Stinson, *Contrast optimal threshold visual cryptography schemes*. Submitted for publication (1998).

[6] C. Blundo and A. De Bonis, *New constructions for visual cryptography*. ICTCS '98. P. Degano, U. Vaccaro, G. Pirillo, (Eds.), 290–303, World Scientific.

[7] C. Blundo, A. De Santis, and D. R. Stinson, *On the contrast in visual cryptography schemes*. Journal of Cryptology **12**, 261–289 (1999). Available also at Theory of Cryptology Library as `ftp://theory.lcs.mit.edu/pub/tcryptol/96-13.ps`. September 1996.

[8] C. Blundo, A. De Santis, and U. Vaccaro, *Randomness in distribution protocols*. Information and Computation **131**, 111–139 (1996).

[9] A. De Bonis and A. De Santis, *Randomness in visual cryptography*. STACS 2000, LNCS **1770**, 626–638.

[10] A. De Bonis and A. De Santis, *On the randomness in visual cryptography*. Submitted for publication.

[11] T. Hofmeister, M. Krause, and H. U. Simon, *Contrast-optimal k out of n secret sharing schemes in visual cryptography*. COCOON '97, LNCS **1276**, 176–185.

[12] R. Impagliazzo and D. Zuckerman, *How to recycle random bits*. STOC '89, 248–255.

[13] M. Naor and A. Shamir, *Visual cryptography*. Advances in Cryptology – EURO-CRYPT '94, LNCS **950**, 1–12.

[14] D.R. Stinson, *An introduction to visual cryptography*. Presented at Public Key Solutions '97, Toronto, Canada, April 28–30 (1997).
Available as `http://bibd.unl.edu/~stinson/VKS-PKS.ps`.

[15] G. J. Simmons, W.-A. Jackson, and K. Martin, *Decomposition constructions for secret sharing schemes*. Bulletin of the ICA, 1:72–88 (1991).

[16] J. H. van Lint and R. M. Wilson, *A Course in Combinatorics*. Cambridge University Press (1992).

[17] D. Zuckerman, *Simulating BPP using a general weak random source*. FOCS '91, 79–89.

Dipartimento di Informatica ed Applicazioni,
Università di Salerno,
84081 Baronissi (SA), Italy
E-mail address: {debonis,ads}@dia.unisa.it
URL: http://www.unisa.it/~ads

Progress in Computer Science and Applied Logic, Vol. 20

Authentication – Myths and Misconceptions

Dieter Gollmann

Abstract. There is a mantra telling us that authentication is difficult. The failure to design robust authentication protocols is commonly attributed to a lack of good design strategies, and to a lack of verification tools. This paper tells the story of entity authentication arguing that clarity is more important than precision, and that formal methods sometimes even add to the confusion about the meaning of 'authentication'. Verifying claimed identities translates into checking whether a party is alive, or into checking the identity of the party at the other end of a connection. Correspondence properties can capture both aspects, obscuring an important distinction.

1. Introduction

Formal methods are used to analyse security properties in an abstract setting. Abstraction (idealization) creates a model of the protocol and of its execution environment. Properties derived in the model have to be translated back to the original problem domain ('reality'). Crossing the boundary between reality and model is error-prone (Figure 1). Formalization does not remove this boundary, it crosses the boundary. In this respect, number theory, cryptography, and formal methods are easy subjects as there exist precise definitions of the objects and properties under examination. In contrast, grasping informal concepts is difficult. Formality can yield precision, but precision is not a substitute for clarity.

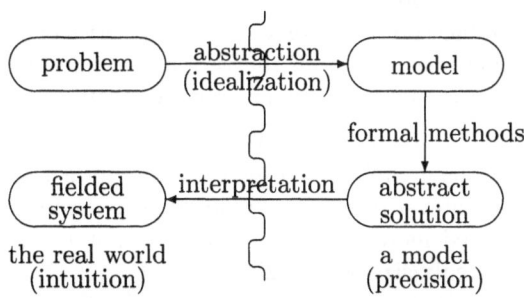

FIGURE 1. Formal methods and mathematics in security analysis

Authenticating has the reputation of being a difficult subject. Stories about broken protocols abound and there exists a considerable body of work on the formal analysis of authentication protocols. The following quote reflects the general mood.

> Many authentication protocols have been published and later found to contain subtle weaknesses or flaws. Two factors contribute to this: (i) the lack of well-established guiding principles for protocol design, and (ii) the use of informal operational reasoning for protocol analysis [30].

Typically, such a quote motivates the presentation of a new formalism for the analysis of authentication protocols, and the biggest prize to be won is the detection of an attack hitherto unreported. We claim, contrarily, that formal analysis more often adds to the problem than helps in its resolution. Perceived problems with authentication are caused by intuitive but imprecise interpretations of the objective of 'authentication', and by neglecting to take into account the environment a protocol is intended to operate in. Attacks in circuit-switched networks, for example, may be meaningless in a packet-switched network. In many cases, new attacks do not expose subtle flaws in a protocol but differences in the assumptions about protocol goals.

1.1. Extensional and Intensional Specifications

To prepare the ground for our discussions, we present Roscoe's distinction between

> *extensional* specifications, which seek to establish what is achieved by the protocol without making detailed analyses of the actual communications nodes have sent, and *intensional* specifications which concentrate on checking that the designer's expectations about these communications are justified [25].

Designers certainly assume, perhaps somewhat optimistically, that protocols meet their stated objectives when there is no interference by an intruder and all steps are executed in the intended order. This view is captured by the following

> *Canonical intensional specification*: A very natural specification we might want a protocol to satisfy is that no node can believe a protocol run has completed unless a correct series of messages has occurred (consistent as to all the various parameters) up to and including the last message the given node communicates.

Of course, it is suspicious if a party can be persuaded to complete a protocol run in circumstances where the intended correspondent did not act according to the rules of the protocol, for whatever reason. Such violations of an intensional specification deserve to be examined more closely, but they do not necessarily constitute an attack, a point repeatedly stressed in [25]:

> the main weakness is that it [an intensional specification] tells us nothing in any abstract sense about what the protocol does.

Strictly speaking, only violations of an extensional specification should be called 'attacks'. A nice example for a violation of an intensional specification that can be described as an attack only with a great stretch of imagination is given in [21]. In this 'attack', a principal A starts to run the Needham-Schroeder public key protocol (see Section 3.3) with itself. The intruder replays A's first message to A, convincing A that it had received a reply to a challenge it had sent to itself.

Successful verification of an intensional specification confirms that a protocol can only be used in the manner intended by its designers, not that it achieves the abstract goals they intended. It is therefore appropriate to say that

> [intensional specifications] provide a separate criterion for judging a protocol to the more abstract extensional specifications one may be able to invent and should not be thought of as a substitute, except perhaps in the area of authentication.

It is interesting to see authentication listed as a possible exception to the general rule. It is tempting to employ intensional specifications in a formal analysis, because they can be formulated in various process calculi in a straightforward fashion, see [28] for an example, and one does not have to think about a more abstract (deeper?) meaning of authentication. However, we hope to show in the following that there exist reasonable extensional definitions of entity authentication.

1.2. Notation

P_X and S_X denote the public encryption key and private signature key of principal X. Symbols like n_X denote a nonce generated by principal X. Encryption of m under key K is written as $eK(m)$, the digital signature of m under K as $sK(m)$.

2. Mutual Authentication

We first focus on authentication protocols where a *verifier* checks the identity of a *claimant* (entity authentication). We do not yet consider protocols that establish shared keys. We will examine various proposals for capturing the meaning of authentication, trying to convince the reader that often the goals of authentication change surreptitiously during the process of formalization. To stay close to our sources, we frequently use quotes when presenting our case. The International Standard ISO/IEC 9798-1 [12] defines entity authentication as follows.

Definition 2.1. *Entity authentication mechanisms allow the verification, of an entity's claimed identity, by another entity. The authenticity of the entity can be ascertained only for the instance of the authentication exchange.*

This view of entity authentication is by definition at odds with any interpretation treating authentication as a step performed at the start of a session. According to Def. 2.1, entity authentication just provides evidence that the claimant is alive, as demonstrated by the following simple challenge-response protocol providing unilateral (one-way) authentication:

$$1. \quad B \to A\colon eP_A(n_B)$$
$$2. \quad A \to B\colon n_B$$

Verifier B sends an encrypted challenge and uses the response to verify the identity
of the responder (claimant) A. If B receives back the challenge n_B it sent encrypted
under A's public key, A has been authenticated to B. In this protocol, A has no
evidence about whom it is replying to and could as well broadcast its response.
This protocol may be used by a network controller to poll nodes in the network
and check their availability, by a security company to check whether a safe has
been tampered with, or by an operating system where users log in with the help of
proximity smart cards. In all cases, the protocol prevents an attacker from replying
to challenges sent to devices that have failed or are no longer present, and thus
deceiving the controller about the true state of the device.

A session-oriented view of entity authentication is given in another International Standard, ISO 7498-2 [11].

Definition 2.2. *Peer entity authentication: The corroboration that a peer entity in
an association is the one claimed. This service is provided for use at the establishment of, or at times during, the data transfer phase of a connection to confirm
the identities of one or more of the entities connected to one or more of the other
entities.*

The two definitions given aim into quite different directions. Given this diver-
gence of goals, it should come as no surprise that research on entity authentication
has been somewhat out of focus.

2.1. Two-way Authentication Is Not Simply Twice One-way

Early proposals for formalizing entity authentication were influenced by the observation that 'two-way (mutual) authentication is not simply twice one-way' [3].
definition proposed in [4] has a distinctly intensional flavour.

Definition 2.3. *Whenever one of the parties completes an execution of the protocol,
it marks the execution as either accepted (in case of successful authentication) or
rejected. The intention is that executions marked as accepted correspond to runs
of the protocol that definitely involved the intended other party.*

Error free executions are once more the yardstick for detecting attacks.

> An error-free history of the protocol runs between A and B is one
> in which all of the executions accepted by both parties ... match
> exactly one-to-one [4].

Errors in a protocol run can be due to an attack or to a communications failure.
The explanation of an attack in [4] is not quite compatible with Def. 2.3:

> The attacker's objective is to cause A or B to erroneously mark
> one of these perverted protocol runs as accepted, even though it
> does not match with any execution accepted by the other party.

The intended other party not only has to be *involved* in the protocol run, it now has to *accept* the protocol run. Moreover, no distinction is made between attacks and protocol runs aborted for some other reason. An attacker could achieve this objective simply by suppressing the last message in a regular protocol run. This can be called a broken connection but hardly a broken protocol.

It may be tempting to interpret any action by a third party as an attack. However, in a network most communications require contributions from intermediate nodes so the fact that such a node was involved in a protocol run accepted by one party (or both) does not necessarily constitute an attack:

> ... a third party can always act as a simple delaying "relay" between the two parties without being considered an attacker. Indeed, such delaying is no different and is indistinguishable from what any network switch does in normal operation.

After these preliminaries, [4] constructs a mutual authentication protocol by taking two separate runs of the one-way authentication protocol given above,

$$
\begin{array}{lll}
1. & A \to B\text{:} & eP_B(n_A) \\
2. & B \to A\text{:} & n_A \\
3. & B \to A\text{:} & eP_A(n_B) \\
4. & A \to B\text{:} & n_B
\end{array}
$$

and merging messages 2 and 3 to save one protocol step:

$$
\begin{array}{lll}
1. & A \to B\text{:} & eP_B(n_A) \\
2. & B \to A\text{:} & n_A, eP_A(n_B) \\
3. & A \to B\text{:} & n_B
\end{array}
$$

Now, it is claimed, an enemy E can launch an oracle session attack by engaging in two protocol runs with A and B, using A as an oracle to respond to B's challenge[1]:

$$
\begin{array}{lll}
1. & E(A) \to B\text{:} & eP_B(n_E) \\
2. & B \to E(A)\text{:} & n_E, eP_A(n_B) \\
3. & E(B) \to A\text{:} & eP_A(n_B) \\
4. & A \to E(B)\text{:} & n_B, eP_B(n_A) \\
5. & E(A) \to B\text{:} & n_B
\end{array}
$$

The description of this sequence of events concludes with this comment on step 5:

> But it [the enemy] then turns around and successfully assumes its faked identity A with respect to B by sending n_B to B.

This 'attack' creates a paradox. If B sends n_A and $eP_A(n_B)$ in separate messages, both parties are authenticated. If B sends the same two items in a single message, mutual authentication fails. How could this happen, or has it happened at all? Before answering this question, we look at a further example.

---

[1]Let $E(A) \to B$ indicate that E sends a message to B that seems to come from A, and $B \to E(A)$ that E receives a message B intended for A.

2.2. The Canadian Attack

The protocol this attack refers to was at one stage [13] considered for inclusion in
ISO/IEC 9798-3, the International Standard for authentication mechanisms using
a public key algorithm. We describe this protocol in simplified form:

$$1. \quad A \to B: n_A$$
$$2. \quad B \to A: n_B, A, n_A, sS_B(n_B, A, n_A)$$
$$3. \quad A \to B: n'_A, B, n_B, sS_A(n'_A, B, n_B)$$

In step 3, A intentionally does not reuse its challenge n_A to avoid signing a string
which is partly defined, and fully known in advance, by B. This protocol survived
in a few revisions of ISO/IEC 9798-3, until an attack was found by the Canadian
member body of ISO [14] (see [8, 23] for further comments on this attack):

$$1. \quad E(A) \to B: n_A$$
$$2. \quad B \to E(A): n_B, A, n_A, sS_B(n_B, A, n_A)$$
$$3. \quad E(B) \to A: n_B$$
$$4. \quad A \to E(B): n'_A, B, n_B, sS_A(n'_A, B, n_B)$$
$$5. \quad E(A) \to B: n'_A, B, n_B, sS_A(n'_A, B, n_B)$$

The description of the attack concludes with the remark 'B now believes to be
connected to A', and the effect of the attack is summarized as follows.

> An intruder who has initiated a session with B can initiate an-
> other session with A and so obtain a signature suitable for use in
> responding to B. In this way, the intruder can successfully mas-
> querade as A in his session with B.

2.3. Resolving the Paradox

Figure 2 shows the Canadian attack in a circuit-switched network. B has a connection with E but is misled into associating the connection with A. This message
had previously been sent by A on another connection. The intruder has succeeded
in deceiving B about the identity of the entity at the other end of a connection. In
circuit-switched networks, we can reason about the entities we are connected to,
and Def. 2.2 is applicable. To masquerade as another party, one sends messages
on a connection the receiver associates with the party impersonated. Mutual authentication of a single connection is not the same as unilateral authentication of
two separate connections. This is the first explanation for our paradox.

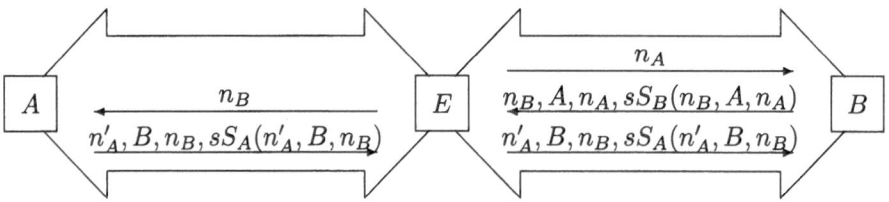

FIGURE 2. The Canadian attack in a circuit-switched network

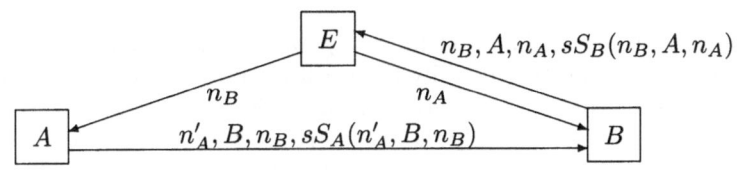

FIGURE 3. The Canadian attack in a packet-switched network

In a packet-switched network, 'E masquerades as A' by intercepting a message from A to B, and forwarding this very message to the intended recipient B. Evidently, E achieves the same effect by doing nothing and letting message 4 pass to B (Figure 3). The attack leaves A with an incomplete protocol run with B. B expects a message containing A's signature on B's challenge n_B. The message sent by A in step 4 and forwarded by E in step 5 fully meets this criterion so no breach has occurred. In packet-switched networks, the Canadian attack counts only if it matters who starts a protocol run. As there are no connections to authenticate Def. 2.2 is inapplicable. Terms like masquerade or impersonate need precise definitions, and 'being connected to' evokes a wrong communications paradigm.

In the attack from Section 2.1, B accepts the protocol run and concludes that A has been authenticated. By Def. 2.3, *the intention [of an authentication protocol] is that executions marked as accepted correspond to runs of the protocol that definitely involved the intended other party.* Party A was indeed involved in this protocol run. It had to serve as an oracle for the intruder. In this interpretation, B has properly authenticated A and there is no attack.

Does the intruder assume a faked identity? In a circuit-switched network, E opens a connection with B that B wrongly associates with A. In a packet-switched network, E has to do a bit more than in the Canadian attack, editing the message from A to B before forwarding it to B. To be precise, E copies part of a message to the intended receiver, and suppresses the rest of the message. Network switches also edit messages, for example, modify headers or reformat messages. Hence, forwarding edited messages to the intended receiver does not constitute an attack in itself. The meaning of words would be stretched considerably if we said that E assumes A's identity when E does nothing more than suppress a part of A's message to B.

If the attacker's objective is *to cause A or B to erroneously mark one of these perverted protocol runs as accepted, even though it does not match with any execution accepted by the other party,* we have an attack but, as said above, we would classify communications failures as security violations. We should thus only ask whether B's protocol run has a one-to-one match with any protocol run involving A. Then, it depends on the definition of matching whether we face an attack.

- There is a match because B issues an encrypted challenge $eP_A(n_B)$ and A replies to the challenge n_B. Hence, there is no attack.
- There is a mismatch because B replies to an encrypted challenge $eP_B(n_E)$ that was not issued by A. Hence, there is an attack.

We have a second explanation for our paradox. In one-way authentication, the claimant is not expected to verify the verifier and will not check the origin of the challenge. In mutual authentication, both parties act as verifiers. Hence, one may be persuaded to check not only the origin of replies but also the origin of challenges. This new condition, which is not explicitly stated in [4] and could be viewed as a typical case of *requirement creep*, may have its justifications but they are not given, and the explanation of the attack hardly refers to the original protocol goals. Finding circumstances where the origin of challenges has to be verified remains an interesting exercise.

2.4. Putting the Coach Before the Horses

A similar story is told in [7], beginning with a familiar high level definition.

> The goal of an authentication protocol is to provide the communicating parties with some assurance about the other's true identity.

To motivate the formalization of this notion, a protocol is introduced, and an attack. This time, we have a challenge-response protocol using digital signatures:

$$1. \quad A \rightarrow B: n_A$$
$$2. \quad B \rightarrow A: sS_B(n_A), n_B$$
$$3. \quad A \rightarrow B: sS_A(n_B)$$

An 'attack' is given before the protocol goals have yet been defined in detail:

$$1. \quad E(A) \rightarrow B: n_E$$
$$2. \quad B \rightarrow E(A): sS_B(n_E), n_B$$
$$3. \quad E(B) \rightarrow A: n_B$$
$$4. \quad A \rightarrow E(B): sS_A(n_B), n_A$$
$$5. \quad E(A) \rightarrow B: sS_A(n_B)$$

After the last message

> Eve [the attacker] can then complete the authentication with Bob and impersonate Alice.

The definition of authentication adopted in [7] turns the above sequence of events into an attack. Following [3] (and [4]) protocol runs are required to match, but the notion of matching protocol runs is made more precise.

Definition 2.4. *Two records of a run match if their messages can be partitioned into sets of matching messages (each set containing one message from each record), the messages originated by one participant appear in the same order in both records, and the messages originated by the other participant appear in the same order in both records.*

Definition 2.5. *A run of an authentication protocol is successful if the following two properties hold:*

- *Both Alice and Bob accept each other's identities.*
- *If Alice and Bob have recorded their exchange, then their records of the run will match.*

The issues raised here are familiar. *Impersonation* has a reasonable explanation only in circuit-switched networks. Protocol runs can fail as the consequence of attacks or of communications failures. To differentiate, a definition should allow for incomplete protocol runs and only ask for a match between certain crucial events in the protocol runs, rather than for a complete match between all messages.

There is also a fundamental conceptual problem. Def. 2.5 is an intensional specification of authentication, phrasing the goals of a protocol so directly in terms of the details of its execution that the execution itself appears as the main reason for running the protocol. Even worse, whilst protocol designers strive to limit the amount of state participants have to keep, this definition explains authentication in terms of artificial state components. In our current protocol, B does not have to store n_A, and has to store n_B only between steps 2 and 3. Why should the success of a protocol run be judged on the basis of data items B does not 'remember' (store) at the end of the run?

Def. 2.5 is certainly sufficient to reach the informal objectives of authentication, but it is by no means necessary. It may be convenient when verifying protocols in a formalism that models protocol executions, but at the same time it becomes more difficult to justify why 'incorrect' executions amount to attacks (see also [22]).

2.5. Simulating Connections

Bellare and Rogaway [2] define entity authentication in terms of *matching conversations*. A conversation of a given entity is a sequence

$$(\tau_1, \alpha_1, \beta_1), (\tau_2, \alpha_2, \beta_2), \ldots, (\tau_m, \alpha_m, \beta_m),$$

where τ_i are time instances, α_i the input to that entity at time τ_i, and β_i its output at time τ_i. Two conversations match if there exist $\tau_0 < \tau_1 < \ldots < \tau_{2r-1}$ and $\alpha_1, \beta_1, \ldots, \alpha_r, \beta_r$ so that the conversations are prefixed by

$$(\tau_0, ., \alpha_1), (\tau_2, \beta_1, \alpha_2), \ldots, (\tau_{2r-4}, \beta_{r-2}, \alpha_{r-1}), (\tau_{2r-2}, \beta_{r-1}, \alpha_r)$$

and

$$(\tau_1, \alpha_1, \beta_1), (\tau_3, \alpha_2, \beta_2), \ldots, (\tau_{2r-3}, \alpha_{r-1}, \beta_{r-1}), (\tau_{2r-1}, \alpha_r, \beta_r)$$

respectively. When matching conversations become the benchmark for successful authentication, protocols are forced to simulate a connection. All messages arrive, they arrive unaltered, and in correct order. Simulating a connection is one of many ways of checking whether an entity is alive, but it is hardly the most efficient. Def. 2.2 does not ask for the simulation of connections either.

2.6. Through the Looking Glass

A global observer of a protocol run sees how the states (beliefs) formed by the participants develop as the run proceeds. At first glance, it seems natural to ask that the beliefs formed by the verifier and by the responder are in agreement.

Definition 2.6. *We say that a protocol correctly achieves authentication if whenever an agent A accepts the identity of another agent B, it must be the case then*

B believes that he has been running the protocol with A, and the records of the messages sent and received at the two ends should match [19].

On second thoughts, this definition does not capture authentication very well. To verify the claimant's identity one does not have to verify the claimant's beliefs.[2] The fact that A wants to corroborate B's identity does not imply that B has to know whom it is responding to. As stated above, in most unilateral authentication protocols the responder has no evidence about the verifier's identity, and in a typical three-step authentication protocol the responder can verify the claimant's identity only after the last message has arrived, at which stage the verifier has already authenticated the responder. Hence, the above condition cannot be met unless the communications system guarantees message delivery. Roscoe's definition of intensional specifications is more careful and avoids this particular pitfall.

An example using the station-to-station protocol [7] may illustrate some of the problems with Def. 2.6. Let p be a suitably chosen prime and g an integer of large multiplicative order modulo p. Parties A and B establish a shared session key $K := g^{xy} \bmod p$ as follows:

1. $A \to B$: A, B, g^x
2. $B \to A$: $B, A, g^y, eK(sS_B(g^y, g^x))$
3. $A \to B$: $A, B, eK(sS_A(g^x, g^y))$

The attack in [19] interleaves two protocol runs, marked by α and β, where the messages from A to B are intercepted by the intruder E:

$\alpha.1.$ $A \to E(B)$: A, B, g^x
$\beta.1.$ $E \to B$: E, B, g^x
$\beta.2.$ $B \to E$: $B, E, g^y, eK(sS_B(g^y, g^x))$
$\alpha.2.$ $E(B) \to A$: $B, A, g^y, eK(sS_B(g^y, g^x))$
$\alpha.3.$ $A \to E(B)$: $A, B, eK(sS_A(g^x, g^y))$

The records of the protocol runs at A and B do not match. In this sense, there is an attack. However, A believes to execute the α-run with B. After message $\alpha.2$, A associates the key K with B, and B and A are indeed the only parties in possession of this key. Party B is left with the incomplete β-run, so B cannot yet associate the key K with any party. Compared to an incomplete run of the protocol where the last message does not arrive, the only difference lies in the yet unconfirmed value of the challenger's identity in the responder's local state.

An attack against the KSL protocol, where the intruder makes A and B establish a shared key (not known to the intruder) while both act as responders in their protocol run with the intruder, is given in [19]. At the end of the 'attack', both parties share a secret key with the party they believe to share a key with, but both believe that the other initiated the protocol run. It remains doubtful whether these two attacks violate any realistic extensional authentication goals.

---

[2]This particular piece of requirement creep was probably caused when the authors of the BAN logic chose 'believes' when asked for a word to pronounce the symbol $|\equiv$, which encouraged others to reason about agents forming beliefs about the beliefs of other agents.

3. Formalizing Entity Authentication

The previous section gave a brief survey of attempts to give a precise meaning to the term authentication, highlighting some major shortcomings.

- Communications can fail. The sender of the last message commits without knowing whether it arrived.
- The attacks described are meaningful in circuit-switched networks but not in packet-switched networks.
- Specifications of protocol goals that keep more state than the protocol description itself can hardly claim to match the designers' intentions.
- Explanations of attacks or detailed protocol goals sometimes deviate from the goals originally given, but these discrepancies are not accounted for.

Too often, the view of authentication is dominated by authentication between people, for example, 'Alice wants to talk to Bob', as seen by human observers. Protocol steps $A \rightarrow B : m$ are subconsciously interpreted within the context of connection-oriented networks. Authors perceive problems in mutual authentication without being able to clearly articulate their discomfort. Can an even higher degree of precision and formality overcome these hurdles? To find out, we return to the history of authentication.

3.1. Correspondence

Woo and Lam match runs of the local programs executed by protocol participants [30]. The high level definition of authentication is familiar and uncontroversial.

> The primary goal of an authentication protocol is for an authenticating principal to be ascertained of the identity of the authenticated principal.

The added explanation

> upon successful termination of protocol execution, an authenticating principal should be assured that it is "talking" to the principal it has in mind

is dangerously anthropomorphic (or connection-oriented). In a network, messages pass through several intermediaries, which may legitimately modify parts of the message. When a party receives a message, is it then "talking" to the originator of the message or to the last intermediary that delivered the message?

> Protocol runs match if *correspondence* between certain steps in the local executions can be established.

> Informally, correspondence means that the execution of the different principals in an authentication protocol proceeds in a lock-stepped fashion. In particular, when an authenticating principal finishes its part of the protocol, the authenticated principal must have been present and participated in its part of the protocol.

This statement is ambiguous. The first sentence suggests very stringent requirements on the synchronization of executions performed by the principals, resembling canonical intensional specifications. The second sentence relaxes the conditions and restates the original informal goal of authentication.

Local programs are sequences of *transitions*, for generating nonces and keys, and sending and receiving messages. The transitions indicating start and end of a local protocol run are:

BeginInit(r)	begin initiator program with responder r,
EndInit(r)	end initiator program with responder r,
BeginRespond(i)	begin responder program with initiator i,
EndRespond(i)	end responder program with initiator i.

Correspondence properties are the formally expressed as assertions of the form

$$(i, \textbf{EndInit}(r)) \quad \hookrightarrow \quad (r, \textbf{BeginRespond}(i)) \qquad (1)$$

$$(r, \textbf{EndRespond}(i)) \quad \hookrightarrow \quad (i, \textbf{BeginInit}(r)) \qquad (2)$$

where i and r denote initiator and responder, and $\hookrightarrow$ is read as 'is preceded by'. The first property fits naturally with challenge response protocols. The initiator issuing the challenge may only commit if the intended responder had started to reply. The second property may just seem to be the dual of the first in mutual authentication (this appears to be the view taken in [30] where no additional explanation is given), but is actually different. As in Section 2.3, the responder is asked to verify the origin of the challenge.

3.2. Agreement

In [20], Lowe gives a hierarchy of specifications of authentication that express different flavours of Def. 2.6. It is again suggested that to verify an agent's identity one has to verify that agent's beliefs:

> An authentication protocol is designed to assure an agent A as to the identity of the other agent B with whom A is running the protocol; therefore, in most cases A should be at least assured that B thought he was running the protocol with A.

There is no definition of the term 'identity', and the focus is clearly on situations where there is just one protocol, and agents engage in several runs of this protocol.

> For example, a typical authentication specification is that when A completes a run of the protocol, apparently with B, then B has been recently running the same protocol.

As an example, we present *agreement*, the strongest form of authentication in [20].

Definition 3.1. *We say that a protocol guarantees to an initiator A agreement with a responder B on a set of data items ds if, whenever A (acting as initiator) completes a run of the protocol, apparently with responder B, then B has previously been running the protocol, apparently with A, and B was acting as a responder in his run, and the two agents agreed on the data values of all the variables in ds, and each run of A corresponds to a unique run of B.*

To capture this definition formally in CSP [10] a channel *signal* is introduced to allow parties to express their beliefs. Two types of events are used,

- *signal.Running.role.A.B.ds* represents that agent A believes to be engaged in a protocol run with B, and A acts in role *role*; ds is the list of data values A and B should agree on. This event is performed immediately before the last message A sends.
- *signal.Commit.role.A.B.ds* represents that agent A, acting in role *role*, commits to a protocol run with B; ds is the list of data values A and B should agree on. This event is performed after A's last event (or after the last message A receives.)

Positioning *Commit* events in a protocol run is uncontroversial. An agent decides whether to commit after receiving all relevant data and evaluating these data locally. The position of the *Running* event is less clear cut. The last message sent by a honest agent A confirms that A has faithfully executed all previous protocol steps. Thus, even when the last message does not provide direct evidence about A's identity, it does so indirectly. However, if A does not follow the protocol rules what can A's last message signal? We return to this question in Section 3.4.

We could require that initiator A (role *INIT*) commits to a protocol run with B, where A and B employ nonces n_A and n_B respectively, only after B has signalled that it is acting as a responder (role *RESP*) in a run with A using the same nonces n_A and n_B. Writing $x \rightarrow y$ to denote that event x happens before event y, this condition becomes

$$signal.Running.RESP.B.A.n_A.n_B \rightarrow signal.Commit.INIT.A.B.n_A.n_B$$

The specification for the responder is accordingly:

$$signal.Running.INIT.A.B.n_A.n_B \rightarrow signal.Commit.RESP.B.A.n_A.n_B$$

These specifications match the correspondence assertions

$$(B, \textbf{EndRespond}(A)) \quad \hookrightarrow \quad (A, \textbf{BeginInit}(B))$$
$$(A, \textbf{EndInit}(B)) \quad \hookrightarrow \quad (B, \textbf{BeginRespond}(A))$$

but allow for more granularity in the description of authentication properties. A hierarchy of authentication specifications is obtained by relaxing the number of parameters the initiator and responder have to agree on, including the identity of the correspondent. If A merely wanted to check that B replied to its challenge, the following specification could be used:

$$signal.Running.RESP.B.C.n_A \rightarrow signal.Commit.INIT.A.B.n_A$$

Section 3.4 sketches an alternative way of formalizing authentication in CSP.

3.3. Analysing the Needham-Schroeder Public-key Protocol

The Needham-Schroeder public key protocol [24] is a more 'secure' version of the mutual authentication protocol from Section 2.1. Essentially, the same values are exchanged but the responses are also encrypted. As challenges are encrypted with the receiver's public key, the intended receiver must process a challenge before it can become available to anyone else. In the full version of the protocol, principals obtain certified public keys from a server S. The simplified version (steps in bold below) omits the interactions with the server.

$$
\begin{aligned}
&1. \quad A \rightarrow S\colon B \\
&2. \quad S \rightarrow A\colon P_B, B, sS_S(P_B, B) \\
&\mathbf{3.} \quad \mathbf{A \rightarrow B\colon eP_B(n_A, A)} \\
&4. \quad B \rightarrow S\colon A \\
&5. \quad S \rightarrow B\colon P_A, A, sS_S(P_A, A) \\
&\mathbf{6.} \quad \mathbf{B \rightarrow A\colon eP_A(n_A, n_B)} \\
&\mathbf{7.} \quad \mathbf{A \rightarrow B\colon eP_B(n_B)}
\end{aligned}
$$

Lowe postulated the following protocol goals [18]:

> The initiator commits to a session only if the responder is indeed present and taking part in the protocol. The responder B commits to a session with the initiator A only if A is indeed trying to connect with B.

These goals explicitly state that the responder verifies the source of the challenge it replied to, rather than the source of the reply to its own challenge. The analysis in [17, 18] concludes that the protocol meets the first of the goals while an attacker can launch an attack that violates the second goal (Figure 4).

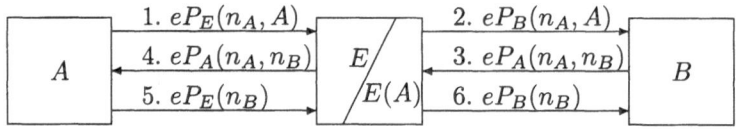

FIGURE 4. Lowe's attack on the Needham-Schroeder public key protocol

After the last step of this attack, the responder B commits to a protocol run with A although A was running the protocol with E, in violation of the responder's authentication specification

$$signal.Running.INIT.A.B.n_A.n_B \rightarrow signal.Commit.RESP.B.A.n_A.n_B$$

The initiator A signals $signal.Running.INIT.A.E.n_A.n_B$ at step 5 of this execution. It is interesting to check why the specification for the initiator,

$$signal.Running.RESP.E.A.n_A.n_B \rightarrow signal.Commit.INIT.A.E.n_A.n_B$$

is satisfied in this attack. The event *signal.Running.RESP.E.A.n_A.n_B* is signaled at step 4, when E intercepts a message from B that is intended for A, and forwards this very message unchanged to A. In a circuit-switched paradigm, the placement of the event is justified because message 4 arrives on A's connection with E.

In a packet-switched network, A cannot detect whether $eP_A(n_A, n_B)$ arrived directly from B or from E. A 'person-in-the-middle-attack' becomes a 'triangle' attack without a place for inserting an event *signal.Running.RESP.E.A.* Now, the interception of this message is not necessary for the attack to succeed, but for verification to succeed. The enemy E fakes the interception, not the message. Even more peculiarly, the event *signal.Running.RESP.E.A.n_A.n_B* includes the value n_B, which is not known to the intruder E at that time. Thus, a proof based on this event would argue that the party E is authenticated because it intercepted a message whose content remains partially unknown to it, rather than because it had used its private key to decrypt A's initial challenge.[3]

These remarks expose a fundamental problem. To capture authentication, an artificial channel and artificial events were introduced. Rules for inserting these events have to be chosen with utmost caution and are an easy source for the kind of abnormalities just noted. In many ways, the introduction of these artificial events resembles idealization in the BAN logic. Both translate informal concepts into a formal framework, and in both cases errors can occur during translation.

Lowe's attack does not violate Def. 2.2. Principal A starts a protocol run with E, and E is indeed active during the attack as it has to decrypt A's message to obtain the challenge n_A. Hence, A's commit is valid as it is preceded by an action by E. Principal B executes a protocol run with A. Principal A is indeed active during the attack as it has to decrypt B's message to obtain the challenge R_B. Hence, B's commit is valid as it is preceded by an action by A.

Meadows' attack on the full Needham-Schroeder public key protocol shows that the initiator may be present at the start of the protocol run, but not when replying to the responder's challenge [21]. The attack assumes that principals do not distinguish between names and nonces. A starts a protocol run with B, which is taken over by the enemy before A replies to B's challenge. A still has to decrypt B's challenge, but does so in another protocol run. The attack works as follows:

$$\alpha.1. \quad A \rightarrow S \ : B$$
$$\alpha.2. \quad S \rightarrow A \ : P_B, B, sS_S(P_B, B)$$
$$\alpha.3. \quad A \rightarrow B \ : eP_B(n_A, A)$$
$$\alpha.4. \quad B \rightarrow S \ : A$$
$$\alpha.5. \quad S \rightarrow B \ : P_A, A, sS_S(P_A, A)$$
$$\alpha.6. \quad B \rightarrow E(A): eP_A(n_A, n_B)$$
$$\beta.3. \quad E \rightarrow A \ : eP_A(n_A, n_B)$$
$$\beta.4. \quad A \rightarrow E(S): n_B$$
$$\alpha.7. \quad E(A) \rightarrow B: eP_B(n_B)$$

---

[3]We are encountering a familiar problem in the analysis of cryptographic protocols. How are messages modelled that contain encrypted fields whose content is not known to the sender?

3.4. Back to Basics

Schneider formalizes entity authentication in CSP avoiding the creation of artificial auxiliary elements [26, 27]. We give the briefest of introductions to his framework. There is a set *USER* of principals. The network is controlled by the enemy, so principals communicate by passing messages to the enemy for delivery and receiving messages from the enemy. User i sends messages to the enemy on a channel *trans.i* and receives messages from the enemy on a channel *rec.i*. Thus, *trans.i.j.m* describes message m being sent from principal i to principal j via the enemy. When a principal receives a message that should contain certain fields, pattern matching is performed and the principal stops if the pattern matching fails. Authentication is expressed as a relation between sets of messages.

Definition 3.2. *A set T of messages authenticates another set of messages R if occurrence of some element in T must have been accompanied by the occurrence of some element of R.*

In this framework, the public-key Needham-Schroeder protocol is described by a process $USER_A$ representing the initiator:

$$USER_A = \quad \square_{i \in USER} \; trans.A!i!eP_i(n_A.A) \rightarrow$$
$$rec.A.i?eP_A(n_A.x) \rightarrow$$
$$trans.A!i!eP_i(x) \rightarrow Stop$$

and process $USER_B$ representing the responder:

$$USER_B = rec.B?j?eP_B(y.j) \rightarrow trans.B!j!eP_j(y.n_B) \rightarrow rec.B.j.eP_B(n_B) \rightarrow Stop$$

We have to pick appropriate authenticating and authenticated messages from the messages in these processes. As the authenticating message, we choose the messages where the respective responses are verified, i.e. messages on *rec* channels. Authenticated messages can be taken either from the *trans* or the *rec* channels. Protocols using public key encryption (rather than digital signatures) illustrate two fundamental alternatives for expressing authentication.

Messages on *rec* channels capture the exercise of the private key during decryption but do not reliably identify the sender. Messages on *trans* channels include the names of the intended participants in a protocol run, giving us the means to verify that 'A wants to talk to B', but do not necessarily demonstrate knowledge of the private key. In many examples, for example, see [18, 27], this will still be the case, but follow only implicitly from assumptions about protocol executions. Meadows' attack sketched above shows that we cannot always rely upon such implicit arguments. This is the major argument against using messages on *trans* channels in protocols using public key encryption. Implicit assumptions are dangerous. Much of the criticism directed at the BAN logic noted that idealization tends to hide implicit assumptions, and that unwarranted hidden assumptions could lead to wrong conclusions.

To demonstrate the alternatives, we discuss three of the nine options listed in [26] for messages that could be authenticated in the Needham-Schroeder public key protocol by the responder's last message, $rec.B.j.eP_B(n_B)$:

- $trans.A.B.eP_B(n_B)$: confirms (implicitly) that A had used its private key to decrypt B's challenge while executing a protocol run with B.
- $trans.A.j.eP_j(n_B)$: confirms (implicitly) that A had used its private key to decrypt B's challenge while executing a protocol run with party j.
- $trans.A.B.eP_B(n_A.A)$: confirms that A initiated a protocol run with B using nonce n_A.

The first property is meaningful if the protocol aims to establish a connection between A and B. However, with interpretations of entity authentication like Def. 2.1 that explicitly exclude session establishment as their goal, the second property is perfectly adequate. The third option captures Lowe's interpretation of entity authentication. Note that neither of the authenticated messages directly captures the stage of the protocol run where A uses its private key. No authentication conditions for the receiver are given in [26]. We pick $rec.A.B.eP_A(n_A.x)$ as the authenticating message and consider the following authenticated messages:

- $trans.B.A.eP_A(n_A.x)$: confirms (implicitly) that B had used its private key to decrypt challenge n_A while executing a protocol run with A.
- $trans.j.A.eP_A(n_A.x)$: confirms (implicitly) that B had used its private key to decrypt A's challenge n_A and somehow had passed n_A to party j.

The second option may appear particularly unsatisfactory because B does not occur at all in the authenticated message. Hence, it may be preferable to use $rec.B.j.P_B(n_A.A)$ as the authenticated message.

3.5. Verifying Identities in Cryptographic Protocols

Specifying authentication by correspondence, we can deal with incomplete protocol runs and be precise about the variables the participants should agree on in a valid protocol run. However, given the conundrum of Figure 5, we can hardly claim to have clarified the meaning of authentication. Depending on the choice of authenticated message, Lowe's attack is a protocol run where none of the parties, only the responder, only the claimant, or both parties are properly authenticated.

initiator A by responder B		responder E by initiator A	
no:	$trans.A.B.eP_B(n_A.A)$		
no:	$trans.A.B.eP_B(n_B)$	yes/no[4]:	$trans.E.A.eP_A(n_A, x)$
yes:	$trans.A.j.eP_j(n_B)$	yes/no[4]:	$trans.E.A.eP_A(n_A, n_B)$
yes:	$rec.A.j.eP_A(n_A.n_B)$	yes:	$rec.E.A.eP_E(n_A.A)$

FIGURE 5. Authentication in the Needham-Schroeder public key protocol

---

[4]The answer is no if the sender is required to have constructed the message.

All choices are plausible in their own way, but lead to very different conclusions about the security of an authentication protocol. This confusion is a reflection of the differences between Def. 2.1 and Def. 2.2. Correspondence can capture that the claimant is alive, but correspondence can also model a shared connection. Hence, correspondence by itself does not clarify the meaning of authentication.

There is yet another issue that has to be resolved in our attempt to explain the phrase 'verifying claimed identities'. We finally have to declare what we mean by 'identity'. In cryptographic protocols, the translation of 'identity' into 'cryptographic key' is quite natural. We follow ISO/IEC 9798-3 [15], the International Standard on entity authentication using a public key algorithm, which states:

Definition 3.3. *In the authentication mechanisms specified in this part of ISO/IEC 9798 an entity to be authenticated corroborates its identity by demonstrating its knowledge of its secret signature key. This is achieved by the entity using its signature to sign specific data. The signature can be verified by anyone using the entity's public verification key.*

The authenticated message should thus be linked to the operation *demonstrating knowledge of the claimant's private key*. Protocol descriptions like to focus on the message exchange, while cryptographic keys are used in computations by the local programs. It is important that our abstract model considers all aspects of a protocol execution. We could thus explicitly add *ApplyKey* events to our model, or identify those events in a given protocol specification that capture exercise of the key implicitly. Taking as an example the CSP model discussed in Section 3.4, it will depend on the cryptographic primitives used in an authentication protocol whether events on *trans* channels or events on *rec* channels are better suited to capturing exercise of the private key.

4. Breaking Assumptions

We now turn to authentication protocols that establish shared session keys. The two protocols examined are due to Needham and Schroeder, who made two fundamental assumptions in the design of their protocols [24].

1. Secrets are never disclosed.
2. The parties involved in the protocol, i.e. the insiders, follow the rules. Danger comes from outside intruders.

These assumptions also underpin the BAN logic of authentication [5], where they are stated quite explicitly.

4.1. Compromised Session Keys

In the Needham-Schroeder protocol, two parties A and B establish a shared session key K_{AB} with the help of a trusted server S. In the symmetric key variant of the protocol, each of the parties shares a key with the server, denoted by K_{AS} and K_{BS} respectively. The following messages are sent in the protocol.

1. $A \rightarrow S: A, B, n_A$
2. $S \rightarrow A: eK_{AS}(n_A, B, K_{AB}, eK_{BS}(K_{AB}, A))$
3. $A \rightarrow B: eK_{BS}(K_{AB}, A)$
4. $B \rightarrow A: eK_{AB}(n_B)$
5. $A \rightarrow B: eK_{AB}(n_B - 1)$

In the last two steps B verifies that A is in possession of the session key K_{AB}. Denning and Sacco observed that an attacker who had broken an old session key, say K'_{AB}, and recorded the corresponding message $eK_{BS}(K'_{AB}, A)$ from the authentication server could successfully impersonate A starting the protocol at step 3 and using the broken session key [6]. The literature contains numerous remarks to the effect that Denning and Sacco had found a flaw in the Needham-Schroeder protocol. Here is their own explanation.

> If communication keys and private keys are never compromised (as Needham and Schroeder assume), the protocol is secure (i.e., can be used to establish a secure channel). We will show that the protocol is not secure when communication keys are compromised, and propose a solution using timestamps.

The flaw in the protocol follows from a change in assumptions. Here, this change is acknowledged explicitly. In other papers, for example, see [18], changes in assumptions are hidden in the formalization of the attack model.

4.2. Misbehaving Insiders

In the original publication of the Needham-Schroeder public-key protocol [24] and elsewhere [29], the nonces n_A and n_B are used as secret initial values or as components of a session key to establish a secure channel between A and B. In the BAN logic, assertions like 'A and B believe n_A and n_B to be shared secrets' have been proven. However, in Lowe's attack the nonces n_A and n_B become known to three parties. Without doubt the protocol has failed to establish shared secrets. How can we explain the co-existence of the BAN proof and such a blatant counterexample? Of course, the answer lies again in the assumptions. In Lowe's attack, principal A starts a protocol run with E, but E departs from the rule book and uses the 'secret' n_A in a run with B.

A similar attack against an early version of the SSH protocol is given in [1]. In the following, h is a hash function, K_{Bl} and K_{Bs} are the long term and short term public encryption keys of principal B, K_A is the public verification key of principal A, K is the session key, and K' is a key derived from K, n_A, and n_B. The protocol has the following steps.

1. $A \rightarrow B: n_A$
2. $B \rightarrow A: n_B$
3. $B \rightarrow A: K_{Bl}, K_{Bs}$
4. $A \rightarrow B: eK_{Bl}(eK_{Bs}(h(\text{previous messages}), K))$
5. $A \rightarrow B: eK'(A, K_A, sK_A^{-1}(A, n_A, n_B))$

In the attack, principal C (an insider) uses a protocol run initiated by A to impersonate A vis-a-vis a third party B.

$\alpha.1.\quad A \rightarrow C:\ n_A$

$\beta.1.\quad C \rightarrow B:\ n_A$

$\beta.2.\quad B \rightarrow C:\ n_B$

$\alpha.2.\quad C \rightarrow A:\ n_B$

$\beta.3.\quad B \rightarrow C:\ K_{Bl}, K_{Bs}$

$\alpha.3.\quad C \rightarrow A:\ K_{Cl}, K_{Cs}$

$\alpha.4.\quad A \rightarrow C:\ eK_{Cl}(eK_{Cs}(h(previous\ messages), K))$

$\beta.4.\quad C \rightarrow B:\ eK_{Bl}(eK_{Bs}(h(previous\ messages'), K))$

$\alpha.5.\quad A \rightarrow C:\ eK'(A, K_A, sK_A^{-1}(A, n_A, n_B))$

$\beta.5.\quad C \rightarrow B:\ eK'(A, K_A, sK_A^{-1}(A, n_A, n_B))$

The traditional view of cryptographic protocols sees A and B as well behaving principals communicating over an insecure channel. The enemy is in full control of this channel and can delete, insert, replay, and modify messages. Cryptography provides A and B with the means to establish a secure logical channel over this insecure communications channel. This paradigm still strongly influences the general perception of security. Within this paradigm, protocols are analysed assuming that they are always executed between two parties following the rules. The enemy may of course engage in protocol runs too to obtain information that can help to mount an attack.

Today, there is a new paradigm. Parties engaging in electronic commerce have to provide for the eventuality that their business partners misbehave. In this scenario, protocols should be analysed under the assumption that parties do not necessarily follow the rules. In our last two examples, the attackers were insiders re-using a session key with some other party. On occasion, this paradigm shift occurs within a single document. The attacker in [18] starts off as an intruder:

> the intruder can observe and intercept messages, and so learn information – such as the values of nonces – and then use this information to introduce fake messages.

Later, the attacker changes colours and becomes an insider:

> We assume that the intruder is a user of the computer network, and so can take part in normal runs of the protocol, and other agents may initiate runs of the protocol with him.

There is another common theme to these examples. Needham and Schroeder treat session keys exactly in the same manner as long term secret keys. In contrast, Denning and Sacco, and other authors, analyse protocols in situations where short term keys may become disclosed. If a long term secret gets compromised security collapses. This ought not to be the case with session keys so it is a worthwhile exercise to re-examine protocols in this new setting. It would, however, be imprecise to say that a protocol was found to be flawed if the analysis shows that it is not applicable in the new environment but was satisfactory in its original setting.

5. Conclusion

From the early 1990's, claims were made to the effect that mutual authentication is more than unilateral authentication in both directions. We have described some of the attacks that were presented in support of these claims, noting that the attacks, and their explanations, are quite convincing in the context of circuit-switched networks but much less so in packet-switched networks. These claims led to formal definitions of authentication that require protocol participants to have matching states during protocol execution [3, 4, 7, 30, 2]. Some of these definitions are quite close to the canonical intensional specifications proposed in [25].

It is not too far fetched to conclude that the originators of the chain of investigations we have pursued in this paper thought in terms of circuit-switched networks. Mutual authentication of a single connection is materially different from unilateral authentication of two separate connections. In the attempt to capture this fact, definitions were coined that capture the essence of a connection much better than the verification of a claimed identity. Authentication protocols may well be used to *establish* a connection. Still, this does not imply that they should *simulate* a connection.

Formalization led to intensional specifications of authentication, but few attempts were made to relate these specifications to the two distinct extensional specifications for entity authentication. On one hand, we can verify whether an entity is alive (Def. 2.1). In circuit-switched networks, authenticating connections is a second reasonable endeavour and answers the question 'Who speaks on this connection?' (Def. 2.2). If this is the goal, we should explicitly identify the connection that is being authenticated, because *for A to assert [channel] C speaks for A it must be able to name C* [16]. Presently, connections lurk beneath the surface but terms like 'impersonate' and 'masquerade' really need their presence. Two further assumptions are necessary when authenticating connections:

1. We have to know *where* the connection ends. Entity authentication checks *who* is at that end.
2. The entity to be authenticated is honest. Otherwise, we face the paradox of Lowe's analysis and authenticate a party based on data it has not seen.

Of course, there may be a case for further properties like verifying the origin of challenges or other asymmetric definitions of mutual authentication, but the case for such goals has to be made. It is rarely good security practice to ask for as much as possible. Overprotection can reduce actual security.

We are asking for examples where it matters who starts the run of an authentication protocol. Are there delicate diplomatic situations where parties refuse to make the first move but are happy to respond? Is there pride in cyberspace? Maybe, conciliator is a more apt description for an 'attacker' who manages to make two parties talk with each other that otherwise would not.

In general, it is advisable to model the system before modelling the protocol. Cryptographic operations and checks on incoming messages are important steps in a protocol execution, both for describing protocol goals and for defending against attacks. The discussion of unique session identifiers in the analysis of the IKE protocol makes interesting reading in this regard [22]. The model also has to describe the behaviour of principals. Protocol goals are often formalized as if agents could engage in a protocol run only by following the rules of the protocol. The triangle attack in Section 3.3 shows that an agent can deviate from a protocol run whilst the corresponding agent still validly completes its run of the protocol. Insider attacks are a realistic threat scenario in electronic commerce [9].

References

[1] M. Abadi, *Explicit communication revisited: Two new attacks on authentication protocols*, IEEE Transactions on Software Engineering **23** (1997), no. 3, 185–186.

[2] M. Bellare and P. Rogaway, *Entity authentication and key distribution*, Advances in Cryptology – CRYPTO'93, LNCS 773 (D. R. Stinson, ed.), Springer Verlag, 1994, pp. 232–249.

[3] R. Bird, I. Gopal, A. Herzberg, P. Janson, S. Kutten, R. Molva, and M. Yung, *Systematic design of two-party authentication protocols*, Advances in Cryptology – CRYPTO'91, LNCS 576 (J. Feigenbaum, ed.), Springer Verlag, 1992, pp. 44–61.

[4] R. Bird, I. Gopal, A. Herzberg, P. A. Janson, S. Kutten, R. Molva, and M. Yung, *Systematic design of a family of attack-resistant authentication protocols*, IEEE Journal on Selected Areas in Communications **11** (1993), no. 5, 679–693.

[5] M. Burrows, M. Abadi, and R. Needham, *A logic of authentication*, DEC Systems Research Center Report **39** (1990).

[6] D. E. Denning and G. M. Sacco, *Timestamps in key distribution protocols*, Communications of the ACM **24** (1981), no. 8, 533–536.

[7] W. Diffie, P. C. van Oorschot, and M. J. Wiener, *Authentication and authenticated key exchanges*, Designs, Codes and Cryptography **2** (1992), 107–125.

[8] D. Gollmann, *Proving authentication protocols – what do authentication protocols prove?*, Mathematics of Dependable Systems (V. Stavridou C. J. Mitchell, ed.), Clarendon Press, 1995, pp. 95–102.

[9] D. Gollmann, *Insider fraud*, Proceedings of the Cambridge Security Protocols Workshop, LNCS 1550 (B. Christiansen et al., ed.), Springer Verlag, 1999, pp. 213–219.

[10] C. A. R. Hoare, *Communicating sequential processes*, Prentice-Hall International, Englewood Cliffs, NJ, 1985.

[11] International Organization for Standardization, *Basic Reference Model for Open Systems Interconnection (OSI) Part 2: Security Architecture*, Genève, Switzerland, 1988.

[12] International Organization for Standardization, *Information technology – Security techniques – Entity authentication mechanisms; Part 1: General model*, Genève, Switzerland, September 1991, ISO/IEC 9798-1, Second Edition.

[13] International Organization for Standardization, *Information technology – Security techniques – Entity authentication mechanisms; Part 3: Entity authentication mechanisms using a public key algorithm*, Genève, Switzerland, March 1991, ISO/IEC JTC1/SC27/WG2 N51.

[14] International Organization for Standardization, *Information technology – Security techniques – summary of voting on letter ballot No.6, document SC27 N277, CD*

9798-3.3 "Entity authentication mechanisms; Part 3: Entity authentication mechanisms using a public key algorithm", Genève, Switzerland, October 1991, ISO/IEC JTC1/SC27 N313.

[15] International Organization for Standardization, *Information technology – Security techniques – Entity authentication mechanisms; Part 3: Entity authentication mechanisms using a public key algorithm*, Genève, Switzerland, August 1993, ISO/IEC 9798-3.

[16] B. Lampson, M. Abadi, M. Burrows, and E. Wobber, *Authentication in distributed systems: Theory and practice*, ACM Transactions on Computer Systems **10** (1992), no. 4, 265–310.

[17] G. Lowe, *An attack on the Needham-Schroeder public-key authentication protocol*, Information Processing Letters **56** (1995), no. 3, 131–133.

[18] G. Lowe, *Breaking and fixing the Needham-Schroeder public-key protocol using FDR*, Proceedings of TACAS, LNCS 1055, Springer Verlag, 1996, pp. 147–166.

[19] G. Lowe, *Some new attacks upon security protocols*, Proceedings of the 9th IEEE Computer Security Foundations Workshop, 1996, pp. 162–169.

[20] G. Lowe, *A hierarchy of authentication specifications*, Proceedings of the 10th IEEE Computer Security Foundations Workshop, 1997, pp. 31–43.

[21] C. A. Meadows, *Analyzing the Needham-Schroeder public key protocol: A comparison of two approaches*, Proceedings of ESORICS'96, LNCS 1146 (E. Bertino et al., ed.), Springer Verlag 1996, pp. 351–364.

[22] C. A. Meadows, *Analysis of the internet key exchange protocol using the NRL protocol analyzer*, Proceedings of the 1999 IEEE Symposium on Security and Privacy, 1999, pp. 216–231.

[23] C. J. Mitchell and A. Thomas, *Standardising authentication protocols based on public key techniques*, Journal of Computer Security **2** (1993), 23–36.

[24] R. M. Needham and M. D. Schroeder, *Using encryption for authentication in large networks of computers*, Communications of the ACM **21** (1978), 993–999.

[25] A. W. Roscoe, *Intensional specifications of security protocols*, Proceedings of the 9th IEEE Computer Security Foundations Workshop, 1996, pp. 28–38.

[26] S. Schneider, *Verifying authentication protocols with CSP*, Proceedings of the 10th IEEE Computer Security Foundations Workshop, 1997, pp. 3–17.

[27] S. Schneider, *Verifying authentication protocols in CSP*, IEEE Transactions on Software Engineering **24** (1998), no. 9, 741–758.

[28] F. J. Thayer Fábrega, J. C. Herzog, and J. D. Guttman, *Strand spaces: Why is a security protocol correct?*, Proceedings of the 1998 IEEE Symposium on Security and Privacy, 1998, pp. 160–171.

[29] E. Wobber, M. Abadi, M. Burrows, and B. Lampson, *Authentication in the TAOS operating systems*, ACM Transactions on Computer Systems **12** (1994), no. 1, 3–32.

[30] T. Y. C. Woo and S. S. Lam, *A semantic model for authentication protocols*, Proceedings of the 1993 IEEE Symposium on Research in Security and Privacy, 1993, pp. 178–194.

Microsoft Research
Cambridge, United Kingdom
E-mail address: `diego@microsoft.com`

Progress in Computer Science and Applied Logic, Vol. 20
© 2001 Birkhäuser Verlag Basel/Switzerland

A Survey of Hard Core Functions

Maria Isabel González Vasco and Mats Näslund

Abstract. The security of public key protocols relies nowadays on the use of one-way functions. However, even assuming a certain function $f(x)$ is *hard enough* to invert, we should always keep in mind the fact that some information may leak through. A function $b(x)$ that does not leak in this way is said to be a hard core for f; given $f(x)$, $b(x)$ cannot even be computationally distinguished from a random string. In this survey, we review what is known in this area, both from a more theoretical point of view and also for 'practical' choices of f such as RSA.

Keywords: hard core function, bit-security, one-way function, pseudorandom generator

1. Introduction

Most people, whether cryptologists/mathematicians or not, have probably at some point contemplated over the issue of randomness. What is it? Why do some events seem random and unpredictable, whereas others come as no surprise? If one takes the problem of weather forecasting as an example, huge progress has been made thanks to advances in science and computing power. Perhaps there really is no true randomness – it is just that current state of the art in technology does not allow us to predict.

Although the above discussion leads into rather deep philosophical issues that we will not pursue, there is a point to be made – how one defines randomness should preferably not depend on current state of technology, at least not if security of cryptosystems is to rely on it. For instance, if a protocol is proved secure assuming the key is chosen at random (which is most of the times the case) if we use a too weak definition of randomness, we could be using a key generator which output sequences are easy to predict and therefore the protocol might in the end be far from secure.

From a mathematical point of view there are several definitions of randomness. For instance, according to the *Encyclopedic Dictionary of Mathematics*, a sequence of random numbers is

> *a sequence of numbers that can be considered as realizations of independently and identically distributed random variables.*

In the cryptographic context, where we assume randomness to imply unpredictability, we shall also require that the random variables in the former definition are uniformly distributed. Note that otherwise even constant sequences are considered to be random.

With the idea of an analogy between unpredictability and randomness, Shannon used the basic concepts of Information Theory to give a definition of random source that would be suitable for cryptography. In [52] he introduces a way to measure randomness in terms of entropy, that is, a certain source is considered random if its entropy is $\log k$, were there are only k possible outputs. In other words, the average 'complexity' of a string generated by the source is $\log k$ bits. Although this notion is too close to the statistical definition to be useful in practice, his ideas are of great importance, for he already pointed out the importance of measuring randomness in terms of computational complexity. Some other early approaches to the issue of randomness dealt with the problem of giving a definition of random sequence. A key paper on the study of properties of single random sequence is that of Kolmogorov [37]. He also used a complexity measure: a string is random if it cannot be *compressed* into a shorter description. Though both mentioned approaches achieved no results of practical interest, their underlying ideas are basic for understanding the theory of randomness we have nowadays at hand.

In cryptography we would ever so often like to mimic true random sources by deterministic algorithms, so called *pseudorandom generators* (PRG). Statistical and mathematical definitions are not set in a framework that is suitable for cryptographic purposes since they are almost always too strong for the computational model in which we do cryptography. Among the first works trying to give some sufficient conditions for randomness, we mention the work by Knuth, [36]. He tried to device a set of tests that would determine if a source was random according to definitions such as the above. His battery of tests included counting frequencies of specific bit-patterns, distribution of k-tuples in k-dimensional space, etc. But even if a source passes a fixed set of tests, there is no guarantee against someone tomorrow inventing a new test that reveals some non-random behavior of the very same source. For our purposes, it is better to introduce randomness in terms of computing power. Given a certain source, the aim is now to decide how *hard* it is to obtain information about the way it works from its output sequences. Therefore, in defining randomness, time and computational resources (primarily time and space) should always be taken into account.

On the other hand, one must also be careful not to use a too weak definition. Otherwise, performance improvements in computer hardware may disqualify the present pseudorandom generators a few years from now just as weather is more predictable today than it was 50–100 years ago. We know today that the approach taken by Knuth is not the correct one since, as discussed, it does not lead to a sound theory of cryptographic randomness. Such did not exist until the seminal works by Blum-Micali [5], and Yao [60], in the early 80's. Yao manages to unify the concepts of randomness and unpredictability; that predictability implies non-randomness is intuitively clear, but the converse is perhaps not that obvious, and we return to this later. A pseudorandom generator (according to Blum-Micali and Yao) is a device that produces long deterministic strings from a small truly random seed and passes *all* efficient tests (and not just a fixed set of tests). Still, one should also keep mind that these early works, while trying to be very general, did not have very strong practical implications, as their definitions only aim at giving 'asymptotic' results. However, it is possible also to derive quantitative statements for fixed values of the security parameters, for example, see [3].

Let us now give a brief idea of the important role pseudorandom generators play in modern cryptography. The most fundamental cryptographic primitive is the one-way function, a function, f, easy to compute, but hard to invert. The notion of one-way function was first introduced by Diffie-Hellman in [14]. One of the greatest achievement over the last two decades is to show that they are both necessary and sufficient for all secure cryptography. One might perhaps think that this is more or less obvious, but notice that a one-way function f can not be used directly for encryption: even if x (the entire message) is hard to find from $f(x)$, large bits and pieces of x could be very simple to find. A function $b(x)$, which does not leak through $f(x)$ is called a *hard core function* for f. The optimum would of course be to have a cryptosystem f where *any* function $b(x)$ is a hard core for f. If this is the case, $f(x)$ is called *semantically secure*, [26]. As it can be easily seen, no deterministic, public key encryption function, f, can be semantically secure, for we can always build a function b which is not hard core for f; namely $b(x) = f(x)$ itself! However, semantic security can be achieved using a pseudorandom generator g. In order to send a message m we select a short random key, k, feed it to the generator g and encrypt m as $c = g(k) \oplus m$, bitwise XOR. The enemy, seeing c, cannot even tell if what he sees is the encryption, or, just random noise.

With this in mind, how do we create a pseudorandom generator exploiting the cryptographic tools at hand? Suppose that f is a one-way function, that in addition is a permutation, and that the boolean function b is a hard core function for f. Pick x_0 (the key, or seed) at random, let $x_{i+1} = f(x_i)$, and define the generator $g(x_0) = b(x_0), b(x_1), \ldots$. The construction is due to Blum and Micali, [5]. It is worth noticing that the condition that f be a permutation can be dropped, but then a considerably more complicated construction is needed, [31].

Thus, having reduced the existence of semantically secure encryption, via pseudorandom generators, to one-way permutations and hard cores, we are left with a central question: a polynomial time computable function that is not one-way clearly has no hard cores at all, but, is the converse true? Does every one-way function have at least one hard core? The purpose of this paper is to give a survey of what is known in this area. We will among other things see that indeed, *any* one-way function has (several) hard core functions.

Unfortunately, we do not know if there are any one-way functions at all and it is not likely that a formal proof of their existence will be easily found for then the statement $P \neq NP$ would be derived from that proof, and this is still the major open problem in computer science. Still, there are several promising candidates, such as RSA, [51] and discrete exponentiation modulo a prime number. Since

RSA is used in practice, it is of great importance to investigate how secure it is, and being secure involves much more than just being hard to invert. Imagine exchanging a yes/no-message, encoded as $m = 1$ or $m = 0$, using RSA. If the opponent knows this, since we are using a public key system, he can himself encrypt both a 1 and a 0 and compare to the intercepted cipher. To thwart this we may encrypt 0 as $RSA(x)$, where x is a random even number in the RSA domain, and 1 as $RSA(y)$, where y is a random odd number. The intended receiver can then decide if the message was yes or no, by checking if the message is odd or even. Is this secure? Even if RSA is hard to invert as a whole, it could be that a few bits, for example, the least significant bit, leak and therefore this scheme would be broken. In fact, for discrete exponentiation modulo a prime, this bit indeed leaks! So even assuming the function f we are dealing with is one-way, we should always take into account the amount of information (bits of x) that can be 'easily' obtained from $f(x)$.

In summary, the results surveyed in this paper fall into two categories: for general one-way functions, what are their hard cores, and, for specific functions, for example RSA, what can be said about security of individual 'bits'.

The rest of the paper is organized as follows: We give some preliminary notation in §2. In §3 we review general results, applying to any one-way function. Then, in §4 we treat the widely used cryptographic functions; RSA, §4.1, and discrete exponentiation, §4.3. We also look at some results based on combinatorial problems, see §4.7. Finally, we mention some intriguing open problems in §5.

2. Notation and Preliminaries

We let $\{0,1\}^n$ denote the set of binary strings of length n, and $\{0,1\}^*$ the set of all finite length binary strings. In general $|x|$ denotes the length of x. We use $\mathcal{U}_n$ for the uniform distribution on $\{0,1\}^n$.

We use the shortened form pptm for probabilistic polynomial time algorithm (Turing-machine). A function ν is *negligible* if for any constant $c > 0$, $\nu(n) = o(n^{-c})$. A polynomial time computable function, $f : \{0,1\}^* \to \{0,1\}^*$, is called a *one-way function*, if for every pptm A, $\nu(n) = \Pr[f(A(f(x))) = f(x)]$ is negligible, probability taken over $x \in \mathcal{U}_n$, and internal coin flips of A.

Two sequences of distributions $\mathcal{D}^1 = \{\mathcal{D}_n^1\}_{n \geq 1}$, $\mathcal{D}^2 = \{\mathcal{D}_n^2\}_{n \geq 1}$, $\mathcal{D}_n^i$ having support on $\{0,1\}^n$, are said to be (computationally) $\epsilon(n)$-*indistinguishable* if for any pptm A,

$$|\mathrm{E}_{z \in \mathcal{D}_n^1}[A(z)] - \mathrm{E}_{z \in \mathcal{D}_n^2}[A(z)]| \leq \epsilon(n).$$

If this holds for all non-negligible $\epsilon(n)$, we simply say that they are indistinguishable.

Let f be a one-way function and $B = \{B_n\}_{n \geq 1}$ be a family of functions, where $b \in B_n$ maps $\{0,1\}^n \to \{0,1\}^{m(n)}$. We say that B is $\epsilon(n)$-*secure for* f, if the distribution $f(x), b, b(x)$ induced by picking, uniformly, $x \in \mathcal{U}_n$ and $b \in B_n$ is $\epsilon(n)$-indistinguishable from the distribution $f(x), b, r$, induced by $x \in \mathcal{U}_n$, $b \in B_n$, and $r \in \mathcal{U}_{m(n)}$. If B is $\epsilon(n)$-secure for all non-negligible $\epsilon(n)$, we say that B is a family of *hard core functions for* f (or simply hard core for f).

For $z, m \in \mathbb{Z}$, $m \neq 0$, let $z \bmod m$ denote the least non-negative residue of z when divided by m. If $z \in \mathbb{Z}/m\mathbb{Z}$, let $\mathrm{bit}_i(z)$ be the ith bit in the binary representation of z. In particular, $\mathrm{lsb}(z) \triangleq \mathrm{bit}_0(z)$.

Throughout the paper, we ignore the issue of rounding $\log n$ and some other similar expressions to the nearest integer; thus sometimes we talk about $(\log n)^{1/2}$ bits.

3. General Hard Core Functions

By a set of *general hard core functions* we mean a set of functions, B, such that B is hard core for *any* one-way function f. Having such set is the most attractive case we can imagine, since no specific knowledge on f is needed. Even if we are able to show that the functions we are basing cryptography on today, for example RSA, have hard cores, advances in algorithms for integer factorization may make it necessary to resort to other candidates for one-way functions. These may not have the same hard cores as RSA.

3.1. General Proof Technique

We will for the moment only consider boolean hard core functions, that is, $m(n) = 1$. It is easy to see that in this case, B is $\epsilon(n)$-secure for some f, if and only if for any pptm $\mathfrak{O}$, random $b \in B$, $\Pr[\mathfrak{O}(f(x), b) = b(x)] \leq (1 + \epsilon(n))/2$. (To be more precise, a presumed distinguisher with $\epsilon(n)$ advantage can be turned into a predictor, $\mathfrak{O}$, as above and vice versa.) How does one show that some set of boolean functions, $B = \{b_i\}$ is hard core for some one-way function, f? The most obvious is of course a proof by contradiction – an algorithm, $\mathfrak{O}$, being able to non-trivially predict $b(x)$, for random b, could be used to retrieve x, that is, to invert f. Therefore, either b is hard core for f or f is not a one-way function. We call $\mathfrak{O}$ the *oracle* for B. Though this might not be the only way to prove that a set of function is hard core for a given f, as far as we know is the only one that has actually been used.

By picking random b_is, one hopes the oracle's predictions to the set $\{b_i(x)\}$ will uniquely determine x. Since $b_i : \{0,1\}^n \to \{0,1\}$, one $b_i(x)$ does not determine x, but one might hope that knowing $b_i(x)$ for several different b_i would determine x, at least if the b_is are chosen wisely. Note though that only slightly more than half of the predictions are correct, and we do not know which. Furthermore, if the b_i needs to be specifically chosen to enable reconstruction of x, then the oracle could perhaps have no advantage at all on this set of special b_is.

The normal way to get around this is to do random sampling followed by a majority decision. To decide some $b_i(x)$ we must then, somehow, ask $\mathfrak{O}$ about some random set of $\{b_j(x)\}$, that are related to the $b_i(x)$ we are after. That is, each $b_j(x)$ gives one vote for $b_i(x)$. It is clear that the existence of relations, relating $b_i(x)$ to $b_j(x)$ makes the set of possibilities for B rather limited. Indeed, as we

shall see in the following sections, all general hard cores that are known have lots
of similarities.

As will be apparent in the sequel, the results are asymptotic in nature, that is,
there is a 'for all sufficiently large n ... ' involved. This may seem a little trouble-
some, and one perhaps worries whether the results have any practical significance
at all. Furthermore, the proofs are *reductions*, meaning that if cryptographic primitive A is secure, then primitive B is almost as secure. Hence it is desirable to make
the proofs as simple as possible, thereby preserving the most of the security. For
instance, a PRG, whose security on seeds on length 10^6 is only that of RSA with
a 10-bit modulus would have no practical use. Fortunately, digging deeper into
the proofs than we do here, will reveal that *most* results are strong enough to
have at least *some* practical implications. As a specific example, it can be seen
that the Goldreich-Levin theorem, presented in the next section, together with the
aforementioned construction by Blum and Micali, gives a generator, whose security on seeds of length n, is essentially the same as that of the underlying one-way
permutation on 'keys' of size $n/2$.

On the other hand, we note that for many of the surveyed results, there is no
indication that the existing proofs are 'optimal' in any sense, and finding simpler
proofs is a very interesting and open-ended research topic.

3.2. Going Beyond One Bit

The Blum-Micali construction of a pseudorandom generator given in the introduction iterates a one-way permutation f, and outputs a single bit on each iteration.
For efficiency reasons, we would like to produce more output bits at each stage,
as the computation of f, though polynomial time, could be rather costly. Thus,
we would like a set of, say k bits, that are simultaneously hard in the sense that
given $f(x)$, they cannot be distinguished from a random string of length k.

If f is a one-way function and we have a set $\{b_i\}$ of hard core functions for f,
how many bits can each b_i output? Let f_2 be a one-way function, and define f by
$f(x) = f(x_1, \ldots, x_n) = f_2(x_1, \ldots, x_k), x_{k+1}, \ldots, x_n$, that is, f applies f_2 to k of
the bits, and outputs the rest unaffected. From a theoretical point of view, there is
nothing that prevents k from being as small as, say $\Theta(\log^{1+\delta} n)$, $\delta > 0$; f is still one-
way if f_2 is. Therefore, there always is a one-way function $f(x_1, \ldots, x_n)$ that can be
inverted in polynomial time if k bits of *extra* information is available. If b_i reveals
at least k bits, it is therefore conceivable that this enables inversion of f. Hence,
we can in general never hope to have more than $k = \mathcal{O}(\log n)$ (simultaneously)
hard core bits. That is, without any information on f, besides its one-wayness. As
we shortly will see, any one way function *has* these many hard core bits.

As discussed, in the one-bit case, being able to *distinguish* $b(x) \in \{0, 1\}$
from a random bit is precisely the same thing as being able to *predict* $b(x)$. If b
outputs more than 1 bit, this correspondence is no longer clear. For instance, let
h be a function that outputs $m(n)$ bits. Knowing that 'the 17th bit of $h(x)$ is
biased towards 0' is probably enough to distinguish $h(x)$ from random strings, but
may not imply ability to compute $h(x)$. As it turns out, there are basically two

techniques at our disposal to show that functions outputting more than 1 bit, are hard cores. Both in some sense reduce the pseudorandomness of many bits to that of a single bit.

First we have the following lemma by Vazirani and Vazirani, [58]. We recall that a function $f(x)$ is called *length regular* if $|x| = |x'|$ implies $|f(x)| = |f(x')|$ and $|f(x)|$ is non-decreasing with $|x|$.

Lemma 1 (The Computational XOR-Lemma). *Let f and h be arbitrary, polynomial time computable, and length-regular functions defined on $\{0,1\}^n$ with $h(x) \in \{0,1\}^{m(n)}$. Suppose that there is a pptm, D, for which*

$$|\Pr\left[D(f(x), h, h(x)) = 1\right] - \Pr\left[D(f(x), h, r) = 1\right]| \geq \epsilon(n),$$

the probability taken over $x \in \mathcal{U}_n$, $r \in \mathcal{U}_{m(n)}$, and D's random choices. Then there is a pptm T so that

$$\Pr\left[T(f(x), h, s) = \sum_{i=0}^{m(n)-1} s_i \operatorname{bit}_i(h(x)) \bmod 2\right] \geq \frac{1}{2} + \frac{\epsilon(n)}{2^{m(n)} - 1}$$

the probability taken over $x \in \mathcal{U}_n$, $s \in \mathcal{U}_{m(n)} \setminus \{\vec{0}\}$ and T's random choices.

Since a converse statement is obvious, the lemma states that if one is able to distinguish $h(x)$ from the uniform distribution, then it is only because one is able to compute a (non-trivial) XOR over the individual bits of $h(x)$. (Not to confuse the reader, we note that there are also other results known as 'the XOR-lemma', see [20].)

Proof sketch. For every $f(x), h$, a distinguisher D as above, computes a function $d : \{0,1\}^{m(n)} \to \{0,1\}$. As such, we can expand d using a Fourier series over the orthonormal basis consisting of all $\chi_s(y) = (-1)^{\langle s,y\rangle \bmod 2}$ where $s \in \{0,1\}^{m(n)}$. Averaging over s, D must thus have a correlation with $\{\chi_s(y)\}$, which by the isomorphism between the groups $(\{0,1\}, \oplus)$ and $(\{-1,1\}, \cdot)$, corresponds precisely to XORs. Refer to [21] for details.   □

The lemma has the drawback that it is only useful for $m(n) = \mathcal{O}(\log n)$ bits. Fortunately, we also have a result about the *universality of the next-bit-test*, due to Yao [60]. For $j \geq i$, let $\mathrm{B}_i^j(x) \triangleq \operatorname{bit}_i(x) \operatorname{bit}_{i+1}(x) \cdots \operatorname{bit}_j(x)$.

Theorem 2 (Universality of the next bit test). *Let f and h be arbitrary, polynomial time computable, and length-regular functions defined on $\{0,1\}^n$ with $h(x) \in \{0,1\}^{m(n)}$. Then, if there exits a pptm, D, for which*

$$|\Pr\left[D(f(x), h, h(x)) = 1\right] - \Pr\left[D(f(x), h, r) = 1\right]| \geq \epsilon(n),$$

with the probability taken over $x \in \mathcal{U}_n$, $r \in \mathcal{U}_{m(n)}$, and D's random choices, then there is $i \in \{0, \ldots, m(n) - 1\}$, and a pptm T_i such that

$$\Pr\left[T_i(f(x), h, \mathrm{B}_0^{i-1}(h(x))) = \operatorname{bit}_i(h(x))\right] \geq \frac{1}{2} + \frac{\epsilon(n)}{m(n)}$$

the probability taken over $x \in \mathcal{U}_n$ and T_i's random choices. A converse statement also holds.

Proof sketch. Suppose we have a pptm D satisfying the above condition. Assume without loss of generality that most of the times $D(f(x), h, r) = 1$ if r is a random value. Let us now define probability distributions q_i over $\{0,1\}^{m(n)}$, for $i = 0, 1, \ldots, m(n) - 1$, induced by the following construction of sequences:

- for $x \in \mathcal{U}_n$, take the first i bits of $h(x)$ as the first bits of the sequence
- the last $m(n) - i$ bits are chosen at random

It is now easy to check that for at least one $i \in \{1, \ldots, m(n) - 1\}$

$$|\operatorname{E}_{z \in q_{i-1}}[D(f(x), h, z)] - \operatorname{E}_{z \in q_i}[D(f(x), h, z)]| \geq \frac{\epsilon(n)}{m(n)},$$

and we can now build T_i in the following way. On input $f(x)$, h, and $z = \mathrm{B}_0^{i-1}(h(x))$:

- choose at random $r \in \mathcal{U}_{m(n)-i}$
- $T_i(f(x), h, z) = D(f(x), h, z \circ r) + \mathrm{bit}_i(z \circ r) \bmod 2$ ($\circ$ is concatenation).

It is easy to check T_i satisfies the condition stated in the theorem.

For the converse statement, suppose we have a pptm T_i such that

$$\Pr\left[T_i(f(x), h, \mathrm{B}_0^{i-1}(h(x))) = \mathrm{bit}_i(h(x))\right] \geq \frac{1}{2} + \epsilon_2(n).$$

We define D as follows;

$D(f(x), h, z) = 1$ if $T_i(f(x), h, \mathrm{B}_0^{i-1}(z)) = \mathrm{bit}_i(z)$, else, $D(f(x), h, z) = 0$. It is again easy to see that

$$|\Pr[D(f(x), h, h(x)) = 1] - \Pr[D(f(x), h, r) = 1]| \geq \epsilon_2(n).$$

$\square$

The above results give us some hints in order to decide whether a certain function is hard core or not for a given one-way function f. We will therefore in the sequel concentrate on the results for single bits. In cases where simultaneous hardness can be shown, it is essentially always by applications of one of the two results above, and a reduction to the one-bit case.

As we already remarked, it is most interesting for us to have examples of universal hard cores which are easy to construct.

3.3. The Goldreich-Levin Theorem

The first to think of the notion of hard core functions seems to have been Yao, [60]. He argues that if a one-way function is hard to invert, then even if almost all bits are easy to find, there should be a small subset in which each bit cannot be predicted with probability exceeding $\frac{1}{2}(1 + \epsilon)$ for some small ϵ. Suppose the ith bit can be predicted with probability $\frac{1}{2}(1 + \epsilon_i)$. We may not know which the hard bits are, but consider the function

$$b(x) = b(x_0, \ldots, x_{n-1}) = x_0 \oplus x_1 \oplus \cdots \oplus x_{n-1},$$

depending on *all* bits. How well can one predict this b, given $f(x)$? One would expect that, statistically, we should not be able to do better than $\frac{1}{2}(1 + \prod_i \epsilon_i)$. Yao showed that indeed, this was essentially the case also in a computational setting.

The concept of hard core function was, as mentioned, introduced by Blum and Micali, [5] (though different terminology was used). Goldreich and Levin [24], 1989, were the first to show that any one-way function has a hard core function:

Theorem 3 (Goldreich and Levin '89). *Let f be a one-way function and for $x, r \in \{0,1\}^n$, define*

$$b_r(x) \triangleq \sum_{i=0}^{n-1} r_i x_i \bmod 2.$$

Then, the family of functions: $\{b_r(x)\}_{n \geq 1}$ for $r \in \mathcal{U}_n$ is a family of hard core functions for any one-way function.

Instead of the original proof, we sketch a proof credited to Rackoff, which will also demonstrate the proof techniques behind basically all the results in the area. It proceeds as follows. Details can be found in [21, 22].

Proof sketch. Suppose $\mathfrak{O}$ predicts $b_r(x)$ for random r, x with probability $(1 + \epsilon(n))/2$. Let $e^i = 00\ldots10\ldots0$, be the unit vector in the ith dimension in $\{0,1\}^n$, that is, $e^i_j = \delta_{i,j}$, the Kronecker delta. What we would like to do is to ask $\mathfrak{O}$ about $b_{e^i}(x)$, $i = 0, \ldots, n-1$, since $b_{e^i}(x) = x_i$, the ith bit of x. However, there is no reason that $\mathfrak{O}$ should have any advantage at all on these very special inputs. Instead, suppose that we already know $b_r(x)$, and ask $\mathfrak{O}$ about $b_{r \oplus e^i}(x)$ (bitwise XOR). Notice that by linearity,

$$b_{r \oplus e^i}(x) \oplus b_r(x) = b_{e^i}(x) = x_i. \tag{1}$$

Hence, the oracle's answer on $b_{r \oplus e^i}(x)$ combined with our knowledge on $b_r(x)$ indeed gives a vote for x_i. If r is random, so is $r \oplus e^i$.

The problem is of course how to obtain $b_r(x)$ to begin with. We could ask the oracle about $b_r(x)$, but since we then ask the oracle twice, we effectively double the error probability. Since the error probability is $(1 - \epsilon(n))/2$ to begin with, we need $\epsilon(n) > 1/2$, if the double error probability is to be bounded away from $1/2$.

To avoid this, first pick randomly and independently $r_1, \ldots, r_k \in \{0,1\}^n$. Next guess $b_{r_j}(x)$, $j = 1, 2 \ldots, k$. There are 2^k possibilities for these, so if we can limit $k = \mathcal{O}(\log n)$, this is still polynomially bounded. We can therefore try all possibilities in polynomial time, and one of them *will* be correct, so let us focus on that one. If we know $\{b_{r_j}(x)\}$, by linearity, we also know

$$b_{c_1 r_1 \oplus \cdots \oplus c_k r_k}(x)$$

for every non-zero linear combination with $c_i \in \{0,1\}$. Now let R_j be the jth (in lexicographic order) linear combination of the r_is. By asking the oracle on each $b_{R_j \oplus e^i}(x)$, we now get, via (1), $2^k - 1$ votes for x_i. Notice that the R_js are not independent, but they are *pairwise independent*. Hence, the error probability when making a majority decision for x_i can be bounded from above by Chebyshev's inequality, and it is at most $\frac{1}{2n}$ if $k \sim \log_2(n/\epsilon(n)^2)$.   □

This proof technique, which we will have reason to return to, is called *k-point based, pairwise independent sampling.*

Generalizing, Goldreich and Levin also showed that a hard core function outputting $m = \mathcal{O}(\log n)$ bits is obtained by multiplying the binary vector x by a random $m \times n$ 0/1-matrix. (In fact, a random Toeplitz matrix, determined by its first row and column, and constant on diagonals, is sufficient.)

The Goldreich-Levin has been generalized to p-ary functions, that is, for $r, x \in (\mathbb{Z}/p\mathbb{Z})^n$, functions of form

$$b_{p,r}(x) = \sum_i r_i x_i \bmod p$$

are indistinguishable form the uniform distribution on $\mathbb{Z}/p\mathbb{Z}$, see [25]. In fact, multivariate polynomials of bounded degree d can also be used.

3.4. Other General Constructions

3.4.1. NÄSLUND '95, '96. In [44, 45], Näslund shows that certain functions based on affine functions (functions of type $x \mapsto \mathrm{bit}_i(ax + b)$) on $\mathbb{Z}/p\mathbb{Z}$, $|p| = \Omega(|x|)$, and on $\mathbb{F}_{2^n}$ are hard cores for any one-way function. It is shown that each individual bit is a hard core, and that for both types of functions, $\mathcal{O}(\log n)$ bits can be output simultaneously. Observe that the $\mathcal{O}(\log \log p)$ most significant bits in a uniformly distributed element in $\mathbb{Z}/p\mathbb{Z}$ are not hard core by previous definition, since they might be non-negligibly biased towards 0. However, changing the definition of hard core to suit this scenario, accounting for the bias, completely analogous results can be shown. This general definition of unpredictability for biased sources was introduced by Schrift and Shamir, [55].

The proof technique for the first case is similar to that used for the RSA bits, for example see [1, 32], and we postpone discussion until then. In the second case, proof follows by a reduction to the Goldreich-Levin result above.

The result for affine functions on $\mathbb{Z}/p\mathbb{Z}$ was in [46] generalized to hold for polynomials, that is, functions of type $\mathrm{bit}_i(\sum_{j=0}^d a_i x^i \bmod p)$ where $d = O(\log p)$, and also for non-prime p, provided p is not 'smooth' (only has small prime factors). In the more recent paper [32] (see §4.1.7), the bound $|p| = \Omega(|x|)$ was lowered to $|p| = \Omega(\log |x|)$.

3.4.2. BONEH '96 In his thesis, [6], Boneh gives a quite general group theoretic version of the Goldreich-Levin theorem, based on representation theory of finite groups. A representation of a group, G, is a a homomorphism, $\rho : G \to \mathrm{GL}(V)$, that is, each $g \in G$ is mapped to a invertible linear transformation on some vector space V. In the case of the (abelian) additive group on $\mathbb{F}_{2^n}$, the 'interesting' V always has dimension 1 and, in fact, the representations turns out to be precisely the set of functions of form $\rho_r(x) = (-1)^{b_r(x)}$, where b_r is as in Theorem 3 above.

What Boneh shows is that in any group (possibly non-abelian), admitting certain 'suitable' representations, one can show analogues of the Goldreich-Levin theorem.

3.4.3. NÄSLUND AND RUSSELL '99. In [47], a set of so called universal hash functions, introduced by Helleseth and Johansson [29], is considered. We recall that a family of *(strong) universal hash functions* (UHF's) on $\mathbb{F}_{p^n}$ is a set of functions, $\mathcal{H}_{n,m}$, with each $h \in \mathcal{H}_{n,m}$ such that $h : \mathbb{F}_{p^n} \mapsto \mathbb{F}_{p^m}$, $m|n$, and for any $x_1 \neq x_2 \in \mathbb{F}_{p^n}$ and any $y_1, y_2 \in \mathbb{F}_{p^m}$:

$$\Pr_{h \in H_{n,m}} [h(x_1) = y_1 \wedge h(x_2) = y_2] = p^{-2m}.$$

Universal hash functions were introduced by Carter and Wegman, [11].

Consider the fields $\mathbb{F}_{p^n}$ and its subfield $\mathbb{F}_{p^m}$, $m|n$, where p is a fixed prime. Each h in the set of hash functions, $\mathcal{H}$, maps $\mathbb{F}_{p^n}$ to $\mathbb{F}_{p^m}$, as follows. Let $\mathrm{Tr}_{n/m}(\alpha)$ be the usual *trace* function mapping $\mathbb{F}_{p^n}$ to $\mathbb{F}_{p^m}$, that is,

$$\mathrm{Tr}_{n/m}(\alpha) = \alpha + \alpha^{p^m} + \alpha^{p^{2m}} + \cdots + \alpha^{p^{n-m}}.$$

(Recall that Tr is $\mathbb{F}_{p^m}$-linear and hence also $\mathbb{F}_p$-linear.) $\mathcal{H} = \{\mathcal{H}_{n,m}\}_{n=tm,t\geq 1}$, is defined by

$$\mathcal{H}_{n,m} = \{h_g(x) = \mathrm{Tr}_{n/m}(g(x)) \mid g \in \mathbb{F}_{p^n}[x], \deg(g) \leq d\}.$$

Viewing $\mathbb{F}_{p^m}$ as a vector space over $\mathbb{F}_p$, the 'bits' of these functions (which are p-ary values, corresponding to projection onto the basis elements, $p = 2$ being a special case) are shown to be hard cores for any one way function, and $\mathcal{O}(\log n)$ bits can be used as hard core functions.

3.5. Complexity Lower Bounds for General Hard Cores

If we are to use hard core functions for, say, pseudorandom generation, we would like them to be as efficiently computable as possible. How 'simple' can a set of general hard core functions be?

First note that no *fixed* function, b can be a hard core for every one way function. If f is an arbitrary one-way function and we define $f_2(x)$ by $f_2(x) = f(x), b(x)$, we have a counter example for this b, as $b(x)$ is easily recovered from $f_2(x)$ and b. In general, we need to choose b randomly.

To determine the simplicity of general hard core functions some measure of simplicity/complexity is needed. In the paper by Goldmann and Näslund, [17], the model of *small depth circuits* is used. That is, how large/deep is a circuit of boolean AND/OR/NOT-gates that is needed to compute the hard core function. All of the general constructions mentioned above can be computed by circuits of logarithmic (in n) depth, polynomial size, and constant fan-in. What is shown in [17] is that these are the simplest possible: reducing the size would make the depth go up, and reducing the depth would make the size more or less explode. The proof uses the so called Håstad Switching Lemma [30].

The intuition behind the proof is once again that one can imagine one-way functions outputting most of the bits unchanged. Suppose that f outputs the bit x_1 unaffected. If the candidate hard core function is, say

$$b(x_1, \ldots, x_n) = (x_1 \wedge x_{17}) \vee b'(x_1, \ldots, x_n),$$

(where $\wedge, \vee$ denotes logical and/or) then if we observe $x_1 = 0$ (in the output of f), we can immediately simplify the function b and attack b' instead. The Håstad Switching Lemma shows precisely that if a function is too simple, then knowing a few of the input bits are often enough to deduce the function value.

In the recent paper by Goldmann and Russell, [18], another approach is taken by classifying complexity in terms of bounds on Fourier spectra. They can by this show similar results for a more general class of functions. Summarizing these two results, [17, 18], it is shown that general hard core functions cannot be computed by

- polynomial size boolean circuits of sub-logarithmic depth
- monotone functions
- generalized threshold gates (for example, the majority function).

4. Hard Bits of Specific Functions

As discussed above, for general one-way functions we can never hope that single bits of x, for example $\mathrm{lsb}(x)$, will be hard cores, nor that more than $\mathcal{O}(\log n)$ bits are simultaneously hard. For *specific* functions, such as the group- and number theoretic functions below where more known structure is available, the situation is completely different.

Some quite delicate issues are, however, now also introduced. For instance, let $f_1(x)$ be the RSA function, and $f_2(x)$ discrete exponentiation modulo a prime p and consider the candidate hard core: $b(x) = \mathrm{lsb}(x)$. As we shall see, $b(x)$ is a hard core for f_1, but not for f_2.

Since it is still open whether these functions really are one-way, the reductions from inverting the respective function to distinguishing certain bits from randomness, will therefore show that 'certain parts are as hard as the whole'. As noted by others, this is a doubly edged sword, since the results demonstrate that the cryptographic functions we are using in practice today are either very secure, or, very insecure.

We shall in the following sections concentrate on how to decide whether certain bits of x obtained from $f(x)$ can enable reconstruction of x. Hence we normally assume that our oracle is perfect and always answers correctly. Only in a few important special cases will we discuss sampling techniques, that is, how an imperfect oracle is 'amplified' to a good one.

4.1. RSA

Definition 4. *The RSA function is defined as follows. Let $N = pq$ where p, q be primes and e an integer, relatively prime to $\phi(N)$. We define*

$$\mathrm{RSA}_{N,e}(x) \triangleq x^e \bmod N.$$

The inverse is given by $y^{e^{-1} \bmod \phi(N)} \bmod N$.

We have already mentioned that no deterministic, public key system can achieve semantic security. In the case of RSA, for instance, the value $\mathrm{RSA}_{N,e}(x)$ does reveal information – namely the value of the RSA function, applied to x. In fact, even less trivial, potentially useful information such as the Jacobi symbol, $(\frac{x}{N})$, leaks through RSA.

Using the RSA function, a widely studied pseudorandom generator is built. The construction is exactly the one we already mentioned due to Blum and Micali. Pick a random seed $x_0 \in \{1, \ldots, N-1\}$, let $x_{i+1} = \mathrm{RSA}_{N,e}(x)$, and define the generator as $g(x_0) = \mathrm{lsb}(x_0)\,\mathrm{lsb}(x_1)\ldots$. Studying how predictable the output of the former generator is, can be reduced to deciding whether certain bits of the input x are hard core for the RSA function or not.

The security of the least significant bit in an RSA encrypted message has gained a lot of attention. A historic overview follows. From now on, we will use the standard notation n for $\log_2 N$.

4.1.1. GOLDWASSER ET AL. '82. The first to investigate the security of the least significant bit in RSA were Goldwasser, Micali and Tong [27]. They proved that any oracle that, when queried on an RSA-image $\mathrm{RSA}_{N,e}(x)$, outputs a guess for the least significant bit of the corresponding x which is correct with probability $1 - \frac{1}{n}$, could be used to invert RSA efficiently. Note that, unfortunately, very reliable lsb-oracles should be used for the inversion.

However, their results on how partial information about the hidden x can be obtained from $\mathrm{RSA}_{N,e}(x)$ are of great interest. They proved that given $y = \mathrm{RSA}_{N,e}(x)$, any algorithm that computes $\mathrm{lsb}(x)$ or $\mathrm{half}(x)$, where $\mathrm{half}(x) = 0$ if $0 \le x \le \frac{N}{2}$ and 1 otherwise, can be used as an oracle to recover x. Note that computing $\mathrm{lsb}(x)$ is polynomially equivalent to computing $\mathrm{half}(x)$, for

$$\mathrm{half}(x) = \mathrm{lsb}(2x \bmod N)$$

and

$$\mathrm{lsb}(x) = \mathrm{half}(2^{-1}x \bmod N),$$

see also the figure below. Since RSA has the (for these results) important multiplicative property,

$$\mathrm{RSA}_{N,e}(u)\,\mathrm{RSA}_{N,e}(v) = \mathrm{RSA}_{N,e}(uv) \bmod N,$$

it is possible to derive for instance $\mathrm{RSA}_{N,e}(2x)$ from $\mathrm{RSA}_{N,e}(x)$, without knowing x. Observe now that

$$\mathrm{half}(x) = 0 \iff x \in \left[0, \frac{N}{2}\right),$$

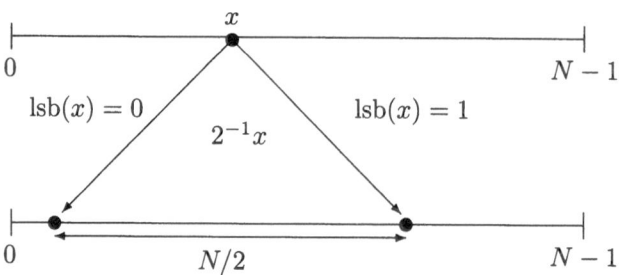

FIGURE 1. Division by 2 in $\mathbb{Z}_N$. Values that only differ in their lsb's are mapped to points $\frac{N}{2}$ apart.

$$\text{half}(2x) = 0 \iff x \in \left[0, \frac{N}{4}\right) \cup \left[\frac{N}{2}, \frac{3N}{4}\right),$$

$$\text{half}(4x) = 0 \iff x \in \left[0, \frac{N}{8}\right) \cup \left[\frac{N}{4}, \frac{3N}{8}\right) \cup \left[\frac{N}{2}, \frac{5N}{8}\right) \cup \left[\frac{3N}{4}, \frac{7N}{8}\right),$$

and so on. Hence, if we can either compute $\text{lsb}(x)$ or $\text{half}(x)$, x is easily recovered by a binary search technique. For further details, see [27].

4.1.2. BEN-OR ET AL. '83. In [4] Ben-Or, Chor and Shamir describe a new method for inverting the RSA function for which less reliable lsb-oracles can be used. They prove that inverting RSA is polynomially reducible to obtaining the parity of some specific encrypted messages, and then design a method of doing so using an lsb-oracle, that gives a correct answer with probability $\frac{3}{4} + \frac{1}{\text{poly}(n)}$.

Their reduction consists of two parts; they first describe an algorithm that inverts RSA using a *parity subroutine* and a gcd-technique. Then, they give a method of constructing the parity subroutine using the lsb-oracle.

Given an integer x, define

$$\text{abs}_N(x) = \begin{cases} x \bmod N & \text{if } x \bmod N < N/2 \\ N - x \bmod N & \text{otherwise.} \end{cases}$$

The *parity* of x, $\text{par}_N(x)$ is the least significant bit of $\text{abs}_N(x)$. Let us now describe briefly the Ben-Or, Chor and Shamir reduction.

1) Inversion algorithm:

Given $y = \text{RSA}_{N,e}(x)$, the 'hidden' x is reconstructed as follows:

1. Choose at random $a, b \in \mathbb{Z}/N\mathbb{Z}$.
2. Apply the Brent-Kung gcd procedure in [9] to $ax \bmod N$ and $bx \bmod N$. Although the values $ax \bmod N, bx \bmod N$ are unknown, this process only needs their RSA images. This procedure uses a *parity subroutine* and gives a 'correct' representation of $\gcd(ax \bmod N, bx \bmod N) = lx \bmod N$ in the form l and $\text{RSA}_{N,e}(lx)$.

3. If $ax \bmod N$ and $bx \bmod N$ are relatively prime (that is, if $lx \bmod N = 1$) then $x = l^{-1} \bmod N$. Otherwise, go back to 1.

2) Parity subroutine:

This subroutine determines with overwhelming probability the parity of a 'small' multiple of $x \bmod N$. It works as follows:

1. It takes as input $\mathrm{RSA}_{N,e}(x)$ and $d \in \mathbb{Z}/N\mathbb{Z}$, where $\mathrm{abs}_N(dx) < \epsilon_n N/2$, ($\epsilon_n$ is

   the same as the oracle's prediction advantage) that is, dx is 'small'. The inputs of the subroutine are believed to satisfy this last assumption (in computing the expected running time of the algorithm we take into account this source of error).

   It picks at random r.

   It calculates:

   $$\mathrm{RSA}_{N,e}(rx) = \mathrm{RSA}_{N,e}(r)\,\mathrm{RSA}_{N,e}(x) \bmod N$$

   and

   $$\begin{aligned} \mathrm{RSA}_{N,e}(rx + dx) &= \mathrm{RSA}_{N,e}((r+d)x) \\ &= \mathrm{RSA}_{N,e}(r+d)\,\mathrm{RSA}_{N,e}(x) \bmod N. \end{aligned}$$

   It queries the oracle for the lsb of $rx \bmod N$ and the lsb of $(rx + dx) \bmod N$. This is called a *dx-measurement*.

   Since $\mathrm{abs}_N(dx)$ is 'small', with very high probability ($\geq 1 - \epsilon_n/2$) no 'wrap' around 0 occurs when $dx \bmod N$ is added to $rx \bmod N$. (If $dx \bmod N = \mathrm{abs}_N(dx)$ then wrap around 0 means $rx \bmod N + \mathrm{abs}_N(dx) > N$, while if $dx \bmod N = N - \mathrm{abs}_N(dx)$ then wrap around 0 means $rx \bmod N - \mathrm{abs}_N(dx) < 0$).

   If no wrap around 0 occurs then:

   $$\mathrm{par}_N(dx \bmod N) = 1 \Leftrightarrow \mathrm{lsb}(rx \bmod N) \neq \mathrm{lsb}(rx + dx \bmod N).$$

Details can be found in [4].

4.1.3. VAZIRANI AND VAZIRANI '84. In [57], Vazirani and Vazirani introduce some changes in the BCS method. They suggest a better use of the oracle answers and their method is proved to succeed when given access to a lsb-oracle, with success probability $0.741 + \epsilon$. Later they improved this result so that the probability required drops to $0.732 + \epsilon$. They also show that some algorithms that invert RSA querying an oracle for the lsb of the hidden x, can be adapted to build inverting algorithms which use an oracle that guesses the jth least significant bit of x from the $j - 1$ previous ones.

4.1.4. GOLDREICH '84. Goldreich introduced a better combinatorial procedure for the Vazirani and Vazirani algorithm to yield a $0.725 + \epsilon$ result. His construction can be seen in [19]. He also showed that the techniques used up to this point had limitations, and could not be used to prove full, $0.5 + \epsilon$, security.

4.1.5. SCHNORR AND ALEXI '84. By some novel sampling techniques, Schnorr and Alexi in [54] improved the above mentioned results. They prove that given an oracle which correctly predicts the k-least significant bits with probability at least $\frac{1}{2} + \epsilon$, it is possible to invert RSA in polynomial time (more precisely in time $n^{\mathcal{O}(\epsilon^{-2}+k)}$).

The former result was not too satisfactory, for the reliability of the oracle needed for the inversion is independent of n, and the ambition was to prove that for larger n, less reliable oracles were needed for the inversion. One would normally expect that the security of the least significant bit in the RSA scheme would grow with the modulus, as it happens with the security of the scheme itself.

4.1.6. ALEXI ET AL. '88. Based on [54] and a subsequent paper by Chor and Goldreich [12], Alexi, Chor, Goldreich and Schnorr wrote together a definitive paper [1] in which they prove that an lsb-oracle that is correct with probability $\frac{1}{2} + \frac{1}{\text{poly}(n)}$ is enough to invert RSA. We give a brief outline of the improvements they made on the previous schemes.

The Ben-Or, Chor, and Shamir parity subroutine has two different sources of error:

1. Wrap around 0 in dx-measurements. The probability of a wrap around 0 is small (bounded by $\frac{1}{\text{poly}(n)}$), since $\text{abs}_N(dx) \leq \frac{1}{\text{poly}(n)} N/2$.
2. The oracle errors: this is the main source of error in each dx-measurement (cf. step 1, p. 241). A dx-measurement is wrong if either the oracle answer for $\text{RSA}_{N,e}(rx)$ or $\text{RSA}_{N,e}(rx + dx)$ is wrong (this is known as 'predicting the end points'). So the error probability of a dx-measurement is 'twice' the oracle error. This is the 'error-doubling phenomena' previously discussed.

Alexi-Chor-Goldreich-Schnorr improve the parity subroutine, introducing two new ideas:

- Avoiding the error-doubling phenomena querying the oracle only for the lsb of *one* end point: $dx + rx \bmod N$, while the other significant bit is known beforehand.
- Generating many such end points with known lsb, in a way that guarantees them 'random' enough to be used as a good sample of the oracle.

Therefore, the Alexi-Chor-Goldreich-Schnorr parity subroutine uses a method for generating $m = \text{poly}(n)$ random points, that are represented as multiples of $x \bmod N$ with the following properties:

1. The points are uniformly distributed in $\mathbb{Z}/N\mathbb{Z}$.
2. The points are pairwise independent.
3. The lsb of each point is known, with probability no less than $1 - \frac{1}{\text{poly}(n)}$.

The technique is interesting. Basically, one chooses random $u, v \in \mathbb{Z}/N\mathbb{Z}$, and sets the jth point as $r_j x = (ux \bmod N) + j(vx \bmod N)$, $j \leq m$. (Note: $\text{RSA}_{N,e}(r_j x) = \text{RSA}_{N,e}(u + jv)\,\text{RSA}_{N,e}(x)$.) If one guesses $(ux \bmod N)/N$ and $(vx \bmod N)/N$ to within $1/\text{poly}(n)$, then except with a small error probability, also $r_j x/N$ is known within $1/\text{poly}(n)$. We in addition now also guess $\text{lsb}(ux \bmod N)$ and $\text{lsb}(vx \bmod$

N). It is not difficult to see that if the guesses are correct (and note that we can afford to try them all), then also $\text{lsb}(r_j x \bmod N)$ will be known with high probability. This is hence a 2-point based, pairwise independent sampling technique.

The points, $\{r_j x \bmod N\}$, generated in such a way are used in the new parity subroutine as follows:

1. The generation of $\{r_j x \bmod N\}$ is performed once per each gcd invocation, as a part of the initialization step of the algorithm.

2. Then, for each j, the parity subroutine only queries the oracle for the lsb of $(r_j x + dx) \bmod N$. Therefore, the probability of error is not doubled in each dx-measurement.

If the oracle predicts the lsb with probability $0.5 + \epsilon$, the use of that parity subroutine yields the overall expected running time of the RSA inversion algorithm:

$$\mathcal{O}(\epsilon^{-8} n^3).$$

More recently, Fischlin and Schnorr, [16] gave a more efficient algorithm for the inversion of the RSA-scheme. The basic idea of this method is inverting $\text{RSA}_{N,e}$ by iteratively tightening approximations of random multiples $ax \bmod N$ with known multiplier a via binary division. Using this method, the number of oracle calls, at most $9n\epsilon^{-2} \log n$, is minimal up to a factor $\mathcal{O}(\log n)$. We give a brief description of the process:

1. Choose $a \in \mathbb{Z}/N\mathbb{Z}$ at random. Use the oracle to guess $\text{lsb}(ax)$ from $\text{RSA}_{N,e}(x)$. Guess rational integer u, satisfying

$$|ax \bmod N - uN| \leq \frac{\epsilon^3}{8} N.$$

(u is chosen at random, and the algorithm succeeds if u satisfies the previous condition.)

2. For $a_t = a2^{-t} \bmod N$, $t = 1, \ldots, n$, iteratively construct rational approximations $u_t N$ such that

$$|a_t x \bmod N - u_t N| \leq \frac{\epsilon}{4 \cdot 2^t} N \quad (u_0 = u).$$

Those approximations are constructed by the following method (*binary division*)

$$u_t = \frac{1}{2}(u_{t-1} + l(a_{t-1} x))$$

where $l(a_{t-1} x)$ is a 'guess' for $\text{lsb}(a_{t-1} x)$ constructed using the oracle

3. From the approximation, $u_n N$, to $a_n x \bmod N$ we get the message

$$x = a_n^{-1} \left[u_n N + \frac{1}{2} \right] \bmod N.$$

Note that the algorithm succeeds if step 2 guesses correctly and the values of $l(a_t x)$ are all correct. The expected running time of this algorithm (considering oracle calls at unit cost) is

$$\mathcal{O}(n^2 \epsilon^{-2} + n^2 \epsilon^{-6}).$$

The proof techniques sketched above also yield the *simultaneous security* of the $\log(n)$ least-significant bits. A full proof can be found in [1]. The j least-significant bits are called *simultaneously secure* if inverting $\text{RSA}_{N,e}$ is polynomial time reducible to distinguishing, given $\text{RSA}_{N,e}(x)$, between the string of j least-significant bits of x and a j − bit string obtained at random. That is, if the j least-significant bits are *simultaneously secure*, they are hard core for RSA. What they do is, first to show that the jth least-significant bit of x is *hard* to get, then use the previously mentioned result by Yao on the next-bit-test to obtain the desired result. Note that once again, we use that 'distiguishability from random' and 'predictability' are equivalent properties of strings, as Yao first stated.

4.1.7. HÅSTAD AND NÄSLUND '98. As seen, the results for the least significant bit generalizes to any of the $\mathcal{O}(\log n)$ least significant bits. For the internal bits of RSA however, the early results appearing in the papers [27, 4, 1], discussed above, were not very strong. The first result appeared in [27], where it was shown that for each i, there *are* N of very special form, for which the ith bit of x cannot be computed without errors.

Recall that the previously discussed equivalence between the half function and the lsb really says that if we have an oracle, predicting *any* boolean predicate such that the oracle is more likely to answer "1" on the interval $[0, N/2)$ than on $[N/2, N-1]$, then this oracle can in fact be used to determine the lsb and therefore to invert RSA. The key idea behind all subsequent works is to find correlations between an oracle for the ith bit and certain 'intervals', $J \subset \mathbb{Z}/N\mathbb{Z}$.

In [4], it was proved that an oracle for the ith bit of RSA can be converted into an lsb-oracle, increasing the error probability by $\frac{1}{4}$ in the worst case. The reduction is interesting and can intuitively be described as follows. Write the modulus $N = N_1 2^{i+1} + N_0$ where $N_0 < 2^{i+1}$. Notice that $\frac{N}{2} \approx N_1 2^i$, where the quality of the approximation depends on $N_0 = N \bmod 2^{i+1}$. If an oracle, $\mathfrak{O}$, is to have any advantage in predicting the ith bit, then it must be the case that

$$\Pr[\mathfrak{O}(\text{RSA}(x)) = 1 \mid \text{bit}_i(x) = 1] - \Pr[\mathfrak{O}(\text{RSA}(x)) = 1 \mid \text{bit}_i(x) = 0] > \delta \quad (2)$$

for some (small) δ. Suppose now that using the multiplicativity of RSA, we ask the oracle about the ith bit of $\text{RSA}_{N,e}(N_1^{-1}x)$. By the observations above, values $y, y+N/2$ will by this transformation be mapped to $N_1^{-1}y = z \bmod N$ and $N_1^{-1}(y + N/2) \approx z + 2^i \bmod N$. Since the ith bits of $z, z + 2^i$ (almost always) are different, observing $\text{bit}_i(N_1^{-1}x)$ intuitively gives an approximation for whether $x > N/2$ or $x < N/2$, that is, for the half($\cdot$)-function discussed previously. Since half($\cdot$) in this case is equivalent to lsb($\cdot$), an oracle for the ith bit can to some extent (depending on $N \bmod 2^{i+1}$), be used as an lsb-oracle.

Ben-Or et al. could prove that for every second bit-position i, the error introduced by this reduction could be bounded by $\frac{3}{16}$. Hence, from their own result for the lsb, a $\frac{7}{8}$-security for half of the individual bits followed. The best result for a long time was the $\frac{3}{4} + o(1)$-security that follows from the work in [1] and the above reduction. The provable security obtainable by this reductions thus depends on N

and i (the bit-position considered), and for worst case N and i, results better than $\frac{3}{4} + o(1)$ are impossible by this method.

The proof of full security for the internal bits came in the work of Håstad and Näslund, [32]. It combines in a novel way the above reduction with an idea from the paper [45] (showing security for bits in functions of form $x \mapsto ax+b \bmod p$). It is shown that at least one of the two approaches [45, 4] works. We briefly outline the ideas behind [45].

Using an oracle for the ith bit of x, we try to decide *two* bits: the $(i+1)$th and the least significant bit. To this end, consider asking the oracle about the ith bit of $2^{-1}x$, feeding it $\mathrm{RSA}_{N,e}(2^{-1}x)$. Observe that since $2^{-1} \bmod N = \frac{N+1}{2}$,

$$2^{-1}x \bmod N = \frac{x - \mathrm{bit}_{i+1}(x)2^{i+1} - \mathrm{lsb}(x)}{2} + \mathrm{bit}_{i+1}(x)2^i + \mathrm{lsb}(x)\frac{N+1}{2}. \quad (3)$$

Assuming we (somehow) already know approximations to the rough size of x and to $x \bmod 2^{i+1}$, two important things happen: the $(i+1)$th bit falls directly into the oracle's bit position, and the lsb indirectly affects more or less all bits by the occurrence/non occurrence of the term $\frac{N+1}{2}$. If we temporarily neglect the effect of the lsb it is clear that we could decide the $(i+1)$th bit by this method since its effect is highly correlated with the oracle's answers on the ith bit. In particular, by (2), there exists an interval (of consecutive integers) $J_0 \subset \mathbb{Z}/N\mathbb{Z}$ such that the oracle's behavior, in terms of the fraction of 1-answers differ on J_0 and the 'shifted' interval J_0+2^i. (We allow intervals to wrap around N.) Suppose now that we also have intervals J_1, J_2, J_3 of non-negligible size, so that the oracle's behavior on $(J_1, J_1 + (N+1)/2)$, $(J_2, J_2 + 2^i + (N+1)/2)$, $(J_3, J_3 + (N+1)/2 - 2^i)$, within each pair is different. Then by randomly sampling in these intervals one should be able to distinguish between the four possibilities for the two bits; bit_{i+1} and lsb. This sampling can be achieved by techniques similar to that used is [1], generating two-point based, pairwise independent, random sample points, whose approximate location in $\mathbb{Z}/N\mathbb{Z}$ is known in advance.

The critical issue is therefore the existence of such intervals. If they do exist, a two-by-two bit decision method from [45] can be used. If such intervals do not exist, then roughly speaking, it is shown that a small (and odd) integer multiple of $\alpha = 2^i + N/2$ is close to an integer multiple of 2^{i+1}, that is,

$$\left| \frac{\alpha}{2^{i+1}} - \frac{r}{s} \right|$$

is 'small', for some small $s \in \mathbb{Z}$. Furthermore, it is shown that the oracle must also be strangely correlated with the value α, essentially always giving the same answer to z as to $z + \alpha$. If this is the case, the simultaneous effects of the $(i+1)$th bit and the lsb when using (3) 'cancel' each other. The good news is that for N, i for which such a relation exist, Håstad and Näslund can show that it is possible to find a number N_2, depending on N_1, s, r and the quality of the above diophantine approximation, such that $N_2^{-1}x \bmod N$ maps values at distance $(N+1)/2$ more or less exactly to values at distance 2^i. That is, a refinement of the reduction from [4] is applied. Furthermore by the correlation between the oracle and the value α, intervals at distance $(N + 1)/2$ can be found where the oracle behaves differently. Hence the Fischlin-Schnorr or gcd-technique for the lsb can be used. In summary, this is a more careful application of the reduction method by Ben-Or et al. [4], using the gathered information about N, i and the oracle's behavior.

Notice that for this method to succeed, the value N_2 must have an inverse modulo N. However, if this is not the case, then $\gcd(N_2, N)$ gives a factorization of N, and we retrieve x immediately. We have a reason to return to this issue in the section on the discrete log.

Finally, the paper also shows security for blocks of $\mathcal{O}(\log n)$ bits.

4.2. The Rabin Function

The Rabin encryption function is RSA with exponent 2, that is, $x \mapsto x^2 \bmod N$.

All the subsequent RSA results above have carried over to Rabin with minor adoptions. The main issue with the Rabin function is that it is a four-to-one map, rather than one-to-one. Given $y = x^2 \bmod N$, it is therefore not well-defined what $\mathrm{lsb}(\sqrt{y} \bmod N)$ is. For this reason, different techniques to make it 1–1 are often introduced. One way is to restrict the domain to certain subsets of $\mathbb{Z}/N\mathbb{Z}$ on which it is one-to-one. By doing this, some technical difficulties arise that we omit here.

4.3. Discrete Exponentiation – Groups Based on Modular Arithmetic

The discrete exponentiation function can be defined on any group. We begin with the case when the group operation is based on ordinary modular arithmetic. Specifically, we first consider $\mathbb{Z}/p\mathbb{Z}$ under multiplication.

Definition 5. *Discrete exponentiation mod p (a prime) is defined by*

$$f_{g,p}(x) \triangleq g^x \bmod p,$$

where $|p| = n$, and g is a generator for $\mathbb{Z}/p\mathbb{Z}^*$.

First of all, note that writing $p = 2^k q + 1$, q odd, all the k least significant bits of x are 'easy' to find given $g^x \bmod p$. More precisely, they can be found by the Pohlig-Hellman algorithm, [50]. In particular, the least significant bit of x can always be found by determining if $y = f_{g,p}(x)$ is a quadratic residue. In the sequel, when we say 'every bit', we do not include these k bits.

The main techniques to show hardness of the bits have been similar to those used for RSA: one views a predictor (or distinguisher) for the ith bit as a window, and by manipulating x, using multiplicative properties of $f_{g,p}$, one can make the different bits of x slide through that window, effectively deciding them one after the other. For instance, left shifts of x can be performed by transforming $y = f_{g,p}(x)$ into $y^2 \bmod p$, being careful about the possible wrap around that is caused by leftmost bits of x when reduction modulo the group order, $p - 1$, occurs. Besides this multiplicativity, however, we now also have a property that RSA lacks: *additivity*. This means that individual bits can bet set/reset once their value has been determined transforming y into $y' = yg^{\pm 2^i} \bmod p$. A small review on the work done follows. In the sequel, we denote by $\mathrm{bit}_i(x)$ the ith bit in the binary representation of x (as we established for the case $x \in \mathbb{Z}/m\mathbb{Z}$).

4.3.1. BLUM AND MICALI '86. In [5], Blum and Micali consider the function

$$\text{umsb}_p^{(k)}(x) \triangleq \lfloor 2^k x/p \rfloor$$

(the unbiased most significant bits). Blum and Micali [5] show, for $k = 1$, that this is a hard core. Note that for $k = 1$, this function coincides with the 'half' function we discussed in the section on RSA. As for RSA, we also here have a relation between this function and the least significant bits of x. Each quadratic residue, $y = g^{2s} \bmod p$, has two roots, g^s and $g^{s+(p-1)/2} \bmod p$. The former one is called the *principal root*. The main result they obtain is that the discrete log problem reduces to finding the principal square root of $f_{g,p}(x)$ (that is, $f_{g,p}((x \bmod 2^k q)/2)$). Obviously, this root can not be obtained if $f_{g,p}(x)$ is not a quadratic residue modulo p, but then we know $\text{lsb}(x) = 1$. We can divide $f_{g,p}(x)$ by g and set this bit to 0. Then, x can be divided by two by computing the principal square root. This way we have shifted x to the right, therefore its second lsb can be now determined by checking whether the remaining value is a quadratic residue or not. All we have to do now is keep on shifting until we get all bits of x.

4.3.2. LONG AND WIGDERSON '88. They generalize Blum and Micali's proof to hold for $k = \mathcal{O}(\log n)$. The proofs are again a kind of binary search method to retrieve x, see [40].

4.3.3. PERALTA '85. He shows that the $\mathcal{O}(\log n)$ least significant bits are hard cores. To do that, he constructs a polynomial time algorithm that obtains x from an oracle that guesses $\text{bit}_{k+l+1}(x)$ on input $(p, g, f_{g,p}(x), \text{bit}_{k+1}(x), \ldots, \text{bit}_{k+l}(x))$, where $l = \mathcal{O}(\log n)$. Assuming the oracle's guess is correct with probability $1/2 + 1/2^{\mathcal{O}(\log n)}$, his algorithm computes x in probabilistic polynomial time. For further details, see [49].

The underlying idea behind his construction is again that the discrete log problem reduces to the principal square root problem. For instance, assuming that $p - 1 = 4r + 2$, and that $f_{g,p}(x)$ is a quadratic residue, notice that $\text{bit}_1(2x \bmod p - 1) = 0$ if and only if $x < p/2$. In general, from $\text{bit}_{k+1}(x)$ the principal square root of $f_{g,p}(x)$ can be easily obtained. Therefore, the idea is to use the known bits and the oracle guesses to obtain principal square roots of different integers y_i obtained from x in the following way: $y_0 = x$ and y_{j+1} is constructed from y_i, first setting $\text{bit}_{k+1}(y) = 0$, and then doing $y_{j+1} = y_j/2$. Note that this division means in fact for us a principal square root computation. It is clear that $\text{bit}_{k+l+1}(y_i) = \text{bit}_{k+l+i}(x)$. This process only recovers x of 'controlled' size. However, Peralta proposes several ways to make it work for arbitrary x.

4.3.4. LONG '84. Long shows in [39] that each individual bit cannot be computed without errors.

4.3.5. PATEL AND SUNDARAM '98. In [48] it is shown that essentially all the bits of x are simultaneously secure, under the assumption that $g^x \bmod p$ is a one-way function, even if x is restricted to be 'small'. The proof is similar to that in [33] (see the following section). Notice that if x is small, x can be shifted many steps to the left, without causing wrap around, and this is a key to making the proof work.

4.3.6. HÅSTAD AND NÄSLUND '99. They prove in [32] that for almost all primes p, any single bit, as well as blocks of $\mathcal{O}(\log n)$ bits of x are hard. This follows by applying the techniques used for the RSA bits in [32] mentioned above. By 'almost all primes' it is to be understood that if p is chosen at random, the result holds with probability $1 - o(1)$. Recall that the method from [32] relies on the existence of the multiplicative inverse of a certain integer $N_2 \in \mathbb{Z}/N\mathbb{Z}$. However, when this does not exist, we argued that we could factor the modulus, retrieving x anyway. When applying methods to the bit security of $f_{g,p}$, it turns out that we need a multiplicative inverse of a number P_2, completely analogous to the N_2-quantity in the RSA case (cf. §4.1.7), but now the inverse is modulo the group order, $p-1$. If P_2 does not have an inverse we now have $\gcd(P_2, p-1) > 1$, but in this case it does not help us in retrieving x in polynomial time. It is therefore necessary that for every P_2' in a set of size $\text{poly}(n)$, $\gcd(p-1, P_2') \leq \text{poly}(n)$. It is shown that if p is chosen at random, this happens with probability close to 1. Still, it could be that there is some completely different method that applies to *all* primes. However, an indication that this is not the case is also given in [32] where examples of primes are given for which no transformation based on the available multiplicative/additive properties of $f_{g,p}$ seem to enable reconstruction of x.

Next, we turn to the case of the ring $\mathbb{Z}/N\mathbb{Z}$.

4.3.7. HÅSTAD, SCHRIFT, AND SHAMIR, '93. Let $f_{g,N}(x) = g^x \mod N$, where N is a Blum-integer, that is, $N = pq$, p, q primes of form $4k + 3$. In [33], it is proven that the low- as well as the high-end $\sim \log N/2$ bits of x are hard (both individually and simultaneously) with respect to $f_{g,n}$. That is, if these bits could be distinguished from random bits, it would be feasible to invert $f_{g,N}$. Since inverting $f_{g,N}$ in turn is easily seen to imply being able to factor N (by computing $\text{dlog}(f_{g,N}(N)) = p + q - 1$ and combining this with the fact that $N = pq$), a quite strong result is obtained.

This is as far as we know the only result showing (simultaneous) hardness for $\mathcal{O}(\log N)$ bits, based on standard intractability assumptions such as integer factoring. The reason it can be done is due to an observation stating that with relatively high probability, an element $g \in \mathbb{Z}/N\mathbb{Z}$ has so called high order. This intuitively means that the discrete logs we compute have leading zeros, and so they can be left-shifted by the multiplicative properties (see above) of the exponentiation function, without causing wrap around.

4.4. Discrete Exponentiation – Elliptic Curves and other Groups

Let $(G, \otimes)$ be a cyclic (and thus abelian) group of order n. For $g \in G$, a generator, one can consider the 'exponentiation' defined by applying $\otimes$ to g, that is,

$$\underbrace{g \otimes g \otimes \cdots \otimes g}_{x \text{ times}} \triangleq [x]g.$$

This gives an analog to the normal discrete log problem: given g and $y = [x]g$, find x. We denote this inverse function glog_g.

Just as we did above, we can define a function $\mathrm{umsb}_g^{(t)}(y)$ to be the j such that $\mathrm{glog}_g(y) \in [jn2^{-t}, (j+1)n2^{-t})$.

4.4.1. KALISKI '86. In [35] it is demonstrated that the function $\mathrm{umsb}_g^{(t)}(y)$ for $t = \mathcal{O}(\log\log n)$ is a hard core for $f(x) = [x]g$. The proof is (as above) to use a kind of binary search to decide $\mathrm{glog}_g(y)$ using an assumed oracle for $\mathrm{umsb}_g^t(y)$. He shows how to construct a pseudorandom generator in an elliptic curve group based on this result.

4.4.2. SCHNORR '98. Using so called shift-bits, $\mathrm{lsb}(2^{-i}x)$ (in groups of odd order), Schnorr proved in [53] security for all bits in this representation under an assumption similar to that of Patel and Sundaram above – that the discrete log is hard, even if the most significant bits of it is given. The key theorem in this paper shows that any arbitrary j consecutive bits of x are simultaneously secure if and only if certain restrictions of f are one way. This can also be applied to the j least significant bits of x as well as the j most significant bits of x/q. Later, he shows that each individual bit in the normal (binary) representation is hard if the the group order is $2^t q$, where q is prime.

4.5. ElGamal

ElGamal encryption is defined by

$$f(x) = (g^r \bmod p, xg^{ar} \bmod p),$$

where g^a is the public key, a the secret key, and where r is randomly chosen.

With $k \in \Omega(\sqrt{\log p})$, the function $\mathrm{umsb}_p^{(k)}(x)$ as above is at least slightly inapproximable in the sense that *computing* this function implies breaking El-Gamal, see Boneh and Venkatesan, [7]. The techniques used are lattice basis reductions, see the discussion below.

4.6. Diffie-Hellman

Let p be an n-bit prime and $g \in \mathbb{F}_p^*$ a primitive element. The problem of computing g^{ab} from g^a and g^b is known as the Diffie-Hellman problem. A widely used key agreement scheme and several other cryptographic protocols rely on the unproven conjecture that any algorithm that solves the Diffie-Hellman problem also solves the discrete log problem in $\mathbb{F}_p$.

When g is a primitive root, Boneh and Venkatesan [7] have proposed a method of recovering a 'hidden' element $\alpha \in \mathbb{F}_p$ from about $n^{1/2}$ most significant bits of $\alpha g^{x_i} \bmod p$, $i = 1, \dots, d$, for $d = 2n^{1/2}$ integers $x_1, \dots, x_d$, chosen uniformly and independently at random in the interval $[0, p-2]$. Their algorithm is based on certain rounding lattice techniques that can be found in [2]. For the randomly chosen integers $x_1, \ldots, x_d$ they consider the lattice L spanned by the rows of the matrix

$$\begin{pmatrix} p & 0 & 0 & \cdots & 0 & 0 \\ 0 & p & 0 & \cdots & 0 & 0 \\ & \vdots & & & & \vdots \\ 0 & 0 & 0 & \cdots & p & 0 \\ t_1 & t_2 & t_3 & \cdots & t_d & 1/p \end{pmatrix}$$

where $t_i = g^{x_i} \bmod p$, $i = 1, \ldots, d$. Let $\mathrm{umsb}_p^{(k)}$ be as above, that is,

$$\mathrm{umsb}_p^{(k)}(t) = j \Leftrightarrow t \bmod p \in [jp2^{-k}, (j+1)p2^{-k})$$

Roughly speaking, $\mathrm{umsb}_p^{(k)}$ is the integer defined by the k most significant bits of $t \bmod p$. Assuming $\mathrm{umsb}_p^{(k)}(\alpha g^{x_i} \bmod p)$ are known for all $i = 1, \ldots, d$, they describe a probabilistic polynomial time algorithm to find a certain vector v_α in L such that $v_\alpha = (r_1, \ldots, r_d, \alpha/p)$. Therefore α can be recovered in polynomial time.

Some statements related to bit-security follow from the mentioned result. In particular, the security $(\log p)^{1/2} + \log \log p$ most significant bits of the private key g^{ab} in the Diffie-Hellman key agreement scheme is proved.

In [28], González Vasco and Shparlinski remove the restriction that g needs to be a primitive root. Using new bounds of exponential sums the above mentioned results are in the aforementioned work extended to the case of elements g of arbitrary multiplicative order T, provided that $T \geq p^{1/3+\epsilon}$; $T \geq p^\epsilon$ is enough for almost all primes. The mentioned results can also be applied to several other cryptosystems based on exponentiation in finite fields, like the Okamoto conference key sharing scheme and ElGamal cryptosystem. Slightly modifying the technique used for the Diffie-Hellman case similar results for the Shamir message passing scheme are obtained.

Finally, there has also been some recent results by Verheul [59], and Shparlinski [56], in the case of composite fields. These results have applications to the security of the cryptosystem proposed in [10, 38].

4.7. Functions Based on Combinatorial Problems

4.7.1. SUBSET SUM. Let $A = \{a_1, \ldots, a_n\}$ be a set of $l(n)$-bit integers and $s = (s_1, \ldots, s_n) \subset \{0, 1\}^n$. The *Subset Sum* function is defined by

$$f_A(s) = (A, \sum_{i=1}^{n} s_i a_i \bmod 2^{l(n)}).$$

This has been conjectured to be a one-way function if $l(n) \sim (1 + \epsilon)n$ for a small constant ϵ.

Note that $\sum_{i=1}^{n} s_i a_i$ is an inner product. Indeed, this is precisely the idea used in [34] by Impagliazzo and Naor to show that if f_A is one-way, then $\sum_{i=1}^{n} s_i a_i \bmod 2^{l(n)}$ is indistinguishable from $\mathcal{U}_{l(n)}$. The proof is thus a reduction to the Goldreich-Levin theorem: if an algorithm, D, distinguishes $f_A(s)$ from random bits, then it is possible to predict $\sum_i r_i s_i$ mod 2, given $r \in \mathcal{U}_n$ and $f_A(s)$. On input $f_A(s), r$, do as follows.

1. guess $k = \sum_i r_i s_i$ (there are n choices)
2. pick a random integer z
3. for $i = 1, \ldots, n$, let $b_i = a_i + r_i z$
4. run D on input $(\{b_i\}, f_A(s) + kz \bmod 2^{l(n)})$
5. if D answers 'non-random', we output k mod 2, else $1 - (k \bmod 2)$

After handling some technicalities arising from the fact that $2^{l(n)}$ is not a prime, one can easily verify that the returned value indeed is a prediction for the inner product.

4.7.2. SYNDROME DECODING PROBLEM. In [15] Fischer and Stern proposed a pseudorandom generator based on the theory of error correcting codes. (The idea of using decoding problems for this purpose actually predates the Goldreich-Levin theorem, see [23].)

Recall that a (n, k, d) binary code is a k-dimensional subspace of $\{0, 1\}^n$, such that any non-zero vector in it has weight at least d. Such space is often defined as the kernel of a $n \times (n - k)$ matrix that is usually referred to as the parity check matrix of the code. Given any vector in $\{0, 1\}^n$, its syndrome is defined as its product by this matrix.

The underlying problem which their generator is based on is that of finding a vector of given weight from its syndrome. This is known as the *Syndrome Decoding Problem* and is known to be NP-hard.

Taking $p \in (0, 1)$ and an integer $\delta \leq n$ (which has to fulfill certain conditions), they consider the set of functions $SD(p, \delta) = \{f_n\}$, such that, $f_n : D_n \longrightarrow \{0, 1\}^{pn(n+1)}$, where

$$D_n = \{(M, x) \mid M \in \{0, 1\}^{pn \times n}, x \in \{0, 1\}^n\}$$

and $f_n(M, x) = (M, Mx)$. Assuming the above collection is a family of one-way functions, they construct a generator which, from a seed (M, x), builds a pseudo-random string using a polynomial time algorithm A that computes a vector of size n and weight δ from a $\log_2 \binom{n}{\delta n}$ bit number.

Given $(M, x) \in D_n$;

1. Do $y = Mx$.
2. Output y_2, where $y = (y_1, y_2)$, and

$$|y_1| = \log_2 \binom{n}{\delta n}$$

3. Do $x = A(y_1)$, go back to 1.

In [15], they prove that given a distinguisher which accepts with different probability $f_n(M, x)$ than a random string (that is, provided that the former generator is predictable) such device can be used to predict the inner product of two bit strings r and s from r and $f_n(s)$. Thus, following the idea of Impagliazzo and Naor, they prove the existence of such device would, through the Goldreich-Levin theorem, contradict the one-wayness of the defined set of functions. For further details, see [15].

5. Summary and Open Problems

We have demonstrated the usefulness of hard core functions. For instance, we have seen that from such predicates pseudorandom generators can be built, which in turn can be used to design semantically secure cryptosystems, bit-commitment schemes etc.

There are some interesting open problems that we would like to mention. The *quadratic residuosity problem* is defined as follows. Given N, a Blum-integer, let $\left(\frac{x}{N}\right) \in \{-1, 1\}$ be the Jacobi symbol of $x \mod N$. For x with $\left(\frac{x}{N}\right) = 1$, let $Q_N(x) = 1$ if x is a square mod N and 0 otherwise. Computing Q_N is no more difficult than factoring, but is the converse also true? If so, Q_N would be a hard core for RSA and similar systems. Based on this assumption, Goldwasser and Micali constructs a semantically secure cryptosystem, [26]. In [13], Damgård conjectures that for a prime p, the family $\{h_a(x) = \left(\frac{x+a}{p}\right) \mid a \in \mathbb{Z}/p\mathbb{Z}\}$ gives hard core functions.

There could be a connection between universal hash functions (see [11]) and hard core functions. First notice that all the general construction of hard core functions mentioned in §3 are, in fact, based on universal hash functions. In §3.5, we reviewed a complexity lower bound for general hard cores. It is interesting to note that precisely the same lower bound holds for universal hash functions, see the paper by Mansour et al., [42].

As far as we know, no results similar to the one mentioned in §4.6 have been obtained for the elliptic curve case. Therefore, studying the bit security of the private key in the Diffie-Hellman elliptic curve scheme could be an interesting topic for further research. Also the reliability of the Naor-Reingold generator [43] could be evaluated through the security of certain bits, as it was done for RSA.

The reader who wishes to look deeper into the complexity theoretic models for secure encryption and pseudorandomness is referred to the comprehensive books by Goldreich, [21, 22], and Luby, [41].

Acknowledgment. We thank Igor Shparlinski and the anonymous reviewers for comments, greatly improving on earlier versions of this paper.

References

[1] W. Alexi, B. Chor, O. Goldreich, and C. P Schnorr, *RSA and Rabin functions: certain parts are as hard as the whole*, SIAM Journal on Computing **17** (1988), 194–209.

[2] L. Babai, *On Lovasz lattice reduction and the nearest lattice point problem*, Combinatorica, **6** (1986), 11–13, 1986.

[3] M. Bellare and P. Rogaway, *Optimal asymmetric encryption*, Proceedings of Eurocrypt '94, LNCS 950, pp. 92–111, Springer-Verlag.

[4] M. Ben-Or, B. Chor, and A. Shamir, *On the cryptographic security of single RSA bits*, Proceedings of the 15th ACM STOC, 1983, pp. 421–430.

[5] M. Blum and S. Micali, *How to generate cryptographically strong sequences of pseudorandom bits*, SIAM Journal on Computing, **13** (1986), 850–864.

[6] D. Boneh, *Studies in computational number theory with applications to cryptography*, PhD Thesis, Dept. of Computer Science, Princeton Univ., 1996.

[7] D. Boneh and R. Venkatesan, *Hardness of computing the most significant bits of secret keys in Diffie-Hellman and related schemes*, Proceedings of Crypto '96, LNCS 921, pp. 129–142, Springer-Verlag.

[8] D. Boneh and R. Venkatesan, *Rounding in lattices and its cryptographic applications*, Proceedings of the 8th Annual ACM-SIAM Symp. on Discr. Algorithms, ACM, NY, 1997. pp 675–681.

[9] R. P. Brent and H. T. Kung, *Systolic VLSI arrays for linear time GCD computation*, VLSI 83, IFIP, F. Anceau and E. J. Aas, eds., Elsevier Science Publishers. 1983. pp. 145–154.

[10] A. E. Brouwer, R. Pellikan, and E. R. Verheul, *Doing more with fewer bits*, Proceedings of Asiacrypt '99, LNCS 1716, pp. 321–332, Springer-Verlag.

[11] J. L. Carter and M. N. Wegman, *Universal classes of hash functions*, Journal of Computer and System Sciences, **18** (1979), pp. 265–278.

[12] B. Chor and O. Goldreich, *RSA/Rabin least significant bits are $\frac{1}{2} + \frac{1}{\text{poly}(\log n)}$ secure*, Proceedings of Crypto '84, LNCS 196, pp. 303–313, Springer-Verlag.

[13] I. B. Damgård, *On the randomness of Legendre and Jacobi sequences*, Proceedings of Crypto '88, LNCS 403, pp. 163–172.

[14] W. Diffie and M. E. Hellman, *New directions in cryptography*, IEEE Transactions on Information Theory **22** (1976), 644–654.

[15] J.-B. Fisher and J. Stern, *An efficient pseudorandom generator provably as secure as syndrome decoding*, Proceedings of Eurocrypt '96, LNCS 1070, pp. 245–255, Springer-Verlag.

[16] R. Fischlin and C. P. Schnorr, *Stronger security proofs for RSA and Rabin bits*, Journal of Cryptology, **13** (2000), no 2, pp. 221–244.

[17] M. Goldmann and M. Näslund, *The complexity of computing hard core predicates*, Proceedings of Crypto '97, LNCS 1294, pp. 1–15, Springer-Verlag.

[18] M. Goldmann and A. Russell, *Spectral bounds on general hard core predicates*, Proceedings of STACS '2000, LNCS 1770, pp. 614–625, Springer-Verlag.

[19] O. Goldreich, *On the number of close-and-equal pairs of Bits in a string (with applications on the security of RSA's L.S.B)*, Proceedings of Eurocrypt '84, LNCS 209, pp. 127–141, Springer-Verlag.

[20] O. Goldreich, *Three XOR lemmas – an exposition*, Theory of Cryptography Library, http://philby.ucsd.edu/cryptolib.html.

[21] O. Goldreich, *Foundations of cryptography (Fragments of a book)*, Theory of Cryptography Library, http://philby.ucsd.edu/cryptolib.html.

[22] O. Goldreich, *Modern cryptography, probabilistic proofs and pseudo-randomness*, Springer-Verlag, 1999.

[23] O. Goldreich, H. Krawczyk, and M. Luby, *On the existence of psuedorandom generators*, Proceedings of the 29th IEEE FOCS, 1988, pp. 12–24.

[24] O. Goldreich and L. A. Levin, *A hard core predicate for any one way function,* Proceedings of the 21st ACM STOC, 1989, pp. 25–32.

[25] O. Goldreich, R. Rubinfeld, and M. Sudan, *Learning polynomials with queries: the highly noisy case,* Proceedings of the 36th IEEE Symp. on Foundations of Computer Science, 1995, pp. 294–303.

[26] S. Goldwasser and S. Micali, *Probabilistic encryption,* Journal of Computer and System Sciences **28** (1984), 270–299.

[27] S. Goldwasser, S. Micali, and P. Tong, *Why and how to establish a private code on a public network,* Proceedings of the 23rd IEEE Symp. on Foundations of Computer Science, 1982, pp. 134–144.

[28] M. I. González Vasco and I. E. Shparlinski, *On the security of Diffie-Hellman bits,* Proceedings of the Workshop on Comp. Number Theory and Cryptography, Birkhäuser, 2000.

[29] T. Helleseth and T. Johansson, *Universal hash functions from exponential sums over finite fields and Galois rings,* Proceedings of Crypto '96, LNCS 921, pp. 31–44, Springer-Verlag.

[30] J. Håstad, *Computational limitations of small-depth circuits,* ACM doctoral dissertation award 1987. MIT Press.

[31] J. Håstad, R. Impagliazzo, L. A. Levin, and M. Luby, *Pseudo random number generators from any one way function,* SIAM Journal on Computing, **28** (1999), 1364–1396.

[32] J. Håstad and M. Näslund, *The security of all RSA and discrete log bits,* Manuscript, 1999. (Preliminary version appears in Proceedings of the 39th IEEE Symp. on Foundations of Computer Science, 1998, pp. 510–519.)

[33] J. Håstad, A. W. Schrift, and A. Shamir, *The discrete logarithm modulo a composite hides $O(n)$ bits,* Journal of Computer and System Sciences, **47** (1993), 376–404.

[34] R. Impagliazzo and M. Naor, *Efficient cryptographic schemes provably as secure as Subset Sum,* Journal of Cryptology, **9** (1996), 199–216.

[35] B. S. Kaliski, *A pseudo-random bit generator based on elliptic logarithms,* Proceedings of Crypto '86, LNCS 263, pp. 84–103, Springer-Verlag.

[36] D. Knuth, *Seminumerical algorithms,* (2 ed.), Volume 2 of *The art of computer programming,* Addison-Wesley, 1982.

[37] A. Kolmogorov, *Three approaches to the concept of the amount of information,* Problems on Information Transmision (1965).

[38] A. K. Lenstra and E. R. Verheul, *The XTR public key system,* Proceedings of Crypto '2000, Springer-Verlag, (to appear).

[39] D. L. Long, *The security of bits in the discrete logarithm,* Phd Thesis, Princeton University, 1984.

[40] D. L. Long and A. Wigderson, *The discrete log hides $O(\log n)$ bits,* SIAM Journal on Computing, **17** (1988), 413–420.

[41] M. Luby, *Pseudorandomness and cryptographic applications,* Princeton University Press, 1996.

[42] Y. Mansour, N. Nisan, and P. Tiwari, *The computational complexity of universal hashing,* Theoretical Computer Science **107** (1993), 121–133.

[43] M. Naor and O. Reingold, *Number Theoretic constructions of efficient pseudorandom functions,* Proceedings of the 38th IEEE Symp. on Foundations of Computer Science, (1997), 458–467.

[44] M. Näslund, *Universal hash functions & hard core bits*, Proceedings of Eurocrypt '95, LNCS 921, pp. 356–366, Springer-Verlag.

[45] M. Näslund, *All bits in $ax+b$ mod p are hard*, Proceedings of Crypto '96, LNCS 1109, pp. 114–128, Springer-Verlag.

[46] M. Näslund, *Bit extraction, hard-core predicates, and the bit security of RSA*, PhD Thesis, Royal Inst. of Technology, Stockholm, 1998.

[47] M. Näslund and A. Russell, *Hard core functions: survey and new results*, Proceedings of Nordsec '99, pp. 305–322.

[48] S. Patel and G. S. Sundaram, *An efficient discrete log pseudo random generator*, Proceedings of Crypto '98, LNCS 1462, pp. 304–317, Springer-Verlag.

[49] R. Peralta, *Simultaneous security of bits in the discrete log*, Proceedings of Eurocrypt '85, LNCS 219, pp. 62-72, Springer-Verlag.

[50] S. C. Pohlig and M. Hellman, *An improved algorithm for computing logarithms over $GF(p)$*, IEEE Trans. on Information Theory, **IT-24** (1978), 106–110.

[51] R. L. Rivest, A. Shamir, and L. Adleman, *A method for obtaining digital signatures and public key cryptosystems*, Communications of the ACM, **21** (1978), 120–126.

[52] C. E. Shannon, *Communication Theory of secrecy systems*, Bell System Technical Journal, **28** (1949), 656–715.

[53] C. P. Schnorr, *Security of almost all discrete log bits*, ECCC report TR98-033, 1998. (This report is available on-line from http://www.eccc.uni-trier.de/eccc/)

[54] C. P. Schnorr and W. Alexi, *RSA-bits are 0.5+ϵ secure*, Proceedings of Eurocrypt '84, LNCS 209, pp. 114–128, Springer-Verlag.

[55] A. W. Schrift and A. Shamir, *On the universality of the next bit test*, Proceedings of Crypto '90, LNCS 537, pp. 394-408, Springer-Verlag.

[56] I. E. Shparlinski, *Security of polynomial transformations of the Diffie–Hellman key*, Preprint, 2000.

[57] U. V. Vazirani and V. V. Vazirani, *RSA bits are .732 + ϵ secure*, Proceedings of Crypto '83, pp. 369–375, Plenum Press.

[58] U. V. Vazirani and V. V. Vazirani, *Efficient and secure pseudo-random number generation*, Proceedings of the 25th IEEE Symp. on Foundations of Computer Science, 1984, pp. 458–463.

[59] E. R. Verheul, *Certificates of recoverability with scalable recovery agent security*, Proceedings of the Inter. Workshop on Practice and Theory of Public Key Cryptography '2000, LNCS 1751, pp. 258–275, Springer-Verlag.

[60] A. C. Yao, *Theory and applications of trapdoor functions*, Proceedings of the 23rd IEEE Symp. on Foundations of Computer Science, 1982, pp. 80–91.

Departamento de Matemáticas, Universidad de Oviedo, 33007 Oviedo, Spain
E-mail address: mvasco@orion.ciencias.uniovi.es

Ericsson Research, SE-164 80 Stockholm, Sweden
E-mail address: mats.naslund@era-t.ericsson.se

Progress in Computer Science and Applied Logic, Vol. 20
© 2001 Birkhäuser Verlag Basel/Switzerland

On the Security of Diffie-Hellman Bits

Maria Isabel González Vasco and Igor E. Shparlinski

Abstract. Boneh and Venkatesan have recently proposed a polynomial time algorithm for recovering a "hidden" element α of a finite field $\mathbb{F}_p$ of p elements from rather short strings of the most significant bits of the remainder modulo p of αt for several values of t selected uniformly at random from $\mathbb{F}_p^*$. We use some recent bounds of exponential sums to generalize this algorithm to the case when t is selected from a quite small subgroup of $\mathbb{F}_p^*$. Namely, our results apply to subgroups of size at least $p^{1/3+\varepsilon}$ for all primes p and to subgroups of size at least p^{ε} for almost all primes p, for any fixed $\varepsilon > 0$. We also use this generalization to improve (and correct) one of the statements of the aforementioned work about the computational security of the most significant bits of the Diffie-Hellman key.

1. Introduction

Let p be an n-bit prime and let $g \in \mathbb{F}_p$ be an element of multiplicative order T, where $\mathbb{F}_p$ is the finite field of p elements.

For integers s and $m \geq 1$ we denote by $(s \operatorname{rem} m)$ the remainder of s on division by m. We also use $\log z$ to denote the binary logarithm of $z > 0$.

In the case of $T = p - 1$, that is, when g is a primitive root, Boneh and Venkatesan [1] have proposed a method of recovering a "hidden" element $\alpha \in \mathbb{F}_p$ from about $n^{1/2}$ most significant bits of $(\alpha g^{x_i} \operatorname{rem} p)$, $i = 1, \dots, d$, for $d = \lceil 2n^{1/2} \rceil$ integers $x_1, \dots, x_d$, chosen uniformly and independently at random in the interval $[0, p - 2]$. This result has been applied to proving security of reasonably small portions of bits of private keys of several cryptosystems. In particular, in Theorem 2 of [1] the security of the $\lceil n^{1/2} \rceil + \lceil \log n \rceil$ most significant bits of the private key $(g^{ab} \operatorname{rem} p)$ of the Diffie-Hellman cryptosystem with public keys $(g^a \operatorname{rem} p)$ and $(g^b \operatorname{rem} p)$ with $a, b \in [0, p - 2]$ is considered.

Namely, a method has been given to recover, in polynomial time, the Diffie-Hellman key $(g^{ab} \operatorname{rem} p)$ from $(g^a \operatorname{rem} p)$ and $(g^b \operatorname{rem} p)$, using an oracle which gives only the $\lceil n^{1/2} \rceil + \lceil \log n \rceil$ most significant bits of the Diffie-Hellman key.

Unfortunately the proof of Theorem 2 in [1] is not quite correct. Indeed, in order to apply Theorem 1 of that paper to $h = g^b$ this element must be a primitive root of $\mathbb{F}_p$. Thus the proof of Theorem 2 of [1] is valid only if $\gcd(b, p-1) = 1$ (of course the same result holds in the case $\gcd(a, p-1) = 1$ as well). However, even in the most favourable case when $l = (p-1)/2$ is prime, only 75% of pairs (a, b) satisfy this condition. Certainly breaking a cryptosystem in 75% of the cases is already bad enough (even in 0.75% is) but unfortunately for the attacker (using the above oracle), these weak cases can easily be described and avoided by the communicating parties. The proof of Theorem 3 of [1] suffers from a similar problem.

Here we use new bounds of exponential sums from [7] to extend some results of [1] to the case of elements g of arbitrary multiplicative order T, provided that $T \geq p^{1/3+\varepsilon}$. This allows us to prove that the statement of Theorem 2 of [1] holds for all pairs (a, b). We also prove that for almost all primes p similar results hold already for $T \geq p^{\varepsilon}$.

A survey of similar results for other functions of cryptographic interest has recently been given in [4].

Throughout the paper the implied constants in symbols 'O' may occasionally, where obvious, depend on the small positive parameter ε and are absolute otherwise; they all are effective and can be explicitly evaluated.

2. Distribution of g^x Modulo p

For integers λ, r and h let us denote by $N_{\lambda,g,p}(r, h)$ the number of $x \in [0, T-1]$ for which $(\lambda g^x \operatorname{rem} p) \in [r+1, r+h]$.

We need the following asymptotic formula which shows that $N_{\lambda,g,p}(r, h)$ is close to its expected value Th/p, provided that T is of larger order than $p^{1/3}$.

Lemma 2.1. *For any $\varepsilon > 0$ there exists $\delta > 0$ such that for any element $g \in \mathbb{F}_p$ of multiplicative order $T \geq p^{1/3+\varepsilon}$ the bound*

$$\max_{0 \leq r,h \leq p-1} \max_{\gcd(\lambda,p)=1} \left| N_{\lambda,g,p}(r, h) - \frac{Th}{p} \right| = O\left(T^{1-\delta}\right)$$

holds.

Proof. We remark that $N_{\lambda,g,p}(r, h)$ is the number of solutions $x \in \{0, \ldots, T-1\}$ of the congruence

$$\lambda g^x \equiv y \pmod{p}, \qquad y = r+1, \ldots, r+h.$$

Using the identity (see Exercise 11.a in Chapter 3 of [19])

$$\sum_{c=0}^{p-1} \exp\left(2\pi i c u / p\right) = \begin{cases} 0, & \text{if } u \not\equiv 0 \pmod{p}; \\ p, & \text{if } u \equiv 0 \pmod{p}; \end{cases}$$

we obtain

$$N_{\lambda,g,p}(r, h) = \frac{1}{p} \sum_{x=0}^{T-1} \sum_{y=r+1}^{r+h} \sum_{c=0}^{p-1} \exp\left(2\pi i c \left(\lambda g^x - y\right)/p\right)$$

$$= \frac{1}{p} \sum_{c=0}^{p-1} \sum_{x=0}^{T-1} \exp\left(2\pi i c \lambda g^x / p\right) \sum_{y=r+1}^{r+h} \exp\left(-2\pi i c y / p\right).$$

Separating the term Th/p corresponding to $c = 0$ we obtain

$$\left| N_{\lambda,g,p}(r,h) - \frac{Th}{p} \right| \leq \frac{1}{p} \sum_{c=1}^{p-1} \left| \sum_{x=0}^{T-1} \exp\left(2\pi i c \lambda g^x/p\right) \right| \left| \sum_{y=r+1}^{r+h} \exp\left(-2\pi i c y/p\right) \right|$$

$$= \frac{1}{p} \sum_{c=1}^{p-1} \left| \sum_{x=0}^{T-1} \exp\left(2\pi i c \lambda g^x/p\right) \right| \left| \sum_{y=r+1}^{r+h} \exp\left(2\pi i c y/p\right) \right|.$$

We estimate the sum over x by using the bound

$$\max_{\gcd(c,p)=1} \left| \sum_{x=0}^{T-1} \exp\left(2\pi i c g^x/p\right) \right| = O\left(B(T,p)\right), \tag{1}$$

where

$$B(T,p) = \begin{cases} p^{1/2}, & \text{if } T \geq p^{2/3}; \\ p^{1/4}T^{3/8}, & \text{if } p^{2/3} > T \geq p^{1/2}; \\ p^{1/8}T^{5/8}, & \text{if } p^{1/2} > T \geq p^{1/3}; \end{cases} \tag{2}$$

which is essentially Theorem 3.4 of [7]. Using the estimate

$$\max_{0 \leq r,h \leq p-1} \sum_{c=1}^{p-1} \left| \sum_{y=r+1}^{r+h} \exp\left(2\pi i c y/p\right) \right| = O(p \log p),$$

see Exercise 11.c in Chapter 3 of [19], we obtain

$$\max_{0 \leq r,h \leq p-1} \left| N_{\lambda,g,p}(r,h) - \frac{Th}{p} \right| = O\left(B(T,p) \log p\right).$$

It is easy to see that for any $\varepsilon > 0$ there exists $\delta > 0$ such that $B(T,p) = O(T^{1-2\delta})$ for $T \geq p^{1/3+\varepsilon}$ and the result follows. $\square$

In the next statement we show that for almost all primes the lower bound $T \geq p^{1/3+\varepsilon}$ can be brought down to $T \geq p^\varepsilon$.

Lemma 2.2. *Let Q be a sufficiently large integer. For any $\varepsilon > 0$ there exists $\delta > 0$ such that for all primes $p \in [Q, 2Q]$, except at most $Q^{5/6+\varepsilon}$ of them, and any element $g \in \mathbb{F}_p$ of multiplicative order $T \geq p^\varepsilon$ the bound*

$$\max_{0 \leq r,h \leq p-1} \max_{\gcd(\lambda,p)=1} \left| N_{\lambda,g,p}(r,h) - \frac{Th}{p} \right| = O\left(T^{1-\delta}\right)$$

holds.

Proof. The proof is analogous to the proof of Lemma 2.1 using in this case Theorem 5.5 of [7] instead of (1) and (2). For each prime $p \equiv 1 \pmod{T}$ we fix an element $g_{p,T}$ of multiplicative order T. Then Theorem 5.5 of [7] claims that for any $U > 1$ and any integer $\nu \geq 2$, for all primes $p \equiv 1 \pmod{T}$ except at most $O(U/\log U)$ of them, the bound

$$\max_{\gcd(c,p)=1} \left| \sum_{x=0}^{T-1} \exp\left(2\pi i c g_{p,T}^x/p\right) \right| = O\left(Tp^{1/2\nu^2}\left(T^{-1/\nu} + U^{-1/\nu^2}\right)\right),$$

holds. We remark that the value of the above exponential sum does not depend on the particular choice of the element $g_{p,T}$.

Taking

$$\nu = \left\lfloor \frac{1}{\varepsilon} \right\rfloor + 1 \qquad \text{and} \qquad U = Q^{1/2+\varepsilon/2},$$

after simple computation we obtain that there exists some $\delta > 0$, depending only on ε, such that for any fixed $T \geq Q^{\varepsilon/2}$ the bound

$$\max_{\gcd(c,p)=1} \left| \sum_{x=0}^{T-1} \exp\left(2\pi i c g_{p,T}^x/p\right) \right| = O\left(T^{1-2\delta}\right), \tag{3}$$

holds for all except $O(Q^{1/2+\varepsilon/2})$ primes $p \equiv 1 \pmod{T}$ in the interval $p \in [Q, 2Q]$. As it follows from (1) and (2), a similar bound also holds for $T \geq Q^{1/3+\varepsilon/2}$. So the total number of exceptional primes p for which (3) does not hold for at least one $T \geq p^\varepsilon > Q^{\varepsilon/2}$ is $O\left(Q^{5/6+\varepsilon}\right)$.

Using the bound (3) in the same way as we have used (1) and (2) in the proof of Lemma 2.1 we derive the desired result.   □

Certainly in both Lemma 1 and Lemma 3 the dependence of δ on ε can be made explicit (as a linear function of ε).

3. Lattices

As in [1], our results rely on rounding techniques in lattices. We therefore review a few related results and definitions.

Let $\{\mathbf{b}_1, \ldots, \mathbf{b}_s\}$ be a set of linearly independent vectors in $\mathbb{R}^s$. The set of vectors

$$L = \{\mathbf{z} \ : \ \mathbf{z} = \sum_{i=1}^{s} t_i \mathbf{b}_i, \quad t_1, \ldots, t_s \in \mathbb{Z}\}$$

is called an s-dimensional full rank lattice. The set $\{\mathbf{b}_1, \ldots, \mathbf{b}_s\}$ is called the *basis* of L.

It has been remarked in in Section 2.1 of [12] and then in Section 2.4 of [13] that the following statement holds which is somewhat stronger than that usually used in the literature.

Lemma 3.1. *There exists a polynomial time algorithm which, for given a lattice L and a vector $\mathbf{r} = (r_1, \ldots, r_s) \in \mathbb{R}^s$, finds a lattice vector $\mathbf{v} = (v_1, \ldots, v_s)$ satisfying* *the inequality*

$$\sum_{i=1}^{s} (v_i - r_i)^2$$

$$\leq \exp\left(O\left(\frac{s \log^2 \log s}{\log s}\right)\right) \min\left\{\sum_{i=1}^{s} (z_i - r_i)^2, \quad \mathbf{z} = (z_1, \ldots, z_s) \in L\right\}.$$

Proof. The statement is a combination of the Schnorr modification [16] of the lattice basis reduction algorithm of Lenstra, Lenstra and Lovász [9] with a result of Kannan [6] about reduction of the closest vector problem to the shortest vector problem. □

For integers $x_1, \ldots, x_d$, selected in the interval $[0, T - 1]$, we denote by $L_{g,p}(x_1, \ldots, x_d)$ the $d + 1$-dimensional lattice generated by the rows of the following $(d + 1) \times (d + 1)$-matrix

$$\begin{pmatrix} p & 0 & 0 & \ldots & 0 & 0 \\ 0 & p & 0 & \ldots & 0 & 0 \\ & & \vdots & & & \vdots \\ 0 & 0 & 0 & \ldots & p & 0 \\ t_1 & t_2 & t_3 & \ldots & t_d & 1/p \end{pmatrix} \tag{4}$$

where $t_i = (g^{x_i} \operatorname{rem} p)$, $i = 1, \ldots, d$.

The following result is a generalization of Theorem 5 of [1] (which corresponds to the case $T = p - 1$).

Lemma 3.2. *Let $d = 2 \lceil n^{1/2} \rceil$ and $\mu = n^{1/2}/2 + 3$. Let α be a fixed integer in the interval $[0, p - 1]$. For any $\varepsilon > 0$, sufficiently large p, and any element $g \in \mathbb{F}_p$ of multiplicative order $T \geq p^{1/3+\varepsilon}$ the following statement holds. Choose integers $x_1, \ldots, x_d$ uniformly and independently at random in the interval $[0, T - 1]$. Then with probability $P \geq 1 - 2^{-n^{1/2}}$ for any vector $\mathbf{u} = (u_1, \ldots, u_d, 0)$ with*

$$\left(\sum_{i=1}^{d} ((\alpha g^{x_i} \operatorname{rem} p) - u_i)^2\right)^{1/2} \leq p 2^{-\mu},$$

all vectors $\mathbf{v} = (v_1, \ldots, v_d, v_{d+1}) \in L_{g,p}(x_1, \ldots, x_d)$ satisfying

$$\left(\sum_{i=1}^{d} (v_i - u_i)^2\right)^{1/2} \leq p 2^{-\mu},$$

are of the form

$$\mathbf{v} = ((\beta g^{x_1} \operatorname{rem} p), \ldots, (\beta g^{x_d} \operatorname{rem} p), \beta/p)$$

with some $\beta \equiv \alpha \pmod{p}$.

Proof. As in [1] we define the modular distance between two integers β and γ as

$$\text{dist}_p(\beta, \gamma) = \min_{b \in \mathbb{Z}} |\beta - \gamma - bp| = \min \{((\beta - \gamma) \operatorname{rem} p), p - ((\beta - \gamma) \operatorname{rem} p)\}.$$

Let x be an integer chosen uniformly at random in the interval $[0, T-1]$. It follows from Lemma 2.1 that for any β and γ with $\beta \not\equiv \gamma \pmod{p}$ the probability $P(\beta, \gamma)$ of

$$\text{dist}_p(\beta g^x, \gamma g^x) > p2^{-\mu+1}$$

for an integer x chosen uniformly at random in the interval $[0, T-1]$ is

$$P(\beta, \gamma) = 1 - 2^{-\mu+2} + O\left(T^{-\delta}\right)$$

for some $\delta > 0$, depending only on ε. Hence

$$P(\beta, \gamma) \geq 1 - \frac{5}{2^\mu}$$

provided that p is large enough.

Therefore, for any $\beta \not\equiv \alpha \pmod{p}$,

$$\Pr\left[\exists i \in [1, d] \mid \text{dist}_p(\beta g^{x_i}, \alpha g^{x_i}) > p2^{-\mu+1}\right] = 1 - (1 - P(\alpha, \beta))^d \geq 1 - \left(\frac{5}{2^\mu}\right)^d,$$

where the probability is taken over integers $x_1, \ldots, x_d$ chosen uniformly and independently at random in the interval $[0, T-1]$.

Since for $\beta \not\equiv \alpha \pmod{p}$ there are only $p-1$ possible values for $(\beta \operatorname{rem} p)$, we obtain

$$\Pr\left[\forall \beta \not\equiv \alpha \pmod{p}, \exists i \in [1, d] \mid \text{dist}_p(\beta g^{x_i}, \alpha g^{x_i}) > p2^{-\mu+1}\right]$$

$$\geq 1 - (p-1)\left(\frac{5}{2^\mu}\right)^d > 1 - 2^{-n^{1/2}}$$

because

$$d(\mu - \log 5) > \left\lceil n^{1/2}\right\rceil n^{1/2} + 2\left\lceil n^{1/2}\right\rceil (3 - \log 5) > \log p + n^{1/2}.$$

The rest of the proof is identical to the proof of Theorem 5 of [1], we outline it for the sake of completeness.

Let us fix some integers $x_1, \ldots, x_d$ with

$$\min_{\beta \not\equiv \alpha \pmod{p}} \min_{i \in [1,d]} \text{dist}_p(\beta g^{x_i}, \alpha g^{x_i}) > p2^{-\mu+1}. \tag{5}$$

Let $\mathbf{v}$ be a lattice point satisfying

$$\left(\sum_{i=1}^{d} (v_i - u_i)^2\right)^{1/2} \leq p2^{-\mu}.$$

Clearly, since $\mathbf{v} \in L_{g,p}(x_1, \ldots, x_d)$, there are integers $\beta, z_1, \ldots, z_d$ such that

$$\mathbf{v} = (\beta t_1 - z_1 p, \ldots, \beta t_d - z_d p, \beta/p),$$

where, as in (4), $t_i = (g^{x_i} \operatorname{rem} p)$, $i = 1, \ldots, d$.

If $\beta \equiv \alpha \pmod{p}$, then for all $i = 1, \ldots, d$ we have $\beta t_i - z_i p = (\beta t_i \operatorname{rem} p)$, for otherwise there would be $j \in \{1, \ldots, d\}$ so that $|v_j - u_j| > p2^{-\mu}$.

Now suppose that $\beta \not\equiv \alpha \pmod{p}$. In this case we have

$$\left(\sum_{i=1}^{d} (v_i - u_i)^2 \right)^{1/2} \geq \min_{i \in [1,d]} \operatorname{dist}_p(\beta t_i, u_i)$$

$$\geq \min_{i \in [1,d]} \left(\operatorname{dist}_p(\beta t_i, \alpha t_i) - \operatorname{dist}_p(u_i, \alpha t_i) \right)$$

$$> p2^{-\mu+1} - p2^{-\mu} = p2^{-\mu}$$

that contradicts to our assumption. As we have seen, the condition (5) holds with probability exceeding $1 - 2^{-n^{1/2}}$ and the result follows.   □

For an integer $k \geq 1$ we define $f_k(t)$ by the inequalities

$$(f_k(t) - 1)\frac{p}{2^k} \leq (t \operatorname{rem} p) < f_k(t)\frac{p}{2^k}.$$

Thus, roughly speaking, $f_k(t)$ is the integer defined by the k most significant bits of $(t \operatorname{rem} p)$.

Using Lemma 3.2 in the same way as Theorem 5 is used in the proof of Theorem 1 of [1] we obtain

Lemma 3.3. *Let $d = 2 \lceil n^{1/2} \rceil$ and $k = \lceil n^{1/2} \rceil + \lceil \log n \rceil$. For any $\varepsilon > 0$, sufficiently large p and any element $g \in \mathbb{F}_p$ of multiplicative order $T \geq p^{1/3+\varepsilon}$, there exists a deterministic polynomial time algorithm $\mathcal{A}$ such that for any integer $\alpha \in [1, p-1]$ given $2d$ integers*

$$t_i = (g^{x_i} \operatorname{rem} p) \qquad and \qquad s_i = f_k(\alpha t_i), \qquad i = 1, \ldots, d,$$

its output satisfies

$$\Pr_{x_1, \ldots, x_d \in [0, T-1]} [\mathcal{A}(t_1, \ldots, t_d; s_1, \ldots, s_d) = \alpha] \geq 1 - 2^{-n^{1/2}}$$

if $x_1, \ldots, x_d$ are chosen uniformly and independently at random in the interval $[0, T-1]$.

Proof. We follow the same arguments as in the proof Theorem 1 of [1] which we briefly outline here for the sake of completeness. We refer to the first d vectors in the defining matrix of $L_{g,p}(x_1, \ldots, x_d)$ as p-vectors.

Let us consider the vector $\mathbf{r} = (r_1, \ldots, r_d, r_{d+1})$ where

$$r_i = s_i \frac{p}{2^k}, \quad i = 1, \ldots, d, \qquad and \qquad r_{d+1} = 0.$$

Multiplying the last row vector $(t_1, \ldots, t_d, 1/p)$ of the matrix (4) by α and subtracting certain multiples of p-vectors, we obtain a lattice point

$$\mathbf{u}_\alpha = (u_1, \ldots, u_d, \alpha/p) \in L_{g,p}(x_1, \ldots, x_d)$$

such that

$$|u_i - r_i| < p2^{-k}, \qquad i = 1, \ldots, d.$$

Therefore,

$$\left(\sum_{i=1}^{d+1}(u_i-r_i)^2\right)^{1/2} \le p(d+1)^{1/2}2^{-k}.$$

Now we can use Lemma 3.1 (with a slightly rougher constant $2^{(d+1)/4}$) to find in polynomial time a lattice vector $\mathbf{v}=(v_1,\dots,v_d,v_{d+1})\in L_{g,p}(x_1,\dots,x_d)$ such that

$$\left(\sum_{i=1}^{d}(v_i-r_i)^2\right)^{1/2}$$

$$\le 2^{(d+1)/4}\min\left\{\left(\sum_{i=1}^{d+1}(z_i-r_i)^2\right)^{1/2}, \quad \mathbf{z}=(z_1,\dots,z_d,z_{d+1})\in L\right\}$$

$$\le 2^{(d+1)/4}p(d+1)^{1/2}2^{-k}\le p2^{-\mu-1},$$

where $\mu=n^{1/2}/2+3$, provided that n is sufficiently large. We also have

$$\left(\sum_{i=1}^{d}(u_i-r_i)^2\right)^{1/2}\le pd^{1/2}2^{-k}\le p2^{-\mu-1}.$$

Therefore

$$\left(\sum_{i=1}^{d}(u_i-v_i)^2\right)^{1/2}\le p2^{-\mu}.$$

Applying Lemma 3.2, we see that $\mathbf{v}=\mathbf{u}_\alpha$ with probability at least $1-2^{-n^{1/2}}$, and therefore, α can be recovered in polynomial time. $\square$

Accordingly, using Lemma 2.2 instead of Lemma 2.1, in a similar way we obtain that for almost all primes much smaller values of T can be considered.

Lemma 3.4. *Let Q be a sufficiently large integer. For any $\varepsilon>0$ there exists $\delta>0$ such that for all primes $p\in[Q,2Q]$, except at most $Q^{5/6+\varepsilon}$ of them, and any element $g\in\mathbb{F}_p$ of multiplicative order $T\ge p^\varepsilon$ there exists a deterministic polynomial time algorithm $\mathcal{A}$ such that for any integer $\alpha\in[1,p-1]$ given $2d$ integers*

$$t_i=(g^{x_i}\operatorname{rem}p)\quad and\quad s_i=f_k(\alpha t_i),\quad i=1,\dots,d,$$

its output satisfies

$$\Pr_{x_1,\dots,x_d\in[0,T-1]}[\mathcal{A}(t_1,\dots,t_d;s_1,\dots,s_d)=\alpha]\ge 1-2^{-n^{1/2}}$$

if $x_1,\dots,x_d$ are chosen uniformly and independently at random in the interval $[0,T-1]$.

4. Security of the Most Significant Bits of the Diffie-Hellman Key

We are ready to prove the main results.

Let $g \in \mathbb{F}_p$ be multiplicative order T. probabilistic polynomial time algorithm which for any pair $(a, b) \in [0, T-1]^2$, For each integer k define the oracle $\mathcal{O}_{k,g}$ as a 'black box' which, any pair $(x, y) \in [0, T-1]^2$, given the values of $X = (g^x \operatorname{rem} p)$ and $Y = (g^y \operatorname{rem} p)$ outputs the value of $f_k(g^{xy})$.

Theorem 4.1. *Let $k = \lceil n^{1/2} \rceil + \lceil \log n \rceil$. For any $\varepsilon > 0$, sufficiently large p and any element $g \in \mathbb{F}_p$ of multiplicative order $T \geq p^{1/3+\varepsilon}$, there exists a probabilistic polynomial time algorithm which for any pair $(a, b) \in [0, T-1]^2$, given the values of $A = (g^a \operatorname{rem} p)$ and $B = (g^b \operatorname{rem} p)$, makes $O(n^{1/2})$ calls of the oracle $\mathcal{O}_{k,g}$ and computes $(g^{ab} \operatorname{rem} p)$ correctly with probability $1 + O\left(2^{-n^{1/2}}\right)$.*

Proof. Given a pair $(a, b) \in [0, T-1]^2$ let us select an integer $r \in [0, T-1]$ uniformly at random. We compute

$$g_r = (Bg^r \operatorname{rem} p)$$

thus $g_r \equiv g^{b+r} \pmod{p}$.

The probability that $\gcd(b+r, T) \geq Tp^{-1/3-\varepsilon/3}$ is at most $\tau(T)T^{-1}p^{1/3+\varepsilon/3}$ where $\tau(T)$ is the number of positive integer divisors of T. Indeed, for any divisor $D|T$ with $D \geq Tp^{-1/3-\varepsilon/3}$ there are at most $T/D \leq p^{1/3+\varepsilon/3}$ values of $s \in [0, T-1]$ with $\gcd(s, T) = D$.

Using the bound $\tau(T) = O(T^{\varepsilon/3})$, see Theorem 5.2 of Chapter 1 of [15], we obtain that the probability of $\gcd(b+r, T) \geq Tp^{-1/3-\varepsilon/3}$ is at most

$$O\left(T^{-1}p^{1/3+2\varepsilon/3}\right) = O\left(p^{-\varepsilon/3}\right) = O\left(2^{-n^{1/2}}\right).$$

In the opposite case, when $\gcd(a+r, T) \leq Tp^{-1/3-\varepsilon/3}$, the multiplicative order of g_r is

$$T_r = \frac{T}{\gcd(b+r, T)} \geq p^{1/3+\varepsilon/3}.$$

Let $\alpha_r \equiv g^{a(b+r)} \pmod{p}$. Then

$$f_k(\alpha_r g_r^x) = f_k\left(g_r^{(a+x)}\right) = f_k\left(g^{(a+x)(b+r)}\right).$$

Now we can use the oracle $\mathcal{O}_{k,g}$ with $(g^x A \operatorname{rem} p)$ and $(g^r B \operatorname{rem} p)$ to evaluate $f_k(\alpha_r g_r^x)$ for an integer x chosen uniformly at random in the interval $[0, p-1]$. Because $T_r | p - 1$ the values of $(x \operatorname{rem} T_r)$ are uniformly distributed in the interval $[0, T_r - 1]$ as well, thus Lemma 3.3 can be applied. Therefore, one can construct a probabilistic polynomial time algorithm that:

- Selects a random $r \in [0, T-1]$.
- Applies algorithm $\mathcal{A}$ from Lemma 3.3 (now g_r plays the role of g in the conditions of Lemma 3.3). This algorithm makes $O(n^{1/2})$ calls to the oracle $\mathcal{O}_{k,g}$.

- Outputs the correct value of α_r with probability at least $1 - O\left(2^{-n^{1/2}}\right)$.

Indeed, the only possible source of error is either the case $T_r \leq p^{1/3+\varepsilon/3}$ or the probability error of the algorithm of Lemma 3.3. The probability of both events is $O\left(2^{-n^{1/2}}\right)$.

Remarking that

$$g^{ab} \equiv \alpha_r A^{-r} \pmod{p},$$

we obtain the desired result. $\qquad\square$

It is easy to see that Theorem 4.1 is nontrivial for any $T \geq p^{1/3+\varepsilon}$. In a similar way, Lemma 2.2 produces a result which holds for almost all primes p and is non-trivial for $T \geq p^\varepsilon$.

Theorem 4.2. *Let* $k = \lceil n^{1/2} \rceil + \lceil \log n \rceil$. *For any* $\varepsilon > 0$ *and for all primes* $p \in [2^{n-1}, 2^n - 1]$, *except at most* $2^{(5/6+\varepsilon)n}$ *of them, and any element* $g \in \mathbb{F}_p$ *of multiplicative order* $T \geq p^\varepsilon$ *the following statement holds: There exists a probabilistic polynomial time algorithm which for any pair* $(a,b) \in [0, T-1]^2$, *given the values of* $A = (g^a \operatorname{rem} p)$ *and* $B = (g^b \operatorname{rem} p)$, *makes* $O\left(n^{1/2}\right)$ *calls of the oracle* $\mathcal{O}_{k,g}$ *and computes* $(g^{ab} \operatorname{rem} p)$ *correctly with probability* $1 + O\left(2^{-n^{1/2}}\right)$.

5. Remarks

First of all we note that the constants in above estimates are effective and can be explicitly evaluated.

We have not used the full power of Lemma 3.1 but rather we have applied it with the same constant as in [1]. It is easy to see that in fact the results of [1] as well as our results hold with some

$$k = O\left(\frac{n^{1/2} \log\log n}{\log^{1/2} n}\right)$$

and a slightly large number of oracle calls.

It would be very interesting to replace the condition $T \geq p^\varepsilon$ for the smallest size of the multiplicative order of g in Lemma 2.2 by a weaker condition of the form $T \geq (\log p)^c$ with some constant c. Although a more careful analysis of the proof of Theorem 5.5 of [7] should allow to replace p^ε with a slower growing function, it seems unlikely that the present method can be applied to T as small as a power of $\log p$.

Our results can also be applied to several other cryptosystems based on exponentiation in finite fields, which have been considered in [1], except the *Shamir message passing scheme*, see [1, 2] (this scheme is also described in Protocol 12.22 in [11]). Unfortunately the proof of Theorem 3 in [1] suffers from the same problem as the proof of Theorem 2 of that paper. Namely, for the ElGamal scheme, see [1, 2] as well as Section 8.4 from [11], it produces a result which applies only to at most 50% of the cases and it cannot be applied to the the Shamir message passing scheme at all. Indeed, in this scheme the exponent x of the corresponding multiplier g^x must satisfy the additional condition $\gcd(bx + 1, p - 1) = 1$, with some b, $\gcd(b, p - 1) = 1$, thus g^x runs through some special subset of $\mathbb{F}_p^*$ (even if g is a primitive root) rather than through the whole $\mathbb{F}_p^*$ and thus Theorem 1 of [1] does not apply. Our results in their present form cannot be used for this problem directly, however it has been shown in [5] that a modification of the technique of this paper, combined with some elementary sieve method produce similar results for the Shamir message passing scheme.

Besides the mentioned in [1, 2] cryptosystems several other schemes can be studied as well. For example, very similar results hold for the Matsumoto-Takachima-Imai key-agreement protocol, see Section 12.6 of [11].

The results of [2] can be generalized in a similar way. To do so one can use the bound of exponential sums of Theorem 3.4 of [7] to study the distribution of the sums $(g^{x_1} + \ldots + g^{x_r} \operatorname{rem} p)$ and thus obtain an analogue of Lemma 2.4 of [2].

One can also extend Theorem 4.1 to the case of Diffie-Hellman encryption modulo an arbitrary composite integer $m \geq 2$. Indeed, using the well-known bound

$$\max_{\gcd(c,m)=1} \left| \sum_{x=0}^{T-1} \exp\left(2\pi i c g^x / m\right) \right| \leq m^{1/2},$$

see Theorem 10 of Chapter 1 in [8] or Theorem 8.2 in [14], instead of (1) and (2), one can obtain similar results for elements g, $\gcd(g, m) = 1$, of multiplicative order T modulo m such that $T \geq m^{1/2+\varepsilon}$. In fact, Lemma 3.3 can be extended to elements t_i chosen uniformly and independently at random from any subgroup $\mathcal{G}$ of the group of units modulo m, provided that the cardinality of $\mathcal{G}$ satisfies $\#\mathcal{G} \geq m^{1/2+\varepsilon}$.

As we have mentioned, similar but somewhat more involved technique can be applied to studying the bit security of the Shamir message passing scheme, see [5].

Finally, we remark that somewhat similar problem for extensions of finite fields have been considered in [18]. The results of that paper and some of their improvements in [17] have applications to the security of the new cryptosystem designed in [3, 10].

Acknowledgement. We thank Consuelo Martínez for her interest and for helpful advice. We also thank Mats Näslund for a careful reading of the manuscript and fruitful discussion.

References

[1] D. Boneh and R. Venkatesan, *Hardness of computing the most significant bits of secret keys in Diffie-Hellman and related schemes*, Lect. Notes in Comp. Sci., Springer-Verlag, Berlin, **1109** (1996), 129–142.

[2] D. Boneh and R. Venkatesan, *Rounding in lattices and its cryptographic applications*, Proc. 8-rd Annual ACM-SIAM Symp. on Discr. Algorithms, ACM, NY, 1997, 675–681.

[3] A. E. Brouwer, R. Pellikan, and E. R. Verheul, *Doing more with fewer bits,* Lect. Notes in Comp. Sci., Springer-Verlag, Berlin, **1716**, (1999), 321–332.

[4] M. I. González Vasco and M. Näslund, *A survey of hard core functions,* Proceedings of the Workshop on Comp. Number Theory and Cryptography, Birkhäuser, 2000.

[5] M. I. González Vasco and I. E. Shparlinski, *Security of the most significant bits of the Shamir message passing scheme,* Math. Comp., (to appear).

[6] R. Kannan, *Algorithmic geometry of numbers,* Annual Review of Comp. Sci., **2** (1987), 231–267.

[7] S. V. Konyagin and I. E. Shparlinski, *Character sums with exponential functions and their applications,* Cambridge Univ. Press, Cambridge, 1999.

[8] N. M. Korobov, *Exponential sums and their applications,* Kluwer Acad. Publ., Dordrecht, 1992.

[9] A. K. Lenstra, H. W. Lenstra and L. Lovász, *Factoring polynomials with rational coefficients,* Mathematische Annalen, **261** (1982), 515–534.

[10] A. K. Lenstra and E. R. Verheul, *The XTR public key system,* Lect. Notes in Comp. Sci., Springer-Verlag, Berlin, **1880** (2000), 1–19.

[11] A. J. Menezes, P. C. van Oorschot and S. A. Vanstone, *Handbook of Applied Cryptography,* CRC Press, Boca Raton, FL, 1996.

[12] D. Micciancio, *On the hardness of the shortest vector problem,* PhD Thesis, MIT, 1998.

[13] P. Nguyen and J. Stern, *Lattice reduction in cryptology: An update,* Lect. Notes in Comp. Sci., Springer-Verlag, Berlin, **1838** (2000), 85–112.

[14] H. Niederreiter, *Quasi-Monte Carlo methods and pseudo-random numbers,* Bull. Amer. Math. Soc., **84** (1978), 957–1041.

[15] K. Prachar, *Primzahlverteilung,* Springer-Verlag, Berlin, 1957.

[16] C. P. Schnorr, *A hierarchy of polynomial time basis reduction algorithms,* Theor. Comp. Sci., **53** (1987), 201–224.

[17] I. E. Shparlinski, *Security of polynomial transformations of the Diffie-Hellman key,* Preprint, 2000.

[18] E. R. Verheul, *Certificates of recoverability with scalable recovery agent security,* Lect. Notes in Comp. Sci., Springer-Verlag, Berlin, **1751** (2000), 258–275.

[19] I. M. Vinogradov, *Elements of number theory,* Dover Publ., New York, 1954.

Department of Mathematics, University of Oviedo,
Oviedo, 33007, Spain
E-mail address: mvasco@orion.ciencias.uniovi.es

Department of Computing, Macquarie University,
Sydney, NSW 2109, Australia
E-mail address: igor@comp.mq.edu.au

Progress in Computer Science and Applied Logic, Vol. 20
© 2001 Birkhäuser Verlag Basel/Switzerland

Polynomial Rings and Efficient Public Key Authentication II

Jeffrey Hoffstein and Joseph H. Silverman

Abstract. In a recent paper [3] a highly efficient public key authentication scheme called PASS was introduced. In this paper we show how a small modification in the scheme cuts the size of the public key and the commitment in half while reducing an already minimal computational load.

1. Introduction

In a recent paper [3], a new highly efficient scheme for public key authentication and digital signatures called PASS was introduced. The ideas underlying PASS are related to the ideas originating in [1] and [2]. Each of these three papers used a combination of algebraic and analytic techniques in the context of a commutative ring

$$R = (\mathbb{Z}/q\mathbb{Z})[x]/(x^N - 1), \tag{1}$$

where q and N are moderately sized relatively prime integers.

In order to avoid excessive duplication of exposition, we will assume some familiarity with the previous paper [3]. We will, however, repeat some definitions and concepts when it appears that this would be useful. Thus this paper should be readable without reference to [3].

The general idea in the earlier paper [3] is as follows. Pearl, the prover, wishes to prove her identity to Vinnie, the verifier. Pearl has a secret key (f, f') consisting of a pair of "short" polynomials in R, that is, having coefficients 1, −1, and 0. Pearl's public key is the collection of values $\{f(\alpha), f'(\alpha)\}_{\alpha \in S}$, where α varies over a set S consisting of half the numbers modulo q.

To identify herself, Pearl randomly picks a pair (g, g') of short polynomials in R. She keeps (g, g') secret, but as her commitment, Pearl reveals the set of values $\{g(\alpha), g'(\alpha)\}_{\alpha \in S}$, the collection of values of g and g' at the points in S. The verifier Vinnie sends Pearl a challenge c_0 that Pearl hashes with the commitment to produce a 4-tuple of extremely short polynomials (c_1, c_2, c_3, c_4). Pearl computes and reveals the polynomial

$$h = c_1 * f * g + c_2 * f * g' + c_3 * f' * g + c_4 * f' * g'.$$

(Note all polynomial multiplications take place in the ring R.)

In order to verify Pearl's identity, Vinnie first checks that h is fairly short, and second he checks that the identity

$$h(\alpha) = c_1(\alpha)f(\alpha)g(\alpha) + c_2(\alpha)f(\alpha)g'(\alpha) + c_3(\alpha)f'(\alpha)g(\alpha) + c_4(\alpha)f'(\alpha)g'(\alpha)$$

is true for all $\alpha \in S$. If h passes both of these tests, then Vinnie accepts Pearl's proof of identity, that is, that she has knowledge of the secret short polynomial pair (f, f').

The security of this scheme, and the scheme described in this paper, rests on two hard problems:

Hard Problem 1.1. *Let L be a "reasonably random" lattice of dimension n and discriminant d. Let s be the length of the actual shortest non-zero vector of L and suppose that*

$$C_1 d^{1/n} < s < C_2 d^{1/n}\sqrt{n},$$

where $C_2, C_1 > 0$ are fixed constants. (The gaussian heuristic says that an inequality of this sort will be true.) Suppose further that $d^{1/n}$ lies within a factor of $\sqrt{n}$ of n. Then the time required to find a vector of length less than $C_3 s$, for a fixed positive constant C_3, grows exponentially with n.

Hard Problem 1.2. *Let $f, g \in R$ be chosen to be short, and let $h = f * g$, so h will also be moderately short. With appropriate choices of parameters, it is very difficult to either recover f and g from h, or to find two other polynomials f' and g' such that f' and g' are short and $h = f' * g'$.*

Hard Problem 1.1 is a somewhat refined version of the well-known hard problem of finding a short vector in a lattice of high dimension. Very roughly, if the values of $f \bmod q$ are given at $t < N$ points α in a set S then f corresponds to a point outside a lattice L of dimension N. If f is very short, then f is closer to L than a randomly chosen point should be, and one can ask how easily it can be found through a lattice reduction process. Alternatively, a dimension can be added and f can be related to a shorter than expected vector in a lattice L' of dimension $N + 1$. This polynomial to vector translation is described in Section 3. Hard Problem 1.1 claims that under conditions that correspond to the parameter choices made in PASS and PASS2, the time required to recover f, or to produce a false response h, increases exponentially with N. Further details are given in Section 3.2.

Hard Problem 1.2 is closer to a traditional factorization problem. It's connection with security is discussed in detail in [3].

In this paper we describe a modified version of the above scheme in which the public key and the commitment each consist of a single short polynomial, rather than a pair of short polynomials. This will improve the operating characteristics of the scheme. We call this variation on the PASS scheme PASS2.

The polynomial response h in PASS2 will take a somewhat different form. It is constructed using a pair of challenge polynomials (c_1, c_2), and the check by Vinnie changes to a verification that h is short, followed by a verification that a certain combination of the values $f(\alpha)$, $c(\alpha)$, $g(\alpha)$, and $h(\alpha)$ are squares modulo q for all $\alpha \in S$.

In the following sections we give a precise description of PASS2, propose some specific parameters, and provide security analyses in these cases.

2. The PASS2 Authentication Scheme

2.1. An Outline of the PASS2 Authentication Scheme

We first review some of the PASS notation. Let q be a prime and let $N = q - 1$. A typical element g of R has a representative of the form

$$g = a_0 + a_1 x + a_2 x^2 + \cdots + a_{N-1} x^{N-1}$$

with coefficients $a_i \in \mathbb{Z}/q\mathbb{Z}$. It is useful to define two norms on R. Let g be a polynomial whose coefficients satisfy $|a_i| \leq q/2$ and $\sum_i a_i = 0$. We then define

$$|g|_2 = \sqrt{a_0^2 + \cdots + a_{N-1}^2} \quad \text{and} \quad |g|_\infty = \max a_i - \min a_i. \tag{2}$$

We recall the notion of a short polynomial.

Definition 2.1. *A polynomial f will be called "short" if its norm $|f|_2$ is smaller than a specified constant multiple of $\sqrt{q}$.*

Very roughly, polynomials are called short if their coefficients are sufficiently small with respect to q that no reduction mod q occurs when two of them are multiplied together. We will occasionally find it useful to call a polynomial "moderately" short if its norm is less than a constant times q.

When polynomials are short, the two norms above are related by the rough inequality

$$|g * g'|_\infty \leq c_2 |g|_2 |g'|_2, \tag{3}$$

where c_2 is a constant that varies between 0.3 and 0.5 for parameters in the ranges discussed here. Also we have the approximate relation (for random choices of short polynomials)

$$|g * g'|_2 \approx |g|_2 |g'|_2. \tag{4}$$

The estimate (4) with appropriate choices of constants shows that the product of two short polynomials will have small $| \cdot |_2$ and $| \cdot |_\infty$ norms with respect to q.

For any integer d, we let $\mathcal{L}(d)$ denote the set of polynomials in R that have exactly d coefficients equal to each of 1 and -1, with all other coefficients equal to 0. We fix a set S consisting of $t = N/2$ randomly chosen distinct non-zero elements $\alpha \in \mathbb{Z}/q\mathbb{Z}$. The set S is a system-wide parameter. For technical reasons, we assume that S is chosen so that if $\alpha \in S$, then $\alpha^{-1} \in S$, that is, S is closed under taking inverses.

We further fix four system parameters d_f, d_g, d_c, γ. These are used to define four sets of polynomials:

$$\mathcal{L}_f = \mathcal{L}(d_f), \quad \mathcal{L}_g = \mathcal{L}(d_g), \quad \mathcal{L}_c = \mathcal{L}(d_c), \quad \mathcal{L}_h = \{h \in R : |h|_2 < \gamma q\}.$$

We now describe how an authentication session proceeds in PASS2. Pearl, the prover, has a private key f, known only to her. This private key is chosen by Pearl at random from $\mathcal{L}_f$. Her public key is the associated ordered collection $\{f(\alpha)\}_{\alpha \in S}$ of values mod q. We claim that the following scenario allows Pearl to prove to Vinnie, the verifier, that she possesses the secret key f associated to her public key, without revealing f or information that could help Vinnie, or a third party Irving observing the transaction, to discover f.

- Pearl randomly chooses a commitment $g_1 \in \mathcal{L}_g$ and sends the set of values $\{g_1(\alpha)\}_{\alpha \in S}$ to Vinnie.
- Vinnie chooses an 80 bit challenge c_0 at random and sends c_0 to Pearl. Pearl hashes c_0 with $\{g_1(\alpha)\}_{\alpha \in S}$ to obtain $c_1, c_2 \in \mathcal{L}_c$. Pearl checks that $c_1(\alpha) \not\equiv 0 \mod q$ for all $2 \leq \alpha \leq q - 2$ with $\alpha \notin S$. If this is not the case Pearl rechooses c_1 in a predefined way until c_1 has this property.
- Pearl chooses $g_2 \in \mathcal{L}_g$ and computes and reveals

$$h = (f + c_1 * g_1 + c_2 * g_2) * g_2.$$

- Vinnie verifies that:
  (A) $h \in \mathcal{L}_h$.
  (B) The quantity $(f(\alpha) + c_1(\alpha)g_1(\alpha))^2 + 4c_2(\alpha)h(\alpha)$ is a quadratic residue modulo q for every $\alpha \in S$.

If Pearl passes the two tests, then Vinnie accepts her claim of identity.

Remark 2.2. *One can check that the probability that the c_1 chosen through a hashing process as above will have the desired non-vanishing property is greater than 50%. Thus it will not take long for Pearl to locate a satisfactory c_1.*

As with PASS, or any public key authentication scheme, one must verify, or at least make strong arguments in favor of, several things. First, it must be shown that if Pearl possesses the private key f, than the probability that she will pass the test and be accepted by Vinnie as legitimate can be made arbitrarily high. Second, it must be shown that a potential impostor without knowledge of f or some other false key f' will have a very low probability of passing the test. Finally, it must be shown that even if an impostor knows the public key and has access to an arbitrarily long transcript of genuine authentication transactions using f, he will have a close to zero chance of recovering either the original private key f or an equally useful false key f'.

In the following, we will generally suppress the $*$ in the notation when multiplying polynomials in R.

2.2. Specific Parameter Choices

In this section we give concrete details for the PASS2 scheme described above. Let q be a small prime, for example $q = 769$ or $q = 929$, and let $N = q - 1$. We will explain below why the level of security for $q = 769$ is considerably greater than that of RSA 512, while that of $q = 929$ is greater than RSA 1024.

Let r be a primitive root modulo q, let $t = N/2$, and let J be a collection of t distinct indices j, chosen at random from the collection of integers less than N, with the condition that if $j \in J$, then $q - 1 - j \in J$. Define S by

$$S = \{r^j \bmod q : j \in J\}. \tag{5}$$

Then S consists of t distinct elements $\alpha \bmod q$. As they are non-zero, each has the property that $\alpha^N \equiv 1 \bmod q$. Also, by its definition, S is closed under the taking of multiplicative inverses mod q.

Fix t and a set S with $|S| = t$ as in (5) above. Set the parameters d_f, d_g, d_c, γ as follows:

$$d_f = [1/2 + q/3], \quad d_g = [1/2 + q/6], \quad d_c = 2, \quad \gamma = 1.8. \tag{6}$$

It is simple to check then that for any $q \geq 769$

$$|\mathcal{L}_f| > 2^{160}, \quad |\mathcal{L}_g| > 2^{160}, \quad |\mathcal{L}_c| > 2^{36}, \quad q^t > 2^{160}. \tag{7}$$

In fact these bounds are far exceeded for all spaces except for $\mathcal{L}_c$. Note that the space of challenges is the space of pairs of elements of $\mathcal{L}_c$ and thus has size 2^{72}.

Let us first discuss completeness. We will show that Pearl, knowing the secret key f, can pass Vinnie's test with very high probability.

2.3. On Completeness

Recall how the scenario works. Pearl chooses $g_1 \in \mathcal{L}_g$ and reveals the commitment $\{g_1(\alpha)\}_{\alpha \in S}$. A challenge is sent to Pearl, which she uses to create the pair $c_1, c_2 \in \mathcal{L}_c$. Pearl chooses $g_2 \in \mathcal{L}_g$, then uses her knowledge of f to compute and reveal

$$h = (f + c_1 * g_1 + c_2 * g_2) * g_2.$$

The test $h \in \mathcal{L}_h$ will be passed for the following reason. From (3) and (4), we see that the fact that $|f|_2$, $|g|_2$, $|c_1|_2$, and $|c_2|_2$ are small implies that $|h|_2$ and $|h|_\infty$ must be small. As with PASS, the probability that $|h|_2$ falls into a given range, or that individual coefficients of h fall into given ranges, can be computed theoretically, but it is far easier to do an empirical computation. For example, in the case (6) above with $q = 769$, we found that in $5 \cdot 10^6$ tests of randomly chosen triples (f, g, c_1, c_2) from $\mathcal{L}_f, \mathcal{L}_g, \mathcal{L}_c$,

$$600250 \leq |h|_2{}^2 \leq 1916009$$

for all but one h, for which the value was 1972192. From this we conclude that the probability that $|h|_2 \leq 1.8q$ is roughly $2 \cdot 10^{-7}$ (and even for the exception this inequality held with 1.8 replaced by 1.83). Thus we claim that the probability of a false alarm, that is, that Pearl will fail test (A) despite knowing the secret f, is less than 10^{-6}. If this occurs, the test can simply be repeated, and similarly with a digital signature.

Remark 2.3. *If desired the test can be strengthened by lowering 1.8 to, say, 1.6. Then the chances of a false alarm are somewhat increased, but the security level at a given parameter setting increases dramatically.*

Next consider the test

(B) $(f(\alpha) + c_1(\alpha)g_1(\alpha))^2 + 4c_2(\alpha)h(\alpha)$ is a quadratic residue mod q for every $\alpha \in S$.

This is true because $(f(\alpha) + c_1(\alpha)g_1(\alpha))^2 + 4c_2(\alpha)h(\alpha)$ is a square if and only if the quadratic equation

$$c_2(\alpha)x^2 + (f(\alpha) + c_1(\alpha)g_1(\alpha))x - h(\alpha) \equiv 0 \pmod{q}$$

has a solution. But the construction of h guarantees the existence of a solution, namely $x = g_2(\alpha)$. Thus Pearl will pass this test also and her proof of identity will be accepted by Vinnie.

2.4. Security Discussion

We will now consider the chances that an imposter, Irving, can pretend to be Pearl without knowledge of the secret PASS2 key f. The first few arguments are identical to those in [3]. With the size of the spaces given in (7), the chances of Irving locating f, or an equivalently useful f' by an exhaustive search or meet-in-the-middle attack are, as in PASS, less than 2^{-80}. Since $|\mathcal{L}_c| = 2^{72}$, the chances that a repeat of a previously observed genuine session will help Irving are minimal.

In order to impersonate Pearl, Irving can either choose his h at random satisfying the quadratic constraints and hope that $|h|_2 < 1.8q$, or. Irving can choose h with $|h|_2 < 1.8q$ and hope that h satisfies the quadratic constraints. In the first case, as in PASS, by using Sterling's formula to approximate the volume of an N-sphere one can check that

$$|\mathcal{L}_h| \approx (2\pi e)^{N/2}(1.8q)^N N^{-N/2}. \tag{8}$$

An h chosen to satisfy the quadratic constraints will be uniformly distributed inside a space of volume q^N. Thus by (8) we see that for our parameters, $|\mathcal{L}_h|q^{-N} < 2^{-160}$. This means that this approach will have a less than 2^{-80} chance of success, even including possible meet-in-the-middle off line attacks.

On the other hand, an h picked at random from $\mathcal{L}_h$ will have a 50% chance of satisfying each quadratic constraint, and thus a $2^{-N/2}$ probability of satisfying all of them. For $N > 320$ this is also less than 2^{-160}.

Another potential attack for Irving is to cheat on his choice of g_1, g_2 and pick polynomials far shorter than they should be. In the most extreme case, Irving could choose g_1, g_2 to be simply x^k, x^l for some k, l. If Irving could find a false key f' with $|f'|_2 < 1.8q$ and $f'(\alpha) \equiv f(\alpha)$ mod q for all $\alpha \in S$, then this attack would succeed. The chances of Irving finding such an f' through a random search are covered by (8) above and are less than 2^{-80}. Keys f and f' can also be searched for via lattice reduction methods, which will be discussed below.

2.5. Soundness

We will give a probablistic argument here that for t a bit larger than $N/2$, if Irving can produce a sequence of responses to a single commitment $\{g_1(\alpha)\}_{\alpha \in S}$ and a sequence of challenge pairs c_1, c_2, then he must have knowledge of the secret key f. As in [3], our argument will not be airtight. But we hope it will be convincing.

Suppose that, given $\{g_1(\alpha)\}_{\alpha \in S}$, when confronted by a random challenge pair c_1, c_2 Irving can produce a moderately short polynomial h with the property that

$$(f(\alpha) + c_1(\alpha)g_1(\alpha))^2 + 4c_2(\alpha)h(\alpha)$$

is a quadratic residue mod q for every $\alpha \in S$. It may be the case that Irving does not really have short polynomials g_1, g_2 on hand but has simply selected the collection of values $\{g_1(\alpha)\}_{\alpha \in S}$ by some method. If so, the multiplication by the random c_1 and the inclusion of the random c_2 in the constraint seem to reduce Irving's situation to the general one of finding a moderately short polynomial satisfying a collection of t quadratic constraints. This problem is analysed below in the section on lattice reduction attacks. With high probability there will exist a large number of potential responses h satisfying these constraints. However, the only method available for finding them seems to be lattice reduction methods, and the time estimates for Irving to find a response by this method are quite long.

Let us assume therefore that Irving's response actually has the form

$$h = (F + c_1 g_1 + c_2 g_2)g_2$$

for any challenges c_1, c_2, with g_1, g_2 fixed, but not necessarily short. We also assume that F has the correct values at S but is not necessarily short. If we make the assumption that given any pair of challenges c_1, c_2, Irving has a reasonable chance of producing a valid response, then we should not have to try too many 4-tuples of the form $c_1, c_2, c_1, 1+c_2$ before we find one such that Irving can respond successfully to both c_1, c_2 and $c_1, 1 + c_2$. Then by taking the two responses H and H' to challenge pairs c_1, c_2 and $c_1, 1 + c_2$, Irving can obtain the difference $H' - H = g_2^2$. Unless Irving has solved the problem previously mentioned, of finding general short polynomials whose values are quadratic residues, it is highly probable that $H' - H = g_2^2$ is a square of a short polynomial, that is, that g_2 really must be short. The square root can then be taken, as described in [3], recovering g_2. The short polynomial $c_2 g_2^2$ can then be subtracted from h, yielding a short polynomial $Fg_2 + c_1 g_1 g_2$. Performing this operation with c_1 and $1+c_1$ (while still keeping g_1, g_2 fixed) Irving could obtain the short polynomial Fg_2. This now has the same values at S as fg_2, an actual product of short polynomials. If we knew that $Fg_2 = fg_2$ then Irving could divide by g_2, recovering f. This however, can be seen to be true with high probability, by using the gaussian heuristic as follows.

The difference of the polynomials $H = Fg_2 - fg_2$ satisfies $H(\alpha) \equiv 0 \bmod q$ for all $\alpha \in S$. H is also moderately short, meaning that $|H|_2 < Kq$ for an absolute constant q. The probable existence of a non-zero H satisfying these constraints for large N can be calculated by approximating the volume of an N-sphere using Sterlings formula, as in [3], and applying the gaussian heuristic to a lattice of determinant q^t, described in the next section. One sees that for N large the expected number of such polynomials is on the order of

$$(2\pi e)^{N/2} N^{-N/2} K^N q^N q^{-t}.$$

If $t = N/2 + \epsilon N$ for some small $\epsilon > 0$, then this quantity approaches zero for large N, meaning that with high probability $H = 0$ and $Fg_2 = fg_2$.

3. Lattice Reduction Techniques

Lattice reduction methods can be used by Irving to search for the private key f, or an equally useful false key f'. These methods can also be used in an off line attempt to construct a valid response h to a given challenge. Finally, they can be used in an attempt to recover g_1 from a given commitment and hence f from the corresponding response h. (In fact about 15 different g_1 recoveries would be necessary to recover f.) In this section we will discuss and quantify the difficulty of these questions. First we will discuss an attack on f using the public key $\{f(\alpha)\}_{\alpha \in S}$

3.1. Formulation of a Lattice Attack on the Public Key

This is approached exactly as in [3]. For convenience we will remind the reader of the outline. We begin by constructing a lattice as follows. For any polynomial $F \in R$, associate to F the vector of coefficients $(a_0, a_1, \ldots, a_{N-1})$. Similarly for any such vector or point in $\mathbb{Z}^N$, one can take the polynomial built from these coefficients, reduce mod q, and obtain an $F \in R$.

Let L be the lattice of all points in $\mathbb{Z}^N$ such the corresponding polynomial F satisfies

$$F(\alpha) \equiv 0 \pmod{q} \qquad \text{for each } \alpha \in S.$$

It is easy to check that L is indeed a lattice, and that the determinant of L is equal to q^t.

It is not difficult to find a polynomial $F' \in R$ such that $F'(\alpha) \equiv f(\alpha) \bmod q$ for all $\alpha \in S$. However it is very unlikely that such an F' will have small coefficients. Suppose, instead, that we find an F' with non-small coefficients and then search for a point $F \in L$ close to F'. If such an F is found, set $f' = F' - F$. Then f' will still have the correct valuations at $\alpha \bmod q$, and if F is very close to F', then $|f'|_2$ will be small.

The problem of finding an f' which will give a good impersonation of f is thus reduced to that of finding a point in a lattice which is as close as possible to a given point outside the lattice. This is a non-homogeneous version of the problem of finding a short vector in a lattice. It can also be translated into a homogeneous problem in a similar lattice of one higher dimension. Roughly speaking, an attacker's chance of success in a fixed amount of time improves as the distance of the given point to the lattice decreases. The attacker's chances also deteriorate as the dimension of the lattice increases.

q	N	T_{avg}	Experimental Times T
101	100	171	149, 170, 193
109	108	226	205, 221, 253
113	112	366	317, 347, 433
149	148	3787	3445, 3912, 4005
157	156	7443	3522, 11363
173	172	32760	17437, 48082
181	180	167021	167021

TABLE 1. Time (secs) To Find Original Key f

N	T (seconds)	T (MIPS-years)
640	$8.65 \cdot 10^{20}$	$1.10 \cdot 10^{16}$
768	$2.52 \cdot 10^{25}$	$3.20 \cdot 10^{20}$
928	$9.60 \cdot 10^{30}$	$1.22 \cdot 10^{26}$
1152	$6.24 \cdot 10^{38}$	$7.92 \cdot 10^{33}$

TABLE 2. Estimated Breaking Times For Original f Key

3.2. Some Lattice Reduction Experiments

Consider a list of primes q and $N = q - 1$ with $d_f = [1/2 + q/3]$ as in (6). When $q = 769$, this gives $d_f = 256$. Our experiments used the lattice reduction package provided in version 3.1b of Victor Shoup's implementation of the Schnorr, Euchner and Hoerner improvements of the LLL algorithm. This is distributed in his NTL package, located at http://www.cs.wisc.edu/~shoup/ntl/. Our approach was to obtain results for an increasing sequence of primes q, and $N = q - 1$, and plot the log of the time it took to break a key or find an alternative key against N. We found in all cases that the log time increased linearly with N. We then extrapolated the line we obtained to obtain estimated breaking times for high N.

Table 1 gives the results of experiments to recover the private key f from $\{f(\alpha)\}_{\alpha \in S}$.

The regression line for the average time (in seconds), as a function of N, is

$$\log(T) \approx 0.0803N - 3.1923.$$

The correlation coefficient is 0.9866. We have used the regression line to extrapolate the breaking time for larger values of N. The results are listed in Table 2. We have used the conversion factor 400/31557600 to convert from seconds to MIPS-years, because our experiments were run on 400 MHz Celeron computers.

Now consider a list of primes q and $N = q - 1$ with $d_g = [1/2 + q/6]$ as in (6). When $q = 769$, this gives $d_g = 128$. Table 3 gives the results of experiments to recover g_1 from $\{g_1(\alpha)\}_{\alpha \in S}$. Note that an attempt could be made to recover g_2

q	N	T_{avg}	Experimental Times T
101	100	69	89, 66, 61, 61, 70
109	108	102	95, 103, 100, 118, 95
113	112	108	103, 136, 95, 114, 94
137	136	471	639, 419, 344, 500, 455
149	148	802	689, 773, 1034, 504, 1008
157	156	1712	1743, 2293, 1468, 1976, 1080
173	172	2983	2465, 2598, 1948, 2860, 5045
181	180	6836	4219, 11951, 7541, 3940, 6529
193	192	13227	22919, 4640, 8982, 22598, 6996
197	196	15169	15744, 28535, 20060, 1612, 9892

TABLE 3. Time (secs) To Recover g_1

N	T (seconds)	T (MIPS-years)
640	$1.64 \cdot 10^{15}$	$2.08 \cdot 10^{10}$
768	$2.54 \cdot 10^{18}$	$3.22 \cdot 10^{13}$
928	$2.47 \cdot 10^{22}$	$3.13 \cdot 10^{17}$
1152	$9.42 \cdot 10^{27}$	$1.19 \cdot 10^{23}$

TABLE 4. Estimated Breaking Times For g_1 Recovery

from values given by the solution of the quadratic equation involving f, g_1, c_1, c_2 that g_2 satisfies. However, as the values of f and g_1 are only known in S and each $g_2(\alpha)$ has two possible solutions, this procedure is far more difficult than the problem of recovering g_1.

The regression line for the average time (in seconds), as a function of N, is

$$\log(T) \approx 0.0574N - 1.6850.$$

The correlation coefficient is 0.9978. We have used the regression line to extrapolate the breaking time for larger values of N. The results are listed in Table 4.

Table 5 gives the time required to produce an f' with the property that $f' \equiv f$ mod q for all $\alpha \in S$ and $|f'|_2 < 1.8q$. Such an f' would not equal the original f, but would be sufficient, once discovered, for Irving to have a reasonably good chance of impersonating Pearl. To do this he would cheat on his commitment by choosing g_1, g_2 to be simple powers of x. We give the results of several experiments for each q between 193 and 307, together with the average time required for each q.

The regression line for the average time (in seconds), as a function of N, is

$$\log(T) \approx 0.0487N - 3.9606.$$

q	N	T_{avg}	Experimental Times T
193	192	281	268, 231, 446, 225, 234
197	196	395	284, 326, 287, 507, 573
229	228	956	671, 890, 1602, 989, 630
233	232	1423	1882, 1611, 1056, 1516, 1051
241	240	2369	3620, 1890, 1961, 2177, 2198
257	256	4021	4038, 6990, 2648, 3024, 3405
269	268	7297	10921, 4023, 8235, 8638, 4670
271	270	9949	11562, 7340, 5865, 18289, 6687
277	276	11415	12775, 10393, 20110, 5319, 8477
281	280	15578	24864, 18970, 11842, 9572, 12642
283	282	22022	15506, 14053, 22891, 41195, 16464
293	292	25239	17518, 39748, 18452
307	306	110973	107573, 114373

TABLE 5. Time (secs) To Find False Key f'

N	T (seconds)	T (MIPS-years)
640	$6.46 \cdot 10^{11}$	$8.19 \cdot 10^{6}$
768	$3.28 \cdot 10^{14}$	$4.16 \cdot 10^{9}$
928	$7.92 \cdot 10^{17}$	$1.00 \cdot 10^{13}$
1152	$4.31 \cdot 10^{22}$	$5.46 \cdot 10^{17}$

TABLE 6. Estimated Breaking Times for false PASS2 key f'

The correlation coefficient is 0.9876. We have used the regression line to extrapolate the breaking time for larger values of N. The results are listed in Table 6.

Remark 3.1. *Table 6, the estimated time for recovery of a false key f', gives the smallest breaking times, hence should be regarded as providing a lower bound for the security of the PASS2 scheme.*

For comparison purposes, we note that the estimated time to break RSA 512 is $3 \cdot 10^4$ MIPS-years, and the estimated time to break RSA 1024 is $3 \cdot 10^{11}$ MIPS-years. So according to Table 6, the PASS2 scheme with $N = 640$ should be considerably more secure than RSA 512, and security for $N = 928$ should be greater than RSA 1024, while and $N = 728$ lies in between.

3.3. Zero-Forced Lattices

Alexander May [4] has given an improved method for searching for small vectors when the small vectors have a comparatively large number of coordinates equal to 0. These ideas lead to the notion of zero-forced lattices, in which one guesses that r particular coordinates of the target are 0, forces them to be zero, and thereby reduces the dimension of the lattice. Of course, if r is large, it may take many tries before one makes a correct guess. Full details of how zero-forced lattices work and how to estimate their effectiveness are given in [5]. However, since the polynomials have only $1/3$ of their coefficients equal to 0, in the case of f, and $2/3$ equal to 0, in the case of g_1, it is very difficult to correctly guess many zeros. As it would be necessary to guess considerably more than 100 zero locations correctly in order to reduce the key breaking time for g_1 or f down to even the time estimate for finding a false f', one sees that the use of zero-forced lattices has a negligible effect on security estimates for PASS2.

3.4. Lattice Based Creation of a Response Without the Private Key

Irving faces the following problem. Given a challenge c, he must find a polynomial h with $|h|_2 < 1.8q$ such that $(f(\alpha) + c_1(\alpha)g_1(\alpha))^2 + 4c_2(\alpha)h(\alpha)$ is a quadratic residue mod q for every $\alpha \in S$. There are several different approaches that Irving can take, but none seem to have any chance of success in time less than that estimated in Table 6 for recovery of a false key f'.

For example, Irving could choose his collection of commitment values $g_1(\alpha)$ at random. After receiving the challenges Irving could choose t values for $h(\alpha)$ at random that satisfy the quadratic constraints. He could then use LLL to search for h with $|h|_2 < 1.8q$ satisfying these constraints. For $t = N/2$ the expected size of a vector satisfying these constraints is (see [3]) about $q/\sqrt{2\pi e}$. As this is less than $1.8q$, there is a high probability that such an h exists. However the time required to find such an h by lattice reduction methods is greater than or equal to the time required to find a false key f' as given in Table 6. Thus the security estimate for PASS2 remains unchanged after considering this potential attack.

Another possibility for Irving is to pick a random collection of commitment values $g_1(\alpha)$. He can then choose G_1 very short, (even a power of x), and define $g_2(\alpha)$ by $G_1(\alpha) = g_2(\alpha)f(\alpha)$. Then he can search for a short G_2 such that $|G_2|_2 < 1.8q$ and such that $G_2(\alpha) = (c_1(\alpha)g_1(\alpha) + c_2(\alpha)g_2(\alpha))g_2(\alpha)$ for every $\alpha \in S$. This, however, reduces to the same search as just mentioned, and should be solvable in about the same time.

The last possibility we will consider is that Irving could find relatively short polynomials G_1, G_2, G_3, that is, polynomials satisfying $|G_1|_2, |G_2|_2, |G_3|_2 < 1.8q$, and collections of values $\{g_1(\alpha), g_2(\alpha)\}_{\alpha \in S}$ such that

$$G_1(\alpha) = f(\alpha)g_2(\alpha), \qquad G_2(\alpha) = g_1(\alpha)g_2(\alpha), \qquad G_3(\alpha) = g_2(\alpha)^2.$$

This problem seems to be just as hard as that mentioned in the first two possibilities, but the dimension is tripled, leading to considerably greater breaking times.

3.5. Attacks on a Transcript of Authentication Sessions

Consider the information revealed in a large collection of distinct examples of

$$h = (f + c_1 g_1 + c_2 g_2)g_2$$

for fixed f and varying c_1, c_2 and g_1, g_2. It is important to note that since the polynomials f, g_1, g_2, c_1, c_2 are small, an attacker may assume that no reduction modulo q has occurred in the construction of h, and thus that the coefficients of h are given over $\mathbb{Z}$. Significant reduction, however, has occurred modulo $x^N - 1$.

First, fix some β not in S and let us consider the information revealed from a collection of responses h for which c_1 vanishes at β, that is, $c_1(\beta) \equiv 0 \mod q$. Let QR denote the set of quadratic residues mod q. Since $g_2(\beta)$ is the solution of a quadratic equation, it must be true that

$$(f(\beta) + c_1(\beta)g_1(\beta))^2 + 4c_2(\beta)h(\beta) \in QR.$$

Since we are assuming that $c_1(\beta)$ vanishes, it follows that $f(\beta)^2 + 4c_2(\beta)h(\beta) \in QR$. This constrains $f(\beta)^2$ to lie in the translated set

$$f(\beta)^2 \in QR - 4c_2(\beta)h(\beta) = \{u^2 - 4c_2(\beta)h(\beta) : u \mod q\}.$$

Each response h for which $c_1(\beta) = 0$ will cut the possibilities for $f(\beta)^2$ by approximately 50%, so after little more than $\log_2 q$ such responses, an attacker can determine $f(\beta)^2$.

If this attack is carried out for every β not in S, then the polynomial $f(x)^2$ can be determined, since the values $f(\alpha)$ for $\alpha \in S$ are already public knowledge. It is then easy to extract the small square root and recover $f(x)$ itself, see [3] for details. The attack we have just described is the reason for the requirement in the protocol that $c_1(\beta) \neq 0$ for all $\beta \notin S$ (other than $\beta = 0, \pm 1$, which are not important). This requirement means that the above attack cannot even get started. We are indebted to Don Coppersmith for informing us of this potential attack.

One might ask if the attacker could apply the same approach using an irreducible quadratic factor of c_1 and thus a root of c_1 in a quadratic extension of $\mathbb{Z}/q\mathbb{Z}$. This will not work, because the polynomial h is only given modulo $x^N - 1$, and it is only elements of $\mathbb{Z}/q\mathbb{Z}$ that have the property that $\alpha^N = 1$; elements in extension fields do not have this property. In other words, the evaluation map at a element of an extension field is not a homomorphism from R to that extension field, so the attack using extension fields is not possible.

The collection of all h in a transcript will generate a lattice over $\mathbb{Z}$. However, because of the presence of the non-zero c_i, the full (and thus useless) lattice is generated by this collection.

One can consider the average of many different responses h. As in [3] this does not provide useful information because the expected values of the coefficients of g_1, g_2 are 0. The expected value, incidentally, of the polynomial g_2^2 is $(x^N - 1)/(x^2 - 1)$. This is not quite zero, but highly non-invertible.

An attacker might also consider the product $h\sigma(h)$, the autocorrelation polynomial corresponding to h. Here, for any polynomial $h(x)$, $\sigma(h)(x) = h(x^{-1})$. (This is simply the polynomial h with its coefficients reflected about $x^{N/2}$.)

This is potentially a very powerful attack, due to Burt Kaliski, as after averaging a long transcript, an attacker can hope to obtain the polynomial

$$a_f A_g^{(2)} + A_c^{(1)} A_g^{(1,2)} + A_c^{(2)} A_g^{(2,2)}.$$

Here for any polynomial F, a_F denotes the even autocorrelation polynomial $a_F = F * \sigma(F)$. Also $A_c^{(i)}$ denotes the expected value of $c_i * \sigma c_i$ for $i = 1, 2$ and $A_g^{(i,j)}$ denotes the expected value of $g_i g_j * \sigma(g_i g_j)$. The average of $h\sigma(h)$ will approach this limit as the cross terms of the product will have expected value zero.

If g_1, g_2, c_1, c_2 vary uniformly, the limiting autocorrelation polynomials are simple constants and hence a_f can be recovered. This means in effect that the quantity $f(\alpha)f(\alpha^{-1})$ can be assumed to be known, and thus that once $f(\alpha)$ is known mod q for any α, $f(\alpha^{-1})$ can be found. This is the reason for the original assumption that S is closed under multiplicative inverses, as an attacker gets no additional knowledge from a_f. We refer also to the analysis given in [3] for the conclusion that it is very difficult to factor a_f as a polynomial and obtain f.

We will close this section by remarking briefly on an important observation of Coppersmith. By selecting any fixed 4-tuple of indices i, j, k, l and computing an average of the product of the i, j, k, l individual coefficients h_i, h_j, h_k, h_l, information can be obtained about a combination of second and fourth power moments of f. (In this terminology, a_f is the second power moment of f.) It is then possible to recover f by a process which, while computationally intensive, is still subexponential in N and feasable for the parameter choice $N = 768$. We have conducted a number of computer experiments to determine lower bounds that the length of a transcript must exceed before an attacker has a chance of determining the limiting value of the products h_i, h_j, h_k, h_l. Some experimental evidence is given in the Appendix B. The experiments show that the convergence to the limiting value is extremely slow. Even after averaging 100 million responses, that is, examining 100 million digital signatures produced by a single private key, the variation in each product is still wide enough to allow considerably greater than 2^{160} choices for a sufficiently large (greater than N) limiting collection of 4-tuple products. We thus feel that it is safe to use a single key for at least 100 million authentication sessions or digital signatures.

3.6. Cheating Verifiers

A cheating verifier can pass specially constructed challenges with given expected values to Pearl and extract information from the responses as outlined above. (For example, choosing challenges equal to 0, or those where c_1 has roots consistently in specific places.) However, by the protocol, c_1 is forced to have no roots and both c_1, c_2 are non-zero. Thus the only potential advantages known to us at this time that a cheating verifier might obtain are eliminated.

		$N = 768$	$N = 928$
Public key	$t \log_2 q$	3840	4640
Private key	$2d_f \log_2 N$	1020	1240
Commitment	$t \log_2 q$	3840	4640
Challenge	$\log_2 c_0$	80	80
Response	$\approx N \log_2 |h|_\infty$	4280	5170
Total Bits Transmitted		8,200	9,890

TABLE 7. Key Length and Communication Requirements in Bits

4. Key Length and Communication Requirements

The key lengths and number of bits transmitted for $N = 768$ and $N = 928$ are given in Table 7. (For $N = 928$, we take $d_f = 62$.) It is worth noting that if desired, as in the PASS scheme, the private key can be stored as, or generated from, any random string of 80 bits, as long as a non-linear uniform mapping is provided into the space $\mathcal{L}_f$. The number of bits in the response is an upper bound, based on the fact that most coefficients of h will have a rather small absolute value and hence can be recorded using 5, 6 or 7 bits. On average, one finds that with these parameter choices, about 34% of the coefficients can be recorded with 5 bits, 29.14% with 6 bits, 29.64% with 7 bits. Only about 0.03% will require 8 bits and one or two rare exceptions require 9. Note that the length of a digital signature attached to a message will be the total number of bits transmitted as recorded below, minus the 80 bits required for the challenge. This is because, as usual when constructing a digital signature, the message is hashed with the commitment to produce the challenge. The signature is then the commitment, followed by the response.

5. Final Remarks

Recall that we established above that the security level of PASS2 with $q = 769$ is considerably greater than that of RSA 512, while the security level of PASS2 with $q = 929$ is greater to RSA 1024.

When Vinnie checks that the quadratic condition is fulfilled, he need only do this for a randomly chosen subset of 80 values in S. It will probably be most efficient for Vinnie to use a precomputed table of quadratic residues mod q, but if space is at a premium, then quadratic reciprocity could be used for this test.

Finally, we remark that the evaluation of polynomials by Pearl and Vinnie can be done most efficiently by means of the FFT. This is because the evaluation of a polynomial is simply the association between a vector over $\mathbb{Z}/q\mathbb{Z}$ and its discrete Fourier transform, where a polynomial is identified with the vector of its coefficients. Naive computation of discrete Fourier transforms of vectors of dimension N only takes N^2 steps, so is not an onerous task. However, the suggested

	Sign	Verify
PASS2 768	4.73	2.12
RSA 512	4.34	0.27
DSA 512	4.75	5.56
PASS2 928	10.69	4.25
RSA 1024	27.62	0.70
DSA 1024	15.38	18.16
ECC(p) 168	15.99	27.42
ECC(2^n) 155	20.12	24.51
PASS2 1152	8.31	3.73
RSA 2048	181.82	2.11

TABLE 8. Timing Estimates (Milliseconds Per Operation)

parameter values were selected so that N is divisible by a large power of 2, which means that one can use Fast Fourier Transforms (FFT) to speed the process. Note that one can do these FFT's in $\mathbb{Z}/q\mathbb{Z}$ working entirely with integers, because $\mathbb{Z}/q\mathbb{Z}$ contains a primitive N^{th} root of unity. There is no need to use real or complex numbers.

We note finally that the key length, signature length and communication requirements described above can be substantially reduced. This will be described in detail in a further paper. Our purpose here is to present the basic scheme with as few "tricks" as possible, so that the underlying validity of the approach can be more easily evaluated.

Appendix A. Timing Comparisons

In this section we compare digital signature and verification times for various cryptosystems. We note that the PASS2 times are based on a preliminary non-optimized implementation by Tao Group, Inc. We also note that the extremely fast RSA verification times are due to the use of the very small value $k = 17$ as verification exponent.

The timing data for the RSA, DSA, and ECC signature schemes in Table 8 are taken from the Crypto++ 3.1 Benchmarks page, which may be found at

<http://www.eskimo.com/~weidai/benchmarks.html>.

All were coded in C++ or ported to C++ from C implementations, compiled with Microsoft Visual C++ 6.0 SP2 (optimized for speed, Pentium Pro code generation), and run on a Celeron 450MHz machine under Windows 2000 beta 3. No assembly language was used. The RSA computations were done using the small verification exponent 17. The DSA and ECC values can be improved somewhat (up to a factor of 2 in some cases) by storing precomputed values. The PASS2 times are for the preliminary implementation by Tao Group (run on a 300MHz machine and extrapolated to 450MHz). The reason that PASS2 1152 is faster than PASS2 928 is because 1152 is more highly divisible by 2 than is 928, which allows greater efficiency in the FFT routines.

Appendix B. Transcript Experiments

In this section we describe experiments performed using the PASS2 parameters

$$N = 768, \quad d_f = 256, \quad d_g = 128, \quad d_c = 2.$$

For each experiment we fixed a random polynomial f, and four random indices i, j, k, l. We randomly chose 100 million 4-tuples of polynomials g_1, g_2, c_1, c_2 according to these parameters. For each of these choices we computed

$$h = (f + g_1 * c_1 + g_2 * c_2) * g_2.$$

We then took the four random indices and computed the product

$$h_i h_j h_k h_l$$

of the corresponding four coefficients of h. We kept a running average of these quadruple products.

Results from a typical experiment are given in Table 9. Other experiments gave similar behavior, so we have selected a few pieces of one run to give a feel for the rate of convergence. In this table, the four indices fixed were 55, 105, 537, and 551 and we have recorded the running average, denoted Avg_h, rounded to the nearest integer, for various numbers of trials. As is clear from the table, even after 10^8 trials, the value of the product has not fully settled down, so it would be difficult to guess the correct value. Note that even if the value of each quadruple product is known to within 2 or 3, say, the number of possible values for all of the products $h_i h_j h_k h_l$ would be far greater than 2^{768}, so it would not be possible to perform an exhaustive search.

Remark B.1. After this paper was submitted, it was pointed out to us that a transcript can be analysed more efficiently by using the fact that $h_i h_j h_k h_l$ and $h_{i+r} h_{j+r} h_{k+r} h_{l+r}$ have the same expected value, increasing the effective length of the transcript by a factor of about $\sqrt{N}$. Thus, to be conservative, we will only claim that the present experiments indicate that transcripts of length up to 3 million do not reveal useful information. We expect that further experiments, to be described in a subsequent publication, will show that considerably longer transcripts are still safe from this quadruple product attack.

# of Trials	Avg$_h$	# of Trials	Avg$_h$
29,990,000	−98	99,900,000	49
30,030,000	−108	99,905,000	49
30,070,000	−88	99,910,000	49
30,110,000	−78	99,915,000	49
30,150,000	−74	99,920,000	50
30,190,000	−63	99,925,000	50
30,230,000	−62	99,930,000	52
44,180,000	−166	99,935,000	52
44,220,000	−162	99,940,000	50
44,260,000	−164	99,945,000	51
44,300,000	−166	99,950,000	52
44,340,000	−156	99,955,000	50
44,370,000	−156	99,960,000	49
89,790,000	21	99,965,000	49
89,830,000	26	99,970,000	50
89,870,000	33	99,975,000	51
89,910,000	32	99,980,000	50
89,950,000	34	99,985,000	49
89,990,000	33	99,990,000	47
90,000,000	32	99,995,000	48
		100,000,000	48

TABLE 9. Average Values of Products $h_i h_j h_k h_l$

References

[1] J. Hoffstein, B.S. Kaliski, D. Lieman, M.J.B. Robshaw, Y.L. Yin, *A New Identification Scheme Based on Polynomial Evaluation*, patent application.

[2] J. Hoffstein, J. Pipher, J.H. Silverman, *NTRU: A new high speed public key cryptosystem*, in Algorithmic Number Theory (ANTS III), Portland, OR, June 1998, Lecture Notes in Computer Science 1423 (J.P. Buhler, ed.), Springer-Verlag, Berlin, 1998, 267–288.

[3] J. Hoffstein, D. Lieman, J.H. Silverman, *Polynomial Rings and Efficient Public Key Authentication*, in Proceeding of the International Workshop on Cryptographic Techniques and E-Commerce (CrypTEC '99), Hong Kong, (M. Blum and C.H. Lee, eds.), City University of Hong Kong Press.

[4] A. May, *Cryptanalysis of NTRU*, preprint, February 1999.

[5] J.H. Silverman, *Dimension-Reduced Lattices, Zero-Forced Lattices, and the NTRU Public Key Cryptosystem*, NTRU Technical Note 013, March 2, 1999, available at ⟨www.ntru.com⟩.

NTRU Cryptosystems, Inc.
5 Burlington Woods, Burlington, MA 01803, USA
E-mail address: jhoff@ntru.com, jhs@ntru.com

Progress in Computer Science and Applied Logic, Vol. 20
© 2001 Birkhäuser Verlag Basel/Switzerland

Security of Biased Sources for Cryptographic Keys

Preda Mihăilescu

Abstract. Cryptographic schemes are based on keys that are highly involved in granting their security. It is in general assumed that the source producing these keys has uniform distribution, that is, it produces keys from a given key space with equal probability. Consequently, deviations from uniform distribution of the key source may be regarded a priori as a potential security breach, even if no dedicated attack is known, which might take advantage of these deviations.

We propose in this paper a model for biased key sources and show that it is possible to prove some results about tolerance of biases, that have the property of being inherent to the bias itself and not requiring assumptions about unknown attacks, using these biases. The model is based on comparing the average case complexities of generic attacks to some number theoretical problems, with respect to uniform and to biased distributions.

We also show the connection to information entropy based analysis of biased sources, which was used in earlier works, for suggesting the tolerance of biased sources.

1. Introduction

Cryptographic schemes, both secret and public key ones, use keys or sets of keys upon which their security is based. It is assumed in general that the keys are produced by a random uniform distributed source, so that the probability for every key from a given key space is constant. The perception, that a random uniform distribution of the cryptographic keys is a minimal security constraint to be respected by a safe implementation, is therefore quite widespread. This perception is based on the following reasoning: breaking a scheme based on uniform random distributed keys requires not less than the general state of the art attack against the given scheme. If the key generation has some bias, on the other hand, the bias could make it vulnerable. One may suppose that some – still unknown – algorithm can exist, that takes advantage of the given bias in order to break the scheme considerably faster than the state of the art general solution would.

Purely uniformly distributed random sources are not easy to provide in practice and are also hard to prove. This may be due to the practical difficulty for creating random bits – as the Netscape flaw proved, several years ago – or to the fact that provably secure random sources may involve a considerable computational overhead. Several authors – Bach [1], [2], Peralta, Shoup [12], [14], etc. – therefore explicitly introduced the notion that *random bits are not for free* into the analysis of algorithms. It may also be that some key generation algorithm, which has slight distributional biases, enjoys further advantages that one would prefer to conserve, provided the security of the bias may be ascertained.

The primary conception that bias equals insecurity may therefore be too strong. One would like to have some simple measures allowing one to distinguish between *harmless* biases and possibly dangerous ones. It is the purpose of this paper to provide a simple model for biases, based on complexity theory, which allows defining the tolerability of a bias by an intrinsic measure of the bias distribution. It is then shown that attacks, which would successfully take advantage of tolerable biases, implicitly improve upon the state of the art attack against the same scheme using uniformly distributed key sources. Of course, there is no general protection against such improvements, and avoiding small biases is therefore not the place where to invest for obtaining better security.

Our concepts are naturally related to the powerful theory of *average case complexity*, initiated by Levin [8] and extended by Gurevitch [5], [6] Ben-David, Chor, Goldreich and Luby [3], etc. In terms of this theory, we compare the average case complexity of some unknown algorithm – optimally adapted to take advantage of a given biased distribution – with respect to two different distributions: the uniform and a given biased one. We thus derive bounds which are intrinsic to the biased distribution, either in terms of measure or of entropy. The class of polynomially bounded biases itself will be defined in the spirit of related average case complexity definitions.

By working with the model, it becomes clear that it is rather powerful in dealing with very general key sources. We focus our attention, however, primarily upon public key cryptographic schemes that base their security upon the assumption that one of the following number theoretic problems is hard:

IF The problem of factoring rational integers.

DL The discrete logarithm in the multiplicative group of a finite field.

EDL The discrete logarithm in the group structure of an elliptic curve over a finite field.

We shall thus not load the model with more generality than necessary for treating these cases in an unified way. Possible further generalizations are obvious and may be made by cryptographers when concrete problems that do not fit in our restricted presentation, raise similar distribution questions. We give examples for the problems which we take into consideration. These will show both useful tolerable biases, and biases which are definitely not tolerable, with respect to this model and from the point of view of *common sense*.

2. Key and Token Spaces

We start with some examples illustrating several issues to be taken into account when formalizing notions like key, key space, key token, etc. We want to define these notions in the following sense: Let S denote some source generating *key tokens* $\tau_i, i = 1, 2, \ldots k$ from given *token spaces* $\mathcal{I}_i$. These tokens are combined by a key generation function

$$\kappa : \mathcal{K}_0 = \mathcal{I}_1 \times \mathcal{I}_2 \times \ldots \times \mathcal{I}_k \longrightarrow \mathcal{K},$$

where $\mathcal{K}$ is a key space of some cryptographic scheme.

For the secret key, the combination can be the identity, since the key holder is allowed to know the single tokens out of which his key is built. The key generation function typically *hides* information when building public keys out of their tokens. We will not be concerned with secret keys, since attacks against these keys are physical or organizational and cannot be protected against by use of cryptographic means. Domain parameters are the common data to which individuals add secret tokens in order to produce key pairs. As such, they can display weaknesses concerning the whole domain. For instance if an elliptic curve with Frobenius trace 1 is chosen as a domain parameter for a scheme based on EDL, the curve is prone to the Smart attack [15]. Domain keys are therefore also part of the topic.

Example 2.1. *For the DL problem, the usual tokens are $\tau_1 = p$, a large prime, and $\tau_2 = \alpha \in \mathbb{F}_p^*$, an element generating a subgroup of large size in the multiplicative group $\mathbb{F}_p^*$. In the RSA scheme, $\tau_1 = p, \tau_2 = q$ are two primes so that $\kappa(p, q) = p \cdot q = n$ is one part of the public key. The other part, the public exponent is sometimes fixed, but may in general be regarded as a key token. For a fixed public modulus size of $2m$ bits, the key space may be defined in different ways. It is in most cases contained in the set: $S = \{p, q : 2^{2m-1} < N := p \cdot q < 2^{2m}\}$, with p and q primes, although the condition $N < 2^{2m}$ can be found sometimes relaxed to, $N < 2^{2m+l}$, with, for example $l = 7$. Even for $l = 0$, the primes p and q are subject to additional conditions which can be more or less severe. One set of conditions concerns the sizes of p and q and it ranges from $p, q > 2^{(1-\varepsilon)m}$ for some fixed fraction ε to $2^m > p, q > \sqrt{2} \cdot 2^{m-1}$. Other conditions restrict the $\gcd(p-1, q-1)$. The resulting key spaces therefore have slight differences.*

Finally, for the EL problem, the tokens $\tau_1 = p$, with p prime, $(\tau_2, \tau_3) = (A, B)$ are the parameters of an elliptic curve which is a domain key:

$$\kappa(p; A, B) = \mathcal{E}_{p^h} : Y^2 = X^3 + A \cdot X + B.$$

An important implicit token is $N(p; A, B) = \sharp \mathcal{E}_p$, the number of points of the elliptic curve $\mathcal{E}_p$ that can be evaluated faster by specifying an order $\Lambda = [1, f\omega] \subset \mathcal{O}(\mathbb{K})$ in the integers of an imaginary quadratic extension $\mathbb{K}$, in which $\mathcal{E}_p$ will have complex multiplication. This specification fixes A and B and herewith a possible way to determine a curve consists of specifying $\mathbb{K}$ and f. For computational reasons, the discriminant of $\mathbb{K}$ must be polynomial in $\log p$, and therefore the methods relying on fixing the complex multiplication produce an exponentially thin key subspace of *all the smooth elliptic curves modulo p. This can be avoided by choosing A and B randomly. After that, one counts the number of points on the resulting curve $\mathcal{E}_p$, subject to some restrictions that do not affect the size of the key space.*

Primarily, both keys and tokens are *sets of parameters*. We shall restrict the notion of *key* only by one condition. A key is an organization level of parameters at which an attack against a public key scheme can be conducted. In this sense, both public and domain parameters are considered to be keys. Also, a key *need not* be the *complete public key* of a given algorithm. For instance for the RSA algorithm, the *public modulus* is a key against which an individual factoring attack can be carried out, although it requires one additional *token*, the public exponent, in order to complete the public key. The range of possible attacks may increase by using this token too, but the modulus in itself is definitely a meaningful key, while the public exponent is not. A secret key is equal to a public key plus information about the building tokens. A domain key is not a key in the restricted sense, since it has insufficient information in order to be the object of an attack. Its choice may however influence the possibility of attacking public keys that are built using the domain key. This is the case, for instance, for most supersingular elliptic curves.

We define our notions accordingly:

Definition 2.2. *A* token *is a parameter or a parameter-tuple $\tau \in \mathbb{Z}^k$, for $k \geq 1$. A* key *is a set of parameters $K \in \mathbb{Z}^{\mathbb{N}}$, built by a key generation function acting on a product of token spaces. A key is subject to attacks that aim to reveal the tokens it is built from. Keys are defined by:*

1. *Algorithm and functional type. The algorithm describes a public key scheme for which the key is used. The type defines the keys' functionality – for example, RSA public modulus.*
2. *Size parameters.*
3. *Token dependencies.*

A key space $\mathcal{K}$ is the set of all keys sharing the same Definitions 1.–3. together with a set of conditional token distributions reflecting the token dependencies.

While the first two items completely define the functionality of a key, the token dependencies may, as shown in the example, induce variations of the key spaces for the same functional keys. In particular, token dependencies may be empty, meaning that all tokens are statistically independent, or they may be given by a set of conditional probabilities. It happens that more tokens are bound by a reciprocal condition. For instance, the primes of which an RSA public modulus is built cannot be *too close* to each other, etc. In such cases, the last generated token can depend upon the choice of the previous ones.

We consider generic attacks against the underlying *hard* problem and their average case complexity. It is our purpose to compare the average case complexities in dependency of different distributions of the keys. Firstly, this means that keys shall be defined so that they are in one to one correspondence with instances of the problem. Secondly the key spaces must be invariant under different key generation strategies. Finally, conditional distributions need to be defined in every case. With these, a distribution of the keys can be derived from the distributions of single tokens. Note that our definitions meet these conditions; however, we can only compare sources at the level of identical key spaces. In particular, variations induced by token restrictions on key spaces for the same functional keys, are not object of our approach. They can be addressed by related methods, since the variations of key spaces are relatively small in such cases. We shall not consider this problem here.

3. Distributions

We shall give some definitions of distributions and attack strategies based on average case complexity, using [3], [5].

If $\mathcal{K}$ is a key space, let $s : \mathcal{K} \longrightarrow \mathbb{N}$ be the size parameter and $L = s(\mathcal{K}) \subset \mathbb{N}$. For $n \in L$, let $\mathcal{K}_n$ be the key set of size n. The function s needs not be surjective, but for $n, m \in L$, $n \neq m \Rightarrow \mathcal{K}_m \neq \mathcal{K}_n$. We assume additionally that $\mathcal{K}$ is endowed with an enumeration function $\phi : \mathcal{K} \longrightarrow \mathbb{N}$ with

$$x \in \mathcal{K}_n, y \in \mathcal{K}_m, n < m \Rightarrow \phi(x) < \phi(y).$$

We wish next to define size independent distributions on the key space, in accordance with average case complexity. If $\mu : \mathcal{K} \longrightarrow [0, 1]$ is a probability density function, then $\mu_n := \mu(k|s(k) = n)$ is a conditional density defined on the space of size n keys. Since the key spaces of different sizes are disjoint, the event $s(K) = n$ has a probability function $\rho : \mathbb{N} \longrightarrow [0, 1]$ with

$$
\begin{aligned}
\rho(n) &= 0, \quad \text{for } n \notin L \\
\mu(k) &= \rho(s(k)) \cdot \mu_{s(k)}(k), \quad \text{for } k \in \mathcal{K} \text{ and} \\
\sum_{n > 1} \rho(n) &= 1.
\end{aligned}
\tag{1}
$$

It is also reasonable to assume that there is a polynomial $b \in \mathbb{Z}[x]$ such that $\rho(n) \geq \frac{1}{b(n)}$, $\forall n \in L$ (see [5], Proposition 1.1): In most practical cases, the set L may even be considered to be finite. Note that the function ρ allows building averages over finite and countably infinite sets L. It has no specific empirical correspondence. We may therefore assume that it is shared by *all* probability functions describing key generation sources for $\mathcal{K}$.

For a fixed key size $s(k) = n$, a uniform source $\mathcal{S}_u$ will draw uniform random distributed instances from each token space $\mathcal{I}_i$, so that the probability of drawing any token $\tau_i \in \mathcal{I}_i$ will be the same: $\ell_{n,i} = \frac{1}{|\mathcal{I}_i|}$, where $|\mathcal{I}_i|$ is the number of elements in $\mathcal{I}_i$. If the token spaces are independent, the probability $\ell_n = \prod_i \ell_{n,i}$. Otherwise, ℓ_n is still constant, but depends on the set of conditional probabilities 3 in Definition 1. Note that the constant ℓ_n is the probability of the uniform distribution at each key $k \in \mathcal{K}_n$. We may thus write $\ell_n(k)$ when the accent is on the uniform distribution and ℓ_n when we are performing computations with the (constant) value of its probability function.

We shall assume that the token spaces are independent and let thus $\mu_{n,i}$ be the probability function on token space $\mathcal{I}_i$ for key size n. We write for the number of elements $m(n) := |\mathcal{K}_n|$ and $m_i(n) := |\mathcal{I}_i|$. Thus $\ell_n \cdot m(n) = 1$ and $\ell_{n,i} \cdot m_i(n) = 1$. If $X \subset \mathcal{K}_n$, and $g : \mathcal{K}_n \longrightarrow \mathbb{R}$, we shall write $\sum_X g(k) := \sum_{k \in X} g(k)$. The probability of the event that a key $k \in X$ will be denoted by

$$\mathcal{P}(X) := \sum_X \mu(k) = P(k|k \in X).$$

Let (S, ℓ) be a randomized unique solution search problem [3] – such as all number theoretical problems relevant for cryptography may be considered – and σ_S its standard (or *state of the art*) solution for the uniform distribution ℓ. Let $f_S(k)$ be the operation count of σ_S when applied to the key $k \in \mathcal{K}$ and

$$\Theta_n = \sum_{k \in \mathcal{K}_n} f_S(k)\ell_n(k) \tag{2}$$

the expected operation count of σ_S for keys of size n and $\Theta = \sum_{n \in S} \rho(n)\Theta_n$.

Let μ_n be the probability functions corresponding to some *biased* key generation source S on $\mathcal{K}_n$, $n \in L$. The local bias of S at $k \in \mathcal{K}_n$ is given by $\lambda_n(k) := \mu_n(k)/\ell_n$ and is normed by

$$\sum_{k \in \mathcal{K}_n} \lambda_n(k) = \frac{1}{\ell_n} \sum_{\mathcal{K}_n} \mu_n(k) = m(n).$$

This is an obvious measure of the deviation from uniform distribution. The most likely way to take advantage of biases consists of finding efficient attacks for the region of the token/key space where a given source has extreme bias. We consider briefly the local bias on token space $\mathcal{I}_i$ with size $m_{n,i} = 1/\ell_{n,i} = |\mathcal{I}_i|$. First note that an important consequence of bias is the existence of keys which may never be produced by a given source. Let

$$E_{\mathcal{I}_i} := \{k_i \in \mathcal{I}_i : \mu_{n,i}(k_i) > 0\} \tag{3}$$

be the *effective subspace* of $\mathcal{I}$ for the source S. If E_n is defined consequently for the space $\mathcal{K}_n$, we let $e(n) = |E_n|/m(n)$, the function measuring the relative weight of the effective subspace for different key sizes. We define the complement $Z_{\mathcal{I}_i} = \mathcal{I}_i \setminus E_{\mathcal{I}_i}$ of the effective set to be the *hidden subspace* of $\mathcal{I}_i$, and similarly for $\mathcal{K}_n$ and for $\mathcal{K}$.

As it will show, tolerance of biases is essentially dependent upon the size of the set of tokens with *small* biases. This fact may first surprise, since it is natural that large biases are the ones to be exploited in order to achieve successful bias adapted attacks. Given that the token space is constant and the sum of probabilities $\sum_{k \in \mathcal{K}} = 1$, a surplus of highly biased tokens will induce a set of tokens with low or even 0 bias. It will be convenient to define:

Definition 3.1. *For $b > 0$, let*

$$
\begin{aligned}
A_{\mathcal{I}}(b) &:= \{k_i \in \mathcal{I} : \lambda_{n,i}(k_i) \geq m_i(n)^b\}, \\
B_{\mathcal{I}}(b) &:= \{k_i \in \mathcal{I} : \lambda_{n,i}(k_i) \leq m_i(n)^{-b}\}, \\
C_{\mathcal{I}}(b) &:= \{k_i \in \mathcal{I} : \lambda_{n,i}(k_i) \geq m_i(n)^{-b}\}, \\
D_{\mathcal{I}}(b) &:= C_{\mathcal{I}}(b) \setminus A_{\mathcal{I}}(b).
\end{aligned}
\tag{4}
$$

Note that $\mathcal{Z}_{\mathcal{I}} \subset B_{\mathcal{I}}(b)$. When the token space $\mathcal{I} = \mathcal{I}_i$ is indexed, we shall write for simplicity: $X_i(b) := X_{\mathcal{I}_i}(b)$, for any of $X = A, B, C, E, Z$. Also, when $\mathcal{I} = \mathcal{K}_n$, there will be no subscript, thus $X(\varepsilon) = X_{\mathcal{K}}(\varepsilon)$.

The source S has polynomially bounded bias *on the token space $\mathcal{I}_i$ if there is a $\delta \in \mathbb{R}_+$ such that:*

$$
\forall \varepsilon > 0 \ \exists n_0 \in N : \ |B_i(\varepsilon)| \leq \left(\delta \cdot \frac{m(n)^\varepsilon}{\Theta_n}\right) m_i(n), \quad \forall n > n_0.
\tag{5}
$$

The bias of S is polynomially bounded on the key space, iff (5) holds for each token space uniformly.

Remark 3.2. *There is some freedom in the choice of the factor of $m_i(n)$ in (5). It makes sense to let it depend on Θ_n, which is expected to be superpolynomial (the theory is meaningless for other cases). Note that the factor $\frac{\delta m(n)^\varepsilon}{\Theta_n}$ bounding the fraction of tokens of $\mathcal{I}_i$ with bias $\lambda(k) < m_i(n)^{-\varepsilon}$ is uniform for all token spaces and depends only on $m(n)$ and some constants. This is natural, since the complexity Θ_n is depending on the order of magnitude of $m(n)$ and cannot meaningfully be related to the size of the single token spaces. The dependency on Θ_n can be dropped by replacing Θ_n with $m_i(n)$, resulting in a generally more restrictive condition. But it can also be loosened. If one decides for instance, for some positive $c < 1$, that no improvement of more then Θ_n^c can be expected, it is safe to use this value in the denominator of (5).*

Note that our definition is more restrictive then requiring that the ℓ – average of the bias is polynomial. This condition would in fact allow the probability function to be centered on a small subset of $\mathcal{I}$ with very high local biases, which we do not want to happen.

It follows from (4) and (5), that

$$
\begin{aligned}
|A_i(\varepsilon)| &< m_i(n)^{1-\varepsilon} \quad \text{and} \\
|D_i(\varepsilon)| &> m_i(n)(1 - 2m_i(n)^{-\varepsilon}),
\end{aligned}
\tag{6}
$$

for some $n > n_1(\varepsilon)$. The bound on the size of the set of tokens with small bias implies that the bulk of tokens have moderate bias.

Definition 3.3. *We say the source S is* **polynomially dense,** *if*

$$
\forall \varepsilon > 0, \ \exists n_0 : |C_i(\varepsilon)| > m(n)^{1-\varepsilon}, \quad \forall n > n_0.
$$

It is clear that sources with polynomial bounded biases are polynomially dense, while the converse is false.

4. Strategies

If $f : \mathcal{K} \longrightarrow \mathbb{R}$ is a positive valued function, we define the expectations

$$E(f, \mu_n) := \sum_{k \in \mathcal{K}_n} f(k) \mu_n(k). \tag{7}$$

An attacker who tries to break the scheme for which the keys were generated using the source $\mathcal{S}$, will use some strategy σ based on one or a combination of algorithms. Let σ be given by an oracle acting optimally in accordance to the bias of μ and let $f = f_\sigma : \mathcal{K} \longrightarrow \mathbb{R}$ be a function giving the expected number of operations required by σ for solving (S, μ) for a given key $k \in \mathcal{K}$.

For fixed n, the average running time of the attacker's strategy, when the keys are generated by $\mathcal{S}$ will thus be

$$\theta_n(\mathcal{S}) = E(f, \mu_n).$$

By (2), the state of the art strategy has running time $\Theta_n = E(f_S, \ell_n)$. It is natural to assume that Θ_n is superpolynomial in $\log m(n)$, a measure of the uncertainty about the uniform distributed input key. We can also assume for any strategy σ, without restriction of generality, that

$$f_\sigma(k) \le \Theta_n, \quad \forall\, k \in \mathcal{K}_n. \tag{8}$$

Since the average case run time of the state of the art algorithm σ_S is superpolynomial with respect to the uniform distribution, it is natural to define a biased distribution as tolerable, to be one with respect to which the average case run time does not improve on Θ_n by more than a polynomial amount.

Definition 4.1. *Let $\mathcal{S}, \mathcal{S}'$ be two sources for the key space $\mathcal{K}$ with $m(n) = |\mathcal{K}_n|$, μ, μ' their distribution densities and f be the run time functions for some optimal strategy oracles for $\mathcal{S}$ with respect to μ. We write $\mathcal{S}' \prec \mathcal{S}$: $\mathcal{S}'$ is polynomially bounded by $\mathcal{S}$ if*

$$\forall \varepsilon > 0, \ \exists n_0 : \ m(n)^{-\varepsilon} E(f, \mu'_n) < E(f, \mu_n), \quad \forall n > n_0. \tag{9}$$

The source $\mathcal{S}$ is said to have **tolerable** *bias, iff $\mathcal{S}_u \prec \mathcal{S}$. The bias is* **conditionally tolerable**, *if the source is polynomially dense.*

The definition of tolerable biases can be verified by using only their intrinsic distribution and without making assumptions about individual attacks. In particular, biases that can be proved to be tolerable bring no risks, while nothing can be said about biases which cannot be proved to be tolerable. This is a clear advantage. The condition for the density $e_n(\mathcal{S}) = 1$ is however very restrictive and dropping it to the condition of polynomial bounded density will allow more flexible definitions of bias tolerance.

The main result about security of biased distributions is:

Theorem 4.2. *Polynomially bounded biases are tolerable.*

Proof. Let us consider the single token space case first (for example, DL domain parameters). The proof will take advantage of the fact that the sets of keys with extreme low biases have at most polynomial contribution to the average case complexity. The expectations $E(f, \mu_n)$ are essentially determined by the values taken by f on $C(\varepsilon)$.

We compare $E(f, \mu_n)$ to $E(f, \ell_n)$ for given $\varepsilon > 0$. It is natural to split the evaluation of the expected values in the parts $\sum_X$, for $X = B, C$. Let $\varepsilon' > 0$ be such that $m(n)^{-\varepsilon'} E(f, \ell_n) - \delta > m(n)^{-\varepsilon} E(f, \ell_n)$ and $n > n_0(\varepsilon)$. For such n, the contribution of the set $B(\varepsilon')$ is easily bounded using (8) and (6), so:

$$
\begin{aligned}
\sigma(\mathcal{S}) \;&=\; \sum_{k \in \mathcal{K}_n} \mu_n(k) \cdot f(k) = \left(\sum_{C(\varepsilon')} + \sum_{B(\varepsilon')} \right) (\mu_n(k) \cdot f(k)) \\
&>\; m(n)^{-\varepsilon'} \sum_{C(\varepsilon')} \ell_n f(k) + \sum_{B(\varepsilon')} \mu_n(k) \cdot f(k) \\
&>\; m(n)^{-\varepsilon'} E(f, \ell_n) - m(n)^{-\varepsilon'} \ell_n \sum_{B(\varepsilon')} (1 - \lambda_n(k)) f(k) \\
&>\; m(n)^{-\varepsilon'} \left(E(f, \ell_n) - |B(\varepsilon')| \ell_n \Theta_n \right) \\
&>\; m(n)^{-\varepsilon'} E(f, \ell_n) - \delta > m(n)^{-\varepsilon} E(f, \ell_n).
\end{aligned}
$$

It follows that $E(f, \ell_n) \prec E(f, \mu_n)$ for this case, so $\mathcal{S}$ has tolerable bias.

In general, let $\mathcal{K}_n = \prod_{i=1}^{r} \mathcal{I}_i$, the cartesian product of the token spaces $\mathcal{I}_i$ and let the token distribution densities $\mu_{i,n}$ be statistically independent. We shall also write $\prod_i$ when the cartesian product defines a summation domain. It is understood in these cases that the product extends over the full set of indices $i \in \{1, 2, \dots, r\}$. Let $\varepsilon > 0$ and $R = \max(2, 2r\delta)$. We can choose $\varepsilon' > 0$ so that

$$
m(n)^{-\varepsilon'} / R > m(n)^{-\varepsilon}.
$$

Let $n > n_0(\varepsilon')$ and $M = \max\{m_j(n) : j = 1, 2, \dots, r\}$. Since

$$
\mathcal{K}_n \setminus \left(\prod_{i=1}^{r} C_i(\varepsilon') \right) \subset \bigcup_{j=1}^{r} \left(B_j(\varepsilon') \times \prod_{i=1; i \neq j}^{r} C_i(\varepsilon) \right),
$$

we have

$$
\left| \mathcal{K}_n \setminus \prod_{i=1}^{r} C_i(\varepsilon') \right| \le \sum_{j=1}^{r} \frac{\delta}{\Theta_n} m(n)^{\varepsilon'} \cdot \prod_{i=1}^{r} m_i(n) = \frac{r \delta m(n)^{1+\varepsilon'}}{\Theta_n}.
$$

Let $\overline{E(f, \ell_n)} = \sum_{\mathcal{K}_n \setminus \prod_i C_i(\varepsilon')} \ell_n f(k)$. We gather from the above estimates, together with (8), that

$$
\overline{E(f, \ell_n)} \le \left| \mathcal{K}_n \setminus \prod_{i=1}^{r} C_i(\varepsilon') \right| \ell_n \Theta_n \le \frac{r \delta m(n)^{1+\varepsilon'}}{\Theta_n} \ell_n \Theta_n = r \delta m(n)^{\varepsilon'}.
$$

Suppose that $\overline{E(f, \ell_n)} > E(f, \ell_n)/2$. It is sensible to assume that $f(k) \geq 1, \forall k \in \mathcal{K}_n$, so $E(f, \mu_n) \geq 1$. Then

$$E(f, \ell_n) < 2\overline{E(f, \ell_n)} < 2\delta r m(n)^{\varepsilon'} < m(n)^{\varepsilon} E(f, \mu_n),$$

by the choice of ε'. In this case, the average case complexities are polynomial with respect to both distributions ℓ_n and μ_n. Otherwise we find:

$$
\begin{aligned}
\sigma(\mathcal{S}) &= \sum_{k \in \mathcal{K}_n} f(k) \prod_{i=1}^{r} \mu_{i,n}(k) > \sum_{\prod_i C_i(\varepsilon')} f(k) \prod_{i=1}^{r} \mu_{i,n}(k) \\
&\geq m(n)^{-\varepsilon'} \sum_{\prod_i C_i(\varepsilon')} \ell_n f(k) = m(n)^{-\varepsilon'} \left(E(f, \ell_n) - \overline{E(f, \ell_n)} \right) \quad (10) \\
&\geq E(f, \ell_n) m(n)^{-\varepsilon'}/2 \geq m(n)^{-\varepsilon} E(f, \ell_n).
\end{aligned}
$$

In both subcases, we have $E(f, \ell_n) \prec E(f, \mu_n)$, for $n > n_0(\varepsilon')$, which completes the proof. □

Remark 4.3. *Note that the condition that density $e(n) \to 1$, following from (6), is necessary in the proof of Theorem 4.2. Without it, there is technically no means for bounding the expected value of the uniformly distributed strategy on the hidden space. It is however a strongly restrictive condition and it is sensible to loosen it by allowing more general sources, asymptotically close to $e(n) = 1$, even if the density does not converge to 1. We shall see that the choice made in the definition of conditionally tolerable biases is natural from a further perspective, the one of entropy measures.*

The decision to define tolerance of biases in terms of at most polynomial gains in the average case complexity of attacks, is arbitrary. One can replace $m(n)^{\varepsilon}$ in (9) by any other family of functions of $m(n)$, closed under multiplication by constants. For instance, one may define tolerance in a problem – specific way, by letting Θ_n^b, for some fixed $b > 0$ be the maximal tolerable performance gain.

5. Entropy

It has been argued in earlier research [4], [11], that biases can be tolerated when the *entropy* of the biased source is asymptotically equal to the entropy of the uniform distributed source. We shall investigate in this section the connection between the entropy approach and bias bounds.

According to Shannon's definition, the entropy of a uniform distributed source is:

$$H_n(\mathcal{S}_u) := - \sum_{k \in \mathcal{K}_n} \ell_n \cdot \log \ell_n = \log \ell_n.$$

A biased source $\mathcal{S}$ has the entropy

$$H_n(\mathcal{S}) := - \sum_{k \in \mathcal{K}_n} \mu_n(k) \cdot \log \mu_n(k).$$

When the source S is fixed and $X \subset \mathcal{K}_n$ we shall also write

$$H_n(X) := - \sum_X \mu_n(k) \cdot \log \mu_n(k).$$

The following lemma sets a very useful relation between the size and probability of a subset of the effective set on the one hand, and the entropy of S on that set.

Lemma 5.1. *Let $D \subset \mathcal{K}_n$ be a subset of the key set and*

$$|D| \le cm(n), \quad M := \mathcal{P}(D).$$

Let $H_n(D) := - \sum_D \mu_n(k) \log(\mu_n(k))$. Then

$$H_n(D) \le M \left(\log m(n) - \log(M/c)\right).$$

Furthermore, if for some $b > 0$, $\mu_n(k) \le m(n)^{b-1}$, $\forall k \in D$, then

$$\log(m(n))(1 - 2b) + \log(M/c) \le H_n(D).$$

Proof. We first apply Jensen's inequalities to the concave function $-x \log x$ on $(0, 1)$, using the $|D|$ constant weights $t_n(k) = 1/|D|$.

$$H_n(D)/|D| = - \sum_D t(k)\mu_n(k) \log(\mu_n(k)) \le -\frac{M}{|D|} \log\left(\frac{M}{|D|}\right),$$

so $H_n(D) \le -M \log\left(\frac{M}{cm(n)}\right) = M \left(\log m(n) - \log(M/c)\right).$

We let $t_n(k) := 1/M\mu_n(k)$, a set of weights with $\sum_D t_n(k) = 1$. Now we apply Jensen's inequality to the convex function $- \log x$:

$$H_n(D) = M \sum_D t_n(k) \cdot (-\log \mu_n(k)) \ge -\log\left(\frac{\sum_D \mu_n(k)^2}{M}\right).$$

Using the bound $\mu_n(k) < m(n)^{b-1}$, we have $\sum_D \mu_n(k)^2 < |D|m(n)^{2b-2}$. It follows that

$$
\begin{aligned}
H_n(D) &\ge -\log\left(\frac{|D| \cdot m(n)^{2b-2}}{M}\right) \ge \log(m(n))2(1 - b) + \log(M/|D|) \\
&\ge \log(m(n))(1 - 2b) + \log(M/c).
\end{aligned}
$$

This completes the proof. $\qquad\square$

In general, if $\mathcal{C} \subset \mathcal{K}_n$ is a subset on which $\mu_n(k) = p$ is constant, then its entropy is

$$H_n(\mathcal{C}) = -|\mathcal{C}|p \log p. \tag{11}$$

One may expect that sources with $\lim_{n \to \infty} H_n(S)/H_n(S_u) = 1$ have a controlled bias. This notion is formalized by:

Definition 5.2. *Let S be a source for the key space $\mathcal{K}$, such that:*

$$\forall \varepsilon > 0, \; \exists n_0 \in \mathbb{N}: \quad H_n(S) > H_n(S_u) \cdot (1 - \varepsilon), \quad \forall n > n_0. \tag{12}$$

*If (12) holds, we call the bias of S **entropy bounded**.*

In the following we investigate the relation between entropy bounded and polynomially bounded biases.

Proposition 5.3. *Polynomially bounded biases are entropy bounded.*

Proof. Let $\mathcal{S}$ be a source with polynomial bounded bias and $\varepsilon > 0$. We shall estimate from below the entropy of $\mathcal{S}$ on the set $\mathcal{D}(\varepsilon)$ which generates the bulk of $H_n(\mathcal{S})$. It will be sufficient to show that (12) holds with $H_n(\mathcal{D}(\varepsilon))$ replacing $H_n(\mathcal{S})$. Let $n > n_0(\varepsilon/3)$. Note that $1/2 > \mu_n(k) \geq \ell_n m(n)^{-\varepsilon}$ for $k \in C(\varepsilon)$ – implying local convexity of $-x \log x$.

We apply Lemma 5.1 to the set $D(\varepsilon/3)$, with $b = \varepsilon/3$. By (6) we may set $c = 1 - 2m(n)^{-\varepsilon/3}$. By definition of $D(\varepsilon/3)$, we have $M/c > m(n)^{-\varepsilon/3}$ and the second case of the lemma implies

$$H_n(\mathcal{S}) \geq H_n(D(\varepsilon/3)) \geq H_n(\mathcal{S}_u)(1 - 2\varepsilon/3) + \log(M/c) \geq H_n(\mathcal{S}_u)(1 - \varepsilon).$$

This completes the proof. $\qquad\qquad\qquad\qquad\qquad\qquad\qquad\qquad\qquad\qquad\qquad\qquad$ $\square$

This proposition indicates a connection between (12) and previous criteria for bias bounding. We show that (12) is in fact equivalent to conditional polynomial bounds.

Theorem 5.4. *A source $\mathcal{S}$ has entropy bounded bias iff it is polynomially dense.*

Proof. Let $\mathcal{S}$ be a polynomially dense source and $\varepsilon, \varepsilon' > 0$, so that

$$(1 - m(n)^{-\varepsilon'})(1 - \varepsilon')^2 > (1 - \varepsilon).$$

By definition, $|C(\varepsilon')| > m(n)^{1-\varepsilon'}$ and $\mathcal{P}(C(\varepsilon')) > 1 - m(n)^{-\varepsilon'}$. We apply the previous proposition to the restriction $\mathcal{S}/C(\varepsilon')$, which is a source with distribution density $\mu_n'(k) = \mu_n(k)/\mathcal{P}(C(\varepsilon'))$ and polynomially bounded bias. If $n > n_0(\varepsilon')$, then

$$
\begin{aligned}
H_n(\mathcal{S}) \quad &> \quad H_n(C(\varepsilon')) > \mathcal{P}(C(\varepsilon'))\, H_n(\mathcal{S}/C(\varepsilon')) \\
&> \quad \left(1 - m(n)^{-\varepsilon'}\right)(1 - \varepsilon') \log\left(|C(\varepsilon')|\right) \\
&> \quad \left(1 - m(n)^{-\varepsilon'}\right)(1 - \varepsilon')^2 \log(m(n)).
\end{aligned}
$$

Finally, by choice of ε', it follows that (12) holds. This shows that dense sources are entropy bounded.

We now show that the converse also holds. Suppose that $\mathcal{S}$ verifies (12) and let $\varepsilon > 0$ and $n > n_0(\varepsilon)$, where n_0 is defined by (12). We show that $\mathcal{S}$ is polynomially dense. We have $H_n(\mathcal{S}) = H_n(B(\varepsilon)) + H_n(C(\varepsilon))$. We shall show that $|C(\varepsilon)| > m(n)^{1-\varepsilon}$. This implies that $\mathcal{S}$ is polynomially dense. Note that

$$\mathcal{P}(B(\varepsilon)) \leq m(n)^{-\varepsilon} \ell_n |B(\varepsilon)| < m(n)^{-\varepsilon}.$$

Since by Lemma 5.1, $H_n(B(\varepsilon)) \leq \mathcal{P}(B(\varepsilon)) \log(m(n)) < m(n)^{-\varepsilon} H_n(\mathcal{S}_u)$, it follows that $H_n(C(\varepsilon)) > (1 - \varepsilon) H_n(\mathcal{S}_u)$. But by (11),

$$H_n(C(\varepsilon)) \leq (1 - \mathcal{P}(B(\varepsilon))) \cdot \log\left(m(n) - |B(\varepsilon)|\right) < \log(m(n) - |B(\varepsilon)|).$$

Together, the last two inequalities yield: $\log(m(n) - |B(\varepsilon)|) > \log\left((m(n))^{1-\varepsilon}\right)$, showing, as announced, that $|C(\varepsilon)|/m(n) > m(n)^{-\varepsilon}$.   □

Corollary 5.5. *The bias of a source $\mathcal{S}$ is conditionally tolerable iff it is entropy bounded.*

Proof. This is a consequence of Theorem 5.4 and the definition of conditionally tolerable biases.   □

6. Practical Applications

In the paper [4], Brandt and Darmgård investigate the distribution of primes produced by sequential search starting from a randomly distributed start point. In [11] we investigate the distribution of a source for prime numbers searching sequentially in an arithmetic progression with relatively large ratio. Both distributions have in common a bias due to the sequential search. It stems from the fact that the probability that the source outputs a given prime is directly proportional to the length of the gap between that prime p and its predecessor q. The larger the gap length $p - q$, the higher the probability that the starting point will hit inside the interval (q, p). It is shown in both cases, based upon a classical conjecture of Hardy and Littlewood, that the resulting distribution is closely approximated by a Poisson distribution. The two biases were studied in the papers [4], [11] by using the entropy approach.

The search in arithmetic progressions generates only a polynomially dense subset of all primes with fixed length. In fact, depending on the ratio of the progression, certain primes of given length will not be produced at all. This feature is shared, for different reasons, by the *Gordon strong primes* [7]. Gordon strong primes p are defined by the requirements that there are two primes q and r of size k bits, such that

$$q|p - 1 \quad \text{and} \quad r|p + 1, \tag{13}$$

a condition which was useful for security, prior to the discovery of the elliptic curve factoring algorithm [9]. An additional condition on $q-1$ having a large prime factor of given size was imposed too. We shall drop this condition for simplicity, as it is not relevant to our analysis. The behavior of the different sources of primes is summarized in:

Theorem 6.1. *The biases of the incremental search source $\mathcal{S}_i$ and of the source $\mathcal{S}_p$ for search in arithmetic progressions are entropy bounded. A uniformly distributed source $\mathcal{S}_g$ for Gordon strong primes (that is, a source producing uniformly distributed primes subject to the above conditions) verifies:*

$$H_n(\mathcal{S}_u) - H_n(\mathcal{S}_g) < 2\log k \tag{14}$$

uniformly for all $n > 2k$.

Proof. The first statement is [4], Theorem 3. The second is proved in [11], §4. Note that the statement holds also for procedures of prime generation similar to the one in [11], such as Shawe-Taylor's algorithm [16]. Let G_n be the set of all n bit primes satisfying condition (13) for some uniformly distributed q, r. The statement (14) follows from $H_n(S_g) = \log(|G_n|)$. $\qquad\square$

The bias for Gordon primes is asymptotically tolerable. For sizes of cryptographic primes it is however comparable to the bias of S_p. Note that S_i is dense, while S_p is only polynomially dense, which suggests a lower entropy for the second source. The source S_g is also dense, but has a relatively large hidden subspace for values of n encountered in practice. Its bias is thus most important when the size of the primes generated is relatively small.

In [10], Maurer gives an algorithm for recursive generation of provable primes, which uses the Dickman function in order to *approximate* the uniform distribution of the primes produced. This can be done, as suggested by Bach [10], with the help of a source of random bits whose distribution is related to the Dickman function. Under this condition, the uniform distribution can be approximated asymptotically with arbitrary accuracy. Based on several heuristic assumptions concerning the behavior of the factorization of $p-1$ for random primes p, the entropy of the source S_m for primes generated by Maurer's algorithm has thus $H_n(S_m) = H_n(S_u)$.

In practice, random sources are most often simulated by some computational process and it is tempting to wish to bound biases of computational random sources by using the concepts described here. It is however inherent to the problem setting, that such bounds can only be given based on some *model* of the computational bias and not upon statistical data of the same. Indeed, assuming that all actual computations that can be done are polynomial in the size of input, it is hardly conceivable to achieve a statistical estimation of the entropy of a computational random source beyond such a bound.

We give an example of a biased source for elliptic curves, that is not conditionally tolerable. Let p be a given prime and consider sources producing elliptic curves $\mathcal{E}(a, b) \mod p$, together with $N = \#\mathcal{E}$. The uniform distributed source S will produce uniformly distributed parameters a, b and then compute N with some recent variant of Schoof's algorithm, for example, see [13]. The complex multiplication source S_{CM} chooses a complex multiplication field $\mathbb{K}$ from a list of fields with polynomial discriminant and easily finds $N = p + 1 \pm \mathbf{Tr}(\pi)$, with $\mathbf{Tr}(\pi) = \pi + \overline{\pi}$, provided that p is a norm in $\mathbb{K} : p = \pi \cdot \overline{\pi}$. If $\mathcal{R}$ is the space of curves produced by S_{CM} and $\mathcal{K}$ the space of elliptic curves over $\mathbb{F}_p$, then $\frac{\#\mathcal{R}}{\#\mathcal{K}} = \frac{\mathcal{O}(g(\log p)^2)}{\mathcal{O}(p^2)}$, where g is the polynomial bounding the discriminants of the complex multiplication fields. If the two sources are compared with respect to the full key space $\mathcal{K}$, it is obvious that S_{CM} is far from being *dense* in this space, and its bias is thus not conditionally tolerable in our terminology. Although no EL algorithm dedicated for curves with CM in fields with small discriminant is known or made plausible, the source S_{CM} is obviously incompatible with our general notion of tolerable biases.

7. Conclusions

We have given a model for measuring biases of non uniformly distributed key sources. The model provides the means for upper bounding the run time gain of an algorithm making use of the bias for breaking a scheme faster than by use of the state of the art algorithm in presence of uniform distributed keys. This is done by comparing the average case complexity of *any* algorithm with respect to the two distributions: uniform and biased. The two average complexities are related by functions depending only on the bias. Since there is no implicit assumption about the algorithms involved, the model gives an universal tool for evaluating upper bounds of run time gains that can be expected in presence of some biased sources. If these bounds are considered as irrelevant from the point of view of complexity theory, a sound proof of the security of respective biased sources results.

We have also shown the connection to entropy measure of biases and proved that some sources of prime numbers considered by earlier papers are tolerable in the sense defined here.

Acknowledgments. I thank the anonymous referee for pointing out the connection to the theory of average case complexity and providing valuable references and I thank him and I. Shparlinski for the encouragement to develop the ideas exposed in this paper. Thanks go also to U. Maurer for valuable discussions about the general framework for the exposition of the ideas in this paper. I am grateful to R. Silverman for his careful comments and suggestions.

References

[1] E. Bach: *Realistic analysis of some randomized algorithms*, J. Comput. Sys. Sci. **42** (1992), pp. 30–53.

[2] E. Bach: *Explicit bounds for primality testing and related problems*, Math. Comp. **55** (1990) pp. 355–380.

[3] S. Ben-David, B. Chor, O. Goldreich and M. Luby: *On the theory of average case complexity*, J. of Computer and System Sciences, **44** (1992), pp. 193–219.

[4] J. Brandt and I. Damgård: *On generation of probable primes by incremental search*, Proceedings CRYPTO'92, Lecture Notes in Computer Science, **740** (1992), pp. 358–370.

[5] Y. Gurevich: *Average case complexity*, J. Comp. Sys. Sci., **42** (1991), pp. 346–398.

[6] Y. Gurevich: *Matrix decomposition problem is complete for the average case*, Proceedings 31-st IEEE Symp. on Foundations of Computer Science, (1990), pp. 802–811.

[7] J. Gordon: *Strong primes are easy to find*, Proceedings EUROCRYPT'84, Lecture Notes in Computer Science, **209** (1984), pp. 216–223.

[8] L. Levin: *Average case complete problems*, SIAM J. on Computing, **15** (1986), pp. 285–286.

[9] H.W. Lenstra Jr.: *Factoring integers with elliptic curves*, Annals of Mathematics, **126** (1987), pp. 649–673.

[10] U. Maurer: *Fast generation of prime numbers and secure public-key cryptographic parameters*, Journal of Cryptology, **8 (3)** (1995), pp. 123–155.

[11] P. Mihăilescu: *Fast generation of provable primes using search in arithmetic progressions*, Proceedings CRYPTO'94, Lecture Notes in Computer Science, **839** (1994), pp. 282–293.

[12] R. Peralta and V. Shoup: *Primality testing with fewer random bits*, Computational Complexity, **3** (1993), pp. 355–367.

[13] R. Schoof: *Counting points on elliptic curves over finite fields*, J. de Théorie des Nombres, Bordeaux, **7** (1995), 219–254.

[14] V. Shoup: *Removing randomness from computational number theory*, PhD Thesis, University of Wisconsin – Madison (1989).

[15] N. Smart: *The discrete logarithm problem on elliptic curves of trace one.* Journal of Cryptology **12 (3)**: pp. 193–196 (1999)

[16] J. Shawe-Taylor: *Generating strong primes*, Electronics Letters, **22 (16)** (1986), pp. 875–877.

Institut für Wissenschaftliches Rechnen, ETH, 8090 Zürich
E-mail address: `mihailes@inf.ethz.ch`

Progress in Computer Science and Applied Logic, Vol. 20
© 2001 Birkhäuser Verlag Basel/Switzerland

Achieving Optimal Fairness from Biased Coinflips

Mats Näslund and Alexander Russell

Abstract. We explore the problem of transforming n biased, independently identically distributed, $\{-1, 1\}$-valued random variables, $X_1, \ldots, X_n$, into a single $\{-1, 1\}$ random variable, $f(X_1, \ldots, X_n)$, so that this result is as unbiased as possible. We perform the first quantitative study of the relationship between the bias b of these X_i, $b = \mathbb{E}(X_i)$, and the *rate* at which $f(X_1, \ldots, X_n)$ can converge to an unbiased $\{-1, 1\}$ random variable (as $n \to \infty$). In general, it is not possible to produce a totally unbiased output. Moreover, quite different behavior can be observed, depending on whether the bias is rational or irrational, algebraic or transcendental. A perhaps surprising common property of all our results is that the *magnitude* of the input bias b has only secondary influence on the achievable output bias of f.

For rational biases we explicitly construct a polynomial time (in n) computable function $f(X_1, \ldots, X_n)$, achieving the optimal output bias.

We give a metrical result, stating that for (Lebesgue) almost all biases $b \in [0, 1]$, a perfectly unbiased output can be produced. This result is derived from a new metrical result on multi-dimensional Diophantine approximation and uniform distribution mod 1.

When the bias is an algebraic number, we demonstrate a curious "boundary behavior" which to some extent coincides with whether the bias is an algebraic integer or not.

1. Introduction

The general problem of producing unbiased random bits from an imperfect random source has received enormous attention. This study essentially began with von Neumann [17] in 1951 and, following his work, a variety of models of such "imperfect sources" have been defined and studied. We study the problem of transforming n independent random bits $X_1, \ldots, X_n$, each of fixed bias b, into a single bit $f(X_1, \ldots, X_n)$ which is as unbiased as possible. These "bits," X_i, are $\{-1, 1\}$ random variables, terminology we shall use throughout the article. Such a bit then has bias $-1 < b < 1$ if $\mathbb{E}(X_i) = \Pr[X_i = 1] - \Pr[X_i = -1] = b$. We generally assume that b is positive.

A natural method for producing such a "nearly unbiased" bit is to XOR (multiply) the input bits. These input bits being independent, $\mathbb{E}(\prod_{i=1}^{n} X_i) =$

$\prod_{i=1}^{n} \mathbb{E}(X_i) = b^n$, so the XOR function produces the respectable bias b^n. One curious conclusion of this paper, however, is that XOR is essentially always non-optimal (see below). In general, the bias produced by a specific function $f : \{-1,1\}^* \to \{-1,1\}$ when coupled with n input bits of bias b is denoted

$$\xi_{f,n}(b) \triangleq |\mathbb{E}(f(X_1,\dots,X_n))| .$$

The natural value with which to compare this is

$$\Xi_n(b) \triangleq \min_f \xi_{f,n}(b),$$

this minimum taken over all functions $f : \{-1,1\}^* \to \{-1,1\}$.

Returning to the XOR function, we have $\xi_{\text{XOR},n}(b) = b^n$. From each new b-biased bit, then, XOR "extracts" another (constant) factor b in the output bias. We naturally expect any asymptotically optimal function to extract at least this extra multiplicative factor of b with each new bit. With this in mind, we define the normalized quantity

$$\Xi(b) \triangleq \lim_{n\to\infty} \sqrt[n]{\Xi_n(b)}.$$

This quantity represents the optimal asymptotic multiplicative effect of each new input bit on the resulting bias. Of course, $\Xi(b) \leq \lim_{n\to\infty} \sqrt[n]{\xi_{\text{XOR},n}(b)} = b$. This value, $\Xi(b)$, we call the *extraction rate* of b.

The general problem is that of suitably partitioning the space of possible outcomes of $(X_1,\dots,X_n)$ into two sets (the "-1 set" and the "1 set") so that the probability of $(X_1,\dots,X_n)$ falling into each of these sets is roughly the same. In the case where the bias is rational and one can rely upon the arithmetic structure of $\mathbb{Z}$, such partitions can be explicitly built (see §2) with little difficulty. Specifically, for a bias b with $\frac{1+b}{2} = \frac{r}{s}$ (where $(r,s) = 1$), one can easily see that $\Xi(b) \geq \frac{1}{s}$ (see Theorem 4) and we show, in fact, that polynomial-time computable partitions of the outcomes of $(X_1,\dots,X_n)$ can be found which asymptotically achieve this lower bound, that is, $\Xi(b) = \frac{1}{s}$. It is worth noticing that this offers exponential improvement (in the achieved bias) over XOR (that is, the function $\prod_i X_i$). (For example, when $b = \frac{3}{5}$, so that $\frac{r}{s} = \frac{4}{5} = (1+\frac{3}{5})/2$, XOR produces bias $\xi_{\text{XOR},n}\left(\frac{3}{5}\right) = (\frac{3}{5})^n$, exponentially inferior to the $(\frac{1+o(1)}{5})^n$ offered by our optimal algorithm.)

In general, however, the behavior of Ξ depends on rather deep properties of the bias b. Indeed, the cases for (most) algebraic and transcendental numbers seem to require quantitative bounds on the behavior of sums of "independent" irrationals. The theory of Diophantine approximation seems the natural place to turn for such machinery. Section §3 shall be spent developing that theory in the direction necessary to prove results concerning irrational biases.

Sharply contrasting the case for rational biases (where $\Xi(b) > 0$), in §5 we prove a metrical result concerning multi-dimensional Diophantine approximation which shows (see Theorem 21) that for (Lebesgue) almost all b, $\Xi(b) = 0$.

Algebraic numbers exhibit particularly interesting "boundary" behavior. This is discussed in section §4, where we prove Theorems 16, and 17, which can be viewed as a classification result. Algebraic numbers fall into two classes;

Class I: Those algebraic numbers b for which $\Xi(b) = \epsilon > 0$. Furthermore, $c_1 \leq \Xi(b) \leq c_2$, where c_1 and c_2 are constants depending only on b's algebraic characteristics.

Class II: Those algebraic numbers b for which $\Xi(b) = 0$ and, furthermore, for which $\exists n_0, \forall n > n_0, \Xi_n(b) = 0$.

Notice that this classification excludes the possibility of algebraic biases b with $\Xi(b) = 0$ but for which no unbiased bit can be extracted from any finite collection of X_i. Both classes are rich: it is easy to construct infinite collections of examples of Class 2 (one can take $b = 2\alpha_i - 1$, where $\alpha_i \in \{\frac{1}{\sqrt{2}}, \frac{1}{\sqrt[3]{2}}, \dots\}$, for example), and a result of Feldman, et. al. [7] shows that Class 1 in fact contains all algebraic integers. These results also demonstrate that $\Xi(b)$ depends essentially on b's algebraic properties (and not on its magnitude, as one might expect).

1.1. Related Work

The bulk of the previous work in this area studies problems roughly dual to ours. There are two predominant directions. The first is the case originally considered by von Neumann, [17]. In this scenario, the bias of the X_i is unknown and one wishes to produce totally unbiased output bits. His work has been substantially generalized and refined (see [2, 13, 8, 5, 6, 10, 1, 16]). It is an essential property of the algorithms studied in these papers that they have the option of producing an empty stream of output. These algorithms, where the bias is unknown, provide upper bounds on the *expected* number of input bits necessary to produce an unbiased output bit. The second scenario is that of Itoh [9] and Feldman, et. al. [7], which focuses more closely on properties of the input biases. They show that by *carefully selecting* the input biases, they can always produce an unbiased output bit with a large, but *fixed* number of input bits. Finally, for a sequence of independently identically distributed random variables $\{X_i\}$, Cover [3] has studied procedures for determining certain rationality characteristics of $\mathbb{E}(X_i)$.

It is worth noticing that these "dual" problems discussed above are quite different in nature from the problem we consider, as they focus on conditions necessary for producing perfectly unbiased output distributions. For general biases, one cannot hope to achieve this. The main contribution of this paper is the first systematic approach to bounding the error terms resulting from the combination of such independent, similarly biased bits. In contrast to the above work, this allows us to elucidate the behavior of algorithms designed to produce approximately unbiased output distributions with no requirements on the either the number or bias of the input bits.

1.2. Some Notation

If S is a set, we will let $|S|$ denote its cardinality. For $\gamma \in \mathbb{R}$, let $[\gamma]$ denote the integer part of γ and $\|\gamma\|$ (or $\gamma \bmod 1$) the fractional part. Similarly, by $\langle\gamma\rangle$ we denote the distance from γ to the closest integer, that is, $\langle\gamma\rangle = \min(\|\gamma\|, 1 - \|\gamma\|)$. Furthermore for $\vec{\gamma} = (\gamma_1, \dots, \gamma_k) \in \mathbb{R}^k$ we write $\langle\vec{\gamma}\rangle$ rather than $(\langle\gamma_1\rangle, \dots, \langle\gamma_k\rangle)$ and use $\|\vec{\gamma}\|$ similarly. For an algebraic number α, the *irreducible polynomial of α*

refers to the unique polynomial $g \in \mathbb{Z}[x]$ of minimal degree with relatively prime coefficients and positive leading term for which $g(\alpha) = 0$. Then the *height* of α, denoted $\mathfrak{H}(\alpha)$, is defined as the maximum of the absolute values of the coefficients of this polynomial.

Let $X_1, \ldots, X_n$ be independent $\{-1, 1\}$-valued random variables each with bias $0 < b < 1$, so that $\Pr[X_i = 1] = (1+b)/2 \triangleq p$. We let $\{-1, 1\}^*$ denote the set of all finite length strings over these two symbols. For a function $f : \{-1, 1\}^* \rightarrow \{-1, 1\}$ we define $\xi_{f,n}(b) \triangleq |\mathbb{E}(f(X_1, \ldots, X_n))|$ and $\Xi_n(b) \triangleq \min_f \xi_{f,n}(b)$, where the minimum is taken over all functions f. Finally, the *extraction rate* of b is defined

$$\Xi(b) \triangleq \lim_{n \to \infty} \sqrt[n]{\Xi_n(b)}.$$

(It is easy to see that this limit indeed exists.)

Let $\mathcal{B}_n$ be the boolean lattice $\{-1, 1\}^n$ and for $\vec{x} \in \mathcal{B}_n$, let $\mathrm{wt}(\vec{x})$ denote the Hamming weight of $\vec{x}$. $\mathcal{B}_n^i$ shall denote the ith level of $\mathcal{B}_n$, that is, all $\vec{x} \in \mathcal{B}_n$ with $\mathrm{wt}(\vec{x}) = i$. Then, $|\mathcal{B}_n^i| = \binom{n}{i}$. If $X_1, \ldots, X_n$ are independent bits with bias b we may associate with each outcome of $(X_1, \ldots, X_n)$ a point in $\mathcal{B}_n$ and, for $\vec{x} \in \mathcal{B}_n^i$, the event $(X_1, \ldots, X_n) = \vec{x}$ has probability $P_n^i(p) \triangleq p^i(1-p)^{n-i}$. For a collection of outcomes $C \subseteq \mathcal{B}_n$ we define $w(C) = \Pr[(X_1, \ldots, X_n) \in C] = \sum_{\vec{c} \in C} P_n^{\mathrm{wt}(\vec{c})}(p)$.

The k-dimensional Lebesgue measure of a measurable set $A \subset \mathbb{R}^k$ is denoted $\lambda(A)$.

2. Rational Bias

We begin now by studying the behavior of $\Xi(b)$ for rational biases $b > 0$. Fix b and define $\frac{r}{s} \triangleq \frac{1+b}{2}$, for relatively prime r and s. To start with, let us see that one cannot extract an unbiased bit from *any* collection of b-biased bits and that, in fact, $\Xi(b) \geq \frac{1}{s}$.

Lemma 1. *For a rational bias b as above, $\Xi_n(b) \geq \frac{1}{s^n}$ and hence $\Xi(b) \geq \frac{1}{s}$.*

Proof. Consider a collection $C \subseteq \mathcal{B}_n$ with $w(C) = \frac{1+\delta}{2}$, $-1 \leq \delta \leq 1$ so that $|\delta|$ is the bias produced by the function

$$f(X_1, \ldots, X_n) = \begin{cases} 1 & \text{if } (X_1, \ldots, X_n) \in C, \\ -1 & \text{otherwise.} \end{cases}$$

Suppose, without loss of generality, that $(-1, \ldots, -1) \notin C$. The weight of C is

$$\frac{1+\delta}{2} = \sum_{i=1}^n t_i \left(\frac{r}{s}\right)^i \left(1 - \frac{r}{s}\right)^{n-i} = \frac{1}{s^n} \sum_{i=1}^n t_i r^i (s-r)^{n-i},$$

for some integers $0 \leq t_i \leq \binom{n}{i}$. Multiplying through by $2s^n$ we have

$$s^n(1+\delta) = 2 \sum_{i=1}^n t_i r^i (s-r)^{n-i}.$$

Notice that δ is non-zero: r divides the right-hand side, $r > 1$ (since $b > 0$), and r and s are relatively prime. Furthermore, the right-hand side is always an integer, so that the left-hand side must be also and $\Xi_n(b) = |\delta| \geq \frac{1}{s^n}$. Hence $\Xi(b) \geq \frac{1}{s}$. $\quad\square$

As promised, we proceed with a matching upper bound. We construct a family of functions $f_n : \{-1,1\}^n \to \{-1,1\}$ for which $\lim_{n\to\infty} \sqrt[n]{|\mathbb{E}(f_n(X_1,\dots,X_n))|} = \frac{1}{s}$, showing that $\Xi(b) \leq \frac{1}{s}$.

Lemma 2. *For a rational bias b as above, $\Xi(b) \leq \frac{1}{s}$. Furthermore, for $n > 2r+1$, $\Xi_n(b) \leq \frac{2r(s-r)^2}{s^n}$ and an optimal f, satisfying $\xi_{f,n}(b) \leq \frac{2r(s-r)^2}{s^n}$, can be found (and computed) in polynomial time.*

Proof. Let $q = s - r$ so that $\frac{q}{s} + \frac{r}{s} = 1$. Since we work with positive bias, $r > q$. Considering the Boolean lattice as described above, the probabilities associated with the ith level are $P_n^i\left(\frac{r}{s}\right) = \frac{r^i q^{n-i}}{s^n}$. For convenience, we shall scale all of our probabilities by s^n, and work with quantities $R_n^i = s^n P_n^i(\frac{r}{s})$. Our goal, then, is to build a collection $C_n \subset \mathcal{B}_n$ for which the scaled weight $s^n w(C_n) = \sum_{\vec{c} \in C_n} R_n^{\text{wt}(\vec{c})}$ is close to $\frac{s^n}{2}$. The function

$$f_n(X_1,\dots,X_n) = \begin{cases} 1 & \text{if } (X_1,\dots,X_n) \in C_n, \\ -1 & \text{otherwise} \end{cases}$$

will then have $\mathbb{E}(f_n(X_1,\dots,X_n))$ close to zero.

Form an initial collection $\tilde{C}_n = \mathcal{B}_n^n \cup \mathcal{B}_n^{n-1} \cup \mathcal{B}_n^{n-2} \cup T$ where T is a maximal subset of $\bigcup_{i < n-2} \mathcal{B}_n^i$ for which $s^n w(\tilde{C}_n) < \frac{s^n}{2}$ and $|(\overline{T} \cap \mathcal{B}_n^j)| \geq r-1$ for $1 \leq j \leq n-3$. (Note that for $n > r$, $|\mathcal{B}_n^j| > r$ for $1 \leq j \leq n-3$.) We shall use this extra space left over on each level j, $(1 \leq j \leq n-3)$, to adjust this initial collection, bringing its scaled weight closer to $\frac{s^n}{2}$. Since $r > q$, $R_n^i < R_n^j$ for any $i < j$ so that from the maximality of T we have $\left|s^n w(\tilde{C}_n) - [\frac{s^n}{2}]\right| < R_n^{n-2} = r^{n-2}q^2$.

We now use elements in levels 1 through $n - 3$ to build an approximation of $s^n w(\tilde{C}_n) - [\frac{s^n}{2}]$ modulo $r^{n-2}q^2$. This approximation, we show, can then be lifted to an agreeable adjustment of $\tilde{C}_n$.

Consider the group $\mathbb{Z}/(r^{n-2}q^2)$ and let $\phi : \mathbb{Z} \to \mathbb{Z}/(r^{n-2}q^2)$ be the natural quotient map. Since r and q are relatively prime, the element $\phi(r^{n-i}q^i)$ has order r^{i-2} for any $i \geq 2$, so that we have a series of subgroups of $\mathbb{Z}/(r^{n-2}q^2)$

$$(0) = (\phi(r^{n-2}q^2)) < (\phi(r^{n-3}q^3)) < \cdots < (\phi(rq^{n-1})),$$

where (γ) denotes the (cyclic) subgroup generated by γ. (Here $<$ denotes subgroup containment.) Each group has index r in the next, the last having index rq^2 inside $\mathbb{Z}/(r^{n-2}q^2)$. This last subgroup (generated by rq^{n-1}) can be used to "approximate" any value in $\mathbb{Z}/(r^{n-2}q^2)$ to within an additive error of rq^2. Specifically, defining $\Delta = \phi([\frac{s^n}{2}] - s^n w(\tilde{C}_n))$, there is $\Delta' \in (\phi(rq^{n-1}))$ so that $\Delta - \Delta' \in \{\phi(0), \phi(1), \dots, \phi(rq^2 - 1)\}$.

If we represent Δ' as $c\phi(rq^{n-1})$, we may have difficulty lifting this to an adjustment of $\tilde{C}_n$ because c may be large. Using elements from each of the nested subgroups above leads to an expression for $\tilde{h}$ which we can lift. First record the following observation.

Observation 3. *Let $\{e\} = G_0 < G_1 < \cdots < G_n = G$ be a sequence of groups, written additively, each a subgroup of the next. Let $\Gamma_i \subset G_i$ be a system of representatives for the left cosets of G_{i-1} in G_i (so that $G_i = (\gamma_1 + G_{i-1}) \cup \ldots \cup (\gamma_t + G_{i-1}))$. Then $G = \Gamma_n + \Gamma_{n-1} + \cdots + \Gamma_1 = \{\gamma_n + \cdots + \gamma_1 \;:\; \gamma_i \in \Gamma_i\}$.*

For two cyclic groups $(\gamma) \subset (\delta)$ with $\gamma = t\delta$, say, the set $\{\delta, 2\delta, \ldots, (t-1)\delta\}$ is a system of representatives of the cosets of (γ) in (δ). So, applying the above lemma with $G = (\phi(rq^{n-1}))$ and the subgroups as listed above, we have an equation in G

$$\Delta' = t_1 \cdot \phi(rq^{n-1}) + \cdots + t_{n-3} \cdot \phi(r^{n-3}q^3)$$

where $0 \le t_i < r$ for each i. Since $r > q$, we may lift the modular approximation $\Delta - \Delta' \in \{\phi(0), \ldots, \phi(rq^2 - 1)\}$ to an expression

$$\left[\frac{s^n}{2}\right] = t_1 \cdot rq^{n-1} + \cdots + t_{n-3} \cdot r^{n-3}q^3 - mr^{n-2}q^2 + w(\tilde{C}_n)s^n + E$$

where $m \le nr$ and the error E lies in $\{0, \ldots, rq^2 - 1\}$. Now, adding $t_i < r$ members of $\mathcal{B}_n^i$ to $\tilde{C}_n$ and removing m members of $\mathcal{B}_n^{n-2}$ from $\tilde{C}_n$ (which we can do as long as $m < \binom{n}{n-2}$, that is, $\frac{n-1}{2} > r$), we have a new collection C_n for which $w(C_n)s^n - \left[\frac{s^n}{2}\right] < E$. Dividing through by s^n gives $\Xi_n(b) \le \frac{2rq^2}{s^n}$ and hence $\Xi(b) \le \lim_{n\to\infty} \sqrt[n]{\frac{2rq^2}{s^n}} = \frac{1}{s}$, as desired.

Finally, notice that each step of the above construction can in fact be carried out in polynomial time.   □

Combining these two lemmas yields the theorem discussed in the introduction.

Theorem 4. *For rational b as above, $\Xi(b) = \frac{1}{s}$.*

3. Discrepancy

For the moment, let us again reformulate our general problem as a "partitioning problem": given the probabilities $P_n^i(p) = p^i(1-p)^{n-i}$, we should like to find integers $t_1, t_2, \ldots, t_n$ with $0 \le t_i \le \binom{n}{i}$ so that $\sum_{i=0}^n t_i P_n^i(p)$ is as close to $\frac{1}{2}$ as possible. When p is irrational, the group-theoretic tools of the previous section are no longer available. Instead, we shall study the "richness" of the set of numbers expressible by sums of the above form, showing that it behaves rather like a set of *random* numbers. Intuitively, this seems a favorable state of affairs since a collection of $\prod_i \binom{n}{i}$ random numbers in $[0,1]$ should include some very close to $\frac{1}{2}$. This will lead to very general results about sums of "independent irrationals" which we apply in both the algebraic and the metrical (probability 1) case.

Given an irrational number ζ, the celebrated *Weyl equidistribution theorem* states that as N limits, the sequence of fractional parts of $\zeta, 2\zeta, \ldots, N\zeta$ becomes uniformly distributed[1] in $[0, 1)$. It is this sort of "richness" which we shall study. The rate at which this sequence converges to the uniform distribution is known to depend on the extent to which ζ is approximable by rationals (see [11], for example). This section will prove similar results for sums of k irrationals. That is, we study the rate at which the sequence of fractional parts of

$$t_1\zeta_1 + t_2\zeta_2 + \cdots + t_k\zeta_k, \qquad 0 \le t_i < N, t_i \in \mathbb{Z}$$

converges to uniform distribution on $[0, 1)$. As will be shown, the one-dimensional behavior of this type of sequence is intimately related to the behavior of the vector sequence $\|t\vec{\zeta}\|$, $1 \le t \le N^k$, in $[0, 1)^k$. It is immediately clear that favorably "random" behavior on the part of $\|t\vec{\zeta}\|$ requires that these ζ_i (in addition to being irrational) are "independent". (For example, if $\zeta_i = \zeta_j$ for all i, j, then $\|t\vec{\zeta}\|$ will always lie in the set $\{(x, \ldots, x) : x \in [0, 1)\} \subset [0, 1)^k$ and the set $\{\|t\vec{\zeta}\| : t < N\}$ will then have constant discrepancy (see Definition 2) for any N.) Such issues are, in general, very delicate (see the discussion opening §5), and we refer the interested reader to the comprehensive book of Kuipers and Niederreiter [11]. Since the numbers we must combine are of form $p^i(1-p)^{n-i}$, it is not *a priori* clear that they should be "independent" in this sense. The central notion in the study of such independence is that of *type*[2], defined below.

Definition 1. *The vector $\vec{\zeta} = (\zeta_1, \zeta_2, \ldots, \zeta_k) \in \mathbb{R}^k$, is said to have type at most $\psi : \mathbb{N} \to \mathbb{R}^+$ (or simply type ψ) if there is $q_0 \in \mathbb{N}$ such that for all $q \ge q_0$, and for all integers $0 \le q_1, q_2, \ldots, q_k \le q$ (not all zero):*

$$\langle q_1\zeta_1 + q_2\zeta_2 + \cdots + q_k\zeta_k \rangle \ge \frac{1}{q^k\psi(q)}.$$

In general, for vectors $\vec{\zeta}$ with small type one has graceful convergence of the sequence $\|t\vec{\zeta}\|$ to the uniform distribution. One measure of the *rate* at which such a sequence convergences is the notion of *discrepancy*, defined next.

Definition 2. *For a finite set of points $W = \{\vec{w}_0, \vec{w}_1, \ldots, \vec{w}_{N-1}\} \subset [0, 1)^k$, the discrepancy of W is defined*

$$\mathcal{D}(W) \triangleq \sup_B \left| \frac{|W \cap B|}{N} - \lambda(B) \right|,$$

---

[1] An infinite sequence in $[0, 1)$ is *uniformly distributed* if, for every interval $I \subset [0, 1)$, the fraction of points falling into I is asymptotically $\lambda(I)$. A simple Fourier-analytic proof can be given, which explains the presence of the exponential sums in Theorem 6.

[2] The notion of *type* we shall use is weaker than the traditional one in that the coefficients q_i are required to be positive. (The traditional definition bounds the behavior for all coefficients $q_1, \ldots, q_k$ with $|q_i| < q$, while we work only with positive such q_i.) Then any upper bound on the classical notion is also an upper bound for this weaker notion. Though this difference is predominantly incidental, it is required in the proof of Lemma 18.

where this supremum is taken over all "boxes" $B \subset [0,1)^k$ of form

$$\{(b_1, b_2, \ldots, b_k) \; : \; x_1 \le b_1 < y_1, \ldots, x_k \le b_k < y_k\},$$

for $\vec{x}, \vec{y} \in [0,1)^k$. Occasionally, we will also study one-dimensional discrepancy modulo σ, $\sigma < 1$, that is, the deviation from uniform distribution in $[0, \sigma)$.

The main result of this section will be the following theorem:

Theorem 5. Let $\vec{\zeta} = (\zeta_1, \zeta_2, \ldots, \zeta_k) \in \mathbb{R}^k$ and define

$$P(\vec{\zeta}, N) \triangleq \{\|t_1\zeta_1 + \cdots + t_k\zeta_k\| \; : \; 0 \le t_i < N, t_i \in \mathbb{Z}\}.$$

Then for every $\epsilon > 0$, there is $\delta > 0$ so that if $\vec{\zeta}$ has type $\psi(q) = O(q^\delta)$ then $\mathcal{D}(P(\vec{\zeta}, N)) = O(N^{-k(1-\epsilon)})$.

We shall here only outline the rather technical details behind the proof. The reader who wishes can for the moment simply note Theorem 5, proceed to Section 4, and return here later.

Definition 3. For any $\vec{\zeta} = (\zeta_1, \zeta_2, \ldots, \zeta_k) \in \mathbb{R}^k$, the reduced volume of $\vec{\zeta}$ is

$$\mathfrak{V}\left\langle \vec{\zeta} \right\rangle \triangleq \prod_{j=1}^{k} \langle \zeta_j \rangle.$$

A primary tool for bounding discrepancy is the following theorem. A proof can be found in [11].

Theorem 6 (Erdős-Turan). For any finite set $W = \{w_0, w_1, \ldots, w_{N-1}\} \subset [0,1)$ and any positive integer m:

$$\mathcal{D}(W) \le C \left(\frac{1}{m} + \sum_{h=1}^{m} \frac{1}{h} \left| \frac{1}{N} \sum_{j=0}^{N-1} e^{2\pi i h w_j} \right| \right)$$

where C is a constant independent of W.

Applying the Erdős-Turan theorem to the sets relevant to our study results in the following discrepancy bounds.

Lemma 7. For any $\vec{\zeta} = (\zeta_1, \zeta_2, \ldots, \zeta_k) \in (\mathbb{R} \setminus \mathbb{Q})^k$, the set

$$P(\vec{\zeta}, N) = \{\|t_1\zeta_1 + \cdots + t_k\zeta_k\| \; : \; 0 \le t_i < N, t_i \in \mathbb{Z}\}$$

satisfies

$$\mathcal{D}(P(\vec{\zeta}, N)) \le \frac{C}{N^k} \left(1 + \sum_{h=1}^{N^k} \frac{1}{h\mathfrak{V}\left\langle h\vec{\zeta} \right\rangle} \right) \tag{1}$$

where C is an absolute constant.

The proof is very similar to that of Theorem 2.5 of [11] and therefore omitted. Given Lemma 7, we can bound the discrepancy by controlling sums of the form found in equation (1) above. The primary step in this process is the following bound.

Lemma 8. *Let* $\vec{\zeta} = (\zeta_1, \zeta_2, \ldots, \zeta_k) \in (\mathbb{R} \setminus \mathbb{Q})^k$. *Then for every* $\epsilon > 0$, *there is* $\delta > 0$ *so that if* $\vec{\zeta}$ *is of type* $\psi(q) = O(q^\delta)$ *then*

$$\sum_{j=1}^{N} \frac{1}{\mathfrak{V}\left\langle j\vec{\zeta} \right\rangle} \in O(N^{1+\epsilon}). \tag{2}$$

We only give an outline the proof. Initially, we invoke a lemma (see Lemma 9 below), stating that if the type of $\vec{\zeta}$ is small, then the set

$$S(\vec{\zeta}, N) \triangleq \left\{ \left\| t\vec{\zeta} \right\| : 1 \le t \le N \right\}$$

is roughly uniformly distributed in $[0, 1)^k$. We now partition $[0, 1)^k$ into a family of rectilinear regions, $\{H_s\}$, each region containing points whose $\mathfrak{V}\langle \cdot \rangle$-values are close; $z \in H_{s_i}$ implies $s_i \le \mathfrak{V}\langle z \rangle \le s_{i+1}$. Applying discrepancy results (that is, Lemma 9), we show that each of these H_s intersects roughly the expected number of points from the set S; roughly $N\lambda(H_s)$ points. Regions lying close to the boundaries require special treatment, and for these we observe (see Lemma 10) that the above discrepancy results also offer immediate *lower bounds* on $\mathfrak{V}\langle \cdot \rangle$. This shows that $S(\zeta, \vec{N})$ intersects none of these dangerous regions and there is an s_0 such that we do not need to consider any regions H_s with $s < s_0$. Having carried out this program, we can then rewrite the sum in (2) of Lemma 8 as a sum over regions:

$$\sum_{j=1}^{N} \frac{1}{\mathfrak{V}\left\langle j\vec{\zeta} \right\rangle} \overset{\text{L.10}}{=} \sum_{j>s_0} \sum_{i:\langle i\vec{\zeta} \rangle \in H_{s_j}} \frac{1}{\mathfrak{V}\left\langle i\vec{\zeta} \right\rangle} \overset{\text{L.9}}{\approx} N \sum_{j>s_0} \frac{\lambda(H_{s_j})}{s_j}.$$

Choosing the sets H_s appropriately, we can conclude the statement of Lemma 8.

Lemma 9 (Niederreiter [12]). *For* $\vec{\zeta} \in (\mathbb{R} \setminus \mathbb{Q})^k$, *and, as above, let*

$$S(\vec{\zeta}, N) = \left\{ \left\| t\vec{\zeta} \right\| : 1 \le t \le N \right\}.$$

Then for every $\epsilon > 0$, *there is* $\delta > 0$ *so that if* $\vec{\zeta}$ *has type* $\psi(q) = O(q^\delta)$ *then*

$$\mathcal{D}(S(\vec{\zeta}, N)) = O(N^{-(1-\epsilon)}).$$

Lemma 10. *Suppose that the set* $S(\vec{\zeta}, N)$ *(as in Lemma 9) satisfies* $\mathcal{D}(S(\vec{\zeta}, N)) \in O(N^{-(1-\epsilon)})$. *Then for any* δ *large enough that* $(1 + \delta)(1 - \epsilon) > 1$, *for all but a finite number of* N:

$$\min_{1 \le j \le N} \mathfrak{V}\left\langle j\vec{\zeta} \right\rangle \ge N^{-(1+k\delta)}.$$

In particular, when $\epsilon < \frac{1}{2}$, *we may take* $\delta = 2\epsilon$ *so that* $\mathfrak{V}\left\langle j\vec{\zeta} \right\rangle \ge N^{-(1+2k\epsilon)}$.

The idea behind the proof of this lemma is simply that if extremely tiny regions of $[0, 1]^k$ are "hit" too often by elements from $S(\zeta, \vec{N})$, then the discrepancy of $S(\zeta, \vec{N})$ cannot be good. We note that if $\zeta_1, \ldots, \zeta_k, 1$ are algebraic and linearly independent over $\mathbb{Q}$, the above Lemma also follows from a result by Schmidt, [15].

This concludes the proof sketch of Lemma 8 and we end this section by seeing how Theorem 5 follow from the above.

Proof of Theorem 5. By Lemma 7, $\mathcal{D}(P(\vec{\zeta}, N)) \leq \frac{C}{N^k}\left(1 + \sum_{h=1}^{N^k} \frac{1}{h\mathfrak{V}\langle h\vec{\zeta}\rangle}\right)$.

By the Abel summation formula,

$$\sum_{h=1}^{N^k} \frac{1}{h\mathfrak{V}\left\langle h\vec{\zeta}\right\rangle} = \sum_{h=1}^{N^k} \frac{s_h}{h(h+1)} + \frac{s_{N^k}}{N^k + 1}$$

where $s_h = \sum_{j=1}^{h}\left(\mathfrak{V}\langle j\vec{\zeta}\rangle\right)^{-1}$.

Fix $\epsilon > 0$. Applying Lemma 8, there is $\delta > 0$ so that the assumption that $\vec{\zeta}$ has type $\psi = O(q^\delta)$ yields that $s_h = O(h^{1+\epsilon})$. Then, for some constant B:

$$\sum_{h=1}^{N^k} \frac{1}{h\mathfrak{V}\left\langle h\vec{\zeta}\right\rangle} \leq B\left[\sum_{h=1}^{N^k} \frac{1}{h^{1-\epsilon}} + N^{k\epsilon}\right] = O(N^{k\epsilon}).$$

Thus $\mathcal{D}(P(\vec{\zeta}, N)) = O(N^{-k(1-\epsilon)})$. $\qquad\qquad\qquad\qquad\qquad\qquad\square$

With this machinery at our disposal, we can now proceed to explore the behavior of $\Xi(\cdot)$ for irrational biases.

4. Algebraic Bias

We now consider the case where the bias b is an algebraic number. Let $\alpha = (1+b)/2$ where b is an algebraic number of degree at least $k+1$. As before, our goal is to find a collection C_n of $\mathcal{B}_n$ for which $w(C_n)$, the probability that $(X_1, \ldots, X_n) \in C_n$, is as close to $\frac{1}{2}$ as possible. We shall proceed roughly as in the rational case, selecting an easy initial collection $\tilde{C}_n$ and refining this collection.

Specifically, we begin with a subset of the lattice, taking the entire $\lfloor\frac{n}{2}\rfloor$th level but avoiding levels $\lfloor n/2\rfloor + 1, \ldots, \lfloor n/2\rfloor + k$, for which $w(\tilde{C}_n)$ is fairly close to $\frac{1}{2}$. In fact, one can easily find such a subset which whose weight is within Δ of $\frac{1}{2}$, where $0 < \Delta < p' = P_n^{\lfloor n/2\rfloor}(\alpha)$. We then refine this initial collection by the use of elements on levels $\lfloor n/2\rfloor + 1, \lfloor n/2\rfloor + 2, \ldots, \lfloor n/2\rfloor + k$. Specifically, we shall approximate Δ modulo p' by finding a subset C of these latter k levels whose weight $w(C)$ is very close to Δ when taken modulo p'. Specifically, writing $w(C) = \Delta' + mp'$, for $0 \leq \Delta' < p'$ and $m = \lfloor w(C)/p'\rfloor \in \mathbb{Z}$ we shall have $|\Delta - \Delta'|$

very small. Combining this C with our original collection yields a new set $C \cup \tilde{C}_n$ with

$$w(C \cup \tilde{C}_n) - \frac{1}{2} = mp' + |\Delta - \Delta'| .$$

Provided m is smaller than the number of elements on level $\lfloor n/2 \rfloor$, we can then remove m such elements that are already in $\tilde{C}_n$, producing a final set C_n with $|w(C_n) - 1/2| = |\Delta - \Delta'|$. Hence, the quality of this approximation depends on how close to Δ we can get by forming sums of the probabilities of levels $\lfloor n/2 \rfloor + 1, \lfloor n/2 \rfloor + 2, \ldots, \lfloor n/2 \rfloor + k$, modulo p'. Bounds on the discrepancy of such sums modulo p', then, can be translated into bounds on the quality of such an approximation.

A lower bound for the discrepancy of $\{t_1 \zeta_1 + \cdots + t_k \zeta_k : 0 \le t_j < N\}$ (think of ζ_j as the probability of an element on level $\lfloor n/2 \rfloor + j$ in $\mathcal{B}_n$) is roughly N^{-k}. Wishing to depress this bound as much as possible, we would like N to be as large as possible. In the outline above, N is bounded by the "width" of the Boolean lattice.

Before proceeding, we introduce some more notation.

Definition 4. Let $\alpha \triangleq (1+b)/2$, that is, the probability that a bit from our biased source is 1 and set $\sigma_n \triangleq \alpha^{\lfloor n/2 \rfloor}(1-\alpha)^{n-\lfloor n/2 \rfloor}$, the probability that $(X_1, \ldots, X_n)$ has Hamming weight exactly $\lfloor n/2 \rfloor$. Finally, let $\tau \triangleq \frac{\alpha}{1-\alpha}$. We will focus on the sequences produced by our source of length n and weight j for $j = \lfloor n/2 \rfloor, \lfloor n/2 \rfloor + 1, \ldots, \lfloor n/2 \rfloor + k$ (where $k+1$ is a lower bound on the degree of α). The probabilities associated with such outcomes then form the set $\{\sigma_n, \sigma_n \tau, \sigma_n \tau^2, \ldots, \sigma_n \tau^k\}$. Scaling by $\sigma_n \ne 0$, we define $\Gamma \triangleq \{1, \tau, \tau^2, \ldots, \tau^k\}$.

Next, we need some results on the algebraic properties of the numbers involved. First, it is straight forward to verify that if α is algebraic of degree at least $k+1$, then so is $\tau = \alpha(1-\alpha)^{-1}$. Hence

Claim 11. If α is algebraic of degree $t > k$ then Γ is linearly independent over the rationals.

For notational convenience, for $\vec{r} \in \mathbb{Z}^k$, define $\mathfrak{r}(\vec{r}) \triangleq \prod_{i=1}^{k} \max(|r_i|, 1)$.

Theorem 12 (Schmidt [15]). *Suppose that $\zeta_1, \zeta_2, \ldots, \zeta_k$ are algebraic numbers with $1, \zeta_1, \zeta_2, \ldots, \zeta_k$ linearly independent over $\mathbb{Q}$. Let $\vec{q} = (q_1, q_2, \ldots, q_k) \in \mathbb{Z}^k \setminus \{\vec{0}\}$. Then for any $\delta > 0$, and for all but finitely many $\vec{q}$,*

$$(\mathfrak{r}(\vec{q}))^{1+\delta} \left\langle \sum_{j=1}^{k} q_i \zeta_i \right\rangle \ge 1$$

Such $\vec{\zeta} = (\zeta_1, \ldots, \zeta_k)$ then have type $\psi(q) = q^{k\delta}$.

Collecting Claim 11, the previous theorem, and Theorem 5 we immediately have

Corollary 13. *The set*

$$P((\tau, \dots, \tau^k), N) = \left\{ \|t_1\tau + t_2\tau^2 + \dots + t_k\tau^k\| \ : \ 0 \le t_i < N \right\}$$

has discrepancy $\mathcal{D}(P) \in O(N^{-k(1-\epsilon)})$ for all $\epsilon > 0$.

Recalling the plan outlined at the beginning of the section, we shall start with some nicely structured "initial collection." The structure we need is described in the following observation, the proof of which we omit.

Observation 14. *Let α be as above. For all sufficiently large n, there is a subset $\tilde{C}_n \subset \mathcal{B}_n$ such that*

$$\mathcal{B}_n^{\lfloor n/2 \rfloor} \ \subset \ \tilde{C}_n,$$
$$\tilde{C}_n \cap \mathcal{B}_n^{\lfloor n/2 \rfloor + j} \ = \ \emptyset, \quad 1 \le j \le k, \ and$$
$$w(\tilde{C}_n) \ = \ 1/2 + \Delta(n),$$

where $0 < \Delta(n) < P_n^{\lfloor n/2 \rfloor}(\alpha)$.

We can now prove the main result of this section.

Theorem 15. *Let b be algebraic of degree at least $k + 1$. Then with α defined as above, for all $\epsilon > 0$, for all sufficiently large n,*

$$\Xi_n(b) \le c2^{-(1-\epsilon)kn},$$

for some constant c. Hence $\Xi(b) \le 2^{-k}$.

Proof. Select $\epsilon > 0$. Note that if b is algebraic of degree $d \ge k + 1$, then so is α (and vice versa). Our goal is, once again, to find a collection $C_n \subset \mathcal{B}_n$ so that the probability that $(X_1, \dots, X_n) \in C_n$ is as close to $1/2$ as possible.

Set $\tilde{C}_n$ as the collection found in Observation 14, so that $w(\tilde{C}_n) = \frac{1}{2} + \Delta$, $0 < \Delta < P_n^{\lfloor n/2 \rfloor}(\alpha)$, $\tilde{C}_n \cap (\bigcup_{j=1}^k \mathcal{B}_n^{\lfloor n/2 \rfloor + j}) = \emptyset$ and $\mathcal{B}_n^{\lfloor n/2 \rfloor} \subset \tilde{C}_n$. We now use levels $\lfloor n/2 \rfloor + 1, \dots, \lfloor n/2 \rfloor + k$ to approximate $\Delta \bmod P_n^{\lfloor n/2 \rfloor}(\alpha)$ and then "lift" this to an approximation modulo 1.

Define $N = \left| \mathcal{B}_n^{\lfloor n/2 \rfloor} \right|$ and observe that $\left| \mathcal{B}_n^{\lfloor n/2 \rfloor + j} \right| = 2^{(1-o(1))n}$ for $0 \le j \le k$. Now, let $N_1 = \frac{N}{k \lceil \tau^k \rceil}$ (recall that $\tau = \alpha(1-\alpha)^{-1}$) and consider the set of non-zero linear combinations of form

$$t_1\tau + t_2\tau^2 + \dots + t_k\tau^k, \qquad 0 \le t_i < N_1.$$

By Corollary 13, this set has discrepancy (modulo 1) $O(N_1^{-k(1-\epsilon)})$. Scaling this set by σ_n, we have a set

$$t_1 P_n^{\lfloor n/2 \rfloor + 1}(\alpha) + t_2 P_n^{\lfloor n/2 \rfloor + 2}(\alpha) + \dots + t_k P_n^{\lfloor n/2 \rfloor + k}(\alpha),$$

(for $0 \le t_i < N_1$) also with discrepancy $O(N_1^{-k(1-\epsilon)})$, but now modulo $\sigma_n = P_n^{\lfloor n/2 \rfloor}(\alpha)$. Thus, there is a $\vec{t} = (t_1, \dots, t_k)$ for which the corresponding linear combination gives a Δ' with $|\Delta - \Delta'| \le Q_\epsilon N_1^{-k(1-\epsilon)} \bmod P_n^{\lfloor n/2 \rfloor}(\alpha)$ for a constant

Q_ϵ. Form C by picking t_j elements from level $\lfloor n/2 \rfloor + j$, $j = 1, 2, \ldots, k$. Then with $m = \left\lfloor \frac{w(C)}{P_n^{\lfloor n/2 \rfloor}(\alpha)} \right\rfloor$ we have

$$w\left(\tilde{C}_n \cup C\right) - \frac{1}{2} = m P_n^{\lfloor n/2 \rfloor}(\alpha) + |\Delta - \Delta'|.$$

We now take our final C_n as $\tilde{C}_n \cup C$ except that we remove m elements from level $\lfloor n/2 \rfloor$. The entire $\lfloor n/2 \rfloor$th level is already included in $\tilde{C}_n$ and furthermore, since the bias is positive,

$$
\begin{aligned}
w(C) &= \sum_{j=1}^{k} t_j P_n^{\lfloor n/2 \rfloor + j}(\alpha) < \frac{N}{k \lceil \tau^k \rceil} k P_n^{\lfloor n/2 \rfloor + k}(\alpha) \\
&= \frac{N}{k \lceil \tau^k \rceil} k \tau^k P_n^{\lfloor n/2 \rfloor}(\alpha) < \left| \mathcal{B}_n^{\lfloor n/2 \rfloor} \right| P_n^{\lfloor n/2 \rfloor}(\alpha),
\end{aligned}
$$

so we must have $m < \left| \mathcal{B}_n^{\lfloor n/2 \rfloor} \right|$. Hence, we can find a subset C_n such that

$$\left| w(C_n) - \frac{1}{2} \right| \leq |\Delta - \Delta'| \leq Q_\epsilon \left(\frac{N}{k \lceil \tau^k \rceil} \right)^{-k(1-\epsilon)}$$

with $N = 2^{(1-o(1))n}$. The produced bias implies that

$$\Xi_n(b) \leq 2 Q_\epsilon \left(k \lceil \tau^k \rceil \right)^{k(1-\epsilon)} 2^{-(1-\epsilon)kn}$$

Taking the nth root and the limit as $n \to \infty$, the bound $\Xi(b) \leq 2^{-k}$ follows. $\square$

We proceed by proving that the algebraic biases fall into two distinct classes.

Theorem 16. *Let b be algebraic of degree $k+1$ and set $\alpha = (1+b)/2$. Then b falls into one of the following two classes.*

    Class I: $\Xi(b) > 0$ *and furthermore for all $\epsilon > 0$ and sufficiently large n,* $\Xi_n(b) \geq (2\mathfrak{H}(\alpha))^{-3(1+\epsilon)kn}$. *Hence $\Xi(b) \geq (2\mathfrak{H}(\alpha))^{-3k}$.*
    Class II: $\Xi(b) = 0$ *and there is a number n_0 so that for all $n > n_0$, $\Xi_n(b) = 0$.*

In other words, $\Xi(b)$ can not limit to zero.

Proof. Suppose that b is not in Class 2. Let $g(x) = \sum_{i=0}^{k+1} g_i x^i \in \mathbb{Z}[x]$, $|g_i| \leq \mathfrak{H}(\alpha)$, be the irreducible polynomial of α. Consider a family of partitions, $\{(C_n^{-1}, C_n^1)\}$, where C_n^{-1} and C_n^1 partition $\mathcal{B}_n$ and let $\left| (C_n^{-1} \cap \mathcal{B}_n^j) \right| = q_{nj}$. Then

$$w(C_n^{-1}) = \sum_{j=0}^{n} q_{nj} \alpha^j (1-\alpha)^{n-j} \tag{3}$$

where $q_{nj} \leq \binom{n}{j} \leq \binom{n}{n/2}$.

Recall that the bias produced by C_n^{-1}, C_n^1 is $|2w(C_n^{-1}) - 1|$. Now, reducing powers α^j, $j > k$, modulo g, scaling to clear denominators, enables us to deduce

$$g_{k+1}^{n-k} \left| 2w(C_n^{-1}) - 1 \right| = \left| \sum_{j=1}^{k} 2Q_{nj}\alpha^j - (g_{k+1}^{n-k} - 2Q_{n0}) \right| \geq \left\langle \sum_{j=1}^{k} 2Q_{nj}\alpha^j \right\rangle$$

since it can be shown that $Q_{nj} \in \mathbb{Z}$, and $|Q_{nj}| \leq 8^n \mathfrak{H}(\alpha)^{2n}$. But $1, \alpha, \alpha^2, \ldots, \alpha^k$ are linearly independent over $\mathbb{Q}$, so from Theorem 12, for all $\epsilon > 0$ (and sufficiently large n),

$$\left\langle \sum_{j=1}^{k} 2Q_{nj}\alpha^j \right\rangle \geq \left(\prod_{j=1}^{k} 2Q_{nj} \right)^{-(1+\epsilon)} \geq \frac{1}{2^{k(1+\epsilon)} 8^{nk(1+\epsilon)} \mathfrak{H}(\alpha)^{2nk(1+\epsilon)}}.$$

Hence, the bias produced by C_n^{-1} is bounded from below by

$$|2w(C_n^{-1}) - 1| \geq \frac{1}{g_{k+1}^{n-k} \cdot 2^{k(1+\epsilon)} 8^{nk(1+\epsilon)} \mathfrak{H}(\alpha)^{2nk(1+\epsilon)}} \geq \frac{1}{2^{k(1+\epsilon)} [2\mathfrak{H}(\alpha)]^{3nk(1+\epsilon)}}$$

Slightly adjusting ϵ yields the bound on $\Xi_n(\cdot)$ quoted in the theorem. The bounds on Ξ follow.   □

Combining the results in this section we conclude:

Theorem 17. Let $\alpha = \frac{1 \pm b}{2}$, for an algebraic bias b of degree at least $k + 1$. If $\Xi(b) \neq 0$, then for a constant c, depending essentially on α's algebraic properties, for all $\epsilon > 0$ and sufficiently large n,

$$(2\mathfrak{H}(\alpha))^{-3nk(1+\epsilon)} \leq \Xi_n(b) \leq c2^{-(1-\epsilon)kn}$$

and thus $(2\mathfrak{H}(\alpha))^{-3k} \leq \Xi(b) \leq 2^{-k}$.

5. A Metrical Result

Hoping to follow the program developed in the previous section, we would like to see that most biases $\zeta \in [0,1]$ will produce probabilities $P_n^i(\zeta)$ to which we can apply the discrepancy results of Section 3. As one would expect, the critical issue is the *type* of the vector $(\zeta, \zeta^2, \ldots, \zeta^k)$. Unfortunately, unlike for vectors with algebraic components, this appears to be an open problem when ζ is transcendental. There are strong metrical results (for example, see [14]), but they do not immediately apply to vectors of the above form which have highly correlated entries. Below we prove a metrical result for such vectors.

Lemma 18. Fix $\delta > 0$ and $k \in \mathbb{N}$. For (Lebesgue) almost all $\zeta \in [0,1]$, there are only finitely many $\vec{q} \in (\mathbb{N} \cup \{0\})^k$ for which

$$\left\langle \sum_{i=1}^{k} q_i \zeta^i \right\rangle < \frac{1}{\mathfrak{r}(\vec{q}) \prod_{i=1}^{k} \max(\ln^{1+\delta}(q_i), 1)}.$$

Proof. Fix $\beta \in (0,1)$. For $\vec{q} \in (\mathbb{N} \cup \{0\})^k$ set $D_{\vec{q}} \triangleq (\mathfrak{r}(\vec{q}) \prod_i \max(\ln^{1+\delta}(q_i), 1))^{-1}$ and define $T_{\vec{q}}(x) = q_1 x + \cdots + q_k x^k$, $Q = \sum_{i=1}^k q_i$. Then, for $x \in (0,1]$, we have $T_{\vec{q}}(x) \in [0, Q]$ and

$$T'_{\vec{q}}(x) \triangleq \frac{dT_{\vec{q}}}{dx}(x) \geq \frac{T_{\vec{q}}(x)}{x}.$$

For $S \subset \mathbb{R}$ and $\gamma > 0$, define $B_\gamma(S) \triangleq \{z \ : \ \exists s \in S, |s - z| \leq \gamma\}$. Let $\mathcal{Q}_\beta \subset \mathbb{N}^k$ denote the finite collection of $\vec{q}$ for which $T_{\vec{q}}(\beta) < 1$ or $D_{\vec{q}} \geq \frac{1}{2}$. For $\vec{q} \notin \mathcal{Q}_\beta$, we have

$$\Pr_{x\in[\beta,1]}[\langle T_{\vec{q}}(x)\rangle < D_{\vec{q}}] = \Pr_{x\in[\beta,1]}\left[T_{\vec{q}}(x) \in B_{D_{\vec{q}}}(\{1,\ldots,Q\})\right].$$

For $\vec{q} \notin \mathcal{Q}_\beta$, $T_{\vec{q}}([\beta,1]) \subset [1,Q]$ and, since $T'_{\vec{q}}(x) > 0$ for all $x \in [\beta,1]$, $T_{\vec{q}}$ is invertible on $[\beta,1]$. We define a probability measure $\nu_{\vec{q}}$ on $[1,Q]$ by setting $\nu_{\vec{q}}(S) = \mu(\{x \ : \ T_{\vec{q}}(x) \in S\})$, where μ is the uniform probability measure on $[\beta,1]$. Then

$$\nu_{\vec{q}}(S) = \int_S \frac{g'}{1-\beta} \, d\lambda$$

where $g = T_{\vec{q}}^{-1}$. (See [4, §5.5.4], for example, for a discussion of Radon-Nikodym derivatives.) Furthermore, notice that $T''_{\vec{q}}(x) \geq 0$ on $[\beta,1]$, so that for two intervals $[a,b]$ and $[a+\gamma, b+\gamma] \subset [1,Q]$ (for $\gamma > 0$), we have

$$\nu_{\vec{q}}([a,b]) > \nu_{\vec{q}}([a+\gamma, b+\gamma]). \tag{4}$$

Now, let j_0 be the smallest natural for which $B_{D_{\vec{q}}}(\{j_0\}) \cap T_{\vec{q}}([\beta,1]) \neq \emptyset$. Considering equation (4) above, for intervals of form $[j + D_{\vec{q}}, j + 1 + D_{\vec{q}}]$ (for $j \geq j_0$) we have $\nu_{\vec{q}}(B_{D_{\vec{q}}}(\{j + 1\})) \leq 2D_{\vec{q}} \nu_{\vec{q}}([j + D_{\vec{q}}, j + 1 + D_{\vec{q}}])$, we conclude that

$$\nu_{\vec{q}}\left(B_{D_{\vec{q}}}(\{j_0+1,\ldots,Q\})\right) \leq 2D_{\vec{q}}.$$

Then, since $g' \geq 1/(1-\beta)$ on $[1,Q]$, $\nu_{\vec{q}}(B_{D_{\vec{q}}}(\{j_0\})) < 2D_{\vec{q}}/(1-\beta)$,

$$\Pr_{x\in[\beta,1]}[\langle T_{\vec{q}}(x)\rangle < D_{\vec{q}}] = \nu_{\vec{q}}\left(B_{D_{\vec{q}}}(\{1,\ldots,Q\})\right) \leq \frac{4D_{\vec{q}}}{1-\beta}.$$

Now,

$$\sum_{\vec{q}} \frac{4}{1-\beta} D_{\vec{q}} = \frac{4}{1-\beta} \prod_{i=1}^k \sum_{q_i=0}^\infty \frac{1}{\max(q_i \ln^{1+\delta}(q_i), 1)}$$

which is finite, since each sum converges.

Recall the Borel-Cantelli lemma:

Lemma 19. *Let $E_1, E_2, \ldots$ be events for which $\sum_i \Pr[E_i] < \infty$. Then $\Pr[\bigcap_i \bigcup_{j>i} E_j] = 0$.*

A proof can be found in [4, §8.3.4], for example. Applying this to the events $E_{\vec{q}} = \{x \ : \ \langle T_{\vec{q}}(x)\rangle < D_{\vec{q}}\}$ shows that with probability 1 there are at most finitely many $\vec{q}$ for which $\langle T_{\vec{q}}(x)\rangle < D_{\vec{q}}$, as desired. Since β was arbitrary, the lemma is established. $\qquad\square$

From this easily follows:

Theorem 20. *For every $\delta > 0$, for (Lebesgue) almost all $\zeta \in [0,1]$, $(\zeta, \zeta^2, \ldots, \zeta^k)$ has type $\psi_k(q) = (\ln q)^{(1+\delta)k}$ for all $k \in \mathbb{N}$. Furthermore, for any k and any $\epsilon > 0$, the set $P((\zeta, \ldots, \zeta^k), N) = \{ \|t_1\zeta + t_2\zeta^2 + \cdots + t_k\zeta^k\| : 0 \le t_i < N \}$ satisfies*

$$\mathcal{D}(P((\zeta, \ldots, \zeta^k), N)) \in O(N^{-k(1-\epsilon)}).$$

Following the same ideas as in the proof of Theorem 15, we now have

Theorem 21. *If ζ is randomly chosen in $[0,1]$, with probability 1, $\Xi(\zeta) = 0$.*

6. Concluding Remarks

Some intriguing open problems remain: are there transcendental numbers ζ for which $\Xi(\zeta) > 0$? Is there $\epsilon > 0$, for which $\{b : \Xi(b) > \epsilon\}$ is infinite? We would like to see the gap closed in Theorem 17. The problem which arises from consideration of possibly mixed biases (input bits with perhaps different biases) seems interesting.

Finally, we would like to thank Johan Håstad for interesting discussions and comments concerning this paper. We also thank Harald Niederreiter for several helpful discussions.

References

[1] M. Blum. Independent unbiased coin flips from a correlated biased source: a finite state markov chain. *Combinatorica*, 6:97–108, 1986.

[2] J. Cohen. Fairing of Biased Coins in Bounded Time. *Yale University, Dept. of Computer Science* Technical Report no. 372, March 1985.

[3] T. Cover. On determining the irrationality of the mean of a random variable. *The Annals of Statistics*, 1(5):862–871, 1973.

[4] R. Dudley. *Real Analysis and Probability*. Chapman and Hall Mathematics Series. Chapman and Hall, New York, NY, 1989.

[5] M. Dwass. Unbiased coin tossing with discrete random variables. *Annals of Mathematical Statistics*, 43:860–864, 1972.

[6] P. Elias. The efficient construction of an unbiased random sequence. *Annals of Mathematical Statistics*, 43:865–870, 1972.

[7] D. Feldman, R. Impagliazzo, M. Naor, N. Nisan, S. Rudich, and A. Shamir. On dice and coins: Models of computation for random generation. *Information and Computation*, 104(2):159–174, 1993.

[8] W. Hoeffding and G. Simmons. Unbiased coin tossing with a biased coin. *Annals of Mathematical Statistics*, 41:841–352, 1970.

[9] T. Itoh. Simulating fair dice with biased coins. *Information and Computation*, 126:78–82, 1996.

[10] D. Knuth and A. Yao. The complexity of nonuniform random number generation. In J. Traub, editor, *Algorithms and Complexity, New Directions and Results*, pages 357–428. Academic Press, 1976.

[11] L. Kuipers and H. Niederreiter. *Uniform Distribution of Sequences*. John Wiley & Sons, 1974.

[12] H. Niederreiter. Methods for estimating discrepancy. In S. K. Zaremba, editor, *Applications of Number Theory to Numerical Analysis*, pages 203–235. Academic Press, 1971.

[13] P. Samuelson. Constructing an unbiased random sequence. *Journal of the American Statistical Association*, 63:1526–1527, 1968.

[14] W. Schmidt. A metrical theorem in diophantine approximation. *Canad. J. Math.*, 12:619–631, 1960.

[15] W. Schmidt. Simultaneous approximation to algebraic numbers by rationals. *Acta Math.*, 125:189–201, 1970.

[16] Q. Stout and B. Warren. Tree algorithms for unbiased coin tossing with a biased coin. *The Annals of Probability*, 12(1):212–222, 1984.

[17] J. von Neumann. Various techniques used in connection with random digits. *Applied Math Series*, 12:36–38, 1951. Notes by G. E. Forsythe. National Bureau of Standards.

Ericsson Research, ERA/T/NF, SE-164 80 Stockholm, Sweden
E-mail address: `mats.naslund@era-t.ericsson.se`

Department of Computer Science and Engineering, University of Connecticut, Storrs, CT 06269, USA
E-mail address: `acr@cse.uconn.edu`

Progress in Computer Science and Applied Logic, Vol. 20
© 2001 Birkhäuser Verlag Basel/Switzerland

The Dark Side of the Hidden Number Problem: Lattice Attacks on DSA

Phong Q. Nguyen

Abstract. At Crypto '96, Boneh and Venkatesan introduced the so-called hidden number problem: in a prime field $\mathbb{Z}_q$, recover a number α such that for many known random t, the most significant bits of $t\alpha$ are known. They showed that Babai's LLL-based polynomial-time nearest plane algorithm for approximating the lattice closest vector problem solves the problem with probability at least $\frac{1}{2}$, provided that the number of bits known (for each $t\alpha$) is greater than $\sqrt{\log q} + \log \log q$. That result is often cited as the only positive application known in cryptology of the LLL algorithm, because it enables to prove the hardness of the most significant bits of secret keys in Diffie-Hellman and related schemes. The purpose of this short and elementary note is to highlight the fact that the result also has a dark side. Indeed, we remark that the hidden number problem is an idealized version of the problem which Howgrave-Graham and Smart recently tried to solve heuristically in their (lattice-based) attacks on DSA and related signature schemes: given a few bits of the random nonces k used in sufficiently many DSA signatures, recover the secret key. This suggests to determine what can be achieved in practice, rather than in theory. Since lattice reduction algorithms are known to behave much more nicely than their proved worst-case bounds, we give the number of bits that enables the Boneh-Venkatesan technique to succeed, provided an oracle for the lattice closest vector problem in the Euclidean norm or the infinity norm. An analogous assumption is used in the well-known lattice-based attacks against low-density subset sums. Interestingly, our experiments support our theoretical bounds and improve the experimental bounds of Howgrave-Graham and Smart.

1. Introduction

At Crypto '96, Boneh and Venkatesan [4] proved the hardness of the most significant bits of secret keys in Diffie-Hellman and related cryptographic schemes. Namely, they proved that computing the most significant bits of the secret key in a Diffie-Hellman key-exchange protocol from the public keys of the participants is as hard as computing the secret key itself. This was done by solving the *randomized* or *sampling* version of the so-called hidden number problem: given $t_1, \ldots, t_d$ chosen uniformly and independently at random in $\mathbb{Z}_q^*$, and $\mathrm{MSB}_\ell(\alpha t_i \bmod q)$ for all i, recover $\alpha \in \mathbb{Z}_q$. Here, $\mathrm{MSB}_\ell(x)$ for $x \in \mathbb{Z}_q$ denotes any integer z satisfying $|x - z| < q/2^{\ell+1}$. For instance, $\mathrm{MSB}_1(x)$ indicates whether $x < q/2$ or not. In the rest of the paper, we will simply call that problem the hidden number problem (HNP).

Boneh and Venkatesan transformed HNP into a lattice closest vector problem (CVP), in which one has to output a point in a given lattice which is closest to a given point in the space. More precisely, they constructed a vector $\mathbf{u}$ and a lattice L from the HNP inputs such that any lattice point sufficiently close to $\mathbf{u}$ discloses the hidden number α. To find a sufficiently close lattice point, they applied Babai's nearest plane algorithm [1], an old application of the celebrated LLL lattice reduction algorithm [7] to CVP. This showed that HNP could be solved in polynomial time using $\ell = \sqrt{\log p} + \log\log p$ bits. Using Schnorr's improved reduction algorithms [13], this can be asymptotically improved to $\varepsilon\sqrt{\log p}$.

The LLL-based solution of the hidden number problem is often cited as the only positive application known of the LLL algorithm in cryptology (for the well-known negative applications in cryptanalysis, we refer to the survey [11]). The purpose of this note is to highlight its dark side. More precisely, we notice that HNP happens to be an ideal version of the problem which Howgrave-Graham and Smart recently tried to solve heuristically in their (lattice-based) attacks on DSA and related schemes (see [8]). Although Howgrave-Graham and Smart referenced Boneh-Venkatesan paper [4], they apparently did not notice how close their problem was to the hidden number problem. Recall that, in its simplest form, the DSA signature of a message m is $(r, s) \in \mathbb{Z}_q^2$ where $r = (g^k \bmod p) \bmod q$, $s = k^{-1}(h(m) + \alpha r) \bmod q$, h is the SHA-1 hash function, k is a random element in $\mathbb{Z}_q$, q is a 160-bit prime, p a large prime such that q divides $p - 1$, g is a public element of $\mathbb{Z}_p$ of order q, $\alpha \in \mathbb{Z}_q$ is the user's secret key. It is well known that if the random nonce k is disclosed, one can recover the secret key $\alpha \in \mathbb{Z}_q$. It is also known [3] that if k is produced by a cryptographically weak pseudo-random generator such as Knuth's linear congruential generator with known parameters, then α can also be recovered using only a few signatures. Recently, Howgrave-Graham and Smart [6] noticed that Babai's nearest plane algorithm could heuristically recover α, provided that sufficiently many signatures and sufficiently many bits of the corresponding nonces k are known. Although Howgrave-Graham and Smart referenced Boneh-Venkatesan paper [4], they apparently did not notice how close their problem was to the hidden number problem.

Indeed, assume for instance that for many signatures (r_i, s_i) of messages m_i, the ℓ least significant bits of k_i are known to the attacker: one knows $a_i \in \{0, \ldots, 2^\ell - 1\}$ such that $k_i - a_i$ is of the form $2^\ell b_i$. Then $\alpha r_i \equiv s_i(a_i + b_i 2^\ell) - h(m_i)$ $(\bmod\ q)$, which can be rewritten as:

$$\alpha r_i 2^{-\ell} s_i^{-1} \equiv (a_i - s_i^{-1} h(m_i)) \cdot 2^{-\ell} + b_i \pmod{q}.$$

In that equation, $t_i = r_i 2^{-\ell} s_i^{-1} \bmod q$ and $u_i = (a_i - s_i^{-1} h(m_i)) \cdot 2^{-\ell} \bmod q$ are known and are such that $0 \le (\alpha t_i - u_i) \bmod q = b_i < q/2^\ell$. Hence, one knows the ℓ most significant bits of $\alpha t_i \bmod q$. Recovering the secret key α is therefore a slightly different hidden number problem in which the t_i's are not assumed to be independent and uniformly distributed over $\mathbb{Z}_q$, but are of the form $r_i 2^{-\ell} s_i^{-1}$ where the k_i's are independent and uniformly distributed over $\mathbb{Z}_q$. In other words, the hidden number problem is an idealized version of the problem of breaking DSA (or related signature schemes) when the ℓ least significant bits (or more generally, ℓ consecutive bits) of the random nonce k are known for many signatures. It follows that Boneh-Venkatesan's result, as such, does not directly imply a provable attack on DSA in such settings, which explains why the attack of [6] was only heuristic. In fact, since the distribution of $r_i 2^{-\ell} s_i^{-1} \equiv (g^{k_i} \bmod p) k_i (h(m_i) + \alpha r_i)^{-1} 2^{-\ell}$ (mod q) seems hard to study (this issue is being addressed by [10]), we will restrict ourselves to the hidden number problem. Our experiments show that in practice, the difference of distribution does not matter much.

Besides that technical difference between the two problems, the experimental results of [6] were below the theoretical bounds of [4]: Howgrave-Graham and Smart could recover the secret key in practice provided as few as 8 bits for each nonce of 30 signatures; while $\sqrt{\log q} + \log \log q$ is approximately 20 for a 160-bit prime q. The practical improvements follow from the well-known experimental evidence that lattice reduction algorithms behave much more nicely than their proved worst-case bounds. In fact, Boneh and Venkatesan already noticed in [4] that the theoretical bound they obtained could be much lower in practice. However, they did not give the improved theoretical bound if one assumed access to ideal lattice reduction, that is a CVP-oracle here. Note that the well-known lattice-based attacks [5] on low-density subset sums use a similar assumption: in such attacks, one assumes access to an oracle for the lattice shortest vector problem. We therefore extend the Boneh-Venkatesan technique to the case where a CVP-oracle is available. This allows us to guess more precisely what can be achieved in practice in the case of attacks on DSA. Indeed, despite NP-hardness results for CVP, it is often possible to solve CVP exactly up to moderate dimensions, using standard lattice reduction algorithms. More precisely, with a 160-bit prime q as in DSA, we prove that a CVP-oracle for the infinity norm or the Euclidean norm solves HNP with respectively $\ell = 2$ and $\ell = 6$. Experimental evidences support our bound, and extend Howgrave-Graham and Smart's experimental results. Namely, we were able to solve HNP and break the corresponding DSA with $\ell = 3$.

Finally, it should be pointed out that the study of the security of DSA in such settings might have practical implications. Indeed, Bleichenbacher [2] recently noticed that in AT&T's CryptoLib (a widely distributed cryptographic library), the implementation of DSA suffers from the following flaw: the random nonce k is always , thus leaking its least significant bit. Apparently, this is because the same routine is used in the implementation of the ElGamal signature scheme, for which k must be coprime with $p-1$, and thus necessarily odd. Our results do not show that CryptoLib's DSA implementation can be broken, but they do not rule out such a possibility either, even with the same attack. In fact, they indicate a potential weakness in this implementation.

2. Notation

We use log to denote the logarithm in base 2. $a \bmod b$ will denote the smallest positive residue of a modulo b, while $a \operatorname{cmod} b$ will denote the centered reduction of a modulo b, that is the smallest residue (in absolute value) of a modulo b. $\lfloor x \rfloor$ denotes the integer part of x, that is the integer n such that $n \leq x < n+1$. $\lfloor x \rceil$ denotes the closest integer to x: $\lfloor x \rceil = \lfloor x + \frac{1}{2} \rfloor$. $\lceil x \rceil$ is the integer n such that $n \geq x > n-1$.

We call *lattice* any full-dimensional lattice, that is any subgroup of $\mathbb{R}^d$ which includes at least d vectors linearly independent. A basic lattice problem is the closest vector problem (CVP): given a lattice L in $\mathbb{R}^d$ and a target $\mathbf{a} \in \mathbb{R}^d$, find a lattice point $\mathbf{u} \in L$ which minimizes $\|\mathbf{u} - \mathbf{a}\|$. CVP generally refers to the Euclidean norm, but of course, other norms are possible as well: we denote CVP_∞ the problem corresponding to the infinity norm. We call CVP-*oracle* any algorithm that solves CVP exactly. Babai's nearest plane algorithm [1] is a polynomial-time approximation algorithm for CVP: given as input a lattice L in $\mathbb{Q}^d$ and $\mathbf{a} \in \mathbb{Q}^d$, it outputs $\mathbf{u} \in L$ such that for all $\mathbf{v} \in L$, $\|\mathbf{u} - \mathbf{a}\| \leq 2^{d/4} \|\mathbf{v} - \mathbf{a}\|$. The algorithm is based on the celebrated LLL algorithm [7].

Let q be a prime number, ℓ be a positive integer, and $t_1, \ldots, t_d \in \mathbb{Z}_q$. We will denote $L(q, \ell, t_1, \ldots, t_d)$ the $(d+1)$-dimensional lattice spanned by the rows of the following matrix:

$$\begin{pmatrix} q & 0 & \cdots & 0 & 0 \\ 0 & q & \ddots & \vdots & \vdots \\ \vdots & \ddots & \ddots & 0 & \vdots \\ 0 & \cdots & 0 & q & 0 \\ t_1 & \cdots & \cdots & t_d & 1/2^{\ell+1} \end{pmatrix}$$

That lattice differs from the lattice considered in [4, 6] only in the $(d+1) \times (d+1)$ coefficient. The new choice is more natural, but the difference is marginal: in practice, it should (heuristically) increase slightly the probability of success.

3. Former results

The main result of Boneh and Venkatesan is the following [4, Theorem 1]:

Theorem 3.1. *Let α be some integer $[1, q-1]$ and $n = \log q$. Let $\mathcal{O}$ be a function defined by $\mathcal{O}(t) = \mathrm{MSB}_\ell(\alpha t \bmod q)$ with $\ell = \lceil \sqrt{n} \rceil + \lceil \log n \rceil$. There exists a deterministic polynomial time algorithm $\mathcal{A}$ such that*

$$\Pr_{t_1, \ldots, t_d} [\mathcal{A}(t_1, \ldots, t_d, \mathcal{O}(t_1), \ldots, \mathcal{O}(t_d)) = \alpha] \geq \frac{1}{2},$$

where $d = 2\lceil \sqrt{n} \rceil$ and $t_1, \ldots, t_d$ are chosen uniformly and independently at random from $\mathbb{Z}_q^*$.

Thus the hidden number problem can be solved using $\ell = \sqrt{\log q} + \log\log q$ bits. Using Schnorr's improved lattice reduction algorithms, this can be asymptotically improved to $\varepsilon\sqrt{\log q}$ for any fixed $\varepsilon > 0$. One may also replace the bound $\frac{1}{2}$ by $\frac{1}{2\sqrt{n}}$ and reduce the number of bits required by $\log\log q$. Then, the expected run time goes up by a factor $\sqrt{\log q}$. One can alternately run $\sqrt{\log q}$ copies of the algorithm in parallel.

Theorem 3.1 follows from a translation of the hidden number problem in terms of lattices. Consider an HNP-instance: let $t_1, \ldots, t_d$ be chosen uniformly and independently at random in $\mathbb{Z}_q^*$, and $a_i = \mathrm{MSB}_\ell(\alpha t_i \bmod q)$. Clearly, the vector $\mathbf{t} = (t_1\alpha \bmod q, \ldots t_d\alpha \bmod q, \alpha/2^{\ell+1})$ belongs to the lattice $L = L(q, \ell, t_1, \ldots, t_d)$. We call that vector the *target vector*. The target vector is very close to the row vector $\mathbf{a} = (a_1, \ldots, a_d, 0)$. Indeed, $\|\mathbf{t} - \mathbf{a}\| \leq q\sqrt{d+1}/2^{\ell+1}$. It follows that the distance of $\mathbf{a}$ to the lattice L is bounded by $q\sqrt{d+1}/2^{\ell+1}$.

The following result, which is essentially [4, Theorem 5], says that any lattice point of L sufficiently close to $\mathbf{a}$ is related to $\mathbf{t}$ and discloses the hidden number α:

Theorem 3.2 (Uniqueness theorem). *Set $d = 2\lceil\sqrt{\log q}\rceil$ and $\mu = \frac{1}{2}\sqrt{\log q} + 3$. Let α be a fixed integer in the range $[1, q-1]$. Choose integers $t_1, \ldots, t_d$ uniformly and independently at random in the range $[1, q-1]$. Let L be the lattice $L(q, \ell, t_1, \ldots, t_d)$ and $\mathbf{a} = (a_1, \ldots, a_d, 0)$ be an integer vector satisfying*

$$|(\alpha t_i \bmod q) - a_i| < q/2^\mu.$$

Then with probability at least $\frac{1}{2}$, all $\mathbf{u} \in L$ with $\|\mathbf{u} - \mathbf{a}\| < \frac{q}{2^\mu}$ are of the form:

$$\mathbf{u} = (t_1\beta \bmod q, \ldots t_d\beta \bmod q, \beta/2^{\ell+1}) \text{ where } \alpha \equiv \beta \pmod{q}.$$

The uniqueness theorem is the key to Theorem 3.1. Indeed, Babai's nearest plane algorithm applied to L and $\mathbf{a}$ yields a lattice point $\mathbf{u}$ such that:

$$\|\mathbf{u} - \mathbf{a}\| \leq 2^{(d+1)/4}\mathrm{dist}(L, \mathbf{a}) \leq 2^{(d+1)/4}q\sqrt{d+1}/2^{\ell+1} < q/2^\mu,$$

where $\mu = \frac{1}{2}\sqrt{n} + 3$ as in the uniqueness theorem. The last inequality follows from $\mu < \ell + 1 - \frac{d+1}{4} - \frac{1}{2}\log(d+1)$ for large n. Hence, with probability at least $\frac{1}{2}$, $\mathbf{u}$ discloses the hidden number α.

The lattice attack of Howgrave-Graham and Smart [6] used the same lattice as in [4]. However, apparently, they were not aware that the attack could be proved if the t_i's were supposed to be independent and uniformly distributed over $\mathbb{Z}_q$. Instead, they relied on the heuristic argument that the target vector was the closest lattice point to $\mathbf{a}$, by comparing the lattice volume with the distance of target vector and $\mathbf{a}$.

4. Extension to CVP-oracles

To extend Theorem 3.1 to the case where a CVP-oracle is available, one needs a more precise version of the uniqueness theorem 3.2, which is as follows:

Lemma 4.1. *Let $d \geq 1$ and $\mu \geq 3$. Let α be a fixed integer in the range $[1, q-1]$. Choose integers $t_1, \ldots, t_d$ uniformly and independently at random in the range $[1, q-1]$. Let L be the lattice $L(q, \ell, t_1, \ldots, t_d)$ and $\mathbf{a} = (a_1, \ldots, a_d, 0)$ be an integer vector satisfying*

$$|(\alpha t_i \bmod q) - a_i| < \frac{q}{2^\mu}.$$

Then with probability at least $1 - (q-1)/2^{d(\mu-2)}$, all $\mathbf{u} \in L$ with $\|\mathbf{u} - \mathbf{a}\|_\infty < \frac{q}{2^\mu}$ have their last coordinate of the form $\beta/2^{\ell+1}$ where $\alpha \equiv \beta \pmod{q}$.

Proof. We follow the proof of [4, Theorem 5]. Define the modular distance between two integers β and γ as $\operatorname{dist}_q(\beta, \gamma) = (\beta - \gamma) \operatorname{cmod} q$. Suppose $\beta \not\equiv \gamma \pmod{q}$ where β and γ are both integers in the range $[1, \ldots, q-1]$. Define:

$$A = \Pr_t \left[\operatorname{dist}_q(\beta t, \gamma t) > 2q/2^\mu \right],$$

where t is an integer chosen uniformly at random in $[1, q-1]$. Then, provided $\mu \geq 3$:

$$A = \Pr_t \left[\frac{2q}{2^\mu} < (\beta - \gamma) t \bmod q < q - \frac{2q}{2^\mu} \right] = \frac{\lfloor q - \frac{2q}{2^\mu} \rfloor - \lceil \frac{2q}{2^\mu} \rceil}{q-1} = \frac{q - \lfloor \frac{q}{2^{\mu-2}} \rceil}{q-1}.$$

Now let $\mathbf{u}$ be a lattice point satisfying the assumption of the lemma. $\mathbf{u}$ is of the form:

$$\mathbf{u} = (\beta t_1 - b_1 q, \ldots, \beta t_d - b_d q, \beta/2^{\ell+1}),$$

for some integers $\beta, b_1, \ldots, b_d$. Assume that $\beta \not\equiv \alpha \pmod{q}$. Then:

$$
\begin{aligned}
\Pr \left[\|\mathbf{u} - \mathbf{a}\|_\infty > \frac{q}{2^\mu} \right] & \geq \Pr \left[\exists i : \operatorname{dist}_q(t_i \beta, a_i) > \frac{q}{2^\mu} \right] \\
& \geq \Pr \left[\exists i : \operatorname{dist}_q(t_i \beta, t_i \alpha) > \frac{2q}{2^\mu} \right] \\
& = 1 - (1-A)^d = 1 - \frac{1}{(q-1)^d} \cdot \left(\lfloor \frac{q}{2^{\mu-2}} \rceil - 1 \right)^d \\
& \geq 1 - \frac{1}{(q-1)^d} \left(\frac{q - 2^{\mu-3}}{2^{\mu-2}} \right)^d \geq 1 - \frac{1}{2^{d(\mu-2)}}
\end{aligned}
$$

Since $\beta \not\equiv \alpha \pmod{q}$, there are at most $q - 1$ values of $\beta \bmod q$ to consider, from which the result follows. □

Besides, we know that the target vector is quite close to $\mathbf{a}$: $\|\mathbf{t} - \mathbf{a}\| \leq q\sqrt{d+1}/2^{\ell+1}$ and $\|\mathbf{t} - \mathbf{a}\|_\infty < q/2^{\ell+1}$.

4.1. Infinity norm

Given L and $\mathbf{a}$, a CVP_∞-oracle will output a point $\mathbf{u}$ such that:

$$\|\mathbf{u} - \mathbf{a}\|_\infty \leq \|\mathbf{t} - \mathbf{a}\|_\infty \leq q/2^{\ell+1}.$$

Hence, the choice $\mu = \ell + 1$ satisfies Lemma 4.1. To make the probability in Lemma 4.1 larger than $\frac{1}{2}$, we need: $\frac{q-1}{2^{d(\mu-2)}} \leq \frac{1}{2}$, which reads:

$$d \geq \frac{\log(q-1) + 1}{\ell - 1}.$$

We obtain:

Theorem 4.2. *Let α be an integer in $[1, q-1]$, ℓ an integer ≥ 2. Let $\mathcal{O}$ be a function defined by $\mathcal{O}(t) = \mathrm{MSB}_\ell(\alpha t \bmod q)$. There exists a deterministic polynomial time algorithm $\mathcal{A}$ using a CVP_∞-oracle such that*

$$\Pr_{t_1,\ldots,t_d} [\mathcal{A}(t_1,\ldots,t_d, \mathcal{O}(t_1),\ldots,\mathcal{O}(t_d)) = \alpha] \geq \frac{1}{2},$$

where $d = \lceil (1 + \log q)/(\ell - 1) \rceil$ and $t_1,\ldots,t_d$ are chosen uniformly and independently at random from $\mathbb{Z}_q^*$.

Thus, provided a CVP_∞-oracle, one can solve in polynomial time the hidden number problem with only 2 bits and at most $d = \lceil \log(q) + 1 \rceil$.

4.2. Euclidean norm

Using the Euclidean norm, the result is not as good. Given L and $\mathbf{a}$, a CVP-oracle will output a point $\mathbf{u}$ such that:

$$\|\mathbf{u} - \mathbf{a}\| \leq \|\mathbf{t} - \mathbf{a}\| \leq q\sqrt{d+1}/2^{\ell+1}.$$

Hence, the choice $\mu = \ell + 1 - \log(d+1)/2$ satisfies Lemma 4.1, provided $\mu \geq 3$, that is $\log(d+1) \leq 2(\ell - 2)$. The largest possible value for d is $d = 2^{2(\ell-2)} - 1$, which leads to $\mu = 3$. And to make the probability in Lemma 4.1 larger than $\frac{1}{2}$, we need

$$\frac{q-1}{2^{d(\mu-2)}} \leq \frac{1}{2} \tag{1}$$

that is $2^{2(\ell-2)} - 1 \geq \log(q-1) + 1$, which reads:

$$\ell \geq 2 + \frac{\log(2 + \log(q-1))}{2}.$$

We obtain:

Theorem 4.3. *Let α be an integer in $[1, q-1]$, ℓ an integer $\geq 2 + (\log(2 + \log(q-1)))/2$. Let $\mathcal{O}$ be a function defined by $\mathcal{O}(t) = \mathrm{MSB}_\ell(\alpha t \bmod q)$. There exists a deterministic polynomial time algorithm $\mathcal{A}$ using a CVP-oracle such that*

$$\Pr_{t_1,\ldots,t_d} [\mathcal{A}(t_1,\ldots,t_d, \mathcal{O}(t_1),\ldots,\mathcal{O}(t_d)) = \alpha] \geq \frac{1}{2},$$

where $d = 2^{2(\ell-2)} - 1$ and $t_1,\ldots,t_d$ are chosen uniformly and independently at random from $\mathbb{Z}_q^*$.

Thus, provided a CVP-oracle, one can solve in polynomial time the hidden number problem with a 160-bit prime q using only 6 bits and d at most 255. For such a q and ℓ however, a much smaller value of d is possible, as one can see numerically that $d = 100$ satisfies (1). More generally, the choice $d = 2^{2(\ell-2)} - 1$ in Theorem 4.3 is not optimal. For sufficiently large ℓ, something like $d = O(\log q/(\ell - 1))$ should be possible.

5. Experimental results

The best bound obtained in practice in [6] reports that HNP can be solved in practice with $\ell = 8$ and $d = 30$, using Babai's nearest plane algorithm. It was suggested that $\ell = 4$ was infeasible. We performed experiments with the NTL library [15]. The running time is less than half an hour for d less than a hundred, on a 500 MHz DEC Alpha. For each choice of parameters size, we ran several times the method on newly generated parameters (including the prime q and the multipliers of the HNP). Each trial is referred as a *sample*. We made experiments with both the randomized HNP and the HNP where the multipliers come from DSA signatures. We did not experience any difference of results for the two problems. Using Babai's nearest plane algorithm and Schnorr's Korkine-Zolotarev reduction [13, 14] with blocksize 20, we could solve HNP with ℓ as low as $\ell = 4$ and $d = 70$. More precisely, the method always worked for $\ell = 5$ (a hundred samples). For $\ell = 4$, it worked 90% of the time over 100 samples. For $\ell = 3$, it always failed on about 100 samples, even with $d = 100$.

We made additional experiments with the well-known embedding strategy (see [9]) and Schnorr's improved lattice reduction [13, 14] to solve CVP. The embedding strategy heuristically reduces CVP to the lattice shortest vector problem. More precisely, if the CVP-instance is given by the vector $\mathbf{a} = (a_1, \ldots, a_d)$ and a d-dimensional lattice spanned by the row vectors $\mathbf{b}_i = (b_{i,1}, \ldots, b_{i,d})$ with $1 \leq i \leq d$, the embedding strategy builds the lattice L spanned by the rows of the following matrix:

$$\begin{pmatrix} b_{1.1} & \cdots & b_{1,d} & 0 \\ b_{2,1} & \cdots & b_{2,d} & 0 \\ \vdots & & \vdots & \vdots \\ b_{d,1} & \cdots & b_{d,d} & 0 \\ a_1 & \cdots & a_d & 1 \end{pmatrix}$$

It is hoped that the shortest vector of L (or one of the vectors of the reduced basis) is of the form $(\mathbf{a} - \mathbf{u}, 1)$ where $\mathbf{u}$ is a sufficiently close lattice vector we are looking for. Using that strategy, we were always able to solve HNP with $\ell = 3$ and $d = 100$ (on more than 50 samples). We always failed with $\ell = 2$ and $d = 150$. In our experiments, to balance the coefficients of the lattice, we replaced the coefficient 1 in the lowest right-hand entry by $q/2^{\ell+1}$. When the attack succeeded, the vector $(\mathbf{a} - \mathbf{u}, q/2^{\ell+1})$ (where $\mathbf{u}$ is a lattice point revealing the hidden number) was generally the second vector of the reduced basis.

Our experiments suggest that the bound of Theorem 4.3 regarding ℓ is not tight. This is not surprising, as the proof technique due to Boneh-Venkatesan is more suited to the infinity norm than the Euclidean norm. It would be interesting to improve Theorems 4.2 and 4.3. We note that the experimental bounds are between the theoretical bounds of the two theorems, and very close to the one of Theorem 4.2. We believe it should be possible to reach $\ell = 2$ in practice using a lattice basis reduction algorithm more suited to the infinity norm (see for instance [12]), especially since the lattice dimension is reasonable. In fact, even $\ell = 1$ might be possible in practice, since the proof of Theorem 4.2 does not rule out such a possibility.

Acknowledgements. We thank Daniel Bleichenbacher, Dan Boneh, Nick Howgrave-Graham and Igor Shparlinski for helpful discussions.

References

[1] L. Babai. On Lovász lattice reduction and the nearest lattice point problem. *Combinatorica*, 6:1–13, 1986.

[2] D. Bleichenbacher, 1999. Private communication.

[3] M. Bellare, S. Goldwasser, and D. Micciancio. "Pseudo-random" number generation within cryptographic algorithms: The DSS case. In *Proc. of Crypto '97*, volume 1294 of *LNCS*. IACR, Springer-Verlag, 1997.

[4] D. Boneh and R. Venkatesan. Hardness of computing the most significant bits of secret keys in diffie-hellman and related schemes. In *Proc. of Crypto '96*. IACR, Springer-Verlag, 1996.

[5] M.J. Coster, A. Joux, B.A. LaMacchia, A.M. Odlyzko, C.-P. Schnorr, and J. Stern. Improved low-density subset sum algorithms. *Comput. Complexity*, 2:111–128, 1992.

[6] N.A. Howgrave-Graham and N.P. Smart. Lattice attacks on digital signature schemes. Technical report, HP Labs, 1999. Report HPL-1999-90. To appear in *Designs, Codes and Cryptography*.

[7] A.K. Lenstra, H.W. Lenstra, Jr., and L. Lovász. Factoring polynomials with rational coefficients. *Mathematische Ann.*, 261:513–534, 1982.

[8] A. Menezes, P. Van Oorschot, and S. Vanstone. *Handbook of Applied Cryptography*. CRC Press, 1997.

[9] P. Nguyen. Cryptanalysis of the Goldreich-Goldwasser-Halevi cryptosystem from Crypto '97. In *Proc. of Crypto '99*, volume 1666 of *LNCS*, pages 288–304. IACR, Springer-Verlag, 1999.

[10] P.Q. Nguyen and I.E. Shparlinski. The insecurity of the Digital Signature Algorithm with partially known nonces. Manuscript in preparation, 2000.

[11] P.Q. Nguyen and J. Stern. Lattice reduction in cryptology: An update. In *Algorithmic Number Theory – Proc. of ANTS-IV*, LNCS. Springer-Verlag, 2000.

[12] H. Ritter. Breaking knapsack cryptosystems by max-norm enumeration. In *Proc. of Pragocrypt '96*, pages 480–492. CTU Publishing House, 1996.

[13] C.P. Schnorr. A hierarchy of polynomial lattice basis reduction algorithms. *Theoretical Computer Science*, 53:201–224, 1987.

[14] C.P. Schnorr and M. Euchner. Lattice basis reduction: improved practical algorithms and solving subset sum problems. *Math. Programming*, 66:181–199, 1994.

[15] V. Shoup. Number Theory C++ Library (NTL) version 3.7. Available at http://www.shoup.net/ntl/.

Département d'Informatique,
École Normale Supérieure,
45 rue d'Ulm
75005 Paris, France
URL: http://www.di.ens.fr/~pnguyen/
E-mail address: pnguyen@ens.fr

Progress in Computer Science and Applied Logic, Vol. 20
© 2001 Birkhäuser Verlag Basel/Switzerland

Distribution of Modular Sums and the Security of the Server Aided Exponentiation

Phong Q. Nguyen, Igor E. Shparlinski, and Jacques Stern

Abstract. We obtain some uniformity of distribution results for the values of modular sums of the form

$$\sum_{j=1}^{n} a_j x_j \quad (\mathrm{mod}\ M) \qquad (x_1, \dots, x_n) \in \mathcal{B}$$

where $M \geq 1$ is an integer, $a_1, \dots, a_n$ are elements of the residue ring modulo M, selected unformly at random, and $\mathcal{B}$ is an arbitrary set of n-dimensional integer vectors. In some partial cases, for very special sets $\mathcal{B}$, some results of this kind have been known, however our estimates are more precise and more general. Our technique is based on fairly simple properties of exponential sums. We also give cryptographic applications of some of these results. In particular, we consider an extension of a pseudo-random number generator due to V. Boyko, M. Peinado and R. Venkatesan, and establish the security of some discrete logarithm based signature schemes making use of this generator (in both its original and extended forms). One of these schemes, which uses precomputation is well known. The other scheme which uses server aided computation, seems to be new. We show that for a certain choice of parameters one can guarantee an essential speed-up of both of these schemes without compromising the security (compared to the traditional discrete logarithm based signature scheme).

1. Introduction

Let $M \geq 2$ be integers. We denote by $\mathbb{Z}_M$ the residue ring modulo M.

For a given n-dimensional vector $\mathbf{a} = (a_1, \dots, a_n) \in \mathbb{Z}_M$ and a given set of n-dimensional vectors $\mathcal{B} \subseteq \mathbb{Z}_M^n$ we consider the distribution of *modular sums*

$$\mathbf{a} \cdot \mathbf{x} \equiv \sum_{j=1}^{n} a_j x_j \quad (\mathrm{mod}\ M), \qquad (x_1, \dots, x_n) \in \mathcal{B}.$$

More precisely, given a residue $c \in \mathbb{Z}_M$ we denote by $N_{\mathbf{a}}(\mathcal{B}, c)$ the number of solutions of the congruence

$$\mathbf{a} \cdot \mathbf{x} \equiv c \quad (\mathrm{mod}\ M), \qquad \mathbf{x} \in \mathcal{B}.$$

Because for many applications it is technically more useful to work with probabilities we define

$$P_{\mathbf{a}}(\mathcal{B}, c) = \frac{1}{|\mathcal{B}|} N_{\mathbf{a}}(\mathcal{B}, c)$$

as the probability that $\mathbf{a} \cdot \mathbf{x} \equiv c \pmod{M}$ for a random vector $\mathbf{x}$ chosen uniformly from $\mathcal{B}$.

We use exponential sums to show that for almost all vectors $\mathbf{a} \in \mathbb{Z}_M^n$ these quantities take their expected values under some natural restrictions on the cardinality of $\mathcal{B}$.

Although this holds to arbitrary sets $\mathcal{B}$, for our applications we are mainly interested in the following special sets. Given an integer $h \le M - 1$ we define $\mathcal{B}_n(h)$ as the set of integer vectors $\mathbf{x} = (x_1, \dots, x_n)$ with $0 \le x_j \le h - 1$, $j = 1, \dots, n$. Given an integer k, $n \ge k \ge 1$, we also define $\mathcal{B}_{n,k}(h)$ as the subset of $\mathcal{B}_n(h)$ consisting of vectors with precisely k non-zero components. Hence

$$|\mathcal{B}_n(h)| = h^n \qquad \text{and} \qquad |\mathcal{B}_{n,k}(h)| = (h-1)^k \binom{n}{k}.$$

In the important (although not always optimal for our applications) special case $h = 2$ we put

$$\mathcal{B}_n(2) = \mathcal{B}_n \qquad \text{and} \qquad \mathcal{B}_{n,k}(2) = \mathcal{B}_{n,k}.$$

In the case $\mathcal{B} = \mathcal{B}_n$ several results of this kind have been known, see [1, 8, 9, 13, 15]. These results have found several cryptographic applications, in particular, to some cryptosystems proposed in [2], see [15]. However for other sets no similar results seemed to be known. In particular for sets $\mathcal{B}_{n,k}$ this problem has been mentioned in [15].

We also give some cryptographic motivations and applications of our results, in particular in the signature scheme with precomputation [2].

In many discrete logarithm based protocols, one needs to generate pairs of the form $(x, g^x \pmod{p})$ where x is random and g is a fixed base. The El Gamal [5] and DSA [14] (*Digital Signature Algorithm*) signatures as well as the Schnorr [19, 20] and Brickell-McCurley [4] identification and signature schemes are examples of such protocols. The generation of these pairs is often the most expensive operation, which makes it tempting to reduce the number of modular multiplications required per generation, especially for smartcards. There are basically two ways to solve this problem. One way is to generate separately a random x, and then to compute $g^x \pmod{p}$ using a precomputation method [3, 7, 17, 10].

The other way is to generate x and $g^x \pmod{p}$ together by a special pseudo-random number generator which also uses precomputation. Schnorr [19] was the first to propose such a preprocessing scheme. The scheme has much better performances than all other methods but there is a certain drawback: the output exponent x is no more guaranteed to be random, and therefore, each generation might leak information. Indeed, de Rooij [16] showed how to break the scheme. A later modification proposed by Schnorr in [20], was also broken by de Rooij [18].

Boyko, Peinado and Venkatesan [2] have recently proposed a new and very simple generator to produce pairs of the form $(x, g^x \pmod{p})$, where p is a prime number, and $g \in \mathbb{Z}_p^*$ is of multiplicative order M.

We describe a generalization of the **BPV** generator, which combined with a certain exponentiation algorithm from [3], becomes computationally more efficient.

This generator as well as our extension, can naturally be used for a signature scheme with precomputation, and thus provides a substantial speed-up compared to the traditional scheme. However, it should be mentioned that the approach of [10] seems to be more efficient for the above values of M (although the asymptotic behaviour of that method has not been evaluated in that paper).

On the other hand, we also describe another scheme, to which the results of [10] do not apply. In that scheme, instead of precomputing and thus storing a rather substantial set of integers, one uses the server to assist the signature generation. This new scheme is suitable for computation on a device with low computational power and memory (such as a smart-card, for example).

We show that the security of both signature schemes is preserved, when certain conditions on their parameters are met. Our proof is based on the notion of statistical distance.

Throughout the paper $\log z$ and $\ln z$ denote the binary and the natural logarithm of real $a > 0$, respectively.

2. Extended BPV Generator

Let $g \in \mathbb{Z}_p^*$ be of multiplicative order M.

The original generator of Boyko, Peinado and Venkatesan [2], which we call **BPV**, with integer parameters $n \geq k \geq 1$ can be described as follows:

Preprocessing Step: Generate n random integers $\alpha_1, \ldots, \alpha_n \in \mathbb{Z}_M$. For each $j = 1, \ldots, n$, compute $\beta_j \equiv g^{\alpha_j} \pmod{p}$ and store the values of α_j and β_j in a table.
Pair Generation: Whenever a pair (x, g^x) is needed, randomly generate $S \subseteq \{1, \ldots, n\}$ such that $|S| = k$. Compute

$$x \equiv \sum_{j \in S} \alpha_j \pmod{M} \quad \text{and} \quad X \equiv \prod_{j \in S} \beta_j \equiv g^b \pmod{p}.$$

If $x \equiv 0 \pmod{M}$ then start again, otherwise return the pair (x, X).

For any output (x, X), we indeed have $X \equiv g^x \pmod{p}$. The scheme needs to store n elements of $\mathbb{Z}_M$, and n elements of $\mathbb{Z}_p^*$. It requires $k - 1$ modular multiplications to compute X (and the same number of additions to compute x, but this cost is negligible). This can be compared with the cost of direct computation of g^x which is about $1.5 \log M$ modular multiplications on average and about about $2 \log M$ modular multiplications in the worst case. Thus the ratio $k/\log M$ is a natural measure of speed-up of the **BPV** generator.

Recall that for the DSA [14] and Schnorr [19, 20] schemes M has 160 bits, while for the El Gamal [5] and Brickell-McCurley [4] schemes M has at least 512 bits. Each generation requires k modular multiplications. For $M = p - 1$ where p is a 512-bit prime the authors of [2] suggest to take $n = 512$ and $k = 64$.

We now describe an extension of the **BPV** generator which we will call **EBPV**, with the same integer parameters $n \geq k \geq 1$ and another integer parameter $h \geq 1$. The preprocessing step of the generator is the same however the pair generation follows a slightly more general algorithm:

Extended Pair Generation: Whenever a pair (x, g^x) is needed, randomly generate $S \subseteq \{1, \ldots, n\}$ such that $|S| = k$ and for each $j \in S$ select a random integer $x_j \in \{1, \ldots, h - 1\}$. Compute

$$x \equiv \sum_{j \in S} \alpha_j x_j \pmod{M} \qquad \text{and} \qquad X \equiv \prod_{j \in S} \beta_j^{x_j} \pmod{p},$$

thus $X \equiv g^x \pmod{p}$. If $x \equiv 0 \pmod{M}$ then start again, otherwise return the pair (x, X).

It is easy to see that we still have $X \equiv g^x \pmod{p}$ and that for $h = 2$ this is precisely the **BPV** generator.

The cost of computing x is still very low at least if h is small (it involves $k - 1$ modular additions and $k - 1$ modular multiplications but one of the factors is small). It has been shown in Theorem 1 of [3] (see also Theorem 7 of [7]) that there exists a (simple) algorithm which computes X with only $k + h - 3$ modular multiplications.

We finally explain how to use the above technique in a completely different scenario: instead of precomputing and thus storing a rather substantial set of integers, one uses the server to help with random pair generation. Applications that we have in mind are related with devices whose computational power and memory are extremely low (such as a smart-card, for example). Such devices can optionnally be plugged in a base so as to communicate with a server (a PC which may or may not be connected to the Internet). We use the server in the following way:

Loading Step – server side: The server generates n random integers $\alpha_1, \ldots, \alpha_n \in \mathbb{Z}_M$. For each $j = 1, \ldots, n$, it computes $\beta_j \equiv g^{\alpha_j} \pmod{p}$ and broadcasts the values of α_j and β_j.

Loading Step - client side: The server randomly generates $S \subseteq \{1, \ldots, n\}$ such that $|S| = k$ and, while receiving the server's communication, the client stores the k pairs (α_j, β_j) corresponding to the indices j in S.

After this, one applies the above *extended pair generation* procedure and stores the pair (x, X) for further signature generation.

It should be noted that, in the special case where $h = 2$, the two steps that have to be performed by the server can be merged since the required multiplications can be performed on the fly. This definitely reduces the overall memory needed.

We believe that our scheme is realistic and, in Section 6, we will propose suitable choices for the parameters that should allow practical implementations. However, some precautions are in order. Firstly, it should be impossible to detect the choices of indices performed by the client to form S. Otherwise, the pair (x, X) used at signature generation becomes known to the attacker and, as is well known, this allows to disclose the secret signing key. Secondly, the server should be trusted. Otherwise, there are easy strategies that disclose the signing key again: for example, one can feed the device with n identical pairs. On the other hand, we do not require the various pairs (α_j, β_j) to remain secret since we only care about the randomness of the output pair.

Smart cards manufacturers have developped countermeasures against "timing" and "power" attacks, which should hopefully address our first concern. As for the second, it can be handled by various cryptographic means. If there is a communication link from the card to the server, both can perform a key exchange and the server can encipher the communication. There are ways to achieve this and keep the computational overhead low for the server. In a "broadcast" scenario, this becomes impossible and the only way seems to have the server hash-and-sign the broadcast. Again, this can be done by keeping the computing power of the client low, for example, by using low-exponent RSA. Still, it might be the case that the device cannot hash on the fly and keep up with the communication speed. A way to overcome the difficulty is to use hash-trees: once the pairs have been sent, the full hash-tree built from the n pairs is broadcast, together with the signature of its root. The client only needs to capture the k paths requested for checking the hash computations and the signature.

3. Distribution of Modular Sums

Let us define

$$\mathbf{e}(z) = \exp(2\pi i z / M).$$

We make use of the identity

$$\sum_{\lambda=0}^{M-1} \mathbf{e}(\lambda u) = \begin{cases} 0, & \text{if } u \not\equiv 0 \pmod{M}; \\ M, & \text{if } u \equiv 0 \pmod{M}; \end{cases} \tag{1}$$

which holds for any integer u and which follows from the formula for the sum of a geometric progression (see Exercise 11.a Chapter 3 of [21]).

Lemma 3.1. *For any integer $\lambda \not\equiv 0 \pmod{M}$,*

$$\sum_{\mathbf{a} \in \mathbb{Z}_M^n} \left| \sum_{\mathbf{x} \in \mathcal{B}} \mathbf{e}(\lambda \mathbf{a} \cdot \mathbf{x}) \right|^2 = M^n |\mathcal{B}|.$$

Proof. We have

$$\sum_{\mathbf{a} \in \mathbb{Z}_M^n} \left| \sum_{\mathbf{x} \in \mathcal{B}} \mathbf{e}(\lambda \mathbf{a} \cdot \mathbf{x}) \right|^2 = \sum_{\mathbf{x}, \mathbf{y} \in \mathcal{B}} \sum_{\mathbf{a} \in \mathbb{Z}_M^n} \mathbf{e}(\lambda(\mathbf{a} \cdot \mathbf{x} - \mathbf{a} \cdot \mathbf{y})).$$

If $\mathbf{x} = \mathbf{y}$ the contribution of the inner sum is M^n. If $\mathbf{x} \neq \mathbf{y}$ then, assuming without loss of generality that $x_n \neq y_n$, we obtain

$$\sum_{\mathbf{a}\in\mathbf{Z}_M^n} \mathbf{e}\left(\lambda\left(\mathbf{a}\cdot\mathbf{x}-\mathbf{a}\cdot\mathbf{y}\right)\right)$$

$$= \sum_{(a_1,\dots,a_{n-1})\in\mathbf{Z}_M^{n-1}} \mathbf{e}\left(\lambda\sum_{j=1}^{n-1} a_j(x_j-y_j)\right) \sum_{a_n=0}^{M-1} \mathbf{e}(\lambda a_n(x_n-y_n)) = 0$$

and the desired result follows. $\qquad\qquad\square$

Using the Cauchy inequality, we derive from Lemma 3.1 that

$$\sum_{\mathbf{a}\in\mathbf{Z}_M^n} \left|\sum_{\mathbf{x}\in\mathcal{B}} \mathbf{e}\left(\lambda\mathbf{a}\cdot\mathbf{x}\right)\right| \leq M^n|\mathcal{B}|^{1/2}. \qquad(2)$$

Theorem 3.2. *The identity*

$$\frac{1}{M^n}\sum_{\mathbf{a}\in\mathbf{Z}_M^n}\sum_{c\in\mathbf{Z}_M}\left(P_{\mathbf{a}}(\mathcal{B},c)-\frac{1}{M}\right)^2 = \frac{M-1}{M|\mathcal{B}|}$$

holds.

Proof. From (1) we have

$$N_{\mathbf{a}}(\mathcal{B},c) = \frac{1}{M}\sum_{\mathbf{x}\in\mathcal{B}}\sum_{\lambda=0}^{M-1} \mathbf{e}\left(\lambda\left(\mathbf{a}\cdot\mathbf{x}-c\right)\right).$$

The contribution of the term corresponding to $\lambda = 0$ is $|\mathcal{B}|/M$.
Therefore

$$\sum_{c\in\mathbf{Z}_M}\left(N_{\mathbf{a}}(\mathcal{B},c)-\frac{|\mathcal{B}|}{M}\right)^2$$

$$= \sum_{c\in\mathbf{Z}_M}\left(\frac{1}{M}\sum_{\lambda=1}^{M-1}\mathbf{e}\left(-\lambda c\right)\sum_{\mathbf{x}\in\mathcal{B}}\mathbf{e}\left(\lambda\mathbf{a}\cdot\mathbf{x}\right)\right)^2$$

$$= \frac{1}{M^2}\sum_{c\in\mathbf{Z}_M}\sum_{\lambda,\eta=1}^{M-1}\mathbf{e}\left(-c(\lambda+\eta)\right)\sum_{\mathbf{x},\mathbf{y}\in\mathcal{B}}\mathbf{e}\left(\lambda\mathbf{a}\cdot\mathbf{x}+\eta\mathbf{a}\cdot\mathbf{y}\right)$$

$$= \frac{1}{M^2}\sum_{\lambda,\eta=1}^{M-1}\sum_{\mathbf{x},\mathbf{y}\in\mathcal{B}}\mathbf{e}\left(\lambda\mathbf{a}\cdot\mathbf{x}+\eta\mathbf{a}\cdot\mathbf{y}\right)\sum_{c\in\mathbf{Z}_M}\mathbf{e}\left(-\lambda c-\eta c\right).$$

The sum over c vanishes if $\lambda + \eta \not\equiv 0 \pmod{M}$ and is equal to M otherwise. Hence,

$$\sum_{c \in \mathbb{Z}_M} \left(N_{\mathbf{a}}(\mathcal{B}, c) - \frac{|\mathcal{B}|}{M} \right)^2 = \frac{1}{M} \sum_{\lambda=1}^{M-1} \sum_{\mathbf{x}, \mathbf{y} \in \mathcal{B}} \mathbf{e}\left(\lambda \left(\mathbf{a} \cdot \mathbf{x} - \mathbf{a} \cdot \mathbf{y} \right) \right)$$

$$= \frac{1}{M} \sum_{\lambda=1}^{M-1} \left| \sum_{\mathbf{x} \in \mathcal{B}} \mathbf{e}\left(\lambda \mathbf{a} \cdot \mathbf{x} \right) \right|^2 .$$

Applying Lemma 3.1, we obtain the desired result. $\square$

In particular, using the Cauchy inequality we obtain

$$\frac{1}{M^n} \sum_{\mathbf{a} \in \mathbb{Z}_M^n} \sum_{c \in \mathbb{Z}_M} \left| P_{\mathbf{a}}(\mathcal{B}_{n,k}(h), c) - \frac{1}{M} \right| \leq M^{1/2} |\mathcal{B}_{n,k}(h)|^{-1/2}. \qquad (3)$$

4. Statistical Distance

In complexity theory, it is customary to use the notion of statistical distance between two distributions (see for instance [6, 11]). Recall that the (probability) distribution $\mathcal{D}$ associated with a random variable X over a set S is the function mapping any n of S to $\mathcal{D}(n) = \Pr(X = n)$. The *statistical distance* between two distributions $\mathcal{U}$ and $\mathcal{V}$ over a finite set S is defined as:

$$L(\mathcal{U}, \mathcal{V}) = \frac{1}{2} \sum_{n \in S} |\mathcal{U}(n) - \mathcal{V}(n)| .$$

It is indeed a distance.

We will need to consider sampled distributions.

Denote by $\mathcal{D}^m$ the distribution defined over S^m by choosing independently at random m samples of S, according to $\mathcal{D}$. In other words, $\mathcal{D}^m(n_1, \ldots, n_m) = \mathcal{D}(n_1)\mathcal{D}(n_2) \cdots \mathcal{D}(n_m)$. We will use the following elementary result, which shows that, when two distributions are close, the sampled distributions remain close:

Lemma 4.1. *Let $\mathcal{U}$ and $\mathcal{V}$ be two distributions defined over a set S. Then for all integer m:*

$$L(\mathcal{U}^m, \mathcal{V}^m) \leq m L(\mathcal{U}, \mathcal{V}).$$

Proof. Define $m + 1$ hybrid distributions $\mathcal{D}_k$ over S^m by choosing independently k elements of S according to $\mathcal{U}$, and $m - k$ according to $\mathcal{V}$. That is, we let for k in $\{0, \ldots, m\}$:

$$\mathcal{D}_k(n_1, \ldots, n_m) = \prod_{1 \leq i \leq k} \mathcal{U}(n_i) \times \prod_{k < i \leq m} \mathcal{V}(n_i).$$

In particular, the distributions $\mathcal{D}_0$ and $\mathcal{D}_m$ are respectively identical to $\mathcal{U}^m$ and $\mathcal{V}^m$. From the triangle inequality, we have:

$$L(\mathcal{U}^m, \mathcal{V}^m) \leq \sum_{k=0}^{m-1} L(\mathcal{D}_k, \mathcal{D}_{k+1}).$$

338   P. Q. Nguyen, I. E. Shparlinski, and J. Stern

We conclude the proof by noting that each distance $L(\mathcal{D}_k, \mathcal{D}_{k+1})$ is actually equal to $L(\mathcal{U}, \mathcal{V})$. □

5. Security of Signature Schemes with the EBPV Generator

Any signature scheme consists of three algorithms: a key generation algorithm $\mathcal{A}_{\text{key}}$, a signature algorithm $\mathcal{A}_{\text{sig}}$, and a verification algorithm $\mathcal{A}_{\text{ver}}$. When a pre-processing generator is used, the key generation and signature verification algorithms remain the same, but one adds a preprocessing generation algorithm $\mathcal{A}_{\text{pre}}$.

Let $\mathcal{A}_{\text{sig}}^*$ be the new signature algorithm using the generator, which takes as input the signature keys, the message to sign, and the precomputations generated by $\mathcal{A}_{\text{pre}}$.

We restrict the definitions to the cases we are interested in (such as the El Gamal, DSA, Schnorr and similar signature schemes), by clarifying the link between $\mathcal{A}_{\text{sig}}$ and $\mathcal{A}_{\text{sig}}^*$. The security proof is only valid for that model, which we now make precise. At each signature request, the algorithm $\mathcal{A}_{\text{sig}}$ calls a probabilistic oracle outputting a value of the form $(x, f(x))$ where x is uniformly distributed over the set of keys and f is a function given on this set. The value is then used to produce a signature. For instance, with the DSA [14] and Schnorr [20] signature schemes one has $f(x) \equiv g^x \pmod{p}$ and the set of keys is $\mathbf{Z}_M^*$, where M is a 160-bit prime divisor of $p - 1$. The bit-size of p varies from 512 to 1024.

The algorithm $\mathcal{A}_{\text{sig}}^*$ differs from $\mathcal{A}_{\text{sig}}$ only by the oracle called. The new probabilistic oracle, that is the generator, always outputs a value of the form $(x, f(x))$ with the same function f, but this time, the distribution of x is not necessarily uniform. We will call *defect* Δ of the generator the statistical distance between its distribution and the uniform distribution.

The oracle of $\mathcal{A}_{\text{sig}}$ only takes the keys as input, whereas the generator of $\mathcal{A}_{\text{sig}}^*$ takes as input both the keys and the precomputations.

Note that the defect of the **EBPV** generator is equal to

$$\Delta_{\mathbf{a}}(n, k, h) = \frac{1}{2} \sum_{c \in \mathbf{Z}_M} \left| P_{\mathbf{a}}(\mathcal{B}_{n,k}(h), c) - \frac{1}{M} \right| \tag{4}$$

and thus can be estimated using (3).

We now model an attacker, with respect to the most powerful attacks, namely adaptive attacks (see [6]). We call an *adaptive attack* with m chosen messages any algorithm which makes exactly m requests to the signature algorithm on messages of his choice, during an execution. Such an attack is said to succeed an *existential forgery* if it produces a valid signature of a message of which it had not requested a signature. For any adaptive attack $\mathcal{T}^*$ against the speeded-up signature scheme, one can associate an adaptive attack $\mathcal{T}$ against the original signature scheme, by replacing each call to $\mathcal{A}_{\text{sig}}^*$ by a call to $\mathcal{A}_{\text{sig}}$, with the same parameters. The attacks therefore have the same complexity (remember that each call to the signature algorithm contributes only by one unit to the overall complexity of the attack).

Theorem 5.1. *Let T^* be an adaptive attack with m chosen messages against the signature scheme using the **EBPV** generator with parameters n, k and h. If T is the adaptive attack corresponding to the original signature scheme, then the probabilities $P_\mathbf{a}(n, k, h, m, T^*)$ and $P(m, T)$ of success of existential forgery for identical choices of the keys for both attacks satisfy*

$$\frac{1}{M^n} \sum_{\mathbf{a} \in \mathbf{Z}_M^n} |P_\mathbf{a}(n, k, h, m, T^*) - P(m, T)| \leq mM^{1/2}|\mathcal{B}_{n,k}(h)|^{-1/2}.$$

Proof. The attack T^* requests at each execution exactly m signatures. For each of those m signatures, the signature algorithm uses a pair of the form $(s_i, f(s_i))$ outputted by the generator, where s_i has values over a certain set S of keys. All these values $s_1, \ldots, s_m$ are pairwise independent. Therefore, the success probability of the attack T^* is equal to:

$$\sum_{(w_1, \ldots, w_m) \in S^m} \Pr{}_{\text{generator}} (\forall i, s_i = w_i) \times \Pr(T^* \text{ succeeds} \mid \forall i, s_i = w_i).$$

Similarly, the success probability of the corresponding adaptive attack T against the original scheme is equal to:

$$\sum_{(w_1, \ldots, w_m) \in S^m} \Pr{}_{\text{oracle}} (\forall i, s_i = w_i) \times \Pr(T \text{ succeeds} \mid \forall i, s_i = w_i),$$

with a uniform distribution for the $s_1, \ldots, s_m$. But the link between $\mathcal{A}_{\text{sig}}$ and $\mathcal{A}^*_{\text{sig}}$ ensures us that the conditional probabilities appearing in the two formulas are identical, and less than 1 by definition. Hence, the difference of the success probabilities of the two attacks T are T^* is bounded from the above by:

$$\sum_{(w_1, \ldots, w_m) \in S^m} \left| \Pr{}_{\text{generator}} (\forall i, s_i = w_i) - \Pr{}_{\text{oracle}} (\forall i, s_i = w_i) \right|.$$

This expression is equal to two times the statistical distance between the distribution obtained by sampling m outputs of the generator and the uniform distribution. Applying Lemma 4.1, the identity (4) and the bound (3) we conclude the proof. $\square$

It follows that when the distribution of the generator output is sufficiently close to the uniform distribution, the **BPV** signature scheme is secure against existential forgeries, provided that the original scheme is.

6. Selecting the Parameters

First of all we remark that for $h = 2$, Theorem 5.1 and well-known bounds of binomial coefficients imply that for every $\rho > 0$ and $k = \lfloor \rho \log M \rfloor$, for every $A > 0$ there exists a constant γ such that for $n = \lfloor \gamma \log M \rfloor$ we have

$$\frac{1}{M^n} \sum_{\mathbf{a} \in \mathbf{Z}_M^n} |P_\mathbf{a}(n, k, 2, m, T^*) - P(m, T)| \leq mM^{-A+o(1)}. \tag{5}$$

For bigger (but still polynomial in $\log M$) values of n one can take even smaller values of k. For example, it is easy to see that for any $A > 0$ if one selects $n = \lfloor (A + 0.5) \log^2 M \rfloor$ and $k = \lfloor \log M / \log \log M \rfloor$ then

$$\frac{1}{M^n} \sum_{\mathbf{a} \in \mathbf{Z}_M^n} |P_{\mathbf{a}}(n, k, 2, m, \mathcal{T}^*) - P(m, \mathcal{T})| \le m M^{-A + o(1)}. \tag{6}$$

In particular, the bound (5) means that for any desired ratio $\rho \sim k / \log M$ there is a value of $n = O(\log M)$, that is, linear in the bit size of M such that the corresponding **BPV** generator signature scheme is secure against adaptive attacks (provided the original signature scheme is secure). Taking n quadratic in the bit-size of M, we derive from (6) that growing speed-up can be achieved without compromising the security of the scheme.

We remark that case $h = 2$ has a very important advantage that the computation of $\mathbf{a} \cdot \mathbf{x}$, for $x \in \mathcal{B}_{n,k}$ does not require any storage, it takes only $k - 1$ multiplications.

Now we show that, selecting bigger values of h rather than $h = 2$, one can achieve even better results.

First of all we remark that if $n = k = \lfloor A \log M / \log \log M \rfloor$ and $h = \lceil n / \log n \rceil$ then

$$|\mathcal{B}_{n,k}(h)| = M^{-A + o(1)},$$

thus the bound (6) holds for this selection of parameters as well. And the total computational cost is $n + o(n)$ thus asymptotically less than for repeated squaring and other exponentiation schemes without compromising on the security and with very reasonable storage requirements, namely with storing $k = O(\log M / \log \log M)$ precomputed values.

Some other choices of n, k and h are possible as well, depending of the particular application. For a given choice of n and a given number of multiplications (which is $\mu = k + h - 3$), one can compute the optimal values of k and h that minimize the defect. Table 1 gives such values and the corresponding defect for a 160-bit M, together with the communication time, assuming a speed of 115200 bauds, the choice of a 1024-bit prime p, and a 160-bit hash function as in the DSA (for n pairs and the hash tree with a 160-bit hash function, one needs to send approximately $1.18n + 2 \cdot 0.16n = 1.5n$ Kbits).

7. Remarks

It is easy to see that one can obtain analogues of Theorem 5.1 for identification schemes as well.

One can also obtain similar results for the RSA-based generator with pre-computation which has been introduced in [2] as well, see also [12].

		Number of multiplications, μ			
n	Comm. time	$\mu = 20$	$\mu = 25$	$\mu = 30$	$\mu = 35$
100	1.3s	2^{29} & 17	2^{17} & 21	2^6 & 24	2^{-5} & 28
1000	13s	2^{-1} & 18	2^{-20} & 23	2^{-39} & 27	2^{-58} & 31
10000	2mins 11s	2^{-33} & 19	2^{-59} & 24	2^{-85} & 28	2^{-111} & 32
100000	22mins	2^{-66} & 20	2^{-99} & 24	2^{-132} & 29	2^{-166} & 33

TABLE 1. Value of $\sqrt{M/|\mathcal{B}_{n,k}(h)|}$ and optimal value of k for a 160-bit M.

References

[1] M. Ajtai, *Generating hard instances of lattice problems*, Electronic Colloq. on Comp. Compl., Univ. of Trier, **TR96-007** (1996), 1–29.

[2] V. Boyko, M. Peinado and R. Venkatesan, *Speeding up discrete log and factoring based schemes via precomputations*, Proc. of Eurocrypt'98, Lect. Notes in Comp. Sci., Springer-Verlag, Berlin, **1403** (1998), 221–234.

[3] E. Brickell, D.M. Gordon, K.S. McCurley, and D. Wilson, *Fast exponentiation with precomputation*, Proc. of Eurocrypt'92, Lect. Notes in Comp. Sci., Springer-Verlag, Berlin, **658** (1993), 200–207.

[4] E. F. Brickell and K. S. McCurley, *An interactive identification scheme based on discrete logarithms and factoring*, Journal of Cryptology, **5** (1992), 29–39.

[5] T. El Gamal, *A public key cryptosystem and a signature scheme based on discrete logarithms*, IEEE Trans. Inform. Theory, **31** (1985), 469–472.

[6] O. Goldreich, *Modern Cryptography, Probabilistic Proofs and Pseudorandomness*, Springer-Verlag, Berlin, 1999.

[7] D.M. Gordon, *A survey of fast exponentiation methods*, Journal of Algorithms, **27** (1998), 129–146.

[8] F. Griffin and I. E. Shparlinski, *On the linear complexity of the Naor–Reingold pseudo-random function*, Proc. of 2nd Intern. Conf. on Inform. and Commun. Security, Sydney, 1999, Lect. Notes in Comp. Sci., Springer-Verlag, Berlin, 1999, 301–308.

[9] R. Impagliazzo and M. Naor, *Efficient cryptographic schemes provably as secure as subset sum*, J. Cryptology, **9** (1996), 199–216.

[10] C. H. Lim and P. J. Lee, *More flexible exponentiation with precomputation*, Proc. of Crypto'94, Lect. Notes in Comp. Sci., Springer-Verlag, Berlin, **839** (1994), 95–107.

[11] M. Luby, Pseudorandomness and cryptographic applications, Princeton University Press, 19969.

[12] J. Merkle and R. Werchner, *On the security of server-aided RSA protocols*, Lect. Notes in Comp. Sci., Springer-Verlag, Berlin, **1431** (1998), 99–116.

[13] M. Naor and O. Reingold, *Synthesizers and their application to the parallel construction of pseudo-random functions*, J. Comp. and Sys. Sci., **58** (1999), 336–375.

[14] *National Institute of Standards and Technology (NIST)*, FIPS Publication 186: Digital Signature Standard, May 1994.

[15] P. Nguyen and J. Stern, *The hardness of the hidden subset sum problem and its cryptographic implications*, Proc. of Crypto'99, Lect. Notes in Comp. Sci., Springer-Verlag, Berlin, **1666** (1999), 31–46.

[16] P. de Rooij, *On the security of the Schnorr scheme using preprocessing*, Proc. of Eurocrypt'91, Lect. Notes in Comp. Sci., Springer-Verlag, Berlin, **547** (1991), 71–80.

[17] P. de Rooij, *Efficient exponentiation using precomputation and vector addition chains*, Proc. of Eurocrypt'94, Lect. Notes in Comp. Sci., Springer-Verlag, Berlin, **950** (1995), 389–399.

[18] P. de Rooij, *On Schnorr's preprocessing for digital signature schemes*, Journal of Cryptology, **10** (1997), 1–16.

[19] C. P. Schnorr, *Efficient identification and signatures for smart cards*, Proc. of Crypto'89, Lect. Notes in Comp. Sci., Springer-Verlag, Berlin, **435** (1990), 239–252.

[20] C. P. Schnorr, *Efficient signature generation by smart cards*, Journal of Cryptology, **4** (1991), 161–174.

[21] I. M. Vinogradov, *Elements of number theory*, Dover Publ., New York, 1954.

Département d'Informatique,
École Normale Supérieure,
45, rue d'Ulm
Paris Cedex 05, France
E-mail address: pnguyen@ens.fr

Department of Computing,
Macquarie University,
Sydney, NSW 2109, Australia
E-mail address: igor@comp.mq.edu.au

Département d'Informatique,
École Normale Supérieure,
45, rue d'Ulm
Paris Cedex 05, France
E-mail address: stern@dmi.ens.fr

Progress in Computer Science and Applied Logic, Vol. 20
© 2001 Birkhäuser Verlag Basel/Switzerland

A General Construction for Fail-Stop Signature using Authentication Codes

Rei Safavi-Naini and Willy Susilo

Abstract. Security of an ordinary digital signature relies on a computational assumption. Fail-stop signature schemes provide security for a sender against a forger with unlimited computational power by enabling the sender to provide a proof of forgery, if it occurs. In this paper, we describe a method of constructing fail-stop signature schemes from authentication codes. We also give an example that fits this general construction and prove its security.

1. Introduction

In an *ordinary digital signature* scheme if an enemy can break the underlying hard problem, he can successfully forge a signature and there is no way for the signer to prove that a forgery has occurred. Fail-stop signature (FSS) schemes have been proposed [16, 12, 8] to provide protection against forgeries of an enemy with unlimited computational power. The basic idea is to have many secret key that correspond to the same public key and so prevent the enemy from finding the signer's secret key even if the underlying problem is solved. In a FSS there are a number of participants: (i) a polynomially bounded *signer* who signs a message that is verifiable by everyone with access to the public key and is protected against forgery by an unbounded enemy, (ii) one or more polynomially bounded *recipients* who directly or indirectly through a trusted centre, take part in the key generation process and are protected against repudiation of the signer, and (iii) a *trusted centre* who is trusted by the recipients and only takes part in the key generation phase. There is another group of participants, the so-called *risk-bearers*, such as insurance companies, who will bear a loss if a proof of forgery is accepted and hence a signature is invalidated. For simplicity we do not make any distinction between a recipient and a risk bearer. In the case of forgery, the presumed signer can provide a proof that a forgery has happened. This is by showing that the underlying computational assumption of the system is broken. The system will be stopped at this stage- hence the name *fail-stop*. In this way, a polynomially bounded signer can be protected against a forger with unlimited computational power. It can be shown that (Theorem 3.2 [8]) a secure FSS can be used to construct an ordinary digital signature that is secure in the sense of [5] and so a fail-stop signature scheme provides a stronger notion of security.

A FSS in its basic form is a *one-time digital signature* that can only be used for signing a single message. However, it is possible to extend a FSS scheme to be used for signing multiple messages [2, 15, 10, 1].

1.1. Our Contributions

In this paper, we describe a method of constructing FSSs from authentication codes. The method requires two families of A-codes and two families of collision intractable hash functions, one with bundling property, that satisfy certain conditions. We prove that the security of the receiver relies on the collision intractiblity of the hash functions and security of the sender can be derived from the bundling property of the hash function and structure of the A-codes. This provides an alternative method to [8] for constructing FSS with provable security. We show a construction proposed in [13] can be regarded as an example of this method.

The paper is organised as follows. In the next section, we present the notations and review basic concepts of FSS and authentication codes. We briefly examine the previous works on FSS and recall a general construction proposed in [8]. In Section 3, we propose a general method of constructing a FSS from an authentication code. In Section 4, we give an example of FSS scheme based on our framework and prove its security requirements. Section 5 concludes the paper.

2. Notations and Preliminaries

2.1. Notations

The ring of integers modulo a number n is denoted by Z_n, and the multiplicative subgroup of integers relatively prime to n, by Z_n^*. Let N denote the set of natural numbers.

Notation $x \overset{?}{=} y$, means that equality of x and y must be checked.

2.2. Authentication Codes

Authentication codes are used to provide protection against tampering with the messages communicated between two participants over an insecure channel.

An *authentication code (A-code)* is a set of functions $\mathbf{E}$, each from a set of messages $\mathbf{M}$ to a set of codewords $\mathbf{C}$ (also called *authenticated messages*). $\mathbf{E}$ is indexed by a piece of information, $K \in \mathcal{K}$, called *key*. A key K uniquely determines an encoding function e_k and conversely, an encoding function is associated with a unique key. For the sake of brevity and when it is clear from the context the term key may be used instead of its corresponding function.

In a systematic Cartesian authentication code, a codeword c is the concatenation of a message m and a tag t, that is: $c = m\|t$. A Cartesian A-code with a message space $\mathbf{M}$, a tag space $\mathbf{T}$ and key space $\mathbf{E}$ is denoted by $A(M, T, E)$ and can be described by a $|E| \times |M|$ matrix over T. A row of the matrix labelled by a key k, defines the *encoding function*, e_k, from M to T given by $e_k(m) = t$ if $A(k,m) = t$.

We considered two types of attacks where in both the enemy is an *intruder-in-the-middle*. The attacks are:

- **Impersonation Attack**: The enemy introduces a message (m, t), where $m \in M$ and $t = e_k(m)$, and hopes that this message is accepted as authentic by the receiver.
- **Substitution Attack**: After observing an authentic message (m, t) in the channel, the enemy constructs a message (m', t'), $m \neq m'$, and hopes that it is accepted as authentic by the receiver.

The chance of enemy's success in making the receiver to accept his fraudulent message as authentic when he follows his optimal strategy in impersonation or substitution, is denoted by Pd_0 and Pd_1, respectively.

We assume that the authentication code, the two probability distributions on **M** and **E** are publicly known but the actual value of the key, and so the encoding function is not known.

2.3. Fail-Stop Signature Schemes

Similar to an ordinary digital signature scheme, a fail-stop signature scheme consists of one polynomial time protocol and two polynomial time algorithms.

1. *Key generation*: is a two party protocol between the signer and the centre to generate a pair of *secret key*, s_k, and *public key*, p_k. This is different from ordinary signature schemes where key generation is performed by the signer individually and without the involvement of the receiver.
2. *Sign*: is the algorithm used for signature generation. For a message m and using the secret key s_k, the signature is given by $y = sign(s_k, m)$.
3. *Test*: is the algorithm for testing acceptability of a signature. For a message m and signature y, and given the public key p_k, the algorithm produces an ok response if the signature is acceptable under p_k. That is $test(p_k, m, y) \overset{?}{=} ok$.

A FSS also includes two more polynomial time algorithms:

4. *Proof*: is an algorithm for proving a forgery;
5. *Proof-test*: is an algorithm for verifying that the proof of forgery is valid.

A secure fail-stop signature scheme must satisfy the following properties [14, 11, 8].

1. If the signer signs a message, the recipient must be able to verify the signature (*correctness*).
2. A polynomially bounded forger cannot create forged signatures that successfully pass the verification test (*recipient's security*).
3. When a forger with an unlimited computational power succeeds in forging a signature that passes the verification test, the presumed signer can construct a proof of forgery and convinces a third party that a forgery has occurred (*signer's security*).

4. A polynomially bounded signer cannot create a signature that he can later prove to be a forgery (*non-repudiability*).

To achieve the above properties, for each public key there exist many matching secret keys such that different secret keys create different signatures on the same message. The real signer knows only one of the secret keys, and can construct one of the many possible signatures. An enemy with unlimited computing power, although can generate all the signatures but does not know which one will be generated by the true signer. Thus, it will be possible for the signer to provide a proof of forgery by generating a second signature on the message with a forged signature, and use the two signatures to show the underlying computational assumption of the system is broken, hence proving the forgery.

FSS are studied in two different models. The main difference between the models is the existence of a dealer who is trusted by all the recipients. Schemes with a trusted dealer, for example [15, 3], allow public verification of the signature and use a two-party protocol between the signer and the dealer to generate the required keys. This ensures that the signer cannot later deny his own signature and can provide a proof of forgery when needed. Schemes without a trusted dealer, for example [14], are obtained by giving the role of the trusted dealer to one of the recipients. This results in a more efficient key exchange at the expense of loosing public verifiability property of the signature. The model is useful for applications such as electronic payment where verification is required for only a single receiver.

Security of a FSS is broken if 1) a signer can construct a signature and can later provide a proof of forgery; or 2) an unbounded forger succeeds in constructing a signature that the signer cannot prove that it is forged. These two types of forgeries are independent and so two different security parameters, k and σ, are used to show the level of security against the two types of attacks. More specifically, k is the security level of the recipient against the forgery of the signer, and σ is that of the signer against the unbounded forger. It is proved [8] that a secure FSS is secure against adaptive chosen plain-text attack and for all $c > 0$ and large enough k, success probability of a polynomially bounded forger is bounded by k^{-c}. For a FSS with security level σ for the signer, the success probability of an unbounded forger is limited by $2^{-\sigma}$.

In the following we briefly recall the general construction given in [8] and outline its security properties.

2.4. A General Construction for FSS

The construction is for a single-message fail-stop signature and uses *bundling homomorphisms*. Bundling homomorphisms can be seen as a special kind of hash functions.

Definition 2.1. [8] *A bundling homomorphism h is a homomorphism $h : G \to H$ between two Abelian groups $(G, +, 0)$ and $(H, \times, 1)$ that satisfies the following.*

1. *Every image $h(x)$ has at least 2^τ preimages. 2^τ is called* bundling degree *of the homomorphism.*

2. *It is infeasible to find collisions, that is, two different elements that are mapped to the same value by h.*

To give a more precise definition, we need to consider two families of groups, $\mathbf{G} = (G_K, +, 0)$ and $\mathbf{H} = (H_K, \times, 1)$, and a family of polynomial-time functions, h_K, indexed by a key, K. There is also a probabilistic polynomial-time key generation algorithm, $gen()$, that on input (k, τ) generates K. For each K, h_K is a homomorphism from G_K to H_K and satisfies the properties of Definition 2.1. The two parameters (k, τ), determine the difficulty of finding collision and the bundling degrees of the homomorphisms, respectively.

A bundling homomorphism can be used to construct a FSS scheme as follows. Let the security parameters of the FSS be given as k and σ. The bundling degree of the homomorphism, τ, will be obtained as a function of σ as shown below.

1. *Prekey generation*: The centre computes $K = gen(k, \tau)$ and so determines a homomorphism h_K, and two groups G_K and H_K. Let $G = G_K, H = K_K$ and $h = h_K$.
2. *Prekey verification*: The signer must be assured that K is a possible output of the algorithm $gen(k, \tau)$. This can be through providing a zero-knowledge proof by the centre or by testing the key by the signer. In any case the chance of accepting a *bad* key must be at most $2^{-\sigma}$.
3. *Main key generation Gen_A*: the signer generates her secret key $sk := (sk_1, sk_2)$ by choosing sk_1 and sk_2 randomly in G and computes $pk := (pk_1, pk_2)$ where $pk_i := h(sk_i)$ for $i = 1, 2$.
4. The message space M is a subset of Z.
5. *Signing*: The signature on a message $m \in M$ is,

$$s = sign(sk, m) = sk_1 + m \times sk_2$$

where multiplying by m is m times addition in G.
6. *Testing the signature*: can be performed by checking,

$$pk_1 \times pk_2^m \overset{?}{=} h(s)$$

7. *Proof of forgery*: Given an acceptable signature $s' \in G$ on m such that $s' \neq sign(sk, m)$, the signer computes $s := sign(sk, m)$ and $proof := (s, s')$.
8. *Verifying proof of forgery*: Given a pair $(x, x') \in G \times G$, verify that $x \neq x'$ and $h(x) = h(x')$.

Theorem 4.1 [8] proves that for any family of bundling homomorphisms and appropriate choice of parameters the general construction:

1. produces correct signature;
2. a polynomially bounded signer cannot construct a valid signature and a proof of forgery;
3. if an acceptable signature $s^* \neq sign(sk, m^*)$ is found the signer can construct a proof of forgery.

Moreover for the chosen parameters k and σ, a good prekey K and two messages $m, m^* \in M$, with $m \neq m^*$, let

$$T := \{d \in G | h(d) = 1 \wedge (m^* - m)d = 0\} \tag{1}$$

Theorem 4.2 [8] shows that given $s = sign(sk, m)$ and a forged signature $s^* \in G$ such that $test(pk, m^*, s^*) = ok$, the probability that $s^* = sign(sk, m^*)$ is at most $|T|/2^\tau$ and so the best chance of success for an unrestricted forger to construct an undetectable forgery is bounded by $|T|/2^\tau$. Thus to provide the required level of security σ, we must choose $|T|/2^\tau \leq 2^{-\sigma}$.

This general construction is the basis of all known *provably secure constructions* of FSS. It provides a powerful framework by which proving security of a scheme is reduced to specifying the underlying homomorphism, and determining the bundling degree and the set T.

Other Previous Works

The first construction of fail-stop signature [16] uses a one-time signature scheme (similar to [6]) and results in bit by bit signing of the message and so is very impractical. In [9] an efficient single-recipient FSS to protect clients in an on-line payment system, is proposed. The main disadvantage of this system is that signature generation is a 3-round protocol between the signer and the recipient and so it has high communication cost. The size of the signature is twice the length of the message.

In [15], an efficient FSS that uses the difficulty of the discrete logarithm problem as the underlying assumption is presented. In the case of a forgery, the presumed signer can solve an instance of the discrete logarithm problem, and prove that the underlying assumption is broken. This is the most efficient scheme known so far which requires only two multiplications for signature generation and results in a signature which is twice the size of the message.

In [8, 11], a formal definition of FSS schemes is given and a general construction using *bundling homomorphism* is proposed. The important property of this construction is its provable security. An instance of this construction uses the difficulty of factoring as the underlying computational assumption of the system [14].

It is also proved that for a system with security level σ for the signer, the signature length and the length of secret key required for signing N messages are at least $2\sigma - 1$ and $(N + 1)(\sigma - 1)$, respectively.

3. Fail-stop signatures from A-codes

We are interested in families of A-codes. A family $\mathcal{A} = \{A(M_K, T_K, E_K) : K \in N\}$, of A-codes is defined by a family of message spaces, $\mathcal{M}$, a family of tag spaces, $\mathcal{T}$ and a family of key spaces $\mathcal{E}$. Each of the three families is an infinite collection of sets indexed by $K \in N$. That is, $\mathcal{M} = \{M_K : K \in N\}$, $\mathcal{T} = \{T_K : K \in N\}$ and $\mathcal{E} = \{E_K : K \in N\}$, and $A(M_K, T_K, E_K)$ is an A-code with

$P_d^{(K)} = \max\{pd_0^{(K)}, pd_1^{(K)}\} \leq \epsilon$ where pd_0^K, pd_1^K are probability of success against impersonation and substitution attack in $A(M_K, T_K, E_K)$.

For a function f, $f : X \mapsto X'$, a *collision* is a pair $x, x' \in X$ such that $f(x) = f(x')$. A function f from X to X' is called a *collision intractable hash function* if finding a pair x, x' such that $f(x) = f(x')$ is hard. To make the hardness definition more precise we need a family of functions indexed by a key $K \in N$. Then intractability means that for all $c > 0$, there exists a $K_0 \in N$ such that for all $K > K_0$, and for all probabilistic polynomial time algorithms the chance of finding collision is less than K^{-c}.

A function h from X to X' is called a *collision intractable bundling hash function* of degree 2^τ, if it satisfies the following properties:

1. for every $x' \in X'$ which is the image of some $x \in X$, that is $x' = h(x)$, there are at least 2^τ preimages, $x_1, x_2, \ldots x_{2^\tau}$, such that $h(x_i) = x'$, $i = 1 \cdots 2^\tau$.
2. It is hard to find a pair of elements $x, x' \in X$ such that $h(x) = h(x')$.

Again to make the second requirement more precise we need a family of functions indexed by a key K. There is a probabilistic polynomial time algorithm *gen()*, that on input k and τ, both from N, generates a key K. In this case for each $z = h_K(X)$ there are at least τ preimages and the difficulty of finding a collision under h_K is at least k. That is, for any polynomial Q_τ that determines the growth of τ with k, and for all $c > 0$ there exists k_0 such that for $k \geq k_0$ the chance of finding collision is at least k^{-c}.

This definition has similarities with the definition of bundling homomorphism but is more general and does not require X and X' to have group structure.

Now consider two families of A-codes $\mathcal{A} = \{A(M_K, T_K, E_K) : K \in N\}$ and $\mathcal{A}' = \{A'(M_K, T'_K, E'_K) : K \in N\}$, a family of polynomial time collision intractable bundling hash function $\mathcal{H} = \{h_K : K \in N\}$ where $h_K : E_K \mapsto E'_K$, and a family of polynomial time collision intractable hash functions $\mathcal{H}' = \{h'_K : K \in N\}$, where $h'_K : T_K \mapsto T'_K$.

We require the following property:

- *Property I*: for any choice of key K, and for an arbitrary $e \in E_K$ the following is satisfied for all $m \in M_K$:

  if $e(m) = t$,   and $h_K(e) = e'$, then $e'(m) = t'$ and $h'_K(t) = t'$

Given a six tuple $(\mathcal{A}, \mathcal{A}', \mathcal{E}, \mathcal{E}', \mathcal{H}, \mathcal{G})$ we can construct a FSS as follows.

The index K is the pre-key and is determined by a pre-key generation algorithm $gen(k, k', \tau)$ which takes the following parameters as input: (i) τ, the bundling degree of the hash function, (ii) k the difficulty of finding collision for h, and (iii) k' which is the difficulty of finding collision for h'. The resulting index is the pre-key that determines various parameters of the system. The signer must be sure that the K is a possible output of $gen()$ by performing a *prekey verification* algorithm.

Once K is determined, $A(M_K, T_K, E_K)$, $A'(M_K, T'_K, E'_K)$, h_K and h'_K are fixed and we have the following stages.

350   R. Safavi-Naini and W. Susilo

1. *Main key generation*: the signer chooses $e \in E_K$ as his secret key (encoding function) and constructs $e' = h_K(e)$ as his public key (verification function).
2. *Signing*: the signature for the message $m \in M_K$ is given by $t = e(m)$.
3. *Testing of the signature*: a signature t on a message m is verified if $h'_K(t) = e'(m)$.
4. *Proof of forgery*: given an acceptable signature t_1 on m where $t_1 \neq e(m)$, the signer produces $t = e(m)$ as the proof of forgery.

Theorem 3.1. *The above construction has the following properties:*

1. *Correct signatures pass the test.*
2. *A polynomially bounded signer cannot construct a signature and a valid proof of forgery.*
3. *If t' is an acceptable signature on m' and $t' \neq e(m')$, the signer obtains a valid proof of forgery.*

Proof: Part 1 follows immediately from the definition of A-codes and the two classes of functions, and Property I. To construct a signature and a proof of forgery, the sender must find a message m with two signatures t and t' that both pass the test. That is finding t and t' such that $h'_K(t) = h'_K(t') = e'(m)$ which because of collision intractability of $h'()$ is hard. An alternative approach for the sender is to try to find a key e_1 with $h_K(e) = h_K(e_1)$ and $e_1(m) \neq e(m)$ which is at least as hard as finding collision for h_K and is hard. Finally Part 3 is true because constructing such signature requires either finding collision for one of the two hash functions which is assumed to be hard.

Theorem 3.2. *Let k and σ denote security parameters of the scheme described above. Then for all pairs of secret and public keys, e, e', and given a message and signature pair (m, t) where $t = e(m)$, the probability of a signature t' on a message m' that satisfies $e'(m') = t'$ also satisfies $e(m) = t$ for an enemy with unlimited power, is at most $|V|/|W|$ (V and W defined below).*

Proof: Suppose enemy has a message (m, t) signed by the sender, and a signature t' for a message m' that passes the verification test. His chance of success in being able to use (m', t') as a successful forgery is the same as the probability of t' being the same as $e(m')$ (that is the signature generated by the valid sender on m'.)

Knowing the public key e' allows the enemy to obtain a set $E(e')$ of keys that contains all the keys $\hat{e}$ such that $h_K(\hat{e}) = h_K(e)$, and includes the sender's key. Using the knowledge of (m, t) the enemy can further reduce the set of possible keys for the sender to $W = E(e', (m, t))$ that contains the keys that map to e' and produce the signature t on m. Now enemy's best chance of success in constructing a forged signature (m', t') that cannot be proven to be a forgery, is obtained by finding the the following set:

$$V = \max_{m', t'}\{e : e \in E(e', (m, t)),\ e'(m') = h'(t'),\ e(m') = t'\}$$

The success chance of an unbounded forger in constructing a forged signature is given by $|V|/|W|$. □

Using the above construction, constructing an FSS with security parameters (σ, k) requires choosing σ to satisfy

$$|V|/|W| \leq 2^{-\sigma}$$

4. A Construction

In the following we describe an example construction for the above model. This construction is first proposed in [13]. Here we show that it can be regarded as an example of the general construction proposed in this paper. We will have two families of A-codes with a bundling hash function that satisfy the requirements of the construction above.

Let p and q denote two large primes, $p < q$, $n = pq$, and $P = 2pq + 1$ where P is also prime. Let g be an element of Z_P^* of order p, $ord_P(g) = p$. Define an A-code as follows.

$\mathbf{M} \in Z_n$.

$\mathbf{T} \in Z_n$.

$\mathbf{E} = \{e_{i,j} = (i,j) : 0 \leq i \leq n-1, 0 \leq j \leq n-1\}$ where $e_{ij} : M \mapsto T$ and $e_{ij}(l) = i + jl \pmod{n}$.

Theorem 4.1. *Let $p < q < 2p$. For the above A-code we have* $Pd_0 = \frac{1}{pq}$ *and* $Pd_1 = \frac{1}{p}$.

Proof: See Appendix.

Bundling Hash Function Let H_p be the subgroup of Z_P^* generated by g. We have $|H_p| = p$. We define a mapping $h : Z_n \mapsto H_p$ given by $h(x) = g^x \pmod{P}$. In the following we show that h is a bundling hash function.

1. *For any $x \in H_p$, there are q preimages in Z_n; that is there are q such y such that $h(y) = x$.*
   This is true because for all elements $y' \in Z_n$ where $y' = y + tp$, and $t \in \{0, 1, ..q - 1\}$, we have

   $$h(y') = g^{y+tp} = g^y = x$$

2. *Given $x \in H_p$ it is difficult to find a y such that $h(y) = x$.*
   This is true because finding y that satisfies $g^y = x \pmod{P}$ is equivalent to finding discrete logarithm in group H_p which is known to be hard (In fact, solving discrete logarithm in group H_p is considerably more difficult than factoring n [7]).

3. *Finding a pair $x, x' \in H_p$ such that $h(x) = h(x')$ is as hard as factoring n.*
   This is true because $h(x) = h(x')$ implies $g^x = g^{x'} \pmod{P}$ and because g is of order p, $x = x' \pmod{p}$ and so $x - x'$ is a multiple of p.

The above bundling hash function partitions Z_n into p subsets each of size q. That is, $Z_n = V_0 \cup V_1 \cdots \cup V_{p-1}$ where $V_i = i + tp$, $0 \le t \le q - 1$. From above, we know that an element x of H_p corresponds to a unique V_i. We use $H_p(i)$ to denote x.

Now consider a family of A-codes, $A'(M, T', E')$ where $M = Z_n$, $T' = H_p$, $E' = H_p \times H_p$ and $e'_{IJ}(l) = g^i g^{jl}$ where $I = H_p(i)$ and $J = H_p(j)$ respectively.

Theorem 4.2. *In $A'(M, T', E')$ we have $Pd_0 = \frac{1}{p}$ and $Pd_1 = \frac{1}{p}$.*

Proof: See Appendix.

A FSS scheme based on the above families of A-codes

Our construction is based on $A(M, T, E)$, $A(M, T', E')$, the bundling hash function defined above and assuming h' to be defined as $h'(x) = h(x)$, $x \in Z_n^*$. That is taking the hash function h' to be equal to the same as the bundling hash function h on all its inputs. We note that the above choices for the A-codes, h and h' will satisfy Property I. This is true because for any choice of K and for an arbitrary $e = (i, j)$, $i, j \in Z_n$, and $m = l \in Z_n$, we have $t = e(m) = i + jl \pmod{n}$. The bundling hash function h_K is defined by g, an element of order p, in Z_n^*. That is $e' = h(i, j) = (g^i, g^j)$ and $e'(l) = g^i g^{jl} = g^{i+jl} = h'(t) \pmod{P}$.

Model

There is a polynomially bounded sender $\tilde{S}$, a polynomially bounded receiver $\tilde{R}$ and an enemy $\tilde{E}$ with unlimited computational power. The A-codes are public. We follow the second model of FSS (as in [14]) which does not require a trusted dealer. This scheme can be easily modified to the first model of FSS (as in [15, 3]) by replacing the role of $\tilde{R}$ with a trusted dealer in the prekey generation phase.

Prekey Generation

$\tilde{R}$ chooses two prime numbers p and q, $p < q$, and computes $n = pq$ and $P = 2pq + 1$, where P is also prime. If P is not a prime, $\tilde{R}$ has to choose another set of p and q such that P is prime. He also chooses an element g of Z_P^* with $ord_P(g) = p$. Finally, he publishes n, g and P, and keeps p and q secret.

Key Generation

$\tilde{S}$ chooses a secret key (i, j), and publishes his public key (γ_1, γ_2), where

$$\gamma_1 = g^i \pmod{P}$$

$$\gamma_2 = g^j \pmod{P}$$

Signing a Message ℓ

To sign a message $\ell \in Z_n$, $\tilde{S}$ computes

$$t = i + j\ell \pmod{n}$$

where t denotes the signature or the tag of ℓ. The signed message is (ℓ, t).

Testing a Signature

(ℓ, t) passes the verification test if

$$\gamma_1 \gamma_2^\ell \overset{?}{=} g^t \pmod{P}$$

holds.

Proof of Forgery

If there is a forged signature t' that also passes the verification test, the presumed signer can prove that he has obtained a collision by showing his own signature t together with t'. That is,

$$g^t = g^{t'} \pmod{P}$$
$$t = t' \pmod{p}$$
$$t - t' = kp, \ k \in Z$$

Then, the presumed signer can find the factorisation of n by calculating $gcd(t - t', n)$ which is p.

Security Proof

Theorem 4.3. *The construction described above under DL and factorisation assumptions, is secure for the receiver and signer.*

Proof: The above construction conforms with the general construction described in Section 3 and so using theorem 4.1 provides security for the receiver. The level of security for the sender is obtained by finding $|W|$ and $|V|$.

Let the public key be e' labelled by (I, J) where $I = H_p(i)$ and $J = H_p(j)$; that is $e'(l) = g^{(i+jl)}$. Firstly we note that $|E(e')| = q^2$. This is because the encoding rule labelled by $((i + mp), (j + np))$, where $m, n \in \{0, 1, \ldots q - 1\}$ are mapped to the same e'. Next for an arbitrary $t \in T$ and a message $l \in M$, we need to find

$$E(e', (l, t)) = |\{e \in E(e') : e(l) = t\}|$$

That is find the number of solutions to

$$(i + mp) + (j + np)l = t \pmod{pq}$$

where $i + jl = t' \pmod{pq}$ or equivalently, finding the number of solutions to $(m + nl)p = t - t' \pmod{pq}$, or the number of solution to $m + nl = t - t' \pmod q$. Now for any l and an arbitrary n we can find a unique m that satisfies this equation. So $|W| = |E(e', (l, t))| = |\{e \in E(e') : e(l) = t\}| = q$.

To find V we consider a message l' with tag t' satisfying $h'(t') = e'_{IJ}(l')$. This means that we have $g^{(i+mp)+(j+np)l'} = g^{t'} \pmod{P}$, or $(i + mp) + (j + np)l' = t'$. Combining this equation with the one obtained from $e(m) = t$, and assuming $i + jl = w \pmod{pq}$ and $i + jl' = w' \pmod{pq}$, we obtain two equations $m + nl = t - w \pmod q$ and $m + nl' = t' - w \pmod q$. That is, $n(l - l') = (u - u') + (w - w') \pmod q$, which has a unique solution. Thus, $|V| = 1$ and $|V|/|W| = 1/q$. □

From the above theorem we must choose $1/p \leq 2^{-\sigma}$ or $p = \log_2 \sigma$ and $p < q$.

5. Conclusion

We proposed a general construction for FSS using authentication codes and showed that a previously proposed FSS fits into this construction. It is likely that other known schemes that are based on the general construction in [8] can be also explained through the A-code construction. However it is not known if the two constructions are equivalent.

References

[1] N. Barić and B. Pfitzmann. Collision-Free Accumulators and Fail-Stop Signature Schemes without Trees. *Advances in Cryptology - Eurocrypt '97, Lecture Notes in Computer Science 1233*, pages 480–494, 1997.

[2] D. Chaum, E. van Heijst, and B. Pfitzmann. Cryptographically strong undeniable signatures, unconditionally secure for the signer. *Interner Bericht, Fakultät für Informatik*, 1/91, 1990.

[3] I. B. Damgård, T. P. Pedersen, and B. Pfitzmann. On the existence of statistically hiding bit commitment schemes and fail-stop signatures. *Journal of Cryptology*, 10/3:163–194, 1997.

[4] W. Diffie and M. Hellman. New directions in cryptography. *IEEE IT*, 22:644–654, 1976.

[5] S. Goldwasser, S. Micali, and R. L. Rivest. A digital signature scheme secure against adaptive chosen-message attacks. *SIAM Journal of Computing*, 17/2:281–308, 1988.

[6] L. Lamport. Constructing digital signatures from a one-way function. *PSRI International CSL-98*, 1979.

[7] A. K. Lenstra and E. R. Verheul. Selecting Cryptographic Key Sizes in Commercial Applications. *Price Waterhouse Coopers, CCE Quarterly Journal*, 3, 3-9, 1999. Full version appears in *http://www.cryptosavvy.com*.

[8] T. P. Pedersen and B. Pfitzmann. Fail-stop signatures. *SIAM Journal on Computing*, 26/2:291–330, 1997.

[9] B. Pfitzmann. Fail-stop signatures: Principles and applications. *Proc. Compsec '91, 8th world conference on computer security, audit and control*, pages 125–134, 1991.

[10] B. Pfitzmann. Fail-stop signatures without trees. *Hildesheimer Informatik-Berichte, Institut für Informatik*, 16/94, 1994.

[11] B. Pfitzmann. *Digital Signature Schemes - General Framework and Fail-Stop Signatures*. Lecture Notes in Computer Science 1100, Springer-Verlag, 1996.

[12] B. Pfitzmann and M. Waidner. Formal aspects of fail-stop signatures. *Interner Bericht, Fakultät für Informatik*, 22/90, 1990.

[13] W. Susilo, R. Safavi-Naini, M. Gysin, and J. Seberry. A New and Efficient Fail-Stop Signature Scheme. *Manuscript*, 2000.

[14] E. van Heijst, T. Pedersen, and B. Pfitzmann. New constructions of fail-stop signatures and lower bounds. *Advances in Cryptology - Crypto '92, Lecture Notes in Computer Science 740*, pages 15–30, 1993.

[15] E. van Heyst and T. Pedersen. How to make efficient fail-stop signatures. *Advances in Cryptology - Eurocrypt '92*, pages 337–346, 1992.

[16] M. Waidner and B. Pfitzmann. The dining cryptographers in the disco: Unconditional sender and recipient untraceability with computationally secure serviceability. *Advances in Cryptology - Eurocrypt '89, Lecture Notes in Computer Science 434*, 1990.

Appendix

Proof Theorem 4.1: (sketch)

1. $Pd_0 = \frac{1}{pq}$.

   We need to show that for an arbitrary source state l, and an arbitrary tag value t, the number of keys e_{ij} such that $e_{ij}(l) = t$ is pq.

   Such keys satisfy the following equation

   $$i + jl = t \pmod{pq}$$

   Since there are pq choices for j and for each choice there is a unique i, then we will obtain pq keys that satisfy the condition.

2. $Pd_1 = \frac{1}{p}$.

   We know that for an arbitrary source state l, a tag t, $0 \le i < pq$, occurs exactly pq times. Now, consider two source states l and l', and all the keys that produce the tag t for the message l. That is: $i + jl = t$. From these keys, the number of keys (i, j) that produce the tag t' for the message l' is given by the number of solutions to the equation $i + jl' = t'$, $t' \in Z_{pq}$. Or the number of solutions to,

   $$t - t' = j(l - l') \pmod{pq}$$

   Consider two cases:

   Case 1. $gcd(l - l', pq) = 1$.

   Then, $j = (l - l')^{-1}(t - t') \pmod{pq}$, and so there is a unique key with this property.

   Case 2. $gcd(l - l', pq) \ne 1$.

   So, there can be either

   i. $l - l' = kp$, $k \in Z_q \setminus 0$ (non-zero elements of Z_q), or
   ii. $l - l' = kq$, $k \in Z_p \setminus 0$ (non-zero elements of Z_p).

   We consider each case as follows:

   i. Let $l - l' = kp$. Then, $k = 1, 2, \ldots, p, p+1, \ldots, q - 1$, and so

   a. if $k \ne p$, then we have $gcd(k, p) = 1$. This means: $(t - t') = jkp \pmod{pq}$. For a fixed $t - t'$, the number of solutions (j) for this equation is p. This is true because if j_0 satisfies $t - t' = j_0 kp \pmod{pq}$, then $(j_0 + uq) \pmod{pq}$ will also satisfy the equation. This is:

   $$t - t' = (j_0 + uq)kp = j_0 kp \pmod{pq}$$

   Now, u can take p values, $0, 1, \ldots, p - 1$, and so there are p solutions. This means that a pair (t, t') will occur p times.

        b. if $k = p$, we have $(t - t') = jp2 \pmod{pq}$. For a fixed $(t - t')$, there are also p solutions for this equation.

    ii. Let $l - l' = kq$, then $k = 1, 2, \ldots, p - 1$, and hence $gcd(k, q) = 1$. This means: $(t - t') = jkq \pmod{pq}$. For a fixed $(t - t')$, the number of solutions for this equation is q. This is true because if j_0 satisfies $t - t' = j_0 kq \pmod{pq}$, then $(j_0 + up) \pmod{pq}$ will also satisfy the equation. Now, u can take q values, $0, 1, \ldots, q - 1$, and so there are q solutions. This means that a pair (t, t') will occur q times.

Hence, the maximum number of solutions for $t - t' = j(l - l') \pmod{pq}$ is q (since $p < q$). Therefore, the probability of success in substitution attack is:

$$Pd_1 = \max_{l, l', t, t'} \frac{|\{e_{ij} : e_{ij}(l) = t, e_{ij}(l') = t'\}|}{|\{e_{ij} : e_{ij}(l) = t\}|} = \frac{q}{pq} = \frac{1}{p} \qquad \square$$

Deception Probabilities for $A'(M, T', E')$

Proof of Theorem 4.2 : (sketch)

1. $Pd_0 = \frac{1}{p}$

   For an arbitrary message l and an arbitrary tag value $W \in H_p$, the number of keys e'_{IJ} for which $e'_{IJ}(l) = W$ is p. This is true because $e'_{IJ}(l) = g^i g^{jl} = g^w \pmod{P}$ where $I = H_p(i)$, $J = H_p(j)$, and $W = H_p(w)$. That is the keys $e'_{IJ}(l)$ have to satisfy

$$i + jl = w \pmod{p}, \quad i, j \in Z_p$$

   This means that $Pd_0 = \frac{1}{p}$

2. To prove $Pd_1 = \frac{1}{p}$ using a similar type of argument and for an arbitrary pair of messages l and l', we must find

$$\max_{w, w'} |\{e'_{ij} : e'_{ij}(l) = w, e'_{ij}(l') = w'\}|$$

   That is we need to find the number of (i, j) that satisfy

$$i + jl = w \pmod{p} \qquad \text{and} \qquad i + jl' = w' \pmod{p}$$

   That is we need to find the number of js that satisfy

$$j(l - l') = w - w' \pmod{p}$$

   This is equal to 1 as $j = (l - l')^{-1}(w - w') \pmod{p}$. So for any chosen i exactly one j can be found and so the number of (i, j) pairs is p which gives $Pd_1 = \frac{1}{p}$. $\qquad \square$

Centre for Computer Security Research
School of Information Technology and Computer Science
University of Wollongong
Wollongong 2522, AUSTRALIA
E-mail address: {rei, s05}@uow.edu.au

Progress in Computer Science and Applied Logic, Vol. 20
© 2001 Birkhäuser Verlag Basel/Switzerland

Robust Additive Secret Sharing Schemes over $\mathbf{Z}_m$

Rei Safavi-Naini and Huaxiong Wang

Abstract. In a threshold secret sharing scheme, a dishonest participant can disrupt the operation of the system by submitting junk instead of his/her share. We propose two constructions for threshold secret sharing schemes that allow identification of cheaters where the secret is an element of the ring $\mathbf{Z}_m$. The main motivation of this work is to design RSA-based threshold cryptosystems, such as robust threshold RSA signature, in which additive (multiplicative) threshold secret sharing schemes over Abelian groups with cheater identification play the central role. The first construction extends Desmedt-Frankel's construction of secret sharing over $\mathbf{Z}_m$ to provide cheater detection, and the second construction uses perfect hash families to construct a robust (t, n) scheme from a (t, t) scheme. We prove security of these schemes and assess their performance.

1. Introduction

A (t, n)-*threshold scheme* is a method of splitting a secret piece of information among n participants in such a way that any t participants can recover the secret by pooling together their *shares*. The shares of the secret are produced and securely transmitted to any participant by a trusted *dealer* during the initialisation of the system. Threshold schemes [7, 20] are special forms of *secret sharing schemes* which allow collaboration of participants belonging to an *access set* [22]. Secret sharing schemes, and in particular threshold schemes, are commonly used in applications where distributed trust is desirable. For example, the secret could be an access code to a secure area, or a cryptographic key to be used for either signing or encrypting data. Threshold schemes have become an indispensable cryptographic tool in any security environment where active entities are groups rather than individuals [13]. They are also commonly used as components of other cryptographic protocols such as in *key escrow* and *key recovery* systems.

The main motivation of this work is to design RSA-based threshold cryptographic systems. Consider a threshold RSA signature scheme. The main problem is to construct shares of the secret exponent, d, where $d \in \mathbf{Z}_{\phi(N)}$ and N is the RSA modulus. Since $\mathbf{Z}_{\phi(N)}$ is a ring and not a field, majority of classical secret sharing schemes, such as Shamir's scheme, can not be directly used. In the case of (n, n) schemes, a simple and elegant approach [9, 17] is to share the secret exponent, d, among the n signers $P_1, \ldots, P_n$ using Karnin-Greene-Hellman scheme [18]. That is, participant P_i holds d_i and $d = d_1 + \cdots + d_n$. To sign a message s, each participant gives his partial signature s^{d_i}. The signature is obtained as the product of the n partial signatures, as $s^d = s^{d_1} \cdot \cdots \cdot s^{d_n}$. Although by using cumulative secret sharing schemes [21] the above *additive share* idea can be generalized to any (t, n) threshold scheme but the resulting system is inefficient and each participant may need to store a share whose size is exponential in n. Desmedt and Frankel [15] initiated the study of threshold RSA signature and gave a heuristic solution for it. The basic idea behind their scheme is to generalize Shamir's scheme over $\mathbf{Z}_{\phi(N)}$ by extending Lagrange polynomial interpolation over a finite field, to a module over a ring. The resulting scheme has a share size n times the size of the secret and so is much more efficient than cumulative secret sharing.

In this paper we consider (t, n) secret sharing schemes over the ring $\mathbf{Z}_m$ with two additional properties: *additiveness* and *cheater identification*. Additive (or *multiplicative*) secret sharing schemes over Abelian groups were introduced by Desmedt et al in [15, 14] as a generalization of *homomorphic* secret sharing schemes [3]. In an additive (t, n) secret sharing scheme over $\mathbf{Z}_m$ the secret can be obtained by adding t components where the ith component is obtained by applying a publicly known function to the share of the ith participant. Additive secret sharing schemes over $\mathbf{Z}_{\phi(N)}$ are the basis of RSA threshold decryption and signature schemes (see [15], [14], [6]). We also consider another property of secret sharing schemes: robustness against dishonesty of a participant. In a traditional secret sharing scheme it is assumed that participants submit their correct shares and so a malicious participant may disrupt the system by simply submitting some junk as his share. A (t, n) threshold scheme is called *robust* if it can correctly produce the secret even in the presence of up to $(t - 1)$ arbitrary malicious participants. To achieve robustness, correctness of the shares or the constructed secret must be verifiable. A similar problem is well studied in the context of *verifiable secret sharing scheme* (VSSS)[1] which is an important primitive in the construction of robust threshold signature schemes.

We give two efficient constructions for (t, n) additive secret sharing schemes over $\mathbf{Z}_m$ which allow identification of cheating participants. The first proposed construction uses the Lagrange polynomial interpolation in the module structure developed by Desmedt and Frankel [15], and can be regarded as an extension of their secret sharing scheme to provide cheater identification. The second scheme uses perfect hash families in a recursive construction to obtain a (t, n) scheme over the ring $\mathbf{Z}_m$, from a (t, t) scheme over the same ring. The construction is inspired by a method given in [6], and is particularly efficient when n is much larger than t. The paper is organised as follows. In Section 2 we give the basic definitions, we then describe the two constructions in Sections 3 and 4. Section 5 concludes the paper.

---

[1]In a VSSS, the correctness of shares distributed by the dealer can also be verified by the participants. In other words, a VSSS also prevents cheating against the dealer. In this paper we, however, assume that the dealer is honest.

2. Preliminaries

A *secret sharing scheme* is a method of protecting a *secret* among a group of *participants* in such a way that only specified subsets of participants can reconstruct the secret. A (t, n)-threshold secret sharing scheme allows any t out of n participants to collaboratively recover the secret, while a set of less than t participants learns nothing about the secret.

Let $\mathbf{P} = \{P_1, \ldots, P_n\}$ be a set of n participants, let $\mathbf{K}$ denote the set of secrets, and assume P_i's share is selected from the set $\mathbf{S}_i$. A (t, n)-threshold scheme consists of a pair of algorithms: the (share) *distribution* algorithm and the (secret) *reconstruction* algorithm. For a secret k from $\mathbf{K}$ the dealer (who is not in $\mathbf{P}$) applies the distribution algorithm

$$\mathbf{D} : \mathbf{K} \to \mathbf{S}_1 \times \ldots \times \mathbf{S}_n$$

to assign shares to participants in $\mathbf{P}$. The reconstruction algorithm takes the shares of a subset $\mathbf{A} \subseteq \mathbf{P}$ of participants and returns either the secret, if the set $\mathbf{A} \subseteq \mathbf{P}$ and $|\mathbf{A}| \geq t$, or a string *fails*.

$$\mathbf{C} : \bigcup_{P_i \in \mathbf{A}} \{S_i\} \to \mathbf{K} \bigcup \{fail\}.$$

The classic example of a (t, n) secret sharing scheme is Shamir's scheme [20], which uses Lagrange interpolation on polynomials over finite fields. It has $\mathbf{K} = \mathbf{S}_i = GF(q)$ and works as follows. To construct a (t, n) threshold scheme ($n < q$) protecting $k \in \mathbf{K}$, the dealer constructs a random polynomial $f(x)$ of degree at most $t - 1$ over $GF(q)$ such that $f(0) = k$, then applies the distribution algorithm D to assign share $f(x_i)$ to participant P_i, where $x_1, x_2, \ldots, x_n$ are N distinct non-zero public elements in $GF(q)$. The reconstruction algorithm $\mathbf{C}$ takes as input at least t valid shares and uses Lagrange interpolation to compute $f(x)$ as

$$f(x) = \sum_{i \in B} f(x_i) \prod_{j \in B, j \neq i} \frac{(x - x_j)}{(x_i - x_j)}$$

where $B \subseteq \{1, \ldots, n\}$ and $|B| = t$, hence reconstructing the secret $k = f(0)$.

In a multiplicative threshold secret sharing scheme over a group [15], [14], the set of secrets $\mathbf{K}$ is a finite group with respect to the operation $*$ and for all sets $B = \{i_1, \ldots, i_t\}$ of t participants there exists a family of functions $f_{i_1, B}, \ldots, f_{i_t, B}$ from $\mathbf{S}_{i_1}, \ldots, \mathbf{S}_{i_t}$ to $\mathbf{K}$, respectively, and a public ordering $i_1, \ldots, i_t$ of elements of B that satisfy the following property. For a key $k \in \mathbf{K}$ and the shares $s_{i_1}, \ldots, s_{i_t}$ given to participants in B by the distribution algorithm $\mathbf{D}$ (on input k), we may express k as:

$$k = f_{i_1, B}(s_{i_1}) * \cdots * f_{i_t, B}(s_{i_t}).$$

It should be noted that multiplicative schemes only require a group structure on the set of keys $\mathbf{K}$, which is different from *homomorphic* schemes [3]. In this paper we will only concentrate on multiplicative schemes over the Abelian group $\mathbf{Z}_m$, where k can be expressed as $k = f_{i_1, B}(s_{i_1}) + \cdots + f_{i_t, B}(s_{i_t})$. We will call such schemes *additive* threshold scheme. Shamir's scheme over the finite field $\mathbf{Z}_q$ is an additive scheme, where $f_{i_\ell,B}(s_{i_\ell})$ is defined by

$$f_{i_\ell,B}(s_{i_\ell}) = f(x_{i_\ell}) \prod_{j \in B, j \neq i_\ell} \frac{-x_j}{(x_{i_\ell} - x_j)}.$$

The *information rate* ρ of a secret sharing scheme is defined by

$$\rho = \min_{1 \leq i \leq n} \left\{ \frac{\log |\mathbf{K}|}{\log |\mathbf{S}_i|} \right\}.$$

The efficiency of a secret sharing scheme can be measured by its information rate. It is well known that $\rho \leq 1$ in any secret sharing scheme. A secret sharing scheme is called *ideal* if $\rho = 1$.

3. Extended Desmedt-Frankel's Secret Sharing Schemes

To construct threshold RSA signature scheme, Desmedt and Frankel [15] generalized Shamir's (t, n) threshold secret sharing scheme to any Abelian group. They extended Lagrange polynomial interpolation over finite fields to modules over rings. We will use their module structure for polynomials in two variables to provide protection against cheaters.

Let G be the Abelian group $< \mathbf{Z}_m, + >$. Let p be a prime such that $n < p$ and let u be a root of the cyclotomic polynomial

$$p(x) = \frac{x^p - 1}{x - 1} = \sum_{i=0}^{p-1} x^i.$$

Let $\mathbf{Z}[u]$ denote the ring of the algebraic extension of integers with element u. That is, $\mathbf{Z}[u] \cong \mathbf{Z}[x]/(p(x))$. Define a module $G^{p-1} = \underbrace{G \times G \times \cdots \times G}_{p-1}$ over the ring $\mathbf{Z}[u]$, with addition and scalar product defined as follows:

1. $\forall [g_0, \cdots, g_{p-2}], [g_0', \cdots, g_{p-2}'] \in G^{p-1}$

   $$[g_0, \cdots, g_{p-2}] + [g_0', \cdots, g_{p-2}'] = [g_0 + g_0', \cdots, g_{p-2} + g_{p-2}'];$$

2. $\forall a_0 + a_1 u + \cdots a_{p-2} u^{p-2} \in \mathbf{Z}[u], \forall [g_0, \cdots, g_{p-2}] \in G^{p-1},$

   $$(a_0 + a_1 u + \cdots a_{p-2} u^{p-2})[g_0, \cdots, g_{p-2}]$$

   is defined inductively by using equations (1) and (2)

   $$a[g_0, \cdots, g_{p-2}] = [ag_0, \cdots, ag_{p-2}], \tag{1}$$

   where $a \in \mathbf{Z}$ and $ag_i = \underbrace{g_i + \cdots + g_i}_{a}$, and

   $$u[g_0, \cdots, g_{p-2}] = [-g_{p-2}, g_0 - g_{p-2}, \cdots, g_{p-3} - g_{p-2}]. \tag{2}$$

It then can be verified that the above addition and scalar product turn G^{p-1} into a module over the ring $\mathbf{Z}[u]$.

For an integer i, let $\alpha_i = \sum_{j=0}^{i-1} u^j \in \mathbf{Z}[u]$. It is proved [15] that $\alpha_1, \ldots, \alpha_{p-2}$ are all units of $\mathbf{Z}[u]$ and so are $\alpha_i - \alpha_j (i \neq j)$. With this definition, Desmedt and Frankel [15] generalized Shamir (t, n) threshold scheme as follows. To share a secret $d \in \mathbf{Z}_m$, the dealer randomly chooses a polynomial of degree at most $t - 1$, $f(x) = a_0 + a_1 x + \cdots + a_{t-1} x^{t-1}$ such that $a_i \in G^{p-1}$ and $a_0 = [d, 0, \cdots, 0]$ and gives to participant P_i his share $f(\alpha_i)$. It is proved that any $t - 1$ participants, pooling their shares, have no information about the secret, but t participants can compute the secret by computing $f(0) = [d, 0, \ldots, 0] = \sum \lambda_{i,B} d_i$, where $d_i =$

$$f(\alpha_i) \text{ and } \lambda_{i,B} = \frac{\displaystyle\prod_{j \in B, j \neq i} (-\alpha_j)}{\displaystyle\prod_{j \in B, j \neq i} (\alpha_i - \alpha_j)}, \text{ for any } B \subset \{1, \ldots, n\} \text{ with } |B| = t.$$

3.1. A secret Sharing Scheme with Cheater Identification

In the following we use the above algebraic setting to construct a (t, n) additive secret sharing scheme with cheater identification over the group $G = <\mathbf{Z}_m, + >$. Let $d \in \mathbf{Z}_m$ be the secret (for example, the secret exponent in RSA). The dealer randomly chooses a $t \times 2$ matrix

$$A = (a_{ij})_{0 \leq i \leq t-1, 0 \leq j \leq 1}$$

such that $a_{ij} \in G^{p-1}$ and $a_{00} = [d, 0, \cdots, 0]$. We define a mapping f from $\mathbf{Z}[u] \times \mathbf{Z}[u]$ to G^{p-1} using the following formal polynomial:

$$F(x, y) = (1, x, \cdots, x^{t-1}) A \begin{pmatrix} 1 \\ y \end{pmatrix} \qquad (3)$$

For $(\alpha, \beta) \in \mathbf{Z}[u] \times \mathbf{Z}[u]$, we define $f(\alpha, \beta) = F(\alpha, \beta)$. Note that the computation of $F(\alpha, \beta)$ employs the formal multiplication of matrices which is defined in a natural way. It is easy to see that any polynomial of the form $F(x, y)$, with degrees at most $t - 1$ in x at most 1 in y, and coefficients in G^{p-1} can be uniquely expressed as (3) and the function f is uniquely determined by the matrix A. The dealer computes $F(\alpha_i, y)$ and gives it to P_i, $i = 1, \ldots, n$, as his share.

Let $U(\mathbf{Z}[u])$ be the set of units of $\mathbf{Z}[u]$ and Δ be a maximal subset of $U(\mathbf{Z}[u])$ such that for any $\alpha, \beta \in \Delta$, $\alpha - \beta \in U(\mathbf{Z}[u])$. Such a subset exists and Δ can be chosen such that $p - 1 \leq |\Delta| \leq 2^{p-1}$. Indeed, $\{\alpha_1, \ldots, \alpha_{p-1}\}$ are $p - 1$ elements with the desired property and so the first inequality holds. On the other hand, we show that any such Δ satisfies $|\Delta| \leq 2^{p-1}$. We take $m = 2$ and consider the module $G = \mathbf{Z}_2^{p-1}$ over the ring $\mathbf{Z}[u]$. Choose $a, b \in G$ such that $a \neq [0, \ldots, 0]$ and define a function ϕ from Δ to G by $\phi : c \to ca + b$. It is clear that ϕ is one-to-one, and so $|\Delta| \leq |G| = 2^{p-1}$, as desired. We will show in Theorem 3.1 that $|\Delta|$ determines the security level of the system and its larger values corresponds to the higher security. The dealer then randomly chooses an element c in Δ, calculates a polynomial of degree at most $t - 1$

$$g(x) = F(x, c) \in G^{p-1}[x]$$

and gives c and $g(x)$ to the combiner as his secret information.

Now suppose t participants P_i, $i \in B$, submit their shares to the combiner. Firstly, for the share $F(\alpha_i, x)$ of P_i the combiner calculates $c_i = F(\alpha_i, c)$ and accepts $F(\alpha_i, 0)$ if and only if $c_i = g(\alpha_i)$, otherwise detects P_i as a cheater. If all shares are correct, the combiner can compute the secret $F(0, 0) = [d, 0, \ldots, 0]$ as

$$F(0, 0) = [d, 0, \ldots, 0] = \sum \lambda_{i,B} d_i,$$

where

$$d_i = F(\alpha_i, 0) \text{ and } \lambda_{i,B} = \frac{\prod\limits_{j \in B, j \neq i} (-\alpha_j)}{\prod\limits_{j \in B, j \neq i} (\alpha_i - \alpha_j)}.$$

Theorem 3.1. *The above scheme is a perfect (t, n) additive secret sharing scheme over $\mathbf{Z}_m$ (that is, any t participants can recover the secret, and any $t-1$ participant together with the combiner have no information about the secret). Moreover, for each participant P_i if $F(\alpha_i, 1) - F(\alpha_i, 0) \neq 0$ then the probability that a false submitted share is not detected by the combiner is $1/|\Delta|$.*

Proof. As we have seen any t participants, pooling their shares, can recover the secret, and the scheme is additive. We show that $t-1$ participants and the combiner have no information about the secret. Without loss of generality, assume that $P_1, \ldots, P_{t-1}$ and the combiner together try to reconstruct the secret. Let $F(x, y)$ be the polynomial chosen by the dealer. Consider the polynomial

$$G(x, y) = F(x, y) + g(x - \alpha_1) \cdots (x - \alpha_{t-1})(y - c)$$

where $g \in G^{p-1}$. Clearly, $F(\alpha_i, y) = G(\alpha_i, y)$, $i = 1, \ldots, t - 1$, and $F(x, c) = G(x, c)$. It follows that $G(x, y)$ also satisfies the conditions subject to the secret information of $P_1, \cdots, P_{t-1}$ and the combiner. In other words, based on the secret information of $P_1, \ldots, P_{t-1}$ and the combiner, $G(x, y)$ could also be a possible dealer polynomial. Since

$$G(0, 0) = F(0, 0) + g(\prod_{i=1}^{t-1} -\alpha_i)(-c)$$

and $(\prod_{i=1}^{t-1} -\alpha_i)(-c)$ is a unit in $\mathbf{Z}[u]$, it is easy to verify that the above identity determines a one-to-one correspondence between g and $G(0, 0)$. Thus, when g runs through all the elements of G^{p-1}, so does $G(0, 0)$, proving the desired result.

Next, assume that P_i wants to submit a false share. P_i succeeds only if P_i can correctly guess $F(\alpha_i, c)$. Since P_i knows the polynomial $F(\alpha_i, y)$, we may assume that $F(\alpha_i, y) = ay + b$, for some $a, b \in G^{p-1}$. By assumption, $a = F(\alpha_i, 1) - F(\alpha_i, 0) \neq 0$, the polynomial $F(\alpha_i, y)$ defines a function from Δ to G, which is one-to-one. Since c is randomly chosen from Δ, it follows that the probability that P_i guesses $F(\alpha_i, c)$ correctly is $1/|\Delta|$, and the result follows. $\square$

We assume that the correctness of the shares is only verifiable by the combiner. It is straightforward to generalize the schemes to make the shares verifiable by other participants. Indeed, this can be done by using a polynomials $F(x,y)$ of degree ℓ in variable y, $\ell \geq t-1$, and giving the share $(F(\alpha_i, y), F(x, \beta_i), \beta_i)$, to participant P_i. The partial share $F(\alpha_i, y)$ is used to reconstruct the secret while $(F(x, \beta_i), \beta_i)$ can be used to verify the integrity of other participants' shares.

It is also worth noticing that in the above scheme the identification of cheaters is independent of finding secret. That is, if k participant, $t < k \leq n$, can find the secret and then attempt to deceive the combiner (to construct the false secret) their success probability is only $1 - (1 - 1/|\Delta|)^k$. The cost of this robustness is that the combiner must store some secret information for verification purposes.

4. A Construction Based on Perfect Hash Families

Desmedt-Frankel construction of threshold secret sharing, which is the basis of the construction given above, has a low information rate. That is for a group of size n, the size of a share is at least n times the size of the secret. In our extension the size of the share is doubled and hence it is at least $2n$ times the size of the secret. More efficient schemes that allow higher information rates are studied by a number of authors. Desmedt, Di Crescenzo and Burmester [14] showed the construction of a (t,n) threshold secret sharing scheme over the Abelian group $< \mathbf{Z}_m, + >$ which has information rate $1/\log n$ when $t = 2$. This is much more efficient compared to the Desmedt-Frankel scheme. In [6] a recursive construction for (t,n) threshold secret sharing scheme which has asymptotically, (t is kept constant and n grows), optimal information rate is given. It was shown in [16] the information rate of Desmedt-Frankel scheme can be doubled.

We will follow the approach of [6] and use a perfect hash family to construct a (t,n) secret sharing scheme with cheater detection and large number of participants, from one with the same t and small number of participants. When t (compared to n) is small this construction results in a much better information rate compared to the scheme described in Section 3.

In the remainder of this section, let $U(\mathbf{Z}_m)$ denote the set of the units of $\mathbf{Z}_m$.

4.1. (t,t) Schemes

We start with a (t,t) secret sharing scheme with cheater detection over $\mathbf{Z}_m$ and then use a perfect hash family to increase the number of participants and construct a (t,n) scheme with the same property.

The (t,t) scheme works as follows. To share a secret $d \in \mathbf{Z}_m$ among the t participants, the dealer randomly chooses t elements $a_1, a_2, \ldots, a_t \in \mathbf{Z}_m$ and $b_1, b_2, \ldots, b_t \in U(\mathbf{Z}_m)$ such that

$$a_1 + a_2 + \cdots + a_t = d,$$

and securely gives (a_i, b_i) to the participant P_i, $1 \leq i \leq t$, as his share. The dealer also randomly chooses an element $v \in U(\mathbf{Z}_m)$, calculates

$$c_i = a_i + vb_i, \quad i = 1, 2, \ldots, t,$$

and gives v and $(c_1, c_2, \ldots, c_t)$ to the combiner. Clearly, the t participants together can recover the secret. The combiner can verify the correctness of a share (a_i, b_i), which together with its identity information i is submitted by P_i, by checking if $c_i = a_i + vb_i$.

Lemma 4.1. *The above construction results in a perfect (t, t) additive secret sharing scheme over $\mathbf{Z}_m$ with cheater identification, (that is, any t participants can recover the secret while no collaboration of up to $t - 1$ participants has information about the secret). The probability that $t - 1$ participants and the combiner recover the secret is $1/|U(\mathbf{Z}_m)|$, and the probability that a dishonest participant successfully submitting a false share is $1/|U(\mathbf{Z}_m)|$.*

Proof. Clearly, when the t participants pool their shares, they can recover the secret, and the scheme is additive. We show that it is perfect. Without loss of generality, assume that $t - 1$ participants $P_1, \ldots, P_{t-1}$ try to recover the secret. Since a_t is independent of $a_1, \ldots, a_{t-1}$, it is clear that they have no information about $d = \sum_{i=1}^{t-1} a_i + a_t$.

Now assume $P_1, \ldots, P_{t-1}$ and the combiner together try to reconstruct the secret. Since the combiner knows c_t and v, they will try to guess a_t according to the equation

$$c_t = a_t + vb_t.$$

Since v is a unit in $\mathbf{Z}_m$, for the fixed v and c_t the function $\psi(x) = c_t - vx$ is one-to-one from $U(\mathbf{Z}_m)$ to $\mathbf{Z}_m$. But b_t is randomly chosen from $U(\mathbf{Z}_m)$, and so the probability that $P_1, \ldots, P_t$ and the combiner correctly guess a_t is $1/|U(\mathbf{Z}_m)|$.

Next, assume that a participant P_i wants to submit a false share. P_i succeeds only if P_i can correctly guess the value of $c_i = a_i + vb_i$, without knowing v. Since b_i is a unit of $\mathbf{Z}_m$ and v is randomly chosen from $U(\mathbf{Z}_m)$, it results in $|U(\mathbf{Z}_m)|$ possible c_i, and so the probability that P_i succeeds is $1/|U(\mathbf{Z}_m)|$. $\square$

4.2. (t, n) Schemes

To construct a (t, n) secret sharing scheme, we use a perfect hash family and multiple copies of the above (t, t) scheme. This approach is implicitly used in [6] and later made explicit in [4] and [13].[2]

An *(n, q, t)-perfect hash family* is a set of functions $\mathbf{F}$ such that

$$f : \{1, \ldots, n\} \longrightarrow \{1, \ldots, q\}$$

for each $f \in \mathbf{F}$, and for any $X \subseteq \{1, \ldots, n\}$ such that $|X| = t$, there exists at least one $f \in \mathbf{F}$ such that $f|_X$ is one-to-one.

---

[2]Both Blackburn [4] and Desmedt [13] have attributed this observation to Kurosawa and Stinson.

We use the notation $PHF(W; n, q, t)$ for an (n, q, t)-perfect hash family with $|\mathbf{F}| = W$ and write $\mathbf{F} = \{f_1, \ldots, f_W\}$. Perfect hashing families originally arose as part of compiler design [19] (see [11] for a survey of recent results). Later they have found numerous applications such as circuit complexity and to the design of deterministic analogues to probabilistic algorithms; see [1]. Numerous constructions for perfect hashing families using finite geometries, designs theory and error-correcting codes are known; see [2], [4].

Let $\mathbf{F} = \{f_1, \ldots, f_W\}$ be a $PHF(W; n, t, t)$. To construct shares of a secret $d \in \mathbf{Z}_m$ in a (t, n) secret sharing scheme, the dealer runs independently W times the (t, t) share distribution algorithm (Section 4.1) for the same secret d and produces

$$\mathbf{d}^1, \ldots, \mathbf{d}^W \in (\mathbf{Z}_m \times U(\mathbf{Z}_m))^t,$$

where $\mathbf{d}^i$ is a vector whose components are the t shares of the ith run, $1 \le i \le W$, of the (t, t) secret sharing scheme. Thus

$$\mathbf{d}^j = (d_{1,j}, \ldots, d_{t,j}),$$

where $d_{k,j} = (a_{k,j}, b_{k,j}) \in \mathbf{Z}_m \times U(\mathbf{Z}_m)$, $1 \le k \le t$, such that $\sum_{k=1}^{t} a_{k,j} = d$, for all $1 \le j \le W$. The dealer also randomly chooses an element $v \in U(\mathbf{Z}_m)$ and calculates elements

$$\mathbf{c}^1, \ldots, \mathbf{c}^W \in (\mathbf{Z}_m)^t,$$

where $\mathbf{c}^j = (c_{1,j}, \ldots, c_{t,j})$ such that $c_{k,j} = a_{k,j} + v b_{k,j}$ for all $1 \le j \le W$ and $1 \le k \le t$. Then the dealer privately sends to the participant P_i, $1 \le i \le n$ the share

$$\mathbf{d}_i^* = (d_{i,1}^*, \ldots, d_{i,W}^*),$$

where

$$d_{i,j}^* = d_{f_j(i),j} = (a_{f_j(i),j}, b_{f_j(i),j}) \in \mathbf{Z}_m \times U(\mathbf{Z}_m), \ 1 \le j \le W$$

and sends v and $\mathbf{c}^1, \ldots, \mathbf{c}^W$ to the combiner.

Theorem 4.2. *The above construction results in a perfect (t, n) additive secret sharing scheme with cheater detection. Moreover, the probability that up to $t - 1$ participants and the combiner together can reconstruct the secret is bounded by $W/|U(\mathbf{Z}_m)|$, and the probability that a participant succeeds in submitting a false share to the combiner is at most $1/|U(\mathbf{Z}_m)|$.*

Proof. First, we show that any t participants can recover the secret. Let B be a subset of $\{1, \ldots, n\}$ and $|B| = t$. Assume that the group of participants P_i's, $i \in B$ want to recover the secret. Since $\mathbf{F}$ is a $PHF(W; n, t, t)$, there exists a function, say $f_\ell \in \mathbf{F}$, such that f_ℓ is one-to-one on B. It is then easy to see that the participants P_i's, $i \in B$, know $\mathbf{d}^\ell$, and so they can recover the secret.

Secondly, any $t - 1$ participants have no information about the secret. We know there are W runs of (t, t) secret sharing schemes and any $t - 1$ out of n

participants know at most $t-1$ shares for each of them which will not give them any information about the secret because the original (t,t) schemes were perfect, and the W schemes are chosen independently. Thirdly, without loss of generality suppose that $t-1$ participants $P_1,\ldots,P_{t-1}$ and the combiner together try to reconstruct the secret. They succeed only if they can correctly guess one of the $a_{t,j}, 1 \le j \le W$. In the same manner as in Lemma 4.1, it is straightforward to show that the probability that they succeed is less than or equal to $W/|U(\mathbf{Z}_m)|$.

Finally, assume that a participant P_i wants to submit a false share. P_i succeeds only if P_i can correctly guess $c_{f_1(i),1},\ldots,c_{f_W(i),W}$. By Lemma 4.1, we know that for each $1 \le j \le W$, the probability that P_i correctly guesses $c_{f_j(i),j}$ is not greater than $1/|U(\mathbf{Z}_m)|$, and so the result follows. □

To assess the efficiency of the above scheme, we examine its information rate ρ. We have

$$\rho = \frac{\log m}{W(\log m + \log |U(\mathbf{Z}_m)|)} = 1/(W + W\frac{\log \phi(m)}{\log m})$$

and so $1/(2W) \le \rho \le 1/W$. This means that the information rate of the resulting scheme is mainly determined by the size W of the perfect hash family $PHF(W;n,t,t)$. It is proved [19] that for any t there exists a perfect hash family $PHF(W;n,t,t)$ with $W = O(\log n)$. In particular, for $n \ge 2$ there is a construction for perfect hash family $PHF(W;n,2,2)$ with $W = \lceil \log n \rceil$. This confirms the result in [14]. The general existence result for perfect hash family with $W = O(\log n)$ is not constructive and finding construction for perfect hash families with small W is an interesting problem. In [2], the authors gave several constructions for $PHF(W;n,q,t)$ from known combinatorial objects in which W is a polynomial function of $\log n$ (for fixed q and t). Compared to extended Desmedt-Frankel scheme, for n much larger than t (this is required because of the constant terms in the actual expression of W), the second scheme has much better information rate.

The following example clarifies some of these ideas. Assume that we need a $(3,500)$ secret sharing scheme for secrets belonging to $\mathbf{Z}_m$. Using extended Desmedt-Frankel scheme the information rate is at most $1/1000$. Using the second scheme, and employing a perfect hash family $PHF(3 \times 4^j; 5^{2^j}, 3, 3)$, where $j \ge 1$ an integer, constructed by Atici, Magliveras, Stinson and Wei [2], we can construct a $(3,625)$ secret sharing scheme with cheater detection and with a information rate greater than $1/98$ which is much more efficient compared to the previous case.

We note that the use of $PHF(W;n,q,t)$ to construct a (t,n) scheme form a (t,t) scheme is general. Such a family can also be used on any (t,q) secret sharing scheme with cheater detection to construct a (t,n) secret sharing scheme with cheater detection.

5. Conclusions

In this paper we presented two threshold secret sharing schemes over a ring $\mathbf{Z}_m$ that allow efficient identification of cheating participants and so provide robustness. As we noted in the introduction, the main motivation of this work is to design the RSA-based threshold cryptographic systems.

It is important to note that in both constructions the information used for verification purposes are completely independent from the secret and cannot be used by the combiner to learn anything about the secret. This also means that the system's protection against cheating does not depend on the knowledge of the secret. That is, even if t participants collude and construct the secret by pooling their shares without the combiner's assistance, and then attempt to impersonate another participant or submit invalid shares at a later time, the chance of success in their attack will not be better than the case they did not know the secret. This protection is not really necessary and one could expect more efficient systems if this condition is relaxed.

Our proposed schemes require a trusted dealer, who knows the secret and generate/distribute shares to the participants. In some applications it is desirable to eliminate the dealer (for example, Boneh and Franklin [8], Cocks [12], Blackburn, Blake-Wilson, Burmester and Galbraith [5]). An interesting extension of our work would be to construct efficient systems without a trusted dealer.

References

[1] N. Alon and M. Naor, Derandomization, witnesses for Boolean matrix multiplication and construction of perfect hash functions. *Algorithmica*, **16** (1996), 434–449.

[2] M. Atici, S.S. Magliveras, D. R. Stinson and W.D. Wei, Some Recursive Constructions for Perfect Hash Families. *Journal of Combinatorial Designs*, **4** (1996), 353–363.

[3] J. C. Benaloh, Secret sharing homomorphisms: Keeping shares of a secret secret. In *Advances in Cryptology – Crypto '86*, LNCS, **263** (1986), 251–260.

[4] S. R. Blackburn, Combinatorics and Threshold Cryptology. In *Combinatorial Designs and their Applications* (Chapman and Hall/CRC Research Notes in Mathematics), CRC Press, London, 1999, 49–70.

[5] S.R. Blackburn, S. Blake-Wilson, M. Burmester and S. Galbraith, Shared generation of shared RSA keys. *Tech. Report CORR98-19, University of Waterloo*, 1998.

[6] S. R. Blackburn, M. Burmester, Y. Desmedt and P. R. Wild, Efficient multiplicative sharing schemes. In *Advances in Cryptology – Eurocrypt '96*, LNCS, **1070** (1996), 107–118.

[7] G. R. Blakley, Safeguarding cryptographic keys. In *Proceedings of AFIPS 1979 National Computer Conference*, **48** (1979), 313–317.

[8] D. Boneh and M. Franklin, Efficient generation of shared RSA keys. In *Advances in Cryptology – Crypto '97*, LNCS, **1294** (1997), 425–439.

[9] C. Boyd, Digital multisignatures, In *Cryptography and coding*, Clarendon Press, 1989, 241–246.

[10] E. Brickell and D. Stinson, The Detection of Cheaters in Threshold Schemes. In *Advances in Cryptology – Proceedings of CRYPTO '88*, LNCS, **403** (1990), 564–577.

[11] Z. J. Czech, G. Havas and B.S. Majewski, Perfect Hashing. *Theoretical Computer Science,* **182** (1997), 1–143.

[12] C. Cocks, Split knowledge generation of RSA parameters. In *Cryptography and Coding,* 6th IMA Conference, LNCS, **1355** (1997), 89–95.

[13] Y. Desmedt, Some recent research aspects of threshold cryptography. In *Information Security Workshop, Japan* (JSW '97), LNCS, **1396** (1998), 99–114.

[14] Y. Desmedt, G. Di Crescenzo and M. Burmester, Multiplicative non-abelian sharing schemes and their application to threshold cryptography. In *Advances in Cryptology – Asiacrypt '94,* LNCS, **917** (1995), 21–32.

[15] Y. Desmedt and Y. Frankel, Homomorphic zero-knowledge threshold schemes over any finite group. *SIAM J. Disc. Math.,* **7** (1994), 667–679.

[16] Y. Desmedt, B. King, W. Kishimoto and K. Kurosawa, A comment on the efficiency of secret sharing scheme over any finite abelian group. In *Information Security and Privacy,* ACISP'98 (Third Australasian Conference on Information Security and Privacy), LNCS, **1438** (1998), 391–402.

[17] Y. Frankel, A practical protocol for large group oriented networks. In *Advances in Cryptology – Eurocrypt '89,* LNCS, **434** (1989), 56–61.

[18] E. Karnin, J. Greene, and M. Hellman, On Secret Sharing Systems. *IEEE Transactions on Information Theory,* **29** (1983), 35–41.

[19] K. Mehlhorn, *Data Structures and Algorithms.* Vol. 1, Springer-Verlag, 1984.

[20] A. Shamir, How to Share a Secret. *Communications of the ACM,* **22** (1979), 612–613.

[21] G. Simmons, W.-A. Jackson and K. Martin, The Geometry of Shared Secret Schemes. *Bulletin of the Institute of Combinatorics and its Applications (ICA),* **1** (1991), 71–88.

[22] D.R. Stinson, An explication of secret sharing schemes. *Des. Codes Cryptogr.,* **2** (1992), 357–390.

[23] M. Tompa and H. Woll, How To Share a Secret with Cheaters. *Journal of Cryptology,* **1** (1988), 133–138.

School of IT and CS, University of Wollongong, Northfields Ave, Wollongong 2522, Australia
E-mail address: [rei, huaxiong]@uow.edu.au

Progress in Computer Science and Applied Logic, Vol. 20
© 2001 Birkhäuser Verlag Basel/Switzerland

RSA Public Key Validation

Robert D. Silverman

Abstract. With increasing demands for internet security, the demand for distribution of public key certificates and certificate authorities spreads. This brings up the new problem of public key validation. This problem can arise in different contexts and can be adressed in some of them using novel zero knowledge protocols, while in other contexts it can be only poorly solved. The purpose of this paper is to give an overview of the topic of key validation and to discuss the possible solutions to the different instances in which it appears. We also present a new zero knowledge protocol for correctness of an RSA public exponent and a simple protocol (not zero-knowledge) for primality of a discrete logarithm. We use the latter to give a proof protocol (not zero knowledge) that an RSA key is the product of strong primes.

1. Introduction

As public keys become intensively exchanged, a variety of scenarios arise in which the *trust* in a given set of keys needs to be established or reestablished.

A first scenario is that of an individual key holder who wants to make sure that a key generated for him by software or hardware is not corrupt due to some functional error of the generating system. A second scenario is that of a Certifying Authority who wants assurance that a key presented for binding in a certificate is both correct and secure. A third scenario occurs when a transaction partner or a public authority wants to know that some public key is not fake or weak, which would allow the holder to repudiate transactions when desired.

This brief presentation suggests that the range of sources of *mistrust* does not as yet present a closed, pertinent list of attacks to protect against. It is therefore sensible to meet the problem by concentrating on a set of *integrity conditions*, which one would like to have the ability to verify for a given key set. We shall restrict our attention in this presentation on key sets for an RSA scheme. Similar problems must also be met for discrete logarithm or elliptic curve based public schemes.

An RSA key set is built from a public modulus $N = pq$, where p and q are primes, together with an exponent pair d, e such that $de \equiv 1 \mod \lambda(N)$ with $\lambda(N) = \text{lcm}(p-1, q-1)$, the Carmichael function. The public part of the key set is (e, N), while the owner alone may have knowledge of all the key parts. Even this condition may not hold, for instance, when the owner uses a hardware encryption black box which never releases its keys.

The security of the RSA scheme is based on the difficulty to factor N, when p and q are large, although a proof that the two problems are equivalent is lacking. Consequently, the primary integrity concerns aim to ascertain that a key set does not have properties which increase the odds of factoring N using known algorithms. Further, one wants assurance that the key possesses the correct mathematical properties to allow encryption/decryption to work. These are separate problems. Currently, in terms of performance and range of applicability, the fastest factoring algorithms are the Generalized Number Field Sieve (GNFS) [16], the multiple polynomial quadratic sieve (MPQS) [25] and the Elliptic Curve Method (ECM) [20]. All of these methods have sub-exponential running time. The first is asymptotically the fastest and has already been shown effective for integers of 512 bits. Of the three, ECM alone has the property of having a run time which depends upon the size of the smallest prime dividing the number to be factored – in particular, the probability of finding a factor p, depends essentially on the size of p and not on the size of the number to be factored.

Some older algorithms which were in a certain sense generalized and outdated by ECM, use properties of *cyclotomic rings*. Such algorithms are the $P \pm 1$ methods [21] and the Bach-Shallit cyclotomic polynomial method [3]. These methods have exponential running times. The Pollard ρ algorithm uses collision search in a pseudo-random sequence and also runs in exponential time. The running time of all of these depends upon the size of the smallest prime factor of the factoring candidate.

A further special factoring algorithm, Lehman's weighted difference of squares method (WDSM) [15] is successfull for integers $N = pq$ such that $|ap - bq|$ is very small, for some small integer parameters a and b. This method generalizes Fermat's classical difference of squares method which applies when the difference $|p - q|$ is very small. In general, it is impossible to guard against a key being deliberately constructed so that it is vulnerable to this attack. Clearly, even if $|ap - bq|$ is large for all a, b less than B_1, one can construct p, q so that $|ap - bq|$ is small for some a, b lying just beyond any chosen bound B_1. Thus, standards usually just protect against $a = b = 1$.

There is no way of constructing keys to protect against GNFS and MPQS except to make the keys sufficiently large. This condition is trivial to check. One protects against ECM by making the factors p and q sufficiently large. This is achieved by making $p \approx q$ and seems impossible to verify without knowledge of p or q.

While the probability of success of one or all of these exponential attacks is astronomically small for randomly generated $N = pq$, some cryptographic standards mandate that keys be constructed to specifically guard against such attacks. The paper by Rivest and Silverman [23] presents arguments for why these constructions should not be necessary, but such requirements persist for historical/hysterical reasons. Applying an exponential algorithm is futile for 1024-bit N unless N has been deliberately constructed to be vulnerable. The largest factor ever found by Pollard $P \pm 1$ is 34 digits, the largest by Pollard ρ is 19 digits, and the largest by ECM is 54 digits. Prospects of finding a 155 digit factor by ECM are remote. See [26].

2. Validation Requirements

This section describes current *integrity conditions* required for an RSA key set that have appeared within some standards.

2.1 The size of N must be sufficiently large – typically ≥ 1024 bits.

2.2 The factors p and q must be prime and simultaneously large, subject to condition 2.3.

2.3 For integers $a, b < B_1$, the linear combination $|ap - bq| > B_2$ for some given bounds B_1, B_2.

2.4 $p \pm 1$ and $q \pm 1$ must be divisible by large primes, typically greater than 2^{100}. The Bach-Shallit algorithm will succeed if other, higher degree cyclotomic polynomials Φ_n in p are smooth, but we know of no standard which requires guarding against this attack.

2.5 $(e, \phi(N)) = 1$.

Four types of validation scenarios are currently distinguished in the industrial and standardization communities. These are:

A. First Party Validation – Validation by the first party when private key is known. The holder or producer of the key wants to ascertain that the key produced by software or a hardware device is both valid and has required security properties.

B. Second Party Validation – Validation by a second party knowing only the public key. A second party wants to ascertain validity of a key without interaction with the 1st party or information about the secret key. This says nothing about security.

C. Zero-Knowledge Validation – Validation and security verification by a second party in cooperation with the first party. The requirements of this case are similar to case B. However, the second party has the possibility to carry out a protocol with the first party, provided this does not leak sufficient information to recover the private key.

D. Joint Generation – Here a party jointly generates a key with the participation of a trusted third party such as a CA in such a way that the CA can then confirm key validity and security. See [4, 11]

3. Known Results

3.1. First Party Validation

It is relatively straightforward for a first party to recheck the arithmetic used to produce a key. We note that such a recheck should be done using software different

from that used to produce the key. In the case where the key was generated by
a black box that does not allow even first party access to the private key, we are
reduced to case B.

3.2. Second Party Validation

It is possible to do some simple and trivial checks on N and e, but these add little
confidence that N and e have correct arithmetic properties. Such checks include
checking thesize of N and that e is in the correct range, that N is not prime, and
that it has no small divisors (checked by applying a factoring algorithm). However,
this do little to engender confidence that the public key is valid. Indeed, verifying that N has only two prime factors without actually factoring N is an open,
unsolved problem [17], and verifying that $(e, \phi(N)) = 1$ seems equally impossible.

3.3. Zero Knowledge Validation

In [18], Liskov and Silverman give a statistical limited-knowledge protocol for
proving that $p \approx q$. The protocol leaks a small number of the uppermost bits of p
and q, but this is not sufficient for conducting an attack with any known algorithm.
If one conducts 2^L rounds of the algorithm, one can get a statistical estimate for
the topmost L bits.

Advances on this work have been made by Camenisch and Michels [7] and
by Boudot [5]. These methods are *slightly* more efficient and do not leak bits, but
have the disadvantage that a bit commitment must be made on p and q, whereas
the Liskov and Silverman result works *only* from N. The bit commitment schemes
require the use of an additional, external parameter (a group of unknown order)
in which to hide p and q.

A related result has been given by Gennaro, Micciancio, and Rabin [13].
Their protocols prove without bit commitment that N is exactly the product of
two primes and can then prove that p and q are so-called *quasi-safe* primes (but
says nothing about size). These are primes r such that $(r - 1)/2$ is also a prime
power. The Liskov and Silverman protocol uses their method for proving that N
has exactly two prime factors before proving size equivalence.

3.3.1. ZKP FOR $(e, \phi(N)) = 1$.   We now give an honest verifier zero-knowlege
protocol for proving $(e, \phi(N)) = 1$. Following the presententation, we will then
show how this may be transformed into a true zero-knowledge protocol. We designate $u[a, b]$ as the uniform density function on $[a, b]$.

We, the Prover, desire to prove to an outside Verifier that $(e, \phi(N)) = 1$. The
Verifier knows only N.

- The Verifier generates $r \in u[1, N - 1]$ and sends it to the Prover. (Alter-
  natively, the Verifier and Prover may jointly agree upon r but the former
  method is simpler.)
- The Prover computes $s = r^d \bmod N$ and sends it to the Verifier.
- The Verifier computes $s^e \bmod N$ and checks that $s = r$. If this fails, the
  Verifier rejects the claim.

Theorem 3.1 (Soundness). *If $(e, \phi(N)) > 1$ the Prover cannot consistently pass the protocol.*

Proof. The protocol requires that a randomly chosen integer r have an eth root mod N. If $(e, \phi(N)) > 1$ then there are at most $\phi(N)/(e, \phi(N))$ such integers r. Hence, if $(e, \phi(N)) > 1$ the probability that $s^e \equiv r \mod N$ is at most $1/(e, \phi(N))$. Therefore the probability of failure may be made arbitrarily small with enough trials.   □

Theorem 3.2 (Completeness). *If $(e, \phi(N)) = 1$ the Prover can succeed at the protocol.*

Proof. If $(e, \phi(N)) = 1$, then there exist integers, a, b such that $ae + b\phi(N) = 1$. Then $ae = 1 - b\phi(N)$ and $r^{ae} = r \mod N$. By choosing $s = r^a \mod N$ where a is the multiplicative inverse of e, the Prover successfully generates an eth root of r whenever $(e, \phi(N)) = 1$ □

Theorem 3.3 (Knowledge Complexity). *There exists a polynomial-time simulator $\mathbb{S}$, which can produce a transcript of the protocol of arbitrary length which is statistically indistinguishable from a real transcript.*

Proof. We exhibit $\mathbb{S}$ here. The simulator works in the following manner.

- The simulator selects a random integer x, and puts $r = x^e \mod N$. If $(e, \phi(N)) = 1$, then $x^e \mod N$ will be uniformly distributed   mod N. Thus, the distribution of r is the same as in the actual protocol.
- The simulator then selects $s = x$ and easily verifies that $s^e = r \mod N$.

It is clear that this does not require knowledge of $e^{-1} \mod N$ and produces a transcript that is indistinguishable from an actual transcipt.   □

Efficiency of the Protocol

If we require that the probability of failure be no more than 2^{-n} for some pre-selected value of n, the protocol requires k rounds where $2^{-n} = (1/e)^k$, for prime e. Thus, $k = n \log 2/ \log e$. For $n = 100$, this is $(64, 44, 25, 7)$ respectively for $e = 3, 5, 17, 65537$. For non-prime e, the number of rounds is n, where $2^{-n} = (1/e_s)^k$, and e_s is the smallest prime factor of e.

The above protocol is not true zero-knowledge, because the verifier can bias the selection of r and thereby obtain information. However, a result of Damgård, Goldreich, and Okamoto [10] shows that any honest verifier zero-knowledge proof may be simply transformed into a true zero-knowledge proof. We also show here how a simple protocol of Chaum, Fiat, and Naor (CFN) [8] may be applied to prevent the verifier from cheating. The verifier does the following. Rather than send a single value of r, the verifier instead sends multiple values $r_1, r_2, \ldots, r_k$ such that $r_i = s_i^f \mod N$ for some $f \neq e$ with $(f, \phi(n)) = 1$. The prover then randomly demands some of the values s_i. The verifier can not know in advance which values of s_i will be demanded, and thus must generate all the r_i appropriately.

Yet another way to achieve true zero knowledge is to use the protocol of Guillou and Quisquater (GQ) [14]. This protocol allows the verifier to generate r in any way he likes, but then merely exhibits knowledge of the eth root of r without revealing it. In terms of efficiency the method of [10] will be very expensive, while the CFN and GQ schemes are comparable.

3.3.2. ZKP FOR STRONG PRIMES   The paper of [7] gives a statistical zero knowlege protocol for proving that $N = pq$ is the product of safe primes; primes p, q such that $(p-1)/2$ and $(q-1)/2$ are also prime. It also gives a statistical zero knowlege protocol for proving that an integer is a Lehman probable prime to a given base. However, this falls somewhat short of what is desired. It does not prove that $p \pm 1$ and $q \pm 1$ have a large, bounded prime factor, and it only proves that p is a weak form of probable prime. This has two sources of possible error. First is that the protocol might succeed, even when the claimed condition isn't true (albeit with very small probability), and the second is that the probable prime may actually be a pseudoprime. The first condition is trivially true of all zero-knowledge methods, but the second is not.

We give here a non zero-knowledge protocol for demonstrating 2.4. It handles the objections above at the expense of losing zero-knowledge, but has the property that it is extensible to showing security of a key from the Bach-Shallit attacks. It further differs from the Camenisch and Michels result in that the bit commitment is peformed in a group of known, rather than unknown order. We also use this non zero-knowledge protocol in order to motivate a conjecture.

We start the protocol for strong primes with the following:

A Proof for Primality of a Discrete Logarithm

Let P designate a very large prime and let g be a generator of $(\mathbb{Z}/S\mathbb{Z})^*$. These parameters are made public *a priori* for all of the protocols. All exponentiations shall be done mod P, so that when we write g^a we shall mean g^a mod P.

In [19] a zero-knowledge protocol is given which proves that $c = ab$ from public information g^a, g^b, and g^c. This information can be known *a priori* or it can be committed to as the first step in the protocol. We abbreviate this protocol as ZKP_Product(a, b, c). In [9] and in [18] statistical limited-knowledge protocols are given which bound the size of a given g^a. We refer to a protocol which proves the size of a as ZKP_Size(a).

We now introduce a protocol for demonstrating that the value of a discrete logarithm is prime without revealing the actual logarithm.

Protocol Prove_Prime(r):

Having committed to g^r for random prime r, we prove that r is prime without revealing it. This protocol does not prove that r by itself is prime. Instead, it constructs another prime s and jointly proves that r and s are prime.

- The Prover generates prime s randomly and sufficiently large such that $H = rs$ is too difficult to factor with existing methods. Also, s is chosen so that $rs < ord(g)$.

- The Prover publishes g^s and H.
- The Prover and Verifier jointly execute the Disjoint Prime Product protocol of [13].
- The Prover and Verifier jointly execute ZKP_Product(r, s, H). This proves that H is the product of two primes, one of which must be r. □

This protocol is not strictly zero-knowledge, since a simulator can't publish H in step 2: a simulator does not know the value of r. Furthermore, this protocol can only be executed once, since executing two instances will publish $H_1 = rs_1$ and $H_2 = rs_2$ and r can be found from $GCD(H_1, H_2)$.

However, while the above protocol is not true zero-knowledge, we have proved that r is a prime, not merely a probable prime.

We now pose the following:

Conjecture: Suppose P is a large prime and g is a primitive root. Recovering p and q from knowledge of $g^p \bmod P$, $g^q \bmod P$, and pq is hard unless factoring or solving discrete logs is easy.

Proof for Strong Keys

The protocol for strong keys is now as follows:

Protocol StrongKey(N):
Having committed to $N = pq$ we prove that p, q are strong primes.

1. The Prover and Verifier jointly execute the protocol of [18] to prove $N = pq$ is the product of exactly two nearly equal primes.
2. The Prover publishes g^p and g^q.
3. The Prover and Verifier jointly execute ZKP_Product(p, q, N). This proves that the values of the discrete logs in (2) are the factors of N.
4. Suppose $p - 1 = a_1 p_1$, $p + 1 = a_2 p_2$, $q - 1 = a_3 p_3$ and $q + 1 = a_4 p_4$ where p_i are large primes. The Prover and Verifier jointly execute ZKP_Product$(a_1, p_1, p - 1)$, ZKP_Product$(a_2, p_2, p + 1)$, ZKP_Product$(a_3, p_3, q + 1)$, and ZKP_Product$(a_4, p_4, q + 1)$. The values for $g^{p\pm1}$ and $g^{q\pm1}$ are easily computed since g^{-1} is easily computed and g^p, g^q are public.
5. The Prover and Verifier jointly execute Prove_Prime(p_i) for $i = 1, 2, 3, 4$.
6. The Prover and Verifier jointly execute ZKP_Size(p_i) for $i = 1, 2, 3, 4$.
7. We have shown that $N = pq$ and each of $p \pm 1$ and $q \pm 1$ are divisible by large primes. □

3.3.3. BACH-SHALLIT ATTACKS It is a simple extension of the methods in §3.3.2 to show that other cyclotomic polynomials in p and q also have large factors. For example, to show that $p^2 \pm p + 1$ has a large prime factor p_5 simply publish g^{p^2} as well as g^p. This allows one to compute $g^{p^2 \pm p + 1}$ from public information. One now executes the methods in §3.3.2 with $p \pm 1$ replaced with $p^2 \pm p + 1$. The verifier can confirm that the value of g^{p^2} is correct by executing ZKP_Product(p, p, p^2) with the prover. Thus, one could also prove that an RSA modulus is secure against the Bach-Shallit factoring attack. Of course one can not prove that all cyclotomic polynomials in p have large prime factors, but for degrees greater than two, the probability that $\Phi_n(p)$ is a smooth integer is vanishingly small. Indeed, for a 512-bit prime factor, if we select a step one limit of 10^{12}, the probability that $P-1$ or $P+1$ succeed is approximately 6×10^{-15}. The probability that (say) $P^2 + P + 1$ succeeds is approximately 6×10^{-37} and the probability that (say) $P^3 + P + 1$ succeeds is about 8×10^{-62}.

3.3.4. ZKP FOR WDSM If p and q are too close together then N may be easily factored by representing it as $x^2 - y^2$. This technique originated with Fermat. A generalization by Lehman works [15] if p/q is close to the ratio of two reasonably small integers, a and b. It might therefore be desirable to prove that $|ap - bq|$ is reasonably large. The X9.31 and X9.44 standards mandate that $|p - q| > 2^{412}$ for 512-bit p and q. We give here a statistical zero-knowledge proof to show that $|ap - bq|$ is large for published parameters a, b.

Protocol ZKP_LargeDiff(N):
Having committed to $N = pq$ and published a, b, we prove that $|ap - bq|$ is large.

- The Prover publishes g^p and g^q. Suppose without loss of generality that $p > q$.
- The Verifier can compute g^{-q} via the Euclidean Algorithm or any other convenient method. The Verifier then computes $g^{ap-bq} = (g^p)^a (g^{-q})^b$.
- Jointly execute ZKP_Size($ap - bq$). $\square$

3.4. Joint Generation

It is possible for a first party to jointly construct a key with the assistance of an outside, trusted party such as a CA. Techniques for this have been developed by [4], and by [11]. The joint construct is achieved in such a way that the CA can then confirm that the key is valid. This technique only gives assurance that the resulting key has the correct arithmetic properties for encryption and decryption to work. It gives no assurance that the key is secure from any of the various attacks under discussion here.

4. Unsolved problems

The discussion in this papers shows that solutions to the following problems would be desirable.

- A true ZKP for primality, not just probable primality of discrete log in a prime field.
- A true ZKP that an RSA modulus is the product of strong primes. It is would be especially desirable to have one which works from N only, and does not require a separate bit commitment to p and q.
- A ZKP for $p \approx q$ using only N that docs not leak bits.

- One security property not previously discussed is the following. It is well known that if d is small, then the modulus can be recovered. A ZKP that d is large working only from N would be useful.
- Techniques for how a second party can validate a key and determine its security without help from a first party. This does not seem possible, for example, if it were possible to verify that N is the product of two primes and that $(e, \phi(N)) = 1$ working from knowledge of N only, then it should be possible to construct an efficient factoring algorithm from this. A proof of this last hypothesis would be useful. Further, such checks do not assure that the key is secure. The risk associated with an invalid key is no worse than the risk associated with one which is valid, but insecure. Therefore, even if such checks were rigorously possible, they would not ameliorate risk or resolve issues involving non-repudiation. Indeed, if it were possible to bound p from knowledge of only N, then one can take good advantage of this via binary search on the size of p [17].

Acknowledgments. The author gratefully acknowledges the assistance of the anonymous referees in drawing particular attention to the fact that the original version of 3.3.1 required an honest verifier, as well as other inciteful comments. The author also thanks Burt Kaliski for drawing attention to [8, 10].

References

[1] *ANSI X9.31-1998 Digital Signatures Using Reversible Public Key Cryptography*, American Bankers Association, (1998).

[2] *ANSI X9.44-1998 Key Transport Using Reversible Public Key Cryptography*, American Bankers Association, (2000).

[3] E. Bach and J. Shallit: *Factoring with cyclotomic polynomials*, Math. Comp. **52**, (1989) pp. 201–219.

[4] D. Boneh and M. K. Franklin: *Efficient Generation of Shared RSA Keys*, Proceedings of Crypto '97, LNCS **1233**, Springer-Verlag, (1997), pp. 425–439.

[5] F. Boudot: *Efficient Proofs that a Committed Number Lies in an Interval*, Proceedings of Eurocrypt 2000, LNCS **1807**, Springer-Verlag (2000), pp. 431–444.

[6] J. Boyar, K. Friedl, and C. Lund: *Practical zero knowledge proofs: Giving hints and using deficiencies*, J. Cryptology 4 1991, pp. 65–85.

[7] J. Camenisch and M. Michels: *Proving in Zero-Knowledge that a Number is the Product of Two Safe Primes*, Proceedings of Eurocrypt '99, LNCS **1592**, Springer-Verlag (1999), pp. 107–22.

[8] D. Chaum, A. Fiat, and M. Naor: *Untraceable Electronic Cash* Proceedings of Crypto '88 LNCS **403**, Springer-Verlag, 1993, pp. 319–317.

[9] I. Damgård: *Practical and Provably Secure Release of a Secret and Exchange of Signatures*, Proceedings of Eurocrypt '93, LNCS **765**, Springer-Verlag (1994), pp. 200–217.

[10] I. Damgärd, O. Goldreich, and T. Okamoto: *Honest Verifier vs Dishhonest Verifier in Public Coin Zero-Knowledge Proofs*, Proceedings of Crypto '95, LNCS **963**, Springer-Verlag, 1995, pp. 325–338.

[11] Y. Frankel, P. MacKenzie and M. Yung: *Robust Efficient Distributed RSA Key Generation*, STOC (1998), pp. 663–72.

[12] Z. Galil, S. Haber, M. Yung: *A Private Interactive Test of a Boolean Predicate and Minimum-Knowledge Public-Key Cryptosystems* (Extended Abstract), FOCS 1985, pp. 360–371.

[13] R. Gennaro, D. Micciancio and T. Rabin: *An Efficient Non-interactive Statistical Zero-knowledge Proof System for Quasi-safe Prime Products*, Proceedings 5th ACM Conference on Computer and Communications Security, (1998), pp. 67–72.

[14] L. Guillou and J.-J. Quisquater: *A Practical Zero-Knowledge Protocol Fitted to Security Microprocessor Minimizing Both Transmission and Memory*, Proceedings of Eurocrypt '88, LNCS **330**, Springer-Verlag (1988), pp. 123–128.

[15] R. S. Lehman: *Factoring Large Integers*, Math. Comp. **28**, (1974), pp. 637–46.

[16] A. K. Lenstra and H. W. Lenstra Jr.: (eds), *The Development of the Number Field Sieve*, Lecture Notes in Mathematics **1554**, Springer-Verlag (1991).

[17] A.K. Lenstra and H. W. Lenstra Jr.: *Algorithms in Number Theory*, Handbook of Theoretical Computer Science, Elsevier Science Pulishers (1990), pp. 674–715.

[18] M. Liskov and R. Silverman: *A Statistical Limited-Knowledge Proof for Secure RSA Keys*, submitted to J. Cryptology.

[19] W. Mao: *Verifiable Partial Sharing of the Factors of an Integer*, Proceedings of SAC '98, Queen's University, pp. 95–109.

[20] P. Montgomery: *Speeding the Pollard and Elliptic Curve Methods of Factorization*, Math. Comp., (1987), **48**, pp. 243–64.

[21] P. Montgomery and R. Silverman: *An FFT Extension to the P-1 Factoring Algorithm*, Math. Comp. **54**, (1990), pp. 839–54.

[22] H. Riesel: *Prime Numbers and Computer Methods for Factorization*, Birkhäuser, Boston, Basel, Stuttgart (1985).

[23] R. Rivest and R. Silverman: *Are 'Strong' Primes Needed for RSA?*, Preprint, http://theory.lcs.mit.edu/ rivest/publications.html.

[24] C.-P. Schnorr and H. W. Lenstra Jr.: *A Monte-Carlo Factoring Algorithm with Linear Storage*, Math. Comp., **43**, (1984), pp. 289–311.

[25] R. Silverman: *The Multiple Polynomial Quadratic Sieve*, Math. Comp. **48**, (1987), pp. 329–39.

[26] R. Silverman and S. S. Wagstaff Jr.: *A Practical Analysis of the Elliptic Curve Factoring Algorithm*, Math. Comp. **61**, (1993), pp. 445–62.

[27] J. van de Graaf and R. Peralta: *A Simple and Secure Way to Show the Validity of Your Public Key*, Advances in Cryptology – Crypto '89 (1989), pp. 128–134.

RSA Laboratories, 20 Crosby Dr. Bedford, MA 01730, USA
E-mail address: rsilverman@rsasecurity.com